THE ROUTLEDGE HANDBOOK OF CRIME FICTION AND ECOLOGY

The Routledge Handbook of Crime Fiction and Ecology is the first comprehensive examination of crime fiction and ecocriticism. Across 33 innovative chapters from leading international scholars, this *Handbook* considers an emergent field of contemporary crime narratives that are actively responding to a diverse assemblage of global environmental concerns, whilst also opening up 'classic' crime fictions and writers to new ecocritical perspectives. Rigorously engaged with cutting-edge critical trends, it places the familiar staples of crime fiction scholarship – from thematic to formal approaches – in conversation with a number of urgent ecological theories and ideas, covering subjects such as environmental security, environmental justice, slow violence, ecofeminism and animal studies. *The Routledge Handbook of Crime Fiction and Ecology* is an essential introduction to this new and dynamic research field for both students and scholars alike.

Nathan Ashman is Lecturer in Crime Writing at the University of East Anglia and the author of *James Ellroy and Voyeur Fiction* (2018). His research spans the fields of crime fiction, contemporary American fiction, and ecocriticism, with a particular specialism in the works of James Ellroy. He has published articles on numerous writers including Ross Macdonald, E.C. Bentley, Don DeLillo, Megan Abbott and Walter Mosley. His second book, *James Sallis: A Companion to the Mystery Fiction*, is forthcoming.

ROUTLEDGE LITERATURE HANDBOOKS

Also available in this series:

THE ROUTLEDGE HANDBOOK OF STAR TREK
Edited by Leimar Garcia-Siino, Sabrina Mittermeier, and Stefan Rabitsch

THE ROUTLEDGE HANDBOOK OF SHAKESPEARE AND INTERFACE
Edited by Clifford Werier and Paul Budra

THE ROUTLEDGE HANDBOOK OF ECOFEMINISM AND LITERATURE
Edited by Douglas A. Vakoch

THE ROUTLEDGE HANDBOOK OF NORTH AMERICAN INDIGENOUS MODERNISMS
Edited by Kirby Brown, Stephen Ross and Alana Sayers

THE ROUTLEDGE HANDBOOK OF VICTORIAN SCANDALS IN LITERATURE AND CULTURE
Edited by Brenda Ayres and Sarah E. Maier

THE ROUTLEDGE HANDBOOK OF REFUGEE NARRATIVES
Edited by Evyn Lê Espiritu Gandhi and Vinh Nguyen

THE ROUTLEDGE HANDBOOK OF COFUTURISMS
Edited by Bodhisattva Chattopadhyay, Grace L. Dillon, Isiah Lavender III and Taryne Jade Taylor

THE ROUTLEDGE HANDBOOK OF CRIME FICTION AND ECOLOGY
Edited by Nathan Ashman

For more information about this series, please visit: www.routledge.com/Routledge-Literature-Handbooks/book-series/RLHB

THE ROUTLEDGE HANDBOOK OF CRIME FICTION AND ECOLOGY

Edited by Nathan Ashman

Routledge
Taylor & Francis Group
NEW YORK AND LONDON

Designed cover image: Getty

First published 2024
by Routledge
605 Third Avenue, New York, NY 10158

and by Routledge
4 Park Square, Milton Park, Abingdon, Oxon, OX14 4RN

Routledge is an imprint of the Taylor & Francis Group, an informa business

Library of Congress Cataloging-in-Publication Data
A catalog record for this book has been requested

ISBN: 978-0-367-55085-1 (hbk)
ISBN: 978-0-367-55086-8 (pbk)
ISBN: 978-1-003-09191-2 (ebk)

DOI: 10.4324/9781003091912

Typeset in Times New Roman
by Apex CoVantage, LLC

CONTENTS

FIGURES

NOTES ON CONTRIBUTORS

Katrin Althans is a DFG-funded research fellow at the Postcolonial Studies Section of the Department of Anglophone Studies, University of Duisburg-Essen, Germany. Her main research interest is in Australian studies, and she has published widely in this area. Here, her focus is on Aboriginal Australian literature, which she has approached from a variety of angles, including genre (the Gothic), ecocritical readings, and geocriticism. Katrin also works in the area of law and literature and has both published on and taught a number of subjects, including crime fiction and representations of trials in literature. For her post-doc project, she is currently working on a second book on the representation of refugees in law and literature.

Caitlin Anderson is an instructor at Auburn University. She graduated with her PhD from Auburn University in the Spring of 2020 with her dissertation, "From Tonic to Toxin: the Shifting and Gendered Social Role of Medicinal Herbs in Long Nineteenth-Century British Literature", which traced the social role of herbs over the course of the long nineteenth century. She now continues that same research as she begins work on her book project based on her dissertation research.

Rafael Andúgar is a PhD candidate at the University of California, Santa Barbara. His current project focuses on the relations between environmentalism and Latin American contemporary crime fiction, particularly how landscape and its affects play an important role in the noir genre as it enables reflection on environmental issues as well as complex emotions that arise from socio-environmental polarization. Some of his work is forthcoming in different collections including Gisela Heffes and Arndt Niebisch (ed.), *Uncontained Toxicity: The Dialectics of Loss and Control* or Victoria Ketz and J. Manuel Gómez (ed.), *Ecocriticism: the Ibero-American Experience.*

Nathan Ashman is Lecturer in Crime Writing at the University of East Anglia and the author of *James Ellroy and Voyeur Fiction* (2018). His research spans the fields of crime fiction, contemporary American fiction and ecocriticism, with a particular specialism in the works of James Ellroy. His second book, *James Sallis: A Companion to Mystery Fiction*, is forthcoming with McFarland.

Nicola Bishop is a senior lecturer in English at Manchester Metropolitan University. She has previously published on Edwardian middlebrow fiction, the works of Agatha Christie, popular history on television and contemporary nostalgia.

Tomas Buitendijk, MPhil, is currently writing his doctoral thesis on the topic of twenty-first century representations of seascapes. He has a particular interest in tracing emergent plurispecies patterns of interaction, both in marine contexts and beyond. He was lead organiser of the 2019 interdisciplinary workshop 'Planet Ocean'.

Alicia Carroll has long worked in nineteenth-century British literature. She is the author of a book on George Eliot and her work on Victorian fiction has recently appeared in the recent *George Eliot in Context* (Cambridge UP, 2013) and in the journals *NOVEL*, *JEGP*, *Women's Studies*, *Nineteenth-Century Literature*, *Dickens Studies Annual*, *Victorian Review* and *Green Letters*. She has recently completed a book, *New Woman Ecologies: From Arts and Crafts to the Great War*, and she has a chapter on Ouida and rivers forthcoming in a book from Palgrave entitled *Victorian Environmental Nightmares*.

Megan Cole Lyle is an English PhD candidate at the University of California, Irvine, with a designated emphasis in critical theory. Her primary research interests include ecocriticism, energy humanities, critical prison studies, infrastructure studies, political economy and transatlantic modernism.

David Conlon is a lecturer in Latin American literature and culture in the School of Languages, Literatures and Cultures at Maynooth University, where he teaches modules on "The Short Story in Latin America", "Latin American Literature and Culture", "Latin American Crime Cinema" and "Crime and Popular Culture in Latin America", and he is also subject leader in Spanish. His research focuses primarily on Latin American crime fiction and Latin American film. He is the author of articles and book chapters on Ricardo Piglia, Rodolfo Walsh, and Antonio Di Benedetto and has articles and chapters forthcoming on Brazilian cinema, Norah Lange, and Jorge Luis Borges. He is currently completing a monograph on Latin American crime fiction on screen.

Rachel Fetherston is a sessional academic in literary studies at Deakin University whose research investigates the representation of the nonhuman in Australian ecofiction and the potential impact that such fiction has on the reader's relationship with nature. Her research includes considerations of speculative and science fiction, crime fiction, multispecies studies and the intersection of reader response and nature connection. She is also a freelance writer and co-founder of Remember The Wild, a non-profit focused on engaging Australians with the natural world.

Alec Follett holds a PhD in literary studies from the University of Guelph where he studied Canadian and Indigenous environmental writer-activists and the politics of reconciliation. He has recently published on short story writer Alice Munro and poet Rita Wong. He is currently co-editor of *The Goose: A Journal of Arts, Environment, and Culture* in Canada.

David Geherin is Professor Emeritus of English at Eastern Michigan University, where he taught courses in modern and contemporary literature and in crime and mystery fiction for 40 years. He is the author of nine books on a wide range of crime writing, including the first book on Elmore Leonard (1989) and two others – *The American Private Eye: The Image in Fiction* (1985) and *Scene of the Crime: The Importance of Place in Crime and Mystery Fiction* (2008) – that were finalists for the Mystery Writers of America's Edgar Allan Poe Award. Among his other books are *The Dragon Tattoo and its Long Tail: The New Wave of European Crime Fiction in America* (2012) and *Carl Hiaasen: Sunshine State Satirist* (2019).

Terry Gifford, a co-founder of British ecocriticism, is Visiting Research Fellow at the Research Centre for Environmental Humanities, Bath Spa University, UK, and Professor Honorifico at the

Universidad de Alicante, Spain. He is the author of *D. H. Lawrence, Ecofeminism and Nature* (2023), *Pastoral* (2nd edn 2020), *Green Voices* (2nd edn 2011), *Reconnecting with John Muir: Essays in Post-Pastoral Practice* (2006) and author, editor or co-editor of seven books on Ted Hughes, most recently *Ted Hughes in Context* (2018). He is on the editorial board of *Green Letters: Studies in Ecocriticism*, *The Journal of Ecocriticism* and *Ecozon@*. His eighth collection of poetry is *A Feast of Fools* (2018).

Colette Guldimann has lived, studied and taught in the United Kingdom, Germany, South Africa, Lebanon and the Sultanate of Oman. She has published across cross-disciplinary boundaries, drawing on a variety of theoretical approaches (film theory, psychoanalysis, postmodernity, semiotics) to illuminate a diverse range of subjects: Southern African literature and culture, cross-cultural symbols and contemporary British poetry. Her current research projects focus on interpretations of popular cultural forms within postcolonial contexts, particularly the postcolonial detective genre.

Malinda Hackett is a PhD student in English and Cultural Studies at Claremont Graduate University. She holds an MA in English from California State University, Northridge, along with a BA in Cinema Studies from San Francisco State University. Areas of interest include Post-1900 American Literature, Gothic Studies, Noir Studies, Transnational Detective Fiction, Feminist Crime Fiction, Urban and Spatial Theory, Feminist Geography, Culture and Media Studies and Digital Humanities.

Felicity Hand is senior lecturer in the English department of the Autonomous University of Barcelona. She teaches postcolonial literature and history and culture of Britain and the United States. She has published articles on various Indian Ocean writers including M. G. Vassanji, Abdulrazak Gurnah, Yasmin Alibhai-Brown and Lindsey Collen and has contributed to the volume *100 Greatest Literary Detectives* (ed. Eric Sandberg, 2018). She is the co-director of the research group Ratnakara (http://grupsderecerca.uab.cat/ratnakara), which explores the literatures and cultures of the South West Indian Ocean. The group's current project is Rhizomatic Communities: Myths of Belonging in the Indian Ocean World (PGC2018–095648-B-I00). Felicity is also the editor of the electronic journal *Indi@logs, Spanish Journal of India Studies*, http://revistes.uab.cat/indialogs.

Ruth Hawthorn is Senior Lecturer in American literature at the University of Lincoln. She is currently completing a monograph on American detective fiction for the BAAS Paperbacks series with Edinburgh University Press. Her research interests include crime fiction, the literature of LA and ecocriticism.

Michael Hinds is a senior lecturer at Dublin City University's School of English. He has a strong interest in popular culture and media, with recent research excursions having had a particularly environmental flavour. He was lead organiser of the 2011 summer school 'Shipwrecks and Desert Islands' and co-organised the 2019 interdisciplinary workshop 'Planet Ocean'. He is author of *Johnny Cash International: How and Why Fans Love the Man in Black* (University of Iowa Press, 2020).

Hsuan Hsu is the author of *Geography and the Production of Space in Nineteenth-Century American Literature* (Cambridge, 2010) and *Sitting in Darkness: Mark Twain, Asia, and Comparative Racialization* (NYU, 2015). His most recent book, *The Smell of Risk: Atmospheric Disparities and the Olfactory Arts* (NYU, 2020), considers olfactory aesthetics as a mode of engaging with environmental injustice in literature, art, memoir and law. His recent courses have examined topics such as geographies of risk, transnational American literature, medical humanities, the aesthetics of atmosphere, the aesthetics of chemosensation and race and realism.

Priscilla Jolly is a PhD candidate in environmental humanities at Concordia University. Her research interests include landscape studies and science fiction. She is particularly interested in bodily transformations effected by landscape as well as spatial representations in literary texts.

Ian Kenny (MA) is a PhD candidate at the Amsterdam School for Cultural Analysis writing about narrative forms and the Anthropocene. He has written about the role of landscape in constituting lieux-de-mémoirs, and his current work investigates the role of mythological thinking as a narrative resource for coming to terms with the braided ecological crises and posthuman realities of the present.

Anna Kirsch completed her English studies MA at Durham University. Her thesis was on environmental ethics and morality in Carl Hiaasen's crime fiction. Her research interests include crime fiction, satire, American studies, gender studies and environmental history.

Anita Lam is Associate Professor of Criminology at York University, Canada. Her research has appeared in *Time & Society*, *Law Text Culture*, *Canadian Journal of Law and Society* and in several edited collections.

Valerie McGuire earned her PhD in Italian studies from New York University. Her research interests integrate Italian studies and history and investigate issues of national identity, Italian empire, citizenship and migration from the Italian unification to the present. She has been a visiting scholar of Italian and Mediterranean history at the University of Aegean (Rhodes, Greece) while holding a Fulbright Scholarship to Greece, as well as while holding a Max Weber postdoctoral position at the European University Institute in Florence. She has taught Italian language and culture courses at San Francisco State University and Chabot Community College, as well as at New York University and Pace University in New York City

Kristopher Mecholsky is Associate Director of Research Advancement at Louisiana State University, where he earned his doctorate in English in 2012. He has taught regularly at LSU, the University of Maryland Global Campus, and Franciscan Missionaries of Our Lady University since 2008. His first book (co-authored with David Madden) is a critical overview of James M. Cain. He regularly publishes essays on the evolution of genre in crime narrative, on adaptation theory and on the South. His work has appeared with McFarland, Palgrave, and Salem Presses, as well as with *South Atlantic Review*, *The Baker Street Journal*, *The Faulkner Journal* and others.

MaKenzie Hope Munson is an independent scholar working in the Chicagoland area; she has plans to begin pursuing her law degree in the Fall of 2024. This chapter, which functioned as the final product of her undergraduate program's senior thesis requirement, is also her first professional publication.

Patrick D. Murphy is Professor and Chair of the Department of English at the University of Central Florida. He has authored *Ecocritical Explorations in Literary and Cultural Studies* (2009), *Farther Afield in the Study of Nature-Oriented Literature* (2000), *A Place for Wayfaring: The Poetry and Prose of Gary Snyder* (2000) and *Literature, Nature, and Other: Ecofeminist Critiques* (1995). He teaches critical theory, modern and contemporary American literature, comparative literature, ecocriticism and ecofeminism.

Leonardo Nolé is a doctoral student in Comparative literature at the Graduate Center – City University of New York and teaching fellow at Hunter College. His research focuses on contemporary American literature, ecocriticism and narrative theory. His most recent publications include

"Understanding the Fabric of the Natural World. The Role of the Collective Protagonist in Annie Proulx's Barkskins" (*JAmIt!*, 2020); "William Faulkner's Go Down, Moses: A Chronicle of Im/Mobilities" (*JAAAS*, 2020); "The Short Story Composite and Ernest Hemingway's In Our Time" (*CoSMo*, 2017).

Bill Phillips is a senior lecturer in English literature and culture at the University of Barcelona where he lectures on poetry, crime fiction and other contemporary fiction. He has published widely on poetry, particularly of the Romantic period, ecocriticism, ecofeminism, postcolonial studies, gender studies and popular fiction including detective fiction, science fiction and zombies. From 2013–2017 he was head of POCRIF (Postcolonial Crime Fiction: a global window into social realities), a Ministerio de Economía y Competividad financed research project on postcolonial crime fiction. The project's team are members of the Australian Studies Centre, based at the University of Barcelona, and the group's research forms part of the wider academic and investigative work carried out by the Centre. His most recent publications have included articles on American TV series *True Detective*, Rudyard Kipling's war fiction and (as editor and contributor) a collection of essays titled *Family Relationships in Contemporary Crime Fiction*.

Marta Puxan-Oliva is a postdoctoral researcher at the Universitat Oberta de Catalunya and Assistant Lecturer at the Universitat de Barcelona. She is a specialist on narrative theory, ecocriticism, world literature, postcolonial studies and racial comparative studies. Her first book, *Narrative Reliability, Racial Conflicts and Ideology in the Modern Novel*, has just been published by Routledge. She has co-edited the special issues "Rethinking World Literature Studies in Latin American and Spanish Contexts" for the *Journal of World Literature* 2(1) (2017) with Annalisa Mirizio and "Historicizing the Global: An Interdisciplinary Perspective" for the *Journal of Global History* 14(3) (2019) with Neus Rotger and Diana Roig-Sanz. She is currently working on oceans, crime and literature, a topic on which she has published the article "Colonial Oceanic Environments, Law, and Narrative in Herman Melville's Benito Cereno and Juan Benet's Sub rosa" in *English Studies* 99(4) (2018). She is a member of the research group GlobaLS.

Damini Ray is currently working as an assistant professor of English literature and language at Syamaprasad College (University of Calcutta). She completed her MPhil research titled "Reimagining Female Identity and Selfhood in Late Twentieth Century Feminist Science Fiction" in 2020, from University of Calcutta, Kolkata. She is presently pursuing PhD research on Victorian and Neo-Victorian crime fiction at Presidency University, Kolkata. Her areas of interest include genre fiction, adaptation studies and women studies.

Irina Souch is a researcher in comparative literature and cultural analysis in the Department of Modern Foreign Languages and Cultures of the University of Amsterdam and Affiliated Fellow at the Amsterdam School for Cultural Analysis (ASCA). She is author of *Popular Tropes of Identity in Contemporary Russian Television and Film* (Bloomsbury 2017) and co-editor of the contributed volume *Heterotopia and Globalisation in the Twenty-First Century* (Routledge 2020). Her current research interrogates the importance of literary and cinematic representations of climate change and ecological catastrophes in shaping perceptions about phenomena which are slow-moving and often unseen.

Kevin Andrew Spicer has a background in medieval literature, Shakespeare, and post-Kantian continental philosophy – with especial focus on Lacanian psychoanalysis, Martin Heidegger and Jacques

Derrida. He has also written and published in the fields of environmental philosophy and ecological thinking as well.

Rebecca Tillett is Senior Lecturer in American Studies and the University of East Anglia, UK. Her recent book publications include *Howling for Justice: New Perspectives on Leslie Marmon Silko's Almanac of the Dead* (2014), and *Indigenous Bodies* (2013). She is founding member of the Native Studies Research Network, UK.

Aina Vidal-Pérez holds a degree in Hispanic philology from the Universitat de València and a master's degree in humanities: contemporary art, literature and culture from the Universitat Oberta de Catalunya (UOC). She is currently following her PhD in information and knowledge society at the UOC and is part of the Global Literary Studies Research Group (GlobaLS). She has a PhD Fellowship (FI 2019) funded by the Government of Catalonia and the European Social Fund. Her doctoral thesis, entitled "Global Mediterranean. Representations of the shore in crisis in the contemporary novel (1990–2020)", studies the literary representations of the Mediterranean coast in the global era, with readings of Rafael Chirbes, Zena el Khalil, Philippe Claudel, Ahmed Khaled Towfik and Donna Leon. The research is situated within the new framework of global literary studies and aims to contribute to the recent research on the novel and to the growing discussion in the field of spatial representations and ecocriticism.

Andrew Yallop is a third-year PhD candidate in English and literary studies at The University of Western Australia. Andrew is in the process of completing a thesis in genre studies examining the way in which detective fiction can function as narrative allegory for "investigating" histories of violence, particularly through the locus of memory-work. Andrew's thesis examines Paul Auster's The New York Trilogy, Roberto Bolaño's *Distant Star*, W.G. Sebald's *Austerlitz* and Patrick Modiano's *Dora Bruder* in order to present a close reading and analysis of the unique adaptations made by these authors to the genre detective fiction. Andrew's essay "Beyond Law and Order: Detecting State Violence and the Search for Justice in Roberto Bolaño's Distant Star" was published in the *Crime Fiction Studies Journal* in March 2021.

ACKNOWLEDGEMENTS

I would firstly like to thank Michelle Salyga at Routledge for initially approaching me about the collection and for being so supportive in the early stages. Bryony Reece also deserves a special mention for her amazing editorial support. Thank you also to all of the incredible contributors who feature in this collection. It has been an absolute pleasure reading and editing your work, and it has no doubt strengthened my own.

Finally, thanks to the usual gang – my colleagues Henry Sutton and Tom Benn, my parents Noel and Sharron, my sister Gemma, and my partner Elisse – for putting up with the manic and delirious energy that surrounds me any time I'm working on a book project such as this.

And thanks to Ivy, my cat, who does literally nothing all of the time.

PLACING CRIME FICTION AND ECOLOGY

An Introduction

Nathan Ashman

In his book *Dark Ecology* (2016), Timothy Morton compares the "darkness of ecological awareness" to the "darkness of noir" (9). It is a provocative statement, one that, for Morton, highlights a certain dissonance that accompanies all critical thinking about anthropogenic climate change. In the same way that the noir protagonist is frequently "implicated" in the narrative's criminal plot, ecological awareness forces the contemporary subject into a "strange loop" where they must confront their own dialectical position as both perpetrator and victim. As Morton puts it: "I'm the detective and the criminal! I'm a person. I'm also part of an entity that is now a geophysical force on a planetary scale" (9).

Morton's employment of the language of crime fiction is suggestive here and points not only to the rich potential for crime writing to articulate affective responses to the contemporary climate crisis, but also to the value of reassessing the mode's various traditions, conventions and subgenres for the ways they might yield new and enlightening perspectives on the entangled global and ecological histories of the Anthropocene. Indeed, since its formal emergence in the mid-nineteenth century, crime fiction has been particularly adept at elucidating correlations between transgressive acts and the specific topographies within which they occur. From the sublime, snow-bound settings of Nordic noir to the enclosed, rural enclaves of the Golden Age cosy, all crime fiction is, to some extent, troubled by questions of ecology. I do not necessarily mean this in the sense of offering a conscious or political interrogation of the relationship between living organisms and the natural world (although some do), but rather as being unavoidably enmeshed in certain historical and ideological understandings of the physical environment.

Whilst it is by no means a unique position to suggest that the crime novel is preoccupied with notions of place, much existing criticism tends to engage with the subject in what David Schmid describes as a fairly "passive manner", where "houses, suburbs, cities, and so on are treated merely as background, as setting, rather than as determinative forces" (7–8). An ecological approach to crime fiction shifts our focus towards these elements that are so "often dismissed as backdrops to human activity", prompting an "adjustment of temporalities" and "urging scholars to situate human activity in seasonal, anthropological, evolutionary and deep timescales" (Walton 115). In this way, we can begin to investigate the specific relation between environment and crime, excavating the buried histories, knowledges, and injustices that constitute the very bedrock of place. As Sam Walton puts it: "Every crime novel is set *somewhere* and investigation of that *somewhere*" is a good place to start thinking about "the wider questions posed by ecological reading" (115).

DOI: 10.4324/9781003091912-1

One of the central aims of this collection is to rethink the crime novel's relationship to place, not only as it pertains to the material (and often local) environments that the genre reveals, but to a particular kind of discourse, a way of "seeing, knowing and understanding the world" (Cresswell 11). As Stewart King suggests, crime fiction has contributed significantly to certain cultural understandings of place. Initially emerging as a very response to the rapid urban and industrial developments of the nineteenth century, the genre also offered its readers new ways of mapping and overcoming the perceived "unreadability" of the modern metropolis (King 1237). Key to the production of these knowledges was the figure of the detective, who, through their heightened capacity to interpret material and spatial signs – often via the embracement of new technologies and innovative modes of deciphering the social and geographical body – worked to establish connections between various 'types' of individuals, thus "revealing (and making meaningful for the reader), the social, political, economic, cultural and physical places they inhabit" (King 1238).

Due to this construction of place "as a locus of meaning" in crime fiction, King argues that scholarship on the genre has been "dominated and delimited by specific and distinct national traditions", a critical approach that has only been further solidified in recent years by a fear of "placelessness" engendered by globalisation (King 1238; Dainotto 4). King points to geographer Gary Hausladen as one of many scholars who have identified the genre's focus on 'local' crime scenes as a form of continuing political resistance against the homogenising effects of global capitalism, a way of preserving the specificity, character and uniqueness of place. This critical desire to safeguard the local and to keep cultures "in their place" would ostensibly appear to service what Roberto Dainotto describes as the "multicultural utopia of a coexistence of different cultures" (3). However, as Dainotto continues, what this "contiguous arrangement of cultures in space" actually tends to do is "bracket away the very question of hegemony – the historical process whereby one culture acquires authority over all others and puts them in order" (3). In crime fiction studies, we can see this in the substantiation and reproduction of a certain "grand narrative" of the genre's development, one that has tended to privilege Anglo-American forms, contexts and histories over all others (Allan et al. 2).

In recent years, crime fiction scholars have therefore looked to challenge this specific chronology, advocating for a more comprehensive understanding of genre that emphasises the transnational and comparative connections between works from across the globe. In their book *Globalisation and the State*, Andrew Pepper and David Schmid open by arguing that "the production, circulation, and translation" of crime fiction has in fact "always been an inherently transnational phenomenon" (1). They point to the genre's roots in the "criminal stories circulating in London and Paris from the early eighteenth century", highlighting how crosspollination and mutability have been constitutive to the mode's development and longevity. At the same time, they also acknowledge that "it is only in the last twenty years" that the genre has truly "mushroomed beyond the familiar scenes of its foundational texts (e.g., London, Paris, New York, Los Angeles)", becoming, in the process, "a truly global literary genre" (1). Developments in technology and transport have no doubt been partly responsible for crime fiction's circulation as a form of world or global literature, while its adaptiveness has made it particularly adept at giving textual shape to the "interconnectedness of the local, the national and the global that characterises today's world" (Gulddal & King 1). As Gulddal and King suggest, many of the issues and ideas that perennially preoccupy the crime novel, be it questions of justice, criminality, investigation or the law, are "prime examples of such transnational linkages" and have invariably developed new and more complex connotations and meanings in the context of globalisation (1).

It is precisely the crime novel's augmenting production and circulation as a form of 'world' literature, not to mention the capaciousness of its literary forms, that makes it particularly well placed to map and narrate "the global environmental crisis" (Puxan-Oliva 365). As Amitav Ghosh suggests, the history of the carbon economy is "a tangled global story with many contradictory and counterintuitive elements" and thus requires new modes of collective discourse that move beyond the limited purview of the individual and the local (2016). Of course – and as this collection hopes to demonstrate – this does not necessarily mean discounting or invalidating the value of local stories or perspectives. Indeed, Stewart King argues that the local context must still play "a significant role in helping readers to see, know and understand the world", particularly when it comes to the "magnitude, complexity, and interconnectedness of the climate crisis" (1). The crime novel's ability to move between an interplay of varying "scales" is what is of particular value to the reader and critic, providing an apposite framework through which to conceptualise the variegated temporal and spatial effects of environmental damage. By examining the numerous ways in which the genre "imagines and theorises entangled locations and worlds", we can come to a more comprehensive understanding of how crime fictions are collectively responding to the climate crisis on a global scale (Stougaard-Nielsen 81).

Within this, we must be cautious not to treat the globalisation of crime fiction entirely uncritically or to see it simply as "a one-way process" whereby the genre "moves inexorably to populate the globe" (Pepper & Schmid 3). Instead, the specific "global implications of the crime being depicted (e.g., the link between individual or collective criminal acts and the exigencies of global capitalism) require new forms and new strategies of representation in order to do justice to the changing world" (Pepper & Schmid 3). While a number of studies have certainly started to examine the shifting definitions of crime, truth and justice in a globalised context, none (to date) have offered a sustained examination of how these considerations intersect with environmental and/or ecological anxieties.[1] This is surprising, particularly given that many of the questions confronted by ecological crime fictions are often eminently global in nature. For instance, in a moment where much of the world's environmental damage is being undertaken 'legally' by governments and transnational corporations, to what extent do notions of criminality and agency need to be reconsidered? How can justice be enacted when criminals are no longer deviant/perverse individuals but abstract, global powers? And how can crime writers reconcile the conventional drive towards resolution and certainty with the paralysing uncertainty of our future?

Ecological Crime Fiction: Histories, Definitions and Aims

These are some of the key questions that the chapters in this collection attempt to grapple with, all the while seeking to consolidate and build upon the work of a small but dedicated group of scholars who have, in recent years, sought to rethink the crime novel from an ecocritical perspective. Pinpointing where and when a particular literary or critical tradition originated is never an easy endeavour; the inherent permeability of genres and modes means that the writing and study of ecological crime fiction likely has a longer and more entangled genealogy that can be adequately outlined here. Certainly, the themes of "environmental destruction and human-induced climate change" have been a constitutive part of the landscape of late twentieth and early twenty-first century literature more broadly, with 'Cli-fi' (climate fiction) emerging as a robust field of study in its own right (King 1235). Yet, as Stewart King appropriately notes, much of the critical attention to date has centred on so-called 'literary' forms, despite the fact that the armature of genre fiction (be it science fiction, crime or thriller writing) has frequently been employed by writers as a means of transmitting such anxieties to a mass audience. Critics such as Adeline Johns-Putra and Axel Goodbody have nonetheless expressed scepticism towards the value of popular forms in meaningfully articulating the climate crisis, arguing that when

> the appeal of a novel resides mainly in its status within a particular genre (or even within the oeuvre of a particularly popular writer of genre fiction), [it] can circumscribe readers' understanding of potential solutions to the problems it presents.
>
> *(4)*

This position not only risks substantiating ingrained and spurious hierarchies between 'literary' and 'popular' forms, but it also falls into the trap of situating genre writing as inherently inflexible and conservative or in some way limited by rules and conventions. If the long history and continued popularity of crime writing has taught us anything, it is that it remains the most accommodating and elastic of literary modes.

Patrick D. Murphy's *Ecocritical Explorations in Literary and Cultural Studies* (2009) is often cited as a key source text in the study of ecological crime fictions, in which he calls on ecocritical scholars to look more closely at "nature-oriented mystery novels" as to "better understand the degree to which environmental awareness has permeated popular and commercial fiction" (143). Murphy points to the works of Florida crime writers John D. MacDonald and Carl Hiaasen as key proponents of a form of mystery fiction that has knowingly and critically taken environmental crime as its subject. Murphy's work has since been built upon by several scholars, perhaps most notably Jo Lindsay Walton and Samantha Walton. In their 2018 special issue on "Crime Fiction and Ecology" for the journal *Green Letters*, Walton and Walton bring together a range of essays that "explore detective fiction in which nature plays a prominent role", as well as other works "whose ecological themes are latent within ostensibly anthropocentric plotting" (4). Walton and Walton are particularly preoccupied with the figure of the "contemporary ecological detective", who, they argue, may be forced to "splinter into new distributed and collective perspectives and forms of agency, including the more-than-human and the assemblage" (3). Whilst the collection acknowledges the kind of consciously 'nature-orientated' mysteries that Murphy pinpoints – mysteries that feature typical genre markers such as the detective, the criminal and the crime scene – it also seeks to extend the field of study to texts that *un*consciously lend themselves to ecological interpretations. One might consider Raymond Chandler's *The Big Sleep* and the abandoned oil field that rests at the foot of the Sternwood family estate. Or Conan Doyle's *The Hound of The Baskervilles* and the untamable moor that defies Sherlock Holmes' drive for mastery. What are the material significances of these landscapes beyond a symbolic or allegorical reading? What can they tell us about the relationship between place, history and culture or between the human and non-human?

Walton's and Walton's special issue is still arguably the most significant contribution to the field to date and suggests two different (yet not necessarily oppositional) ways of approaching the idea of ecological crime fiction: that is, as either a distinctive category of texts with shared preoccupations; or, as a particular mode of interpretation. Marta Puxan-Oliva, for instance, is one of several scholars who has attempted to establish ecocrime fiction as something approximating a subgenre, drawing a clear distinction between crime novels where "environmental issues are marginal" to the story's focus and those where environmental crimes are the "core problem" (362). Stewart King takes a similar approach, coining the neologism "crimate fictions" to refer to novels that "both narrate the climate catastrophe through the popular conventions of the crime genre and apply the genre's ideological concerns with culpability and criminality to the climate crisis" (1237). Although this would seem to circumscribe the parameters of the mode, he does note that crimate fiction is merely "a subset" of a larger category of "environmental crime fiction", one that fosters a "different relationship to place than is evident in many other environmental crime stories" (1239). By thinking about crimate fictions in this way, King hopes to shift the study of "environmental" crime fiction away from the "localised" geographical contexts that have dominated scholarship to date and calls instead for a wider embracement of comparative, postcolonial and global theoretical approaches. He points to the work of scholars such as Marta Puxan-Oliva, Nibedita Bandyopadhyay and Stephanie LeMenager who have analysed the relationship between crime novels "beyond a single locale", helping the reader to comprehend the magnitude and complexity of the climate crisis on differing yet "interconnected scales which, when read together, foster the eco-cosmopolitanism needed if humans are to address the slow destruction of the planet" (1240).

Whilst I agree that a broader critical embracement of comparative approaches to the crime novel is vital in understanding and articulating the climate crisis, King's method does still potentially limit the theoretical field to a localised number of 'actively' engaged contemporary texts and thus threatens to overlook a more temporally dispersed and abstracted range of ecological considerations. Moreover, treating ecological crime fiction as a 'category' in its own right might only serve to produce new forms of chronology and canonisation, which risks further ingraining the same hierarchies and methods of privileging that have characterised crime writing (and criticism) to date. Where would we begin to pinpoint the origins of such a mode? How would we begin to draw its conceptual boundaries? What would we include or exclude and how might such delimitations serve our aims? If we are to truly think about ecological crime fiction at different 'scales', we also need to rethink the crime genre in a broader historical (yet equally comparative) sense. After all, it is only by better understanding where we have been that we can start to imagine different ways of moving forward in the future. We might look to Andrew Pepper's excellent study *Unwilling Executioner: Crime Fiction and The State* – which examines the history of the crime genre's relationship with state power – as a blueprint on how to draw "connections between writers and national traditions across time and space" (7). Although ecological contexts and histories do not feature explicitly in Pepper's work, his claim that the "crime story has a much richer, longer and more radical lineage [than] some critics are prepared to cede" remains relevant to this project (7).

The Routledge Handbook of Crime Fiction and Ecology has several key aims: first, it intends to offer the most comprehensive study of crime fiction and ecocriticism to date, building and expanding upon existing scholarship in an effort to materialise ecological crime fiction as a field of study in its own right; second, its seeks to be internationally inclusive and historically expansive in scope, stressing the importance of reading ecological crime fiction not as a genre with a specific set of conventions, but as a broad field of texts from across a range of geographical and historical locales. Whilst the collection certainly contains chapters that analyse cognisantly ecological crime fictions – and how these fictions are responding to the global climate crisis *now* in the contemporary moment – it also includes approaches that 'open up' a diverse assemblage of crime narratives to new ecocritical perspectives; finally, and through its structural arrangement, the handbook also aims to contribute to the critical dismantling of the grand "story of crime fiction", shifting from an emphasis on the genre's subforms and "subcomponents" to the "complexity and idiosyncrasies of the individual text" (Allan et al. 1). This includes a general movement away from "studying crime fiction in the context of separate national literary traditions" towards an examination of genre as "a transnational and global phenomenon" (Allan et al. 1). The examination of distinct, 'local' crime fictions will still feature as part of this, yet the collection is arranged, as will be discussed in more detail shortly, as to place them within a wider critical conversation. I have landed on Walton's and Walton's term "ecological" crime fiction as my preferred nomenclature – as opposed to variations such as "environmental crime" (Puxan-Oliva 362), "crimate fiction" (King 1236), "eco-thrillers" (Moore 98), "nature-orientated mysteries" (Murphy 143) or "ecologically conscious detective fictions" (Bandyopadhyay 70) – largely because it gestures towards a more comprehensive engagement with the terrains of ecocriticism and crime fiction studies, rather than delimiting the scope to particular 'forms' of crime writing, such as the thriller, or specific 'branches' of ecocritical thought, such as climate.

Whilst this collection includes chapters that engage with ecological crime fictions from a range of scales – local, national, regional and global – and from across a variety of geographical contexts – including works from Asia, Europe, Africa, South America, North America and Oceania – there is still an admittedly greater representation of anglophone literatures on display here, particularly those produced in the United States, Britain and continental Europe. There are obvious factors behind this: access to translations, the difficulty of representing multiple languages within a single study, and, as Andrew Pepper notes, the "international circulation of crime fiction criticism in English"

(9). Nonetheless, more can certainly be done to broaden the frame of ecological crime fiction and criticism, particularly in the context of the global south, where climate change has only worked to widen pre-existing inequalities. If this collection does indeed contain limitations or oversights, as any collection of this kind invariably does, it is my hope that it will, at the very least, be part of a larger conversation that moves the study of crime fiction in new and vital directions.

The Structure of the Handbook

The collection is divided into five subsections – "Space and Topography"; "Bodies and Violence"; "Epistemologies"; "Criminality and Justice"; "Energy, Globality and Circulation" – with each subsection consisting of six to seven thematised chapters. As the headings intimate, the chapters collected here are arranged conceptually rather than mapped against any extant 'chronology' of the crime genre (i.e., 'Golden Age, 'Hard-boiled' etc). This is in an effort to shift away from the entrenched, Anglo-American "grand narrative" of the mode, which still continues to dominate academic discourse (Allan et al. 1). In general, mapping the origins, genealogies or histories of crime fiction through an ecological lens is not the major work of this handbook. Instead, by arranging the chapters around shared thematic or theoretical preoccupations, these individual sections can be understood as figurative "contact zones", imaginary spaces "where cultures meet, clash, and grapple with each other", and where potentially new knowledges and modes of understanding might be produced (Pratt 34).

The chapters in Part I of the handbook, titled "Space and Topography", all broadly rethink the relationship between crime fiction, location and place, examining not only the associations that individual characters and/or investigators form with the topographies where violent crimes are committed, but also how buried traumas and concealed histories come to be sedimented in both the literal and figurative materialities of landscape. This necessitates a rethinking of place beyond its common representation as a fixed or "passive locus of social relations" to something more active, agentive and variable (Lefebvre 11). In the opening chapter of the collection, Terry Gifford considers the affective dimensions of Peter May's Lewis Trilogy, a series of crime novels set on the island of Lewis and Harris in the Outer Hebrides. Gifford conducts research on the island's ecology through a mode of "narrative scholarship", rooting himself in a sense of place through first-hand experience of the island's materiality and culture. From this position, he interrogates the extent to which the narrative arrangements of May's work are influenced by his choice of setting, and whether the social ecology and ethics that the texts present are in anyway shaped or defined by a wider natural ecology. Gifford pays particular attention to the affective capacities of weather, land and sea, ultimately questioning the limitations of May's "intimacy" with the "more-than-human" ecology of Lewis and Harris.

Next, Irina Souch's and Ian Kenny's chapter focuses on the latent ecological resonances in Ann Cleeves' *The Crow Trap*, drawing a connection between the text's stylistic properties and the British school of new nature writing. The authors see Cleeves' novel as a rejection of the nostalgic or idealised Northumbrian countryside, as the land itself comes to undermine the oppositions between nature and culture, perpetrator and victim, science and superstition and past and present. This influence of place writing on British crime fiction is similarly explored by Nicola Bishop in her chapter on the Norfolk mystery novels of contemporary crime writer Elly Griffiths. Bishop principally focuses on Griffiths' representation of the saltmarsh, a liminal and shifting topography that emerges as a site of haunting and of long historied memory. In the process, the chapter argues for the significance of Griffiths' work, not only for the intersections it produces between contemporary place and historical time, but as an example of the wider influence of ecocritical trends on a range of contemporary and popular literary forms.

Chapter 4 shifts away from the British Isles, and from literary forms, to consider two contemporary 'Nordic noir' television shows – *Trapped* (2015–) and *Twin* (2019) – for their signalling of a 'turn' towards non and more-than-human actors as agentive protagonists in what Buitendijk and

Hinds term contemporary "cross-species dramas". The authors examine the hostile manifestations of the sea in both narratives, as it shifts from a passive, aesthetic backdrop to something more active and present. Describing it as "the big deep", a riff on Chandler's description of death as "the big sleep", the sea, for Buitendijk and Hinds, ultimately emerges as a "Pluri-Agential Protagonist", one that confounds the scales of human understanding and morbidity. This relationship between Nordic crime narratives and imaginaries of place is picked up again in Priscilla Jolly's essay on the work of Icelandic crime writer Arnaldur Indriðason. Beginning by sketching a brief history of how images of landscape have been constructed and disseminated in the framing of Iceland as a nation, Jolly goes on to examine Nordic crime fiction's participation in the continued building of a touristic imaginary. Forms of place-making are also the focus of the concluding chapter in this section, in which Rachel Fetherston explores the representation of environmental concerns in contemporary Australian crime fiction, with a specific emphasis on questions of settler-colonial and Indigenous belonging.

Section II of the collection centres on considerations of human and non-human bodies, particularly as they intersect with various forms and scales of violence. In her opening chapter on Agatha Christie's short story, "The Blue Geranium", Alicia Carroll analyses how Christie's work encourages readers to rethink ecological violence *as* violence enacted by humans against beings or entities often deemed as 'pests'. Carroll argues that the narrative not only reveals the uneven distribution of justice across species boundaries, but also the brutality and cruelty endemic to human society. These questions of agency are also examined in Caitlin Anderson's chapter on L.T. Meade's illustrated *Sorceress of the Strand* series (1902–1903). Centring particularly on the topic of plant literacy, and on the association between women and plants in the symbolic economy of the *fin de siecle*, Anderson reads the late-nineteenth-century phenomenon of "poison panic" as deeply rooted in late Victorian anxieties regarding the possibility of a newly autonomous and educated female subject.

In the next chapter, Malinda Hackett studies the concept of "colonial recursions" in Native American author Louise Erdrich's *The Round House*. Hackett argues that the novel provides a crucial insight into the junctures between environment, gender, criminality and coloniality, particularly when considering the habituation of violent sexual crimes against Indigenous women as characteristic of a "European settler mindset" which used the gendering of land as the very means through which to rationalise its violent 'civilising' mission. The legacies and continuities of white patriarchal violence are similarly assessed by Andrew Yallop – albeit through a different lens – in his chapter on Olga Tokarczuk's extraordinary ecofeminist crime novel *Drive Your Plow Over the Bones of the Dead*. Yallop reveals how Tokarczuk works both within and against the conventions of the detective genre, *re*imagining environmental harm as an ecological encounter in which non-human animals are positioned as both victims and perpetrators of murder. Questions of animal agency likewise constitute part of Colette Guldimann's examination of crime fiction set in post-apartheid South Africa, which, she suggests, has emerged as the new brand of "politically engaged fiction" due to its triangulation of the intersections between crime, governmental policy and ecology. Within this, Guldimann places the South African crime novel within the wider context of critical conversations regarding globalisation and the state in contemporary crime fiction.

The final two chapters in Part II consider different forms of violence as they appear in texts from Northern Ireland and Ecuador respectively. In his study of Adrian McKinty's Belfast-based novels, Bill Phillips explores the texts' movements between rural and urban locales, and how these present us with different frames and scales of violence. Phillips shows how the urban spaces of the texts' become inextricably associated with "fast" forms of violence (terrorist bombings and shootings), whereas the ostensibly tranquil and restorative rural locales reveal a divergent yet equally destructive type of "slow violence". Depictions of slow violence are also considered in Rafael Andúgar's chapter on Gabriela Alemán's satirical novel *Poso Wells*, in which the destruction of the fragile ecology becomes rooted in the corrupt hegemonic structures of a heteropatriarchal society.

Part III of the handbook carries the title "Epistemologies", referring to a preoccupation with truth that has traditionally energised the crime novel, as well as to the genre's complicity in producing certain modes and hierarchies of knowledge. In the era of climate catastrophe, ecological crime fictions raise profound questions regarding the previously restorative and reassuring functions of the form, not to mention the very possibility of truth. In the first chapter in this section, Kevin Andrew Spicer and MaKenzie Hope Munson directly engage with these anxieties, reading Lydia Millet's dystopian climate novel as a kind of "Oedipal detective game", one where the drive for truth, in the context of ecological disaster, is deferred and ultimately foreclosed. The subject of epistemology is interpreted slightly differently by Kristopher Mecholsky in the second chapter in this section, in which he explores the genesis and conceptual parameters of "ecocrime" fiction as a subgenre of the crime novel proper. Pinpointing the origins of the mode in the work of American crime writer John D. MacDonald, Mecholsky sees the author's shift towards environmentalism as emerging from the coalescing traditions of naturalism, the western, and the American hard-boiled.

Anita Lam's chapter considers Kate Winkler Dawson's true crime novel *Death in the Air* for the way it offers a symmetrical framing of two killers: the serial murderer John Reginald Christie and the Great London Smog of 1952, an environmental disaster that remains one of the deadliest air pollution events in world history. Lam argues that by treating this environmental event in the same analytical way as the notorious murderer Christie, Dawson ultimately pursues answers to epistemological questions that are only usually reserved for human killers: "Who was victimised? How did the killing occur? Who was responsible for the harm?"

In the same way that Lam interrogates Dawson's unique modes of investigating truth, the next chapter by Hsuan Hsu maps out a history of what he terms "hyperosmic" detective fictions; that is, texts that exhibit a preoccupation with heightened olfactory perception. Moving between works by writers such as Spider Robinson, Arthur Conan Doyle, Rudolph Fisher and Chester Himes, Hsu focuses on the figure of the "deodorizing detective", an ambivalent investigative agent who indicates the extent to which "trans-corporeal modes of embodied knowledge and ecological relation" have always informed the genre of detective fiction.

Whereas Hsu concentrates on methods of detection, Patrick D. Murphy's chapter on the work of science fiction writer Paolo Bacigalupi thinks more about criminality, especially how criminality collides with questions of immorality and culpability in the context of environmental harm. The resistance of ecological crime to the demands of knowledge and narrativisation is also explored by David Conlon in his reading of three contemporary Argentinian ecological crime texts. Conlon argues that all three works subsume the enormity of such abstract crimes into the "internal logic" of their plots, deliberately resisting the typical epistemological drives of the genre. As a consequence, these works expose the inherent difficulty of "delineating the ecological crime scene" when existing notions of justice, casualty and culpability are no longer adequate. In the final chapter of this section, Katrin Althans follows Kristopher Mecholsky in taking a narratological approach to the question of ecological crime fiction. Althans employs the term "ecologemes" to refer to a set of formal ecological entities that, she argues, are characteristic of contemporary Australian crime fiction. Althans aims to produce a new way of reading ecology in such works; that is, not only as a matter of theme and setting, but as something constitutive to their narratological operations.

Part IV of the collection centres on the evolving meanings of criminality and justice, not only in the context of environmental violence, but in an increasingly globalised world. Whereas crime fiction has historically expressed a social unease regarding the potential harm that one individual might inflict upon another, ecological crime fiction increasingly redirects this unease towards intangible powers and/or perpetrators. The spectre of climate catastrophe, which blurs boundaries of culpability, creates a dissonance that forces individuals to confront their own fraught, dialectal position as both perpetrator and victim. The interpretive, typological foundations of detection therefore find themselves under threat, confronting the genre with vital yet challenging questions. In the first chapter of this section, Marta Puxan-Oliva argues that the crime genre is contributing significantly to the exposure of the ambivalences that underly many accepted understandings and interpretations of crime. To demonstrate this, she reads Juan Villoro's *Arrecife* (*The Reef*, 2012) as a novel that illustrates the particular problem of defining environmental wrongdoing in the contemporary era.

Crime and justice are similarly the focus of Rebecca Tillett's chapter, which explores how contemporary Indigenous fiction from North America works to expose the connection between the violences of settler colonialism and the violences of climate crisis in the Anthropocene era. Tillett argues that all three writers – Waubgeshig Rice, Cherie Dimaline and Louise Erdrich – engage in a form of "embodied, contextual, and relational practice", exhibiting an experiential understanding of the world, and relationship to landscape, that must be an essential requirement if we are to move beyond our current era of profound climate crisis. This examination of the relationship between violence and environmental racism is continued in Ruth Hawthorn's chapter, which surveys how two post-Katrina crime novels – James Lee Burke's *The Tin Roof Blowdown* (2007) and Sara Gran's *City of the Dead* (2010) – respond to cultural (and political) narratives that position particular communities as "less worthy of care" during moments of ecological crisis. Felicity Hand scrutinises a connected set of relations between environmental justice and social justice in her chapter on Southern African crime fiction, which reflects on how forms of ritual murder connect to a broader range of social inequalities across the intersections of gender, class, race and landscape.

Anna Kirsch analyses representations of what she terms "wild justice" in her chapter on the work of satirical ecological crime writer Carl Hiaasen. Kirsch centres particularly on manifestations of the "carnivalesque" in Hiaasen's fiction, and how this becomes a tool through which he attempts to influence his readers' "moral decision-making". Despite the gleeful "sense of schadenfreudic delight" that Hiaasen clearly takes in his texts' depictions of radical eco-revenge, Kirsch questions whether this mode of carnivalesque protest has lost its potency in an era when governments are increasingly turning to carnivalesque performance *themselves* as a mode of political discourse. Carl Hiaasen is also the focus of the next chapter in the collection by leading Hiaasen scholar David Geherin, who examines how environmental anxieties manifest across the author's journalism and fiction writing.

The final chapter in this section by Leonardo Nolé continues assessing environmental responsibility, but shifts our focus to two Nordic crime series, *Karppi* and *Bedrag*, as an example of the limitations and the possibilities of crime fiction in representing environmental crimes. Nolé argues that the shows are particularly adept at revealing the "the national and international scale of the energy business and its entanglement with global capital". It is precisely their ability to display and connect crimes on these varying scales (national/international, local/global) that, for Nolé, makes them particularly noteworthy instances of contemporary ecological crime fictions, ones that raise pertinent questions about individual and collective accountability.

Part V places a particular emphasis on questions of energy, globality and literary circulation. In his opening chapter, Nathan Ashman undertakes a close reading of Elliott Chaze's *Black Wings Has My Angel* (1953), examining how early noir fiction obliquely reflects on the rapid transformation of twentieth-century American life by the forces of oil capital. For Ashman, we see this not only in the material landscapes that these texts reveal, but also in their evocation of a particular type of desiring yet "alienated post-war subject". Alec Follett's chapter on Canadian writer Thomas King similarly thinks about the relationship between genre and petroleum energy, with a specific focus on how these ideas intersect with the text's wider engagement with environmental racism and the legacies of settler colonialism. Megan Cole Lyle also considers the "omnipresence of carceral petro-capitalism" in her reading of Rachel Kushner's *The Mars Room*, arguing that the novel demonstrates the ideological limitations of the American environmental literary tradition. Through Kushner's conflation of Henry David Thoreau and Ted Kaczynski as examples of an enduring and regressive "masculinist American fantasy of nature-as-freedom", Cole Lyle argues that the novel presents anthropogenic climate change and the "carceral state" as fundamental parts of the same "capitalist matrix".

The final three chapters in this section are all united by a focus on questions of globality and transnationality, beginning with Valerie McGuire's piece on the Venice-based novels of American crime writer Donna Leon, in which she argues that the ostensibly local nature of Leon's work belies the global scales of ecological injustice that the texts frequently reveal. Leon's fiction similarly features as part of Aina Vidal-Pérez's comparative analysis of the local and global readerships of crime fictions set in Italy. Placing Leon's work alongside Francesco Aloe's *Il vento porta farfalle o neve* – a novel published as part of a wider collaboration between crime writers, the Milan-based publisher Edizioni Ambiente and Italy's top environmental organisation, Legambiente – Vidal-Pérez argues that both authors use crime fiction to reach popular audiences, raising environmental awareness at different scales. Vidal-Pérez examines the "disparate scopes of circulation" that the novels encounter, even when dealing with the same or similar global environmental issues, thus revealing the respective authors' unequal access to "mechanisms of internationalisation". In the final chapter of the collection, Damini Ray looks at Rajat Chaudhuri's ecological crime novel, *The Butterfly Effect*, arguing that the text's vast temporal and geographical focus raises crucial concerns about the uneven transnational impact of environmental disasters, as well as emphasising the importance of memory in sustaining an ethical ecological consciousness.

In some form or other, all of the chapters in this collection are united by a shared preoccupation with endings, resolutions and/or futures, particularly as regards to the crime novel's increasing inability to offer closure or reassurance in a moment of such profound uncertainty. Of course, the degree to which we see this as a failing or fatal flaw of contemporary ecological crime narratives (or crime narratives in general) depends entirely on how we interpret the ideological underpinnings of the crime genre as a whole. The impression that all crime fictions must offer resolutions to the mysteries they present, or somehow work to maintain an ingrained status quo, rests on an assumed political conservatism underpinning the history and traditions of the mode, when, in truth, crime fiction has always moved variably between positions of radicalism and orthodoxy (Pepper 2). Moreover, and as Patrick D. Murphy notes in his chapter in this collection, crime novels that offer overly neat resolutions to a complex series of global, environmental and political entanglements may only work to "reinforce cultural inertia rather than challenging it".

Conversely, then, perhaps it is by embracing open-endedness, or by actively working against plot, that we can begin to meaningfully unpick the complexity of our world systems. In her influential study on petroleum culture, *Living Oil*, Stephanie LeMenager refers to Helon Habila's *Oil on Water* and Attica Locke's *Black Water Rising* as examples of contemporary crime fictions that embrace a strategy of "open plotting", gesturing towards the possibility of "alternate worlds from within a genre" traditionally associated with closure (136). These "imperfectly resolved" mystery novels "pursue plot" only to abandon it, negating the drive for clear or complete understanding (136). However, and to return to where we started at the top of this introduction, LeMenager argues that such texts do not leave us with the same lingering anxiety or dread that one might typically associate with the "imperfectly finished" plots of noir fiction and film. Rather, the open or "diminished" plot instead works to invite the reader into a "kind of production", where they become an activate participant in the continuation of the text's critical work (136).

Like the crime fictions discussed by LeMenager – works that demonstrate the continued significance of crime fiction in reflecting upon, and giving shape to, the density of our world – it is my hope that *The Routledge Handbook of Crime Fiction and Ecology* will impel a similar kind of active response in its readers. Rather than seeking to foreclose or delimit the terrain or scope of ecological crime fiction as a field of study – and by working against the 'master plot' of crime fiction so often reproduced in collections of this kind – the chapters in this handbook point excitedly to the future of crime criticism, hopefully paving the way for further innovations and developments.

Note

1. See, for instance, Andrew Pepper's *Unwilling Executioner: Crime Fiction and the State*; Pepper's and Schmid's *Globalization and the State in Contemporary Crime Fiction*; Matzke's and Muehleisen's *Postcolonial Postmortems: Crime Fiction from a Transcultural Perspective*; Pearson's and Singer's *Detective Fiction in a Postcolonial and Transnational World*; Gulddal's, King's and Rolls' *The Cambridge Companion to World Crime Fiction* and Krajenbrink's and Quinn's *Investigating Identities: Questions of Identity in Contemporary International Crime Fiction*.

Bibliography

Allan, Janice M., et al. "Introduction: New Directions in Crime Fiction Scholarship." *The Routledge Companion to Crime Fiction*, edited by Janice M. Allan et al., Routledge, 2020, pp. 1–9.

Bandyopadhyay, Nibedita. "'The Green Sleuth': An Analysis of the Environmentalism in the Selected Detective Fictions of Sunil Gangopadhyay." *Green Letters*, vol. 22, no. 1, 2018, pp. 67–77, doi:10.1080/14688417.2018.1431142.

Cresswell, Tim. *Place: A Short Introduction*. Wiley, 2013.

Dainotto, Roberto Maria. *Place in Literature: Regions, Cultures and Communities*. Cornell UP, 2000.

Ghosh, Amitav. *The Great Derangement: Climate Change and the Unthinkable*. U Chicago P, 2016.

Goodbody, Axel, and Adeline Johns-Putra. "Introduction." *Cli-fi: A Companion*, edited by Axel Goodbody and Adeline Johns-Putra, Peter Lang, 2019, pp. 1–17.

Gulddal, Jesper, and Stewart King. "What is World Crime Fiction?" *The Cambridge Companion to World Crime Fiction*, edited by Jesper Gulddal et al., Cambridge UP, 2022, pp. 1–24.

King, Stewart. "*Crimate* Fiction and the Environmental Imagination of Place." *The Journal of Popular Culture*, vol. 54, no. 6, 2021, pp. 1235–1253, doi:10.1111/jpcu.13083.

Lefebvre, Henri. *The Production of Space*. Wiley, 1991.

LeMenager, Stephanie. *Living Oil: Petroleum Culture in the American Century*. Oxford UP, 2016.

Moore, Ellen E. *Landscape and the Environment in Hollywood Film*. Palgrave Macmillan, 2017.

Morton, Timothy. *Dark Ecology: For a Logic of Future Coexistence*. Columbia UP, 2018.

Murphy, Patrick. *Ecocritical Explorations in Literary and Cultural Studies*. Lexington Books, 2009.

Pepper, Andrew. *Unwilling Executioner: Crime Fiction and the State*. Oxford UP, 2016.

Pepper, Andrew, and David Schmid. "Introduction: Globalization and the State in Contemporary Crime Fiction." *Globalization and the State in Contemporary Crime Fiction: A World of Crime*, edited by Andrew Pepper and David Schmid, Palgrave Macmillan, 2016, pp. 1–20.

Pratt, Mary Louise. "Arts of the Contact Zone." *Modern Language Association*, 1991, pp. 33–40.

Puxan-Oliva, Marta. "Crime Fiction and the Environment." *The Routledge Companion to Crime Fiction*, edited by Janice M. Allan et al., Routledge, 2020, pp. 362–370.

Schmid, David. "From the Locked Room to the Globe: Space in Crime Fiction." *Cross-Cultural Connections in Crime Fictions*, edited by Vivien Miller and Helen Oakley, Palgrave Macmillan, 2012, pp. 7–23.

Stougaard-Nielsen, Jakob. "World Literature." *The Routledge Companion to Crime Fiction*, edited by Janice M. Allan et al., Routledge, 2020, pp. 76–84.

Walton, Jo Lindsay, and Samantha Walton. "Introduction to Green Letters: Crime Fiction and Ecology." *Green Letters*, vol. 22, no. 1, 2018, pp. 2–6, doi:10.1080/14688417.2018.1484628.

Walton, Samantha. "Studies in Green: Teaching Ecological Crime Fiction." *Teaching Crime Fiction*, edited by Charlotte Beyer, Springer, 2018, pp. 115–130.

PART I

Space and Topography

1

AFFECT IN PETER MAY'S LEWIS AND HARRIS NOVELS

Terry Gifford

I am writing the first draft of this chapter on the Isle of Lewis in Scotland's Outer Hebrides during a tempestuous October and November when the island's elements are at their most volatile and its ecology is in a state of seasonal change. A thunderbolt on my first night cut my internet connection for my entire stay. I have watched snow buntings touch down on the Butt of Lewis, its most northernly point, whilst rainbows are a daily unexpected surprise almost anywhere. My son Tom, partner Sally and granddaughters Islay and Elsie, have missed a day of school in Lincoln after half-term because the ferry off the island was port-bound for 2 days by Storm Aiden (see *The Chessmen* 91). In my rucksack I have brought four crime novels by Peter May which are set on Lewis and Harris. I intend to conduct research on the affect of their ever-present awareness of the island's ecology in the mode of narrative scholarship by acquiring for myself a sense of the place, its materiality and culture that Peter May knows so well. My cottage in Ardroil looks across Uig Sands to Uig Lodge where May spent months between 1992 and 1996 filming the Gaelic language series he produced for Scottish television, *Machair*, before eventually moving to live in France. It was when looking at the images of the Isle of Lewis that hung on the walls of his house in South West France that May decided to set his next crime novel here: "A wild, wind-swept corner at the extreme north-west of Europe that no one had ever used as a setting for a crime novel. A place I knew so intimately it was almost part of me" (*Chessmen* 103).

Those two sentences raise several questions concerning affect, ecology and the crime novel for me as I sit in that "wild, wind-swept corner" myself. Central to them is whether the narrative of May's crime novels is itself influenced by its 'setting', whether the social ecology is shaped by its natural ecology, whether character and ethics presented by the author are defined by the affect of weather, land and sea upon him. Is what he thinks of as 'a setting' for a genre narrative actually driving that narrative, even in ways of which he may be unaware? And what are the limitations of his exploration of his 'intimacy' with this distinctive more-than-human ecology that is actually under threat from industrial-scale mining, windfarms and fishing? "Within ecocriticism, the figure of the detective is richly suggestive", write Jo Lindsay Walton and Samantha Walton in their introduction to the "Crime Fiction and Ecology" special issue of *Green Letters* (2). What are the limitations to Peter May's detective's 'apprehending totality', as they put it, on the island of Lewis?

DOI: 10.4324/9781003091912-3

In a Zoom interview with Hugh Topping for Topping's Bookshop on 26 March 2021, May revealed that he became a writer of crime fiction because of an environmental concern in a global context that might exemplify a kind of 'apprehension of totality':

> I didn't choose crime fiction; it chose me. I wanted to write a first novel set in China about genetic food modification, but how to tell it? If I had a body found at the beginning that would be a good start (*The Firemaker*, 1999). I was trying to make a living as a novelist. After that first book my publisher offered a deal if I wrote about the same characters in the same setting. So I fell into the genre.

Asked whether he believed that people were shaped by their natural environment and whether that might be explored to raise his readers' awareness of ecology and even climate change, May confirmed this as a continuing interest in his work:

> Crime fiction is a broad church. I've tried to push the boundaries of crime fiction. I've a great interest in environment and I wrote a novel based upon bee colony collapse (*Coffin Road*, 2016). I was stopped by the pandemic from going to the Norwegian archipelago of Svalbard to research my next novel, a place which has been much affected by climate change. Yes, I believe that people are affected by their environment. In a place like the Outer Hebrides you develop a way of living that is influenced by weather and geography that is influencing your culture. A sense of place is always important to me. I never write about a place I've never been to.

Indeed, in 2023 May published a novel titled *A Winter Grave* set in mainland Scotland ravaged by climate change in the near future.

'Islandness' in *The Blackhouse*

In the first novel of Peter May's trilogy set on the Isle of Lewis, *The Blackhouse* (2011), that features his detective Fin Macleod, Fin says to his former girlfriend, "The world's like the weather, Marsaili. You can't change it. And you can't shape it. But it'll shape you" (153). The challenging and fast-changing weather of the Atlantic seaboard has a continuing dramatic and forceful presence throughout all three novels of the trilogy, *The Blackhouse*, *The Lewis Man* (2012) and *The Chessmen* (2013), together with the stand-alone novel set on Harris that followed them, *Coffin Road* (2016). The way the weather, a force of ecology that May's characters think of as beyond human influence, can 'shape you' is clearly a matter of affect. The weather is one aspect of what May calls 'the elements' in his novels, and reading the trilogy the reader might be forgiven for wondering how many different forms of discourse the author might deploy to characterise the pervasive presence of rain. People in these novels are clearly 'shaped' by rain but are also aware of the state of the tides, the bleak appearance of peat bogs, the resistance of gneiss, the recurrence of rainbows, the different qualities of beaches that appear to be simply of sand and the particular configuration of beaches, bays, cliffs, machair (grassland behind Marram-grassed dunes), moorland and mountains that define the Isle of Lewis and its attached 'isle' of Harris to the south. Quite how that 'shaping' has come about, and its consequences for character formation, home making, working life and social interaction, are at the centre of these narratives.

But in saying this, already a sense of social ecology is at work, as in the statement that 'the world's like the weather'. The world these islanders inhabit is of a sharply defined culture that is full of tensions. In a crime novel such tensions are revealed in a constant implicit reference to ethical positions. In crime novels set on the Isle of Lewis these ethical tensions are located in a culture that is defined, like the ecology, by a particular kind of 'islandness'. One might say that an archipelagic set of ethics is at play in these crime novels that is itself derived from the tensions of growing up, leaving and returning to this distinctive island culture. Inevitably this particular 'islandness' also shares some characteristics and issues discussed by archipelagic studies.

'The world' is enlarged in *The Lewis Man* as the story ranges through 'the long island' of the Outer Hebrides to the island of Eriskay in the south, and ecological differences are crucial to this narrative. And 'the world' beyond these islands is modified in its reception and influences by the world of 'islandness' in the novels. If the narrative moves to Edinburgh, as it does in *The Lewis Man*, its different ambiance is compared to that of the Isle of Lewis (355). The material ecology of Lewis affects the social ecology, which, in turn, affects the ethical positions of the characters in Peter May's Lewis trilogy. Against a sense of determinism that might be implicit in this analysis of the novels' context is the possibility of human agency, of the detective's potential for changing 'the world' as given, by bringing a changed state of 'justice', or at least of an empowering understanding of its complexity. Actually, in May's narratives, a sense of complexity is available from a sensitivity to the material ecology of the islands which plays a key part in the narrative structure – the unfolding of following the clues in the islands' material and social ecology. This is in contrast with the Shetland detective novels of Ann Cleeves (her Northumberland novel *The Crow Trap*, the subject of the next chapter, is a different matter.) Peter May's crime novels are richly and absorbingly driven by his interest in the material and cultural ecology of the Isle of Lewis. In Cleeves' novel *Raven Black* (2006), for example, the issue of overgrazing on Shetland is only mentioned sarcastically by a teenage daughter as a radio "party piece" (192) of her father, a conservationist whose work is not discussed beyond his monthly "beached bird survey" on which he once takes his daughter. (A daughter in May's *Coffin Road* has a father whose "hobby horse" is GMOs (104).) But at bottom their value is in the mysterious connection between place and values, sensitivity to the dynamics between people that, in their turn, are affected by their sensitivity to the dynamics of living on Lewis. This is not to say that the Isle of Lewis is a character in the novels, but that characters in the novels are 'shaped' by all that is Lewis. Battling against the elements, battling in poverty to earn a living between the land and the sea, battling, in May's characterisation, against a puritan culture, are all somehow at one with Fin Macleod's battle for truth and for justice in these social and elemental conditions. These are complex novels written in a simple, accessible style within a familiar genre, that, in a hard-nosed manner, convey a compassion and humanity that is also hard earned. Their achievement as explorations of archipelagic nature-culture is not to be underestimated.

It was a truism of early ecocriticism that what had formerly been regarded as the 'setting' of a novel was now the foreground of critical attention. In his early and influential book *The Environmental Imagination* (1995), the American ecocritic Lawrence Buell's chapter on 'Place' now reads like a historical turning point. Rejecting Eudora Welty's "subordination of place to the role of handmaiden" (255) in her famous essay 'Place in Fiction' (1942), Buell argues that "we do not think about our surroundings, and our relation to them, as much as we ought to" (261). More recently cognitive ecocritic Alexa Weik von Mossner has drawn attention to the neglect of "aspects of the narrative strategies that writers use to create immersive environments for readers" (12). In Peter May's trilogy anyone living on Lewis cannot but be always aware of the subtle shift in the seasons, the present and oncoming weather, the conditions underfoot and on the roads, and the state of the tides. However, what Buell says of American nature writing may well remain true for the plot-driven crime novel: that their ecological "stature is not recognised . . . because we have not learned how to read them" (270). What D. H. Lawrence wrote in the first version of his essay 'The Spirit of Place' in 1918 remains true for May's novels: "All art partakes of the Spirit of Place in which it is produced" (16). In May's novels that spirit combines breathtaking beauty, in rainbows and beaches, with life-threatening storms and seas, and persistent rain and wind. Lawrence went further: "Every great locality expresses itself perfectly, in its own flowers, its own birds and beasts, lastly its own men, with their perfected works" (30). If the "guga" is this place's Gaelic name for young gannets, their annual hunt, which becomes the focus of *The Blackhouse*, might be regarded as the dangerous and skilful "perfected works" of the twelve men from Ness who not only hunt but process the food for their islander neighbours. Some conservationists, who are represented in the novel, would challenge this hunt as 'perfected works'. But the point is that this storm-bound island, which certainly 'expresses itself perfectly' has produced people who struggle, not only for the means of survival, but also for decency, neighbourly compassion, community solidarity and ethical behaviour. Indeed, the 'imperfect work' of a murder that launches a crime novel demands the drive for the 'perfected works' of justice and resolution.

May's own explicit example of the way a "great locality expresses itself perfectly" in the "perfected works of men" and women would be Gaelic psalm-singing, which Fin reflects upon in precisely this way:

> A strange, unaccompanied tribal chanting which could seem chaotic to the untrained ear. But there was something wonderfully affecting about it. Something of the land and the landscape, of the struggle for existence against overwhelming odds. Something of the people amongst whom he had grown up. Good people, most of them, finding something unique in themselves, in the way they sang their praise to the Lord, an expression of gratitude for hard lives in which they had found meaning.
>
> *(*Blackhouse *85)*

The "overwhelming odds" are, of course, the Lewis ecology, and Fin's celebration of the Gaelic psalm-singing of Lewis is undercut by its being an expression of the resilience of "hard lives". Throughout the novels May endorses not just the "struggle for existence", but the economic poverty of most lives engaged in this elemental struggle as crofters. There are pervasive casual references to this heritage in characterisations such as those of "pale, mean faces born of generations of island poverty" (*Blackhouse* 357). Nevertheless, communal improvised singing reflects, in this passage, a sense of lives, hard as they are, "in which they had found meaning". Despite Fin's hatred for the meanness and rivalry of the puritanical churches, of which there are "five different Protestant sects" on Lewis we are told, in this passage he recognises that there is not a spiritual poverty in communal religious life and language on the island (*Lewis Man* 228).

Perhaps one aspect of Lawrence's notion of 'locality' producing distinctive 'works' might be Ralph Crane's and Lisa Fletcher's assertion that "places produce stories". More than an issue of the personification of landscape for symbolic effect, this view of the island signals the agency of place. Not only can islands function as settings, but they can also operate on the "level of character, and influence plot" (Crane & Fletcher 6). There are two ways in which major features of the plot of *The Blackhouse* signal the agency of place. The first is that, for young people especially, the boundedness of an island and its insular culture can seem a limitation to be transcended by leaving. The obverse of this is the lingering, often unconscious, desire to return. For a novelist this is an example of what Pippa Marland calls "an archipelagic-ecological heuristic" ("Literary Archipelagraphies" 223). May's detective, Fin Macleod, was born and brought up in Ness, the northern tip of Lewis, by parents who spoke Gaelic and neglected to teach him any English by the time he went to school where he found it was the language of instruction. A little girl, Marsaili, offers to translate for him. Later, she becomes his teenage girlfriend and eventually shares a flat with him as a fellow student at Glasgow University. Fin treats her badly, by his own later admission, and she returns to Lewis whilst he also drops out and eventually becomes an Edinburgh policeman. In the idyllic summer before Fin leaves the island he has a conversation on a beach with his best friend, Artair, who is depressed about a bleak dead-end future on the island working as a welder at an oil rig construction yard: "I can smell it already. And then there's all the years of travelling that fucking road from Ness to Stornoway, and a hole in the ground at the end of it all" (162). Due to the nature of the land there is no direct road between Ness and Stornoway. Fin tries to cheer his friend by saying, "Hey, look around you. It doesn't come much better than this" (162). Here is a rare moment when May celebrates island ecology as aesthetics. Artair replies, "Yeh, that's why you're in such a fucking hurry to leave". So here, in its role as both trap and release, is the agency of the island at play, as it is in Fin's reception and adjustment on his return to investigate a murder that has strong similarities with one in Edinburgh which Fin has been investigating.

As Fin investigates the murder, his revisiting people and places is charged with a tension derived from his not having returned since he was a teenager. This is a recognisable archipelagic angst, but it lends May's crime novel a degree of tension particular to place which is a key to the plot. The murdered man was a bully who, as a fellow schoolboy, victimised both Artair and Fin. When Marsaili returned to Lewis from Glasgow she married Artair but told him in a moment of bitterness that their son Fionnlagh, born soon after, was actually Fin's, although later she admits that she is not sure. Actually, Artair has committed a murder in imitation of the Edinburgh murder in order to bring Fin back to the island. His intention is to have Fin watch his son thrown from a cliff during the annual guga hunt, partly in revenge for Fin's leaving in the first place. Thus the plot turns towards the second major feature of *The Blackhouse* to signal the agency of place: the traditional killing of young gannets (the guga) for food. Men of Ness have been drawn to the island of An Sgeir (really Sulasgeir – gannet rock) for generations each August to cull two thousand young gannets for food on Lewis. In recent years they have been granted a licence to do so, with the approval of the Royal Society for the Protection of Birds, despite protests from some conservationists. It is a privileged rite of passage for young men chosen to be part of the twelve men of Ness (nowadays actually ten men). In his authoritative book, *The Guga Hunters* (2008), the Ness-born writer Donald S. Murray points out that harvesting birds for food has been common practice in subsistence communities around the world and documented in journeys to Sualsgeir "since at least the sixteenth century" (8). Robert Macfarlane agrees in his chapter following the hunters in *The Old Ways* (2012), although Pippa Marland has some nuanced reservations about Macfarlane's commentary in her book *Ecocriticism and the Islands* (188–196). Half a mile long and two hundred yards at its widest, the ecology of this barren rock is marked by its two hundred feet high cliffs as nesting sites for a variety of seabirds. Apart from its lighthouse, the only structure on the island is the remains of a blackhouse that gives the book its title, walls on which an improvised roof of tarpaulin (turf or thatch roofs on stone walls mark a blackhouse on Lewis; modern roofs on concrete walls define a whitehouse) is made for the two weeks each year by the guga hunters. The current leader of the hunters in the novel, Gigs, explains that in this experience they are "reaching back through the centuries, joining hands with our ancestors" (201). Later, he admits that "It's not the tradition. That might be part of it, aye. But I'll tell you why I do it, boy. Because nobody else does it, anywhere in the world" (216).

The novel's men of Ness are also a closed community within an insular island community. What happens on the rock, and what is said on the rock, stays on the rock (371). Fin's idyllic last summer on Lewis was darkened only by his dread of joining the hunters together with Artair Macinnes and his teacher father who had been coaching both his son and Fin for their exams. In a drunken state Fin was rescued from a ditch by Gigs to whom he confessed that Mr Macinnes had been sexually abusing Artair and himself. On the rock Gigs confronts Macinnes and bans him from the blackhouse to survive in the sea caves below the cliffs. So Artair's anger at Fin's escaping the abuse by leaving Lewis is deepened by Fin's confession having resulted in ten men knowing this appalling secret, with whom he is condemned to live whilst Fin is not. May's continual reminder to the reader of Lewis's island ecology through references to weather, tides, ferries, foghorns, peat-cutting, beaches and bogs reinforces the archipelagic experience on which a key plot motivation turns. But the final twist is contributed by one of the processes of crime investigation. On the last page of the novel, DNA evidence proves that Fionnlagh is actually Fin's son. When Fin had left Edinburgh he had recently lost his eight-year-old son in a hit and run car accident. It is a mark of May's compassion that he ends *The Blackhouse* by restoring an unknown son to his detective.

Culture Shaped by Nature in *The Lewis Man*

May's compassion is perhaps most in evidence in the second novel of the trilogy, *The Lewis Man*, which takes as its central narrative structure the journeys of 'homers' who were 'boarded out' by childless Catholics in South Uist from orphanages on the mainland. Following on chronologically from the first novel, Fin returns to Edinburgh to realise that his marriage had only been sustained by the son he had lost and, resigning from the police force, he accepts his wife's prediction that he will return to Lewis when she says, "You can never escape the island. It was there between us all those years, like an invisible shadow. It kept us apart. Something we could never share" (17). Fin's island identity does indeed draw him back to restore the abandoned croft house of his parents. When a body is discovered by turf-cutters and its DNA links the body to Marsaili's father, Tormod Macdonald, Fin realises that he has a week to establish whether Tormod, as number one suspect, is really the killer, before a detective is sent over from the mainland. But Tormod is suffering from Alzheimer's, and the novel is partly narrated in his voice, recounting the death of one of a family of bullies at the orphanage, the Kellys, in a bridge-walking challenge at which Tormod's younger brother Peter is held responsible by the bully's remaining brothers. They have misread Peter's reaching out to help as a push that killed their brother. Fin's quest to find out more about Tormod leads him south through the archipelago to the island of Eriskay, where Tormod and Peter were taken in as child labour by Catholics on a croft. It emerges that the avenging Kelly brothers eventually found Peter there and killed him. A distraught elder brother stole the crofter's van, loaded Peter's body in a blanket and drove as far north as he could to bury the body in a peat bank outside Ness. There he adopted the name Tormod Macdonald (he had originally been John McBride), came to be married and settled on a croft where Marsaili was born and raised. When Marsaili discovers all this, she says, "I don't know what to feel about him any more, Fin" (371). Fin makes a case for compassionate respect and interest in the elderly that reveals his deep and articulate humanity.

This humanity has been forged, readers are reminded from time to time, by an archipelagic upbringing in an ecology that is ever-present in the novels. In *The Lewis Man*, Fin is returning to Ness passing peat-cutters out in a rare "dry afternoon" that produces a rare lyrical moment when "sun slanting across the bog to the east spun gold in the dead grass" (169). But the shadow of the bleak church at Cross reminds him that he is nearly home.

> Home? Was this really his home now, he wondered. This wind-ravaged corner of the earth where warring factions of an unforgiving Protestant religion dominated life. Where men and women struggled all their lives to make a living from the land, or the sea, turning in times of unemployment to the industries that came, and went again when subsidies ran out, leaving the rusting detritus of failure in their wake.
>
> *(169–170)*

However, an archipelagic upbringing amidst this struggle with the ecology of Lewis produces personal qualities that are derived from that ecology: "But if here wasn't home, where was? Where else on God's earth did he feel such an affinity with the land, the elements, the people?" (170). Culture, formed in part by nature, and a particular Lewis nature engaged by culture, produces 'an affinity' in Fin. This is an essential factor in May's characterisation of his investigator whose complex, island-formed, nature-aware humanity holds the reader throughout the trilogy. What readers witness in Fin's character is the subtle past and present play of affect, from 'spun gold' to 'rusting detritus' to the 'dark shadow' of an 'unforgiving church'.

It would be easy to dismiss this discourse as merely contributing "atmosphere" to May's crime novels. But in these novels physical, elemental atmosphere, produced by both natural and cultural ecologies, has an effect upon characters that might be understood through the ecological aesthetics of Gernot Böhme, as translated and explained by Kate Rigby in *Reclaiming Romanticism* (2020). Böhme moves from suggesting that "spatial atmospheres feature whenever the moods or motivations of characters are shown to be inflected by their surroundings" (65), to a reciprocal experience of agency since "sensual perception means participating in the articulate presence of things" (63), to a social experience "with others, by whom, like it or not, our psycho-physical state of being is inflected" where atmosphere "can also acquire a moral force" (63). This, in turn, is experienced by the reader since such discourse "performs the work of ecological aesthetics not so much by representing the experience of atmosphere but by actually producing it . . . eliciting an actual affective response to the virtual space of feeling" (65). These might seem to be rather over-theorised claims for crime fiction, but May's punchy prose of short direct sentences can easily be seen to be working in "the virtual space of feeling" and not just in the case of Fin. Early in *The Lewis Man* we hear Tormod's voice, sitting in his care home:

> The rain is hammering on the window. It's making some din! When you were out on the moor you never heard it, of course. You heard nothing above the wind. But you felt it all right. Stinging your face when a force ten drove it at you. Horizontal sometimes. I loved that feeling. Out there in the wild, just me and that great big sky, and the rain burning my face.
>
> But they keep me cooped up inside these days. Not to be trusted outdoors, bad Mary says.
>
> *(59)*

It is surely hard for the reader not to have some empathy from the production of elemental exhilaration and the ultimate feeling of entrapment here. Fin's stoicism in the face of such weather is also part of his character as an investigator: "It was a filthy morning . . . He had grown up with this. It was normal" (387). Even in Edinburgh, where Fin and Marsaili go to find her father's orphanage, the island normality is present in its absence: "It was almost dark by the time they stepped back out onto the street, earlier than it would have been up on the islands" (355). When Fin and Marsaili walk on a beach at Ness, where they had made love as teenagers, May makes a conventional use of atmosphere as metaphor that anticipates their kissing for the first time since Fin's return: "The night was filled with the whispering sound of the sea. It sighed, as if relieved by the removal of its obligation to maintain an angry demeanour" (329). Less conventionally, May can be inventively playful in describing ecology. Tormod looks through the car window travelling south through Harris: "It hardly seems familiar at all. Not sure if it's grass bursting through the rock, or the rock bursting through the grass" (380). May is occasionally more explicit in using nature as metaphor, as when Peter's death is imagined by Fin above the very beach where he died: "Almost as if mirroring the moment, nature turned the sea the colour of blood as the sun sank on the horizon" (405).

The attention to nature that these examples reveal in May's Lewis trilogy exemplifies a primary requirement in Böhme's ecological aesthetics: "An aesthetic relation to nature consists in allowing oneself to be spoken to by it" (63). When Fin first went to Eriskay on his quest to discover the real identity of Marsaili's father he remembered a detail from the post-mortem report on Peter's body:

> the pathologist had found fine silver sand in all the abrasions and contusions of the lower body. Not golden sand, as found on the beaches of Harris. But silver sand as found down here, on what Tormod had called Charlie's beach.
>
> *(261)*

May has one of his clues be dependent upon Fin's being spoken to by differences in ecology because he has "allowed himself to be spoken to by it". It is no coincidence that Fin is able to form a sympathetic relationship with the confused Tormod, not only by listening to his partial recall, but also by taking him to a beach where they are both "spoken to" by its ecology. They walk together barefoot in the sea: "The wind blew Tormod's coat around his legs and filled Fin's jacket. It was strong in their faces, and soft, laced with spray, blown uninterrupted across three thousand miles of Atlantic ocean" (161). Tormod is invigorated and "happy, as in childhood, to delight in the simplest of pleasures" (162). Here a corporeal connection with nature temporarily eclipses culture: "The sound of the wind and the sea filled their ears, drowning out everything else. Pain, memory, sadness. Until finally Fin stopped and turned them around for the walk back" (162). They walk back into the narrative of crime fiction, but moments of humanity and ecological connection like this make Peter May's Lewis crime fiction specially moving and reconnective with a nature-culture depth.

Natural and Social Ecologies in *The Chessmen*

The third novel in the trilogy, *The Chessmen*, begins with an unusual ecological event and sustains more of an environmental sense than the previous two novels through the issue of poaching and land ownership. Fin and his old school friend, Whistler, awake after spending the night sheltering from a storm in a mountain "beehive" stone shelter. They look down to where there had been a loch to find that it has disappeared. The heavy storm has broken a long spell of dry weather during which the peat had dried out to form a hard, deep layer lying on the Lewisian gneiss. Whistler helpfully explains that the rain has run down the cracks in the peat to form a layer of sludge beneath it. The pressure of the water in the loch has forced the peat retaining the loch to slide over the sludge in what is known as a "bog burst" (5). So the loch has drained down to a lower level, revealing an undamaged aircraft in which a body has been hidden under water for seventeen years, strapped into the pilot's seat. Thus is initiated the crime to be solved, for a head injury suggests that this man could not have piloted the plane and was probably dead before the plane landed on the water and sank. The ID on the body identifies Roddy Mackenzie who led a band for which Whistler played the flute and Fin was roadie in their teenage years.

Fin has taken a job as head of security on an estate where poachers are threatening the viability of its rod fishery, which is its main source of income from salmon and sea trout, plus brown trout in hundreds of lochs. As a former detective and local man he believes that he can solve this problem within two months. The estate owner knows that Whistler poaches the odd fish and is not part of the organised local gang who are netting, smoking and selling large numbers of salmon with supply lines into Europe, but asks that Fin makes an example of Whistler by stopping his poaching. The five water systems on the estate provide fishing rights shared with five other estates. Fin is told that "if they put us out of business, a lot of local people are going to lose their jobs" (54). Whistler earns a living by carving and selling large replicas of the famous Lewis chessmen carved by Vikings from walrus bone, found not far below his croft, in the dunes at Uig by a cattle herder in 1831. Whistler does not recognise Highland estate ownership of wild moors, rivers, lochs and mountains, and the creatures that live in them:

> A man's entitled to take from the land that the Lord gave us. And he gave it to us all, Fin. You cannae take it with you when you die, so how can anyone think they own it while they're living? *(29)*

This is an argument with which May himself appears to have some sympathy. Evoking the still-remaining memories of cruel clearances on Lewis in the mid-nineteenth century (see Craig 297–303), the narrator says, "a landowner continued to be seen in a strange reluctant way as being superior. A regard landowners also had of themselves" (53). Fin eventually quits his job because of this attitude in the estate owner.

But as teenagers Fin and Whistler discover that they are linked by one of the greatest tragedies not only in the history of Lewis, but of British maritime history. At 2am on the first of January 1919 the HMY *Iolaire* hit rocks yards from Stornoway harbour entrance whilst returning Royal Naval Reservists to the Outer Hebrides. Over two hundred island men who had survived the First World War drowned in the dark and stormy conditions. John Finlay Macleod managed to swim ashore with a line by which Whistler's great-grandfather saved the life of Fin's grandfather. May has an explicit repeat of this rescue between the two boys when Fin steps back off a wooden bridge into a torrent when they are fishing for brown trout and Whistler manages to throw him a line to save his life. In a typically subversive May touch, Whistler steals the Land Rover of the water-bailiff parked nearby to drive them home before returning it to the bailiff's house. "No one ever did find out who took it, or why" (189).

The role of Lewis's ecology in *The Chessmen* is wide-ranging in technique and significance. From casual references ("He had not heard Gunn approach over the noise of the wind" [190]), to the historic tragic storm of the Iolaire disaster, to the famous riparian economy (Ted Hughes made two fishing trips to Lewis, staying at the expensive fishing lodge at Grimersta), to pure aesthetics ("The tide was in, emerald water a foot deep over acres of golden sand, splinters of distant sunlight stabbing through breaks in the cloud, firing light in fast-moving flashes across the far machair" [193–4]), May integrates nature and culture with a delight that goes beyond the vehicle of a crime novel. Indeed, at times the narrative drive slackens for moments, and even chapters, of sheer archipelagic celebration. As though aware of this, May takes the final denouement of his narrative to a villa above the Mediterranean near Malaga in Spain where Roddy reveals why he did not die in the plane discovered in the drained loch. Over a paella – with prawns, Roddy jokes, "probably shipped from Stornoway" (316) – he explains to Fin that, having learned to fly in later life after losing touch with Fin, Roddy had been pressured by his girlfriend's family into making drug runs to land on Solas beach in North Uist. Wanting to escape from this lifestyle he got into a fight on the beach with his girlfriend's brother who was killed. Roddy then flew away with the body to ditch the plane on a remote loch in Lewis. He dragged the body into the pilot's seat with his ID planted on it and then disappeared to Spain, with the loyal help, it turns out, of Whistler. In May's Lewis novels, loyalty to school friends who have been boarding together during the week at the Nicholson Institute or Lews College in Stornoway from disparate parts of the island, features more than family feuds. Writing about the cultural affect of the Iolaire tragedy, John Macdonald says that it "put a new iron into the Lewis soul; inspire[d] young men to press for change, resist manipulation and dead traditionalism" (230). In the late decades of the twentieth century the survivors found an interest in their voices being heard on Lewis, expressing what Macdonald calls "a tenderness of fellowship" (230). But in Fin's case "fellowship" does not override his commitment to fairness and justice, his fiercely secular morality which has developed in the Lewis context of a predominant religious morality.

Ecological Anxieties in *Coffin Road*

A few years after completing the trilogy of novels centred upon Fin Macleod, May turned to the Isle of Harris in what has been characterised by one reviewer as an "eco-thriller . . . making the atmospheric most of his isolated locations" (O'Connell). *Coffin Road* carries an intriguing epigraph from Jeff Ruch, Executive Director of Public Employees for Environmental Responsibility: "Scientists . . . submitting works on neonicotinoids or the long-term effects of GMO crops, trigger corporate complaints . . . and find that their careers are in jeopardy" (np). Neal Maclean has been secretly conducting research on neonicotinoids with eighteen beehives hidden on the "Coffin Road". But in the dramatic opening of the novel he has been washed up on the beach in front of his cottage in Luskentyre with memory loss. As he pieces together the evidence that he finds about himself, he realises that he appears to have been writing a book about the disappearance of the lighthouse keepers on the Flannan Isles. When he borrows a boat to visit the Isles he discovers a body in the old church there with an instinctive fear that he might be the murderer. In what he knows to be his house, he finds a map with a line marked on it which he is told is the old coffin road. When he walks up it, he is drawn aside to find hidden beehives and finds that he knows a lot about bees. On his hands are bee stings, as there were on the hands of the body he found. It is eventually revealed that Neal is actually Professor Tom Fleming whose research in Edinburgh was funded by "a Swiss agrochemical company" (140) hoping to be vindicated in the matter of the damaging effect of neonicotinoids on bees. In fact his research found the opposite, so Fleming has gone rogue to conduct conclusive research with two other scientists at three secret sites. All data is sent to Fleming. When published it will, as one of the scientists says, "blow the agrochem industry out of the water" (303). The Swiss company is seeking this data, which Fleming hides in the Flannan Isles' famous lighthouse, and will commit murder in order to capture the data on the hidden hard drive.

The book acknowledges the research of Dr Christopher N. Connolly at Dundee University's Centre for Environmental Change and Human Resilience, "on whose research the science of my book is based" (391). But this science has been challenged by Dr Rebecca Nesbit in an online magazine that considers representations of science in literature (the 'Lab Lit' genre) called LabLit.com: "The scientific reality is also far more complex than the one May portrays, and the neonicotinoid debate is surrounded by uncertainty". Whilst admitting that "neonicotinoids are almost certainly a factor in the decline of wild bees", Nesbit wants a recognition of other contributory factors. She also objects that scientists have not, so far, been killed by corporate interests. By her own admission, however, her final conclusion is not justified: "This book seems intent on fuelling the powerful environmental lobby by using 'evidence' which is, unsurprisingly in a novel, a work of fiction". The suspicion here of "the powerful environmental lobby" is a giveaway of her sympathy for the corporate production of a chemical which she has to admit contributes to the decline of pollinators in a global environmental crisis. May's ecologically based plot may be a work of fiction, but in *Coffin Road* the environmental concern that he highlights is certainly not.

Against such ecological anxiety, matched in May's narrative strategy by the identity anxiety of the central character, is the aesthetic of landscape and the repeated topophilic trigger in the text of the famously beautiful location of Luskentyre. The Stornoway policeman, George Gunn from the Lewis novels, arrives wondering "if there could be any more beautiful spot on earth" (156), although he knows that generations of local kids have taken it for granted (167). It is significant that this idealisation comes from a character who is not a native of Harris. With the exception of DS Gunn, all the main characters of *Coffin Road* are not natives of the islands. For this reason, although the novel's plot hinges on an environmental issue, there is less sense of Marland's "archipelagic-ecological heuristic" in this novel. May has chosen an international environmental issue for this novel rather than one specifically located in the islands' ecology, such as the proposal for a Harris superquarry at Lingearabhagh, the decline of the Atlantic salmon or the overfishing of the islands' coastal waters. Only in his photographic book, *The Hebrides*, does May comment on current proposals for huge windfarms on Lewis: "If planners have their way, this landscape will shortly be marred by forests of wind turbines" (108). However, it is not the first priority of the crime novelist to be responsible for addressing these environmental issues. One might argue that May's achievement is both a greater and more subtle one that renders his Lewis and Harris novels of importance beyond the confines of genre. But in doing so May also demonstrates that crime fiction can be a vehicle for insights into the role of affect in our understanding of the intertwining of natural and cultural ecologies in archipelagic literature.

Bibliography

Buell, Lawrence. *The Environmental Imagination: Thoreau, Nature Writing, and the Formation of American Culture*. The Belnap Press, 1995.
Cleeves, Ann. *Raven Black*. Macmillan, 2006.
Craig, David. *On the Crofter's Trail*. Jonathan Cape, 1990.
Crane, Ralph, and Lisa Fletcher. *Island Genres, Genre Islands: Conceptualisation and Representation in Popular Fiction*. Rowman and Littlefield, 2017.
Lawrence, D. H. *The Symbolic Meaning*. Centaur Press, 1962.
Macfarlane, Robert. *The Old Ways*. Hamish Hamilton, 2012.
Macleod, John. *When I Heard the Bell: The Loss of the Iolaire*. Birlinn, 2009.
Marland, Pippa. "Literary Archipelagraphies: Readings from the British-Irish Archipelago." *Contemporary Archipelagic Thinking: Towards New Comparative Methodologies and Disciplinary Formations*, edited by Michelle Stephens and Yolanda Martinez-San Miguel, Rowman and Littlefield, 2020, pp. 221–240.
———. *Ecocriticism and the Island: Readings from the British-Irish Archipelago*. Rowman and Littlefield, 2023.
May, Peter. *The Blackhouse*. Quercus, 2011.
———. *The Lewis Man*. Quercus, 2011.
———. *The Chessmen*. Quercus, 2013.
———. *The Hebrides*. Quercus, 2013.
———. *Coffin Road*. Quercus, 2016.
———. *A Winter Grave*. Quercus, 2023.
Mossner, Alexa Weik von. *Affective Ecologies: Empathy, Emotion, and Environmental Narrative*. Ohio State UP, 2017.
Murray, Donald S. *The Guga Hunters*. Birlinn, 2008.
———. *As the Women Lay Dreaming*. Saraband, 2018.
Nesbit, Rebecca. *Fiction as Environmental Lobby*, www.lablit.com/article/923. Accessed Apr. 2021.
O'Connell, John. "The Best Recent Thrillers: Reviews Roundup." *www.theGuardian.com*, 8 Jan. 2016, www.theguardian.com/books/2016/jan/08/best-recent-thrillers-reviews-roundup.
Rigby, Kate. *Reclaiming Romanticism: Towards an Ecopoetics of Decolonization*. Bloomsbury, 2020.
Walton, Jo Lindsay, and Samantha Walton. "Introduction to *Green Letters*: Crime Fiction and Ecology." *Green Letters*, vol. 22, no. 1, 2018, pp. 2–6, doi:10.1080/14688417.2018.1484628.

2

THE GOSHAWK DID IT

Nature Writing and Detection in Ann Cleeves' *The Crow Trap*

Ian Kenny and Irina Souch

> Rachael worked from a large scale map. She had already chosen her survey areas using the natural boundaries shown on the map. Neither sample was on Black Law land. One, a patch close to the burn and a disused lead mine, was heavily grazed. It was farmed by one of the Holme Park tenants, almost denuded of heather. It would be easy for walking but not, she suspected, very interesting for birds. The other was a piece of heather moorland, managed for grouse. It had been leased by the Holme Park Estate to a syndicate of Italian businessmen. She suspected they would not find the shooting so enjoyable with the industrial noise of the quarry in the background, but she presumed that Slateburn Quarries has offered the estate such a tempting deal that income from the shooting rights would hardly be missed.
>
> *(26)*

This passage is from Ann Cleeves' 1999 detective novel, *The Crow Trap*. It is emblematic of the way the author depicts the rugged Northumbrian landscape in which the story is situated, showing nature as deeply imbricated within the interwoven narrative threads. Although visually picturesque, the countryside is far from idyllic: it is not so much concerned with the tourist gaze as it is with geographic isolation and conflicting industrial, political, and environmentalist interests. In Cleeves' book, human lives are not separated from the environment but revolve around and are given shape and meaning by their relationship to it. It is where people conceal their secrets, relive their memories, and earn their livelihood. At the heart of the novel's human-centric dramas stands the quarry: limestone has been dressed and used in the region for generations, and what initially began as a small-scale local operation has grown from humble beginnings, expanding its hold on the landscape. The Langholme's, landed gentry of the great house, Holme Park, and its encompassing territories and reserves, are traditional aristocratic stewards. They practice hunting and trapping (now made into international business), but also control the species that are indigenous to the landscape. Farmers, like those at Black Law Farm, graze cattle on the wide commons of the hills, distinct from village life, but remain the traditional (and waning) lifeblood of the community. The area is also known as a popular ground for ramblers and naturalists. Thus, the novel shows how the strict boundaries of the human and natural worlds fade as they refract various traits and realities.

The Crow Trap (best known from its successful 2011 ITV adaptation to the small screen, which initiated the internationally acclaimed television series *Vera*) starts when three young women – botanist Anne Preece, ornithologist Rachael Lambert, and mammal expert Grace Fulwell – come together at Baikie's Cottage to work on an Environmental Impact Assessment of the proposed quarry. First to

DOI: 10.4324/9781003091912-4

arrive, the team leader Rachael finds her friend Bella, who lives nearby at Black Law Farm, hanged in the shed along with a suicide note. Despite this tragic discovery, the survey goes as planned. Although the holders of the land are eager to see the project approved, the owners of nearby Slateburn Quarries, who would be developing the site, seem less enthusiastic. The survey continues until Grace is found strangled near Baikie's cottage, having recently observed an unexpectedly high number of otters in the stream running through the area. Assigned to the case is Detective Vera Stanhope, who many years earlier, as a local rookie beat-cop, investigated the disappearance of a young boy close to the same spot.

Despite *The Crow Trap*'s essentially anthropocentric plot – the crimes revolve around a knotted family drama and have, in the end, little to do with environmental opposition to the quarry – it can be undeniably considered as an example of "nature-oriented mystery novels . . . [helping] to understand the degree to which environmental consciousness and nature awareness has permeated popular and commercial fiction" (Murphy 143). And it is not only because, as Jo Lindsay Walton and Samantha Walton argue, "crime fiction is a form of specialist knowledge in its own right, with its own distinct contribution to make to cultural understandings of human-nature relations and environmental crisis" (3). The heated controversies throughout the novel around how much damage the development of the quarry could do to the landscape already align Cleeves' narrative with contemporary ecological themes. However, the novel's environmental stance is most convincingly achieved through the incorporation of stylistic properties inherent to the British school of new nature writing known for its commitment to ecological accuracy and meticulous examination of the impact of human action on the environment (Cowley).[1] New nature writing has emerged within the past decade as a literary tradition that builds upon previous generations of authors whose environmentally focused narratives offered insights into the natural world. New nature writing combines these past approaches with new stylistic elements to create ecologically minded narratives that are affective descriptions and detailed encounters with landscapes and ecosystems. Indeed, without being aware of *The Crow Trap*'s genre, one could arguably perceive the quoted paragraph as pertaining to the new nature writing tradition that eschews the easy "pastoralization" of rural life but, instead, is "closely concerned with the mutual inflection of human and non-human in the idea of place" (Smith 11) and "marked by an attentiveness to the relationships that make up the landscape, the places and forms in which they can be found, and the various ways that they can be seen" (Lilley 3).

One can argue that the thorough analysis of unexpected intersections of environment and human legacies is a feature that makes detective fiction remarkably akin to nature writing, and the connection between these two distinct genres, as Cleeves' novel demonstrates, may prove conducive for generating new narratives bearing "a particular potential for reshaping the individual and collective ecosocial imaginary" (Heise 258).

The book's title refers both metaphorically and literally to this particular sort of genre blending. While Rachael happens upon a trap that uses live bait to lure crows into a large cage, Detective Stanhope sets a similar ploy later in the novel, using the isolated surveyors as bait to entice the murderer into her snares. Mimicking a common method of controlling crow populations in the region, Vera's plan and Cleeves' title gesture to the intricate genre intersections that are developed throughout the text. Such productive genre crosspollination allows readers to enjoy the detective's careful unravelling of multiple mysteries while equally acquiring better understanding of our ecological enmeshment through the complex ways in which the "characters' experiences both shape and are shaped by their engagement with aspects of the natural world" (Lehtimäki 137).[2]

Ann Cleeves herself is a remarkable example of such engagement. The author's recollection of how the novel came into being is inseparable from a distressing personal experience she lived through while embedded in the landscape she so vividly depicts in her story:

> *The Crow Trap* . . . was conceived while I walked miles round the Northumberland countryside with my husband, Tim. He had suffered a major psychotic episode and been hospitalised.

> Though he'd been allowed home, he was still very poorly and very restless. Walking was the best remedy. It was autumn. I remember low sunlight, hedgerows loaded with haws and sloes, and to the rhythm of our footsteps, I brought to life the dishevelled, compassionate middle-aged detective who would very soon become part of my life.
>
> *(Cleeves 2020)*

It is not surprising that Cleeves is often compared to new nature writers when it comes to offering precise, locally informed descriptions of natural environments. Thus, for instance, in their digital writing workshop, Norwich Castle Museum compares her *Shetland* book series to Mark Cocker's *Crow Country* as "in both cases the [authentic] settings are almost characters in themselves" (Norwich Castle Museum and Art Gallery). A British ornithologist website, *Fat Birder*, in turn, names Cleeves as the best-known representative of "birding fiction" – novels written from a birder's perspective or depicting a birder's murder ("Birding Fiction: Novels from a Birding Perspective"). Accordingly, Cleeves' personal webpage keenly mentions her previous employment at the Bird Observatory of Shetland to explain her ornithological knowledge and fascination with the island ("Shetland").

In Cleeves' work, the small-scale emplacements characterised by meticulous and tangible descriptions of the local flora and fauna encourage readers to imagine a real place, enmeshed with(in) other real places and ecologies thus counterbalancing an age wherein both often take on abstract and intangible (global) forms. These images are invaluable and, as Tim Ingold warns us, very fragile because they can be

> all too readily crushed by the high-powered impact of a global science more intent on establishing the authority of its own particular view of the environment, and of what human beings are doing to it, than on enhancing our own awareness or powers of observation.
>
> *(Ingold 95)*

Yet, it is exactly the references to "concrete specific places", as Cathy Elliott argues in relation to British new nature writing, that enable contrasting totalising narratives of the whole world as being on one and the same path of doom (35). For Elliott, nature writing possesses a great political potential which lies in the interrupting of standard dystopian stories of the planet's future:

> [W]e are invited to imagine a landscape in which the creatures (including humans) that inhabit it, as well as different times and their legacies, are piled on top of one another in no particular order . . . engaging with and writing about nature in detailed ways resists the politics of complacency and apathy that too readily suffuse engagements with climate change.
>
> *(Elliott 35)*

The Crow Trap demonstrates this kind of engagement with the complex temporalities of concrete places. The land – in its variety of distinct locations that are knit together – is where the traces of different times overlap. While people survey the landscape, looking to see what organisms – what lifeforms and stories – are associated with it, they weave their human understanding and ordering onto the places around them. The places themselves, and the memories of their past uses and forms, all butt-up against one another throughout the narrative. Described in interwoven layers of the physical features, each place is itself often a harbinger of the memories that suffuse them. These hybrid natural/cultural spaces exist across a variety of timelines, both real and imagined, making pasts present and presenting pasts.

In what follows, we will closely look at how, in the novel, people go about their business in and around a multiplicity of places, noticing and naming things that unfold along the way as they come in and out of focus, rekindle memories, and acquire (new) meaning(s). We ponder how these acts of 'stocktaking' create a sense of familiarity and belonging but also how, occasionally, the most familiar settings become uncannily incomprehensible, disturbing the existing modes of knowing, prompting new revelations and discoveries. Importantly, the latter experience also solicits our acceptance that places refuse fixity and tidiness: the constancy of the morphing landscape and its permeable and imbricated boundaries makes *The Crow Trap* an unconventional detective novel. While Vera's quest to restore order upon the human dramas is palpable, albeit at times convoluted, her relationship to the physical spaces negotiated in the novel is more ambiguous, inhabiting the terrain of remembrance and recollection, as she also negotiates the landscape of her past. Vera is ultimately able to solve the crime on account of her affective encounters with the disruptive and untameable landscape that the novel depicts. And so, being acted upon by both natural forces and human agency – always in process – the land itself perpetually undermines the received oppositions between nature and culture, science and intuition, perpetrator and victim, and past and present.

Taking Stock of Things: Landscape Exploration, Aesthetic Pleasure and Detection

Rather than adhering uncritically to the foundational tropes of the detective fiction genre that uphold myths of mastery and knowability proceeding on account of "superficially convincing evidence that is ultimately irrelevant" (Encyclopaedia Britannica), *The Crow Trap* brings to the fore the puzzle-solving nature of detective fiction through its affective engagement with landscape and place through multiple acts of stocktaking undertaken by different characters. While Vera comes across a number of red herrings and circumvents them in time, ultimately mastering the human-centric dramas and solving the constitutive crimes, Cleeves also elicits a subtle critique of the perceived simplicity by which detection narratives proceed. Typical genre themes of control and categorisation effectively fall short of the mark when turned away from the human psyche toward the presence that the landscape exerts throughout the novel. This landscape cannot be adequately contained on account of conventional methods. Instead, it is perhaps better understood as part of what Charles J. Rzepka describes as the nearly geological turn inherent in some of the earliest examples of crime fiction, which propose to disentangle the various agents within the text as an act of puzzling in order "to enable readers not to solve the crime but to exercise their retrospective imaginations. As we read forward, we imagine backward, analeptically" (Rzepka 3). Lingering within the relationships between various puzzle pieces – landscape and the natural world prominent among them – not only opens up a view on a wide variety of phenomena that might otherwise be overlooked but also invites for reflections of the genre itself. In *The Crow Trap* the puzzle pieces are continuously (re)arranged in intricate patterns by the police team, Vera (who in particular engages with the terrain both physically and psychologically), and the affective and deeply attentive work of the female surveyors.

The novel's prologue opens with a description of an Ordnance Survey map of "The North Pennines, Kimmerston and the Surrounding Areas" (1). Intended to provide sufficient information for tourists, the map does not mention Baikie's cottage by name nor does it convey an idea of "the dark shadow of the forest, the grey stone buildings beyond the yard" and the rocky snow-topped silhouette of the Faiburn Crag, which can be observed from the window of the neighbouring Black Law Farm (2). Echoing the quote from the beginning of this chapter, the description points to the inevitable disparity between the Ordnance Survey's uniform design, levelling the topography and the living experience of the land the map is expected to represent.

> Maps may attempt to assert control but . . . they are always open to subversion . . . The meticulous detail of the Ordinance Survey map (a detail that nonetheless is never complete as the nuclear power station, the bunker, disappear from view) make it open to renegotiation,

writes Barbara Bender (307). Time and time again, the novel addresses the insufficiency of abstract administrative methods of exploring the natural territory *and* investigating the crime, setting them against the value of local knowledge and place lore. Just as the tourist maps fail to relay the full range of vernacular features rendering the area distinctive, so to the large-scale maps and satellite landscape surveys used for the fieldwork need "ground-truthing . . . bending close to the soil . . . getting things right" (Cleeves 89). For the botanist Anne, the most rewarding part of her job is

> the detailed investigation, identifying plants within the [randomly placed two-meter square wooden] frames, recording their abundance. She loved teasing through the sphagnum moss for plants like cranberry, bog rosemary, bog asphodel, squatting so close to the ground so she could smell the peat, feel the insects on her fingers.
>
> *(Cleeves 90–91)*

Throughout the narrative, relying on maps is counterpoised with the intense sensory experience of being in the landscape. Survey maps, as feminist geographer Gillian Rose reminds us, are representative of geographic fieldwork which was once considered "a particular kind of masculine endeavour" (154) and "involved mastering the skills of surveying, mapping, sketching and photographing the land" (153). Arguing against the division of geography into the human, concerned with studies of social relations, and the physical, focusing on the natural environment (which tapped into the traditional nature/culture dichotomy), Rose advocates a different kind of geographical imagination, which goes beyond the disembodied, rationalised knowing of the land, and instead requires "the sensibilities of the aesthete as well as the objectivity of the scientist" (158). "For many, there is pleasure and wonder in fieldwork; looking gives the researcher the gorgeous scene, the enlightening detail, the breathtaking view, the beauty of diversity", suggests Rose (157). In *The Crow Trap* the female scientists (and Detective Stanhope, as we will argue later) maintain their professional gaze and analytical distance but also experience an intense pleasure from taking in the complex and magnificent Northumberland scenery. The fieldwork expeditions speak to all their senses attesting that, as Ingold contends, "the environment is, in the first place, a world we live in, and not a world we look at" (95). Ingold elaborates:

> We inhabit our environment: we are part of it; and through this practice of habitation it becomes part of us too. We see with eyes trained by our experience of watching what is going on around us, hear with ears tuned by the sounds that matter to us, and touch with bodies that have become accustomed, by the lives we lead, to certain kinds of movement. Smells, too, excite memories and anticipations. This inhabited world – the world of our perception – includes the earth beneath our feet, the sky arching above our heads, the air we breathe, not to mention the profusion of vegetation, powered by the light of the sun, and all the animals that depend on it, busily absorbed in their own lives as are we in ours.
>
> *(Ingold 95)*

In the novel, the women's "ambulant encounters" (Bender 306) with the landscape, are imagined as "arriving in a different world" (Cleeves 153) and, indeed, affect all their senses. They find themselves "out on the hill . . . [able to] see to the horizon in every direction" (211) and take in "the smell of gorse, damp peat and crushed heather, the sound of skylark, curlew and distant sheep" (383). They notice "primroses in bud and violets" in the sheltered bank of the river ford (89) and the long grass "mixed with buttercups and clover" in the low meadow towards the hill (437); they inhale "the smell of the peat" (157) and watch "cormorants standing on the staithes in the river" (259) and "the goshawk fly out of the forest to swoop onto a young grouse" (270); they walk into "a thicket of gorse in full flower with its sweet scent of roasted coconut" (382) and listen to "a burst of a birdsong" (357).

As previously noted, such naming, and careful description of a large variety of life animating the landscape clearly reverberates with new nature writing's aesthetic repertoire. Apart from enacting a heartfelt enjoyment from being there – inhabiting the environment in Ingold's sense – the ambient details oppose the flattened spatiality of maps and make readers acutely aware of the Northumberland geography with its unique species, patterns, and ecosystems. These scenes also confuse the perceived difference between human figures and landscape, simultaneously opening and grounding the possibility of ecological consciousness.

In concert with Rose's argument, the fieldwork experiences in *The Crow Trap* further illuminate the inseparability of scientific observation, measuring and recording, and the emotional commitment that informs the desire to know and understand nature's dynamic and self-regulating capacity. Assuming a position of the landscape's responsive inhabitant inevitably brings into question the correctness of easy scientific assumptions. The unbiased perusal of the local plant and animal population at the place of the abandoned lead mine does not confirm the prognoses of the survey's outcomes.[3] No calamitous environmental anomalies are found to advise against the prospective quarrying, which complicates the antagonism between the industrial intervention and what used to be the unpolluted wild area. The evidence of the land's multilayered history and never-ending resilience starts early in the narrative when Rachael visits the disused mine to see it colonised by nature and the elements:

> The track crossed the stream and came to the old lead mine. The estate had talked once of doing it up, turning it into a living museum, but nothing had come of it. Soon there would be little left to preserve. There was still a chimney but it was crumbling from the top, eaten into by the weather, so the brickwork seemed to unravel like a piece of knitting . . . There was the smell of stale water and decay.
>
> *(Cleeves 29)*

Parallel to this spontaneous process, there are the Forestry Commission's plantations bordering to the old mine area (Cleeves 519), and the Wildlife Trust's recent initiatives to make one of the nearby worked-out quarries into a new natural reserve by flooding the pits and turning them into ponds, which has already "attracted mallards, coot and moorhen" (109). Although not implying the pristine return to the wild, the organised attempts of "rewilding" along with the ad hoc adaptation of natural life to new circumstances do emphasise the long-lasting and entangled relationship between human and non-human worlds and bring in stark relief nature's perpetual refusal to recede.

The detective plot develops along a similar line of registering the land's patterns and rhythms and unveiling the ways in which all stories are brought together through temporary alliances between human and non-human nature, the echoes of which outlive their immediate purposes. We learn that as a child, Vera often accompanied her father on his visits to Constance Baikie – the artist and local naturalist – and has indelible memories of the cottage and her endless rambles through the adjacent fields and lanes. Vera's father, who was never fond of children and a widower, was a fervent amateur naturalist (Cleeves 235). Over the years, he and Constance developed a deep friendship through their shared hobbies: both stole (rare) bird eggs, forcing Vera to partake when wardens and park rangers were present, and amassed sizeable collections, even after foraging in this way was made illegal.[4] These complex childhood undertakings – and an early exposure to the landscape combined with her adult experience as a police detective – give Vera an intuitive and affective edge over other less locally attuned colleagues, whose traditional methods of inquiry that mimic the regulatory nature of the mine, the survey ordnance maps, and the rigidity of scientific knowing, result in little to no useful knowledge. Just like in the case of the female researcher team, Vera's investigation requires an objective, disinterested gaze, but also an ability to "inhabit" the environment abandoning herself to emotions and reminiscences it inspires, combining her deeply personal associations with the local natural world and the broader machinations at work in the novel. Armed not with fancy gadgetry that might enable a mastery over the environment and subsequent secrets it might be hiding, instead, she is armed with her natural observations, memories, and keen knowledge of the land. As a detective, Vera proceeds by unearthing and negotiating what lies hidden in the psychological and physical body of the landscape rather than attempting to control and organise. Such affective attunement suggests Detective Stanhope is a new type of a bottom-up detective who is sensitive to place-based details, rather than the traditional top-down, patriarchal figure once common in detective literature who is concerned primarily with the objective 'bigger picture'. This both functions as critique of the genre conventions while also forging important progressions: Cleeves develops a different sort of detective within an affective and feminist turn who senses and puzzles with greater variety, melding impartial observation, intuition, and aesthetic engagement into a new detective figure who also further nuances the propositions of Rzepka developed earlier.

In the final part of the narrative, Vera is walking the public footpath from Langholm towards the mine intending to finally apprehend the culprit:

> She . . . followed the well-worn path towards Black Law, walking steadily, only turning her head from time to time to check that no one was following her. The path crossed the hill. On the lower slopes there were dry stone walls. The grass was cropped low by sheep. When she'd walked here in her childhood she'd been fit. . . . It was another clear, hot day and soon she was sweating and dizzy with exertion. . . . She walked through a gap in the last crumbling wall and the path climbed steeply. The ground was more uneven. Bright green bog and tufts of juncous, curlew and skylark. . . . At the tarn she allowed herself to rest. . . . A slight breeze rippled the water and dried the sweat of her face. From where she sat she could look down into the valley, to Baikie's and Black Law farmhouse and the old mine. She stood up and walked on, finding the going easier because it was downhill.
>
> *(518–519)*

Presenting Vera's walk as an embodied, sensory experience, the fragment highlights how one's sense of self and the environment are mutually constituted. This attention to the simultaneity of the human and non-human worlds together with the unhurried description of the smells, sounds, and textures of things considerably postpone the denouement. W.J.T. Mitchell seminally argued that the emphasis on the acts of seeing (and, as we argue, other sensual experiences) and on spatial relationships, often happening in women's writing, undermines the domination of space by linear narrative progression (1989).[5] The dramatic, purposeful, temporal narrative becomes suspended by what in classic narrative theory has a supportive function: namely by space, description, the sight and material qualities of objects, people, and relationships. Interspersing the linear narrative movement of a whodunnit, these moments of stasis impart a different narrative rhythm. They also challenge the restrictive definition of the landscape as the "natural" setting for the "cultural" story. Creating spaces for remembering and reflective distancing, the landscape here emerges as an actor within the narrative that is more-than scene of the crime, undermining the assumption of a detective story itself: that its solutions can resolve cultural complexities or redeem the past.

Affective Encounters With Landscape and Memory

Throughout *The Crow Trap*'s ecologically-oriented narrative, a variety of "clues" make the traces of the past present. Couched in the stark countryside these clues signify not only criminal acts, but often other memories of intricate family and small-town dramas that particular places give rise to. Thus, the landscape the characters walk and survey functions as a sort of living archive that can be unpacked through their close engagement with the natural world. This is not only undertaken by Vera, but by the entire cast of characters as they engage with the spectres of the past and puzzle through the circumstances of the present. From the get-go, the recollections of the surveyors are tessellated within Vera's, the sum total of which eventually help to unveil the perpetrator.

The piece of land around the abandoned mine emerges as a central and multilayered location within the narrative that assists in marking various memories and their associated meanings. These psychosocial strands are woven into the physical landscape, which operates not only geologically, but also, we argue, culturally, as a "store of information that can be accessed by users" (Skaloš and Kašparová 63). Near the end of the story, when Rachael and Anne return to the mine to clear the quadrats from the survey zone, they describe

> Look[ing] down on the site. With the grey block of the mine, the dark moss of the conifer, the pale snake line of the burn, it was like looking down onto a map. They could see the curve in the burn where Grace's body had been found.
>
> *(Cleeves 383)*

Completing their business is interrupted when they come across a particular spot that elicits an immediate affective response within both women, as they register it as the scene of their colleague's murder. While the two women associate the area with both their work and Grace's untimely death, for Barbara Waugh – the murderer herself – the ruined buildings and channelled streambeds are the burial site of a long-dead child whom she interred in the shallow earth under the floor of the decrepit engine house (Cleeves 531). Barbara commemorates this distant crime by visiting regularly and laying a posy of non-indigenous flowers, "lily of the valley and pale narcissi", at the entrance to the ruin (Cleeves 29). Although unaware of the small bouquet's purpose, Rachael remarks upon it when she first visits the mine, depicting Cleeves' entangled conceptualisation of landscape and memories. The posy of flowers, engine house ruins, and gentle curve of the stream all function as memorial anchors that tie both criminal and commemorative acts to the landscape and are emblematic of the inwrought aspect of the memories that the landscape conceals within plain sight.

Developing upon Mitchell's notions explored in the previous section that descriptions of the natural world advance environmental consciousness as well as provide narrative clues, rather than being an unremarkable happenstance of the novel, local engagements with memory and landscape are essential elements that develop a robust ecology of place. Memorials – be they constructed to those ends or less-official sites – are dependent upon what James E. Young has called the "vast array of . . . material, aesthetic, spatial, [and] ideological [forces] converging in one memorial site" (Young x). But memorial sites, as Young goes on to suggest, are not shut-off from the locales in which they are situated. They are in conversation with the wider world around them. Indeed, as Canadian geographer Edward Relph wrote in his 1976 book *Place and Placelessness*, lived-in/living landscapes achieve their meaning through the experiences of those who dwell there. But a sense of place is not determined by human acts alone, and Relph helps to further nuance our notion in the previous section, that self and environment are mutually constitutive: it is rather a complex relationship between a wide variety of human actions and natural processes, settings, and features that mix together to (in)form these collaboratively produced entities and identities. He writes:

> In our everyday lives places are not experienced as independent, clearly defined entities that can be described simply in terms of their location or appearance. Rather they are sensed in a chiaroscuro of setting, landscape, ritual, routine, other people, personal experiences, care and concern for home, and in the context of other places.
>
> *(Relph 29)*

Relph's text, an early frontrunner in the renegotiation and subsequent attunement of the field of geography to affective and philosophical questions, elaborates upon and grounds Rose's later theorisations on moving from rationalised landscapes towards place-based experiences and imaginative relationships between people and the places they inhabit and encounter. The chiaroscuro – the gradient (non) boundaries – that Relph describes are particularly visible in *The Crow Trap*, woven into the narrative at every turn, helping to read the landscape of the North Pennines as more than mere setting. The landscape that emerges throughout the text takes on new agency and identity as the characters negotiate memories that are roused through their engagement with the environment, enlivening the mode of detective fiction. While the grammar of geology – unearthing, digging, sifting – has been present in crime fiction's structure from its earliest moments, reading with Relph helps reinterpret these earthly encounters, moving away from human interventions carried out by detective and reader alone, reframing the role of landscape within the narrative puzzle to realise its affective imbrications.

Tending toward the "the sensibilities of the aesthete" (Rose 158), during one of her first thorough walks of the landscape, Rachael describes the area near the old mine with a local's level of detail (Cleeves 29, cited in the previous section) situating the focal point of the abandoned buildings – and the time to which they gesture – within the broader panorama when she summits Hope Crag. "The land sloped gently in a series of plateaux to the horizon, which was softened by woodland", including strips of managed and burnt heather in various stages of growth: "[t]here was a soft westerly breeze blowing into her face and all around her was the song of meadow pipit, skylark, and curlew" (Cleeves 30). Such descriptive passages permeate *The Crow Trap*, creating the chiaroscuro of setting that help to define that specific place: these natural/cultural features also function as memory markers, occurring on a scalar variety from the massive size of Hope Crag to the mine's engine house, the swollen skirl, and the migratory range of local birds. Thus, Relph's work helps situate the experience of place in the novel within a series of overlapping concentric circles: from the now-abandoned mine that links the Pennines to resource extraction and industrialisation on a global scale, to the professional acts of surveying the landscape to catalogue the flora and fauna (Cleeves 29), and ultimately to the deeply personal experiences of the people that help to develop and articulate a sense of *this* particular place (Cleeves 63).

Utilising this relational framework encourages a co-constitutive view of self and place where both give meaning to each other, moving towards recognising what Anna Tsing has called "polyphonic assemblages" to illuminate how autonomous identities intertwine their non-unified but nonetheless simultaneous and particular narratives and rhythms (24–25). Frequently, the places that encourage this sort of engagement are themselves liminal: in-between, neither entirely natural nor human, neither entirely present nor past, timeless yet bearing time's trace, echoing Relph's contention that "places are emerging or becoming; with historical and cultural change new elements are added and old elements disappear. Thus, places have a distinct historical component" (3).

Cleeves employs a specific grammar of place descriptions that is informed by the tradition of nature writing, allowing for "different times and their legacies" that are piled-up on top of one another "in no particular order" to come to the fore (Elliot 35, cited earlier). The places that jog the character's memories are often decrepit/abandoned sites that were previously dominated by human usage, but have since slipped into new, similarly liminal versions of themselves, in what Relph conceives as neither uniform nor homogenous (5) and where they are "never empty, but [have] content and substance that derive both from human intention and imagination and from the character of the space" itself (10). In *The Crow Trap*, this includes not only the mine and outbuildings there, but also the cared-for spaces where the local nature is woven into human actions and purposes, like the priory garden replete with "overblown blooms" lovingly nurtured by Anne (347), or the parish church, St. Bartholomew's, "separated from the roads by low stone walls" and a "wooden lych-gate", bearing the trace of past generations who have lived and died in Kimmerston and the surrounding region (261). These liminal territories open room for reflection through people engaging with them: the business of human concerns (alone) recedes, and the vastness of the living landscape in its variety fills the atmosphere of the novel.

While all characters find many echoes from the past while walking the landscape, Vera's occupational function as the detective makes her retrospective moments of primary interest: it is Vera's private reminiscences of her visits to Baikie's that ultimately enable her to expose the sinister acts – both past and present – perpetrated near the cottage, farm, and mine. And while uncovering distant memories and events is a standard feature of crime fiction, brought about by an affective engagement with the surrounding landscape, here they culminate in solving the crime(s). As such, we argue that Vera might be read as a restitutive figure, able to navigate the boundaries between past and present that would otherwise obfuscate one another. As a witness to these recollective occurrences, she is not only a detective, but also an archivist working with and through the memory repository that is the landscape: the reward of bringing the perpetrator to justice requires digging into the traces the past leaves upon the present, searching for clues, and correlating webs of stories, actions, and phenomena. This proffers a new kind of epistemology, highlighting the ways in which the scientific and analytical methods of ordnance, survey, and detection would have been inadequate on their own to solve the crime.

Cleeves' novel also engages with what Andreas Huyssen defines as one of the potential pitfalls of memory – "its hypertrophy" which has the possibility to lead to "self-indulgence, melancholy fixations, and problematic privileging of the traumatic dimension of life" (6). Yet, in crime fiction, this possible pitfall proves to be a central aspect of memory-work: ruminating on complex and conflicting memories brings key features of the previously unexamined to light. Vera's fixations in the latter half of the novel as she contemplates her childhood spent at Baikie's – triggered by her contact with the environmentalists and the cataloguing that they have been brought in to perform – are key to her solving the crime. While walking the terrain at Baikie's and mulling over her past, Vera first navigates the landscape in which the crimes take place in the archive of her memory, ultimately utilising this uncovered knowledge – enabled in turn by the landscape itself – to help her catch the killer. One afternoon at the cottage, Rachael is joined by Vera who tells her about her childhood there, and her motivations for doing so.

> "I expect you think I'm odd", she said. "Eccentric. Even that I'm dragging up my past for effect. That's not the case, and if I do have a reputation for eccentricity, I have one too for getting results . . . I'm telling you, so you know I understand what goes on here."
>
> *(Cleeves 235)*

In the final scenes, the narrative crescendos. After stopping at home for sturdy boots and other hiking equipment, Vera takes to the hills, departing from the edge of the village, transitioning through the liminal spaces that border the settlement to the rugged Pennines, taking note of and negotiating the landscape that she knows so well, accompanied all the while by her own childhood recollections of the place as she hunts down the perpetrator of the crimes (Cleeves 518). Upon arriving at the mine's abandoned engine room, out of breath but invigorated, Vera catches the criminal.

> Tonight the moon was covered by the low, dense cloud which had rolled in like fog. From the shell of the engine room came another sound, the scrape of metal against stone and soil . . . The woman was standing with her back to the gap in the wall [and] had loosened a flagstone from the corner of the room and shifted it enough so she could dig out the soil underneath. The grave must have been shallow because already Vera saw a fragment of bone, cream as ivory, waxy in the candlelight. The woman squatted and began to scrabble at the soil with her fingers.
>
> *(Cleeves 525)*

Vera discovers the woman in question, Barbara Waugh, as she is scrabbling unhinged and nervous at the limestone and soil to exhume the bones of the unnamed child whom she killed in a psychotic episode near Baikie's Cottage long ago. The true purpose of the posies of flowers left at the site as well as the motivation for stopping the environmental assessment by any means necessary – including murder – come to light. Barbara's interment and exhumation of the child further elaborate on and add to the chiaroscuro of actions that Relph describes, adding another layer of memory on to the already elaborately woven sense of place developed throughout the novel. Her criminal and ritualistic acts commemorate the body of the murdered child to the earth, mingling the two together, performing a literal covering up of her crimes in an attempt to bury her deeds within the land itself. Later, half-crazed and frightened of being discovered, Barbara is undone as she brings the physical remainder of those memories – the bones of the deceased child – back to the surface. The interactive and often messy framework of memory, landscape, and detection is the engine driving the narrative throughout *The Crow Trap*. Cleeves achieves her innovative plot resolution through tessellating a confluence of human and natural/cultural forces that work together.

Conclusion, or the Goshawk Did It

While crime fiction is not always ecologically focused, it is, we argue, often concerned with constituting an ecology of a particular place. As the developments of the present spin ever onward, blurring distinctions and shrinking the experiential dimension of space and time, in crime fiction, the local and particular open before us again as a repository not only of memory, but certainly in *The Crow Trap*, as a worthy place to be engaged with. That's in part because places to which we are most attached, according to Relph, might be thought of as "fields of care, settings in which we have had a multiplicity of experience and which call forth an entire complex of affections and responses" (Relph 38). Crime fiction of this sort helps to establish an ecology of place through cataloguing what is right around us – the physical elements of the natural/cultural world, which often also feature as markers of human memories. They might be reminders of a deep ecological past brought forth by contemplating the weathered sky-reaching Pennines, the history and course of a meandering river through a gentle vale, or even a series of crimes committed and commemorated by a disturbed mind.

As we have established, Cleeves' novel is doubtlessly a fascinating continuation and exploration of the trends developed in the British School of new nature writing that were (re)kindled in the late nineties and early aughts and are part of a longer tradition that renegotiates the relationships between people and place through precise and often affective description. Passages in *The Crow Trap* are reminiscent of nature writing frontrunners, such as Nan Shepherd's *The Living Mountain* (1977), where attention to and description of minute detail brings the lived-in reality of the natural world prominently to the fore as more than mere setting. Cleeves and Shepherd both share what Robert Macfarlane has called the "localist perspective" whose "closeness serve[s] to intensify rather than to limit [their] vision" and testifies to a local intuition and affinity that counteracts the placelessness and estrangement often charted in contemporary narratives of global capitalism, as their authors attempt to think on a planetary scale (Macfarlane 2011 xiv, foreword in Shepherd 1977). This sort of writing has the ability to permeate a wide variety of genres, and Cleeves employs it in her crime fiction to the effect that the text reads, like Shepherd's own work, as "field-note, memoir, natural history and [sometimes] philosophical meditation" (Macfarlane xiv). In these sorts of texts, the more one reads them, the more they reveal about not only the environments and places described, but also, of the imbrication of human actions within the living physical bodies of the landscapes and the living psychosocial body of memory.

In Cleeves' novel, people form their relationship with land through both criminal and restitutive acts: they kidnap children by a swollen river in the spring; they bury their dead in an abandoned and crumbling building; they pick flowers and lay them next to a shallow grave. The descriptions and realities of the places themselves – and the mediated recollections of their previous forms – intermingle throughout the story, by turns discontinuous and complementary, are also harbingers of the memories that suffuse them. These hybrid natural/cultural spaces exist across many different timelines, both real and imagined, making pasts present and navigable as they imitate the hidden collies and ravines, the skirls and burns, of the narrative and physical landscape.

While *The Crow Trap* promotes at nearly every turn the reading of an interwoven natural/cultural narrative, at some point, it also opens a different reflection while sketching the challenging and often insurmountable human-imposed schism between natural and cultural worlds. In the final pages of the book, Vera recounts the story of the initial crime committed by a disturbed and mentally unstable Barbara Waugh many years before:

> At the same time as one of Barbara's disappearances, a toddler disappeared. His mother and her boyfriend had brought him for an outing to the hills . . . While the pair were otherwise engaged the boy vanished. If you believed the local newspapers he'd been swept away by a large hawk to its eyrie. If you believed me at the time he was drowned in the skirl which was in flood. In fact we were both wrong. The boy was taken by Barbara Waugh on one of her crazy moorland wanderings . . . later she took him to the old mine and kept him as a pet, a toy, a replacement son . . . We don't know yet how he died . . . At some time, either then or later, she buried him under the flagstones of the engine house. She tried to forget him but couldn't quite.
>
> *(Cleeves 531–532)*

When Vera was a young police officer, the unresolved disappearance of the child was attributed to nature: if not a goshawk carrying off the young toddler, symbolically robbed of its young by egg poachers, then a tempestuous and dangerous river that had swollen beyond its banks was responsible for carrying the child away. In either scenario, the agents of nature are employed as scapegoats that inhabit a curious double bind. On the one hand, it appears that they cannot be guilty of committing a crime or held responsible for doing so. While Cleeves' novel shows a thoroughly blended natural/cultural world, the details of the historical case imply that the divide between the natural and the cultural still exists. On the other hand, when something cannot be explained through the pernicious, premeditated, or passionate criminal acts of humans, then nature is substituted as a necessary catch-all, a summary "Act of God" that is decidedly non-human, to blame, and yet seemingly incapable of true culpability. The human-devised resolution to a conundrum that cannot be adequately solved through lack of evidence when no other option is present? The goshawk did it.

Notes

1. New nature writing as a distinctive literary movement was first conceptualised in the 2008 themed issue of the popular literary journal *Granta*, edited by Jason Cowley. The issue included a wide range of pieces on nature, from writers such as Kathleen Jamie, Jonathan Raban, Richard Mabey, Robert Macfarlane, and Lydia Peelle.
2. Recent developments in landscape theory work to affirm this claim (see Elkins and DeLue). The roots of the word in the Nordic and Germanic languages suggest a close relation between physical features of a place and the population that dwells there. This underscores the mutual shaping of people and place not only through physical tools but also through ideology and policy (Spirn 92–93).
3. In the novel, the impartiality of the environmental survey bears a risk of being compromised when it transpires that Grace deliberately augmented the number of otters in the area. What is more, the readers learn that the women's employer Peter Kemp not only has plagiarised Rachael's sampling technology but also attempts to steer the survey in the direction that would better suit his own career plans.
4. Without exonerating individual people from environmental responsibility, Cleeves juxtaposes the small-scale egg collecting with large-scale and officially authorised industrial activities, showing how humans idiosyncratically

define limits of acceptability with regards to environmental intervention, privileging global and expansive operations above the local by making one legal (and financially lucrative) and the other illicit and shameful.

5. Mitchell illustrates this by turning to the classic examples of Jane Austen's *Emma*, Charlotte Brontë's *Jane Eyre*, and Virginia Woolf's *To the Lighthouse*.

Bibliography

Bender, Barbara. "Place and Landscape." *Handbook of Material Culture*, edited by Christopher Tilley et al., SAGE, 2020, pp. 303–314.

"Birding Fiction: Novels Written from a Birding Perspective". *Fat Birder*, 12 Dec. 2020, fatbirder.com/birding-fiction.

Britannica, T. Editors of Encyclopaedia. "Detective Story." *Encyclopaedia Britannica*, 12 May 2020, www.britannica.com/art/detective-story-narrative-genre.

"Castle Writers Online." *Norwich Castle Museum and Art Gallery*, 16 Dec. 2020, http://eprints.glos.ac.uk/8045/1/8045-Monthly-Creative-Writing-Workshop-Inspired-by-Works-in-Norwich-Castle%27s-Collection-%27Castle-writers-goes-digital%27.pdf.

Cleeves, Anne. *The Crow Trap*. Pan Books, 1999.

———. "Stories Have Always Been Healing, So I'm Funding Bibliotherapists." *The Guardian*, Guardian News and Media, 2 Sept. 2020, www.theguardian.com/books/2020/sep/02/stories-healing-funding-bibliotherapists-ann-cleeves.

———. *Website Home Page*, 16 Dec. 2020, https://anncleeves.com/.

Cowley, Jason, editor. *Granta 102: The New Nature Writing*. Granta Publications, 2008.

DeLue, Rachael, and James Elkins, editors. *Landscape Theory*. Routledge, 2008.

Elliott, Cathy. "H is for Heterotopia: Temporalities of the British New Nature Writing." *Heterotopia and Globalisation in the Twenty-First Century*, edited by Simon Ferdinand et al., Routledge, 2020, pp. 34–48.

Heise, Ursula K. "Postcolonial Ecocriticism and the Question of Literature." *Postcolonial Green: Environmental Politics and World Narratives (Under the Sign of Nature)*, onder redactie van Alex Hunt en Bonnie Roos, U Virginia P, 2010, pp. 251–258.

Huyssen, Andreas. *Present Pasts*. Stanford UP, 2003.

Ingold, Tim. *Being Alive: Essays on Movement, Knowledge and Description*. Routledge, 2011.

Lehtimäki, Markku. "Natural Environments in Narrative Contexts: Cross-Pollinating Ecocriticism and Narrative Theory." *Storyworlds: A Journal of Narrative Studies*, vol. 5, 2013, pp. 119–141, doi:10.5250/storyworlds.5.2013.0119.

Macfarlane, Robert. "Introduction." *The Living Mountain*, edited by Nan Shepherd, Canongate, 2019.

Mitchell, W. J. "Space, Ideology, and Literary Representation." *Poetics Today*, vol. 10, no. 1, 1989, pp. 91–102, doi:10.2307/1772556.

Murphy, Patrick. *Ecocritical Explorations in Literary and Cultural Studies: Fences, Boundaries, and Fields*. Reprint, Lexington Books, 2010.

Relph, Edward. *Place and Placelessness*. Pion, 1976.

Rose, Gillian. *Feminism and Geography: The Limits of Geographical Knowledge*. Polity, 1993.

Rzepka, Charles J., and Lee Horsley. *A Companion to Crime Fiction*. Wiley, 2010.

Skaloš, Jan, and Ivana Kašparová. "Landscape Memory and Landscape Change in Relation to Mining." *Ecological Engineering*, vol. 43, 2012, pp. 60–69, doi:10.1016/j.ecoleng.2011.07.001.

Smith, Jos. "An Archipelagic Literature: Re-Framing 'the New Nature Writing'." *Green Letters*, vol. 17, no. 1, 2013, pp. 5–15, doi:10.1080/14688417.2012.750840.

Spirn, Anne Whiston. "The Art Seminar." *Landscape Theory*, edited by James Elkins and Rachel DeLue, Routledge, 2008, pp. 87–156.

Tsing, Anna Lowenhaupt. *The Mushroom at the End of the World*. Princeton UP, 2017.

Walton, Jo Lindsay, and Samantha Walton. "Introduction to Green Letters: Crime Fiction and Ecology." *Green Letters*, vol. 22, no. 1, 2018, pp. 2–6, doi:10.1080/14688417.2018.1484628.

Young, James Edward. *The Texture of Memory*. Yale UP, 1994.

3
THE NORFOLK SALTMARSH
Elly Griffiths and Place in Contemporary Crime Fiction

Nicola Bishop

Elly Griffiths' series of novels about Ruth Galloway, forensic archaeologist and university lecturer, was recommended to me by my mother-in-law, an avid reader of crime fiction. While not an archaeologist, but an historian and lecturer, and knowing that I prefer my crime on the cosy side of noir, the series lent itself to recommendation. As an East Anglian, the series had appealed to her sense of localism, but as I was drawn into the world that Griffiths creates, it was the genre of contemporary place writing as much as Golden Age crime that rendered the work familiar. Griffiths' novels focus on Norfolk saltmarshes, the picturesque seaside village communities of the Norfolk coast and a fictionalised University of North Norfolk in Kings Lynn. Most prominent of these locations is the lonely and isolated cottage the protagonist, Ruth, lives in at the edge of a stretch of saltmarsh close to a nature reserve. The evocation of this setting as a place "where the earth meets the sky" is certainly one that has featured in imaginings of East Anglia in fiction, non-fiction and beyond, for many centuries (*The Crossing Places* 27).

This chapter explores contemporary crime fiction and the influence of place writing on the construction of narrative place. It takes Griffiths' Dr Ruth Galloway series (2009–) and examines the depiction of the eastern coastal fringes – the Norfolk fens and marshland – and the relationship between the main character and the cottage at the edge of marshes where she lives. The chapter considers Griffiths' writing of north Norfolk within the context of other narratives deeply influenced by place to examine the cultural cross-location of edgelands and self that result in a complex spatial diarisation. It looks at the multifaceted construction of the saltmarsh as a site of haunting and long historied memory while also exploring the spiritual and folkloric significance of this liminal space. Finally, it suggests that the dovetailing of the saltmarshes' long history (from Prehistoric archaeology to Victorian corpses and Second World War bone findings), with the character's recent and personal memory marks the significant cross-fertilisation of contemporary place writing within Griffiths' crime fiction storytelling. In doing so, the chapter suggests that crime writing like Griffiths' can be read in the ways it offers an intersection of historical time and contemporary place, but also as a conduit of ecocritical trends and concerns within a wider range of literary forms.

East Anglian Place Writing

East Anglia, and particularly the coastal fringes, has inspired many writings of place; it has a variety of atmospheres – haunting, isolated, quaint, bustling, picturesque – that are explored and pored over

 DOI: 10.4324/9781003091912-5

through art and on the page. Griffiths' work is both a response to this and a part of it. In many ways, it follows in the footsteps (lost to a palimpsest in the shifting sands) of other writers of East Anglian landscapes; from crime fiction aficionados like P. D. James to W. G. Sebald, M. R. James and Graham Swift. Griffiths is equally part of a contemporary group of writers focusing on Norfolk crime – both cosy and violent – writers such as J. M. Dalgliesh, Keith Finney, David Blake and Judi Daykin.

The significance of setting in the crime genre more widely goes beyond the logistics of shaping and naming the location in which the events take place. As Lisa Fletcher puts it: "geography and genre are mutually constitutive" (1) – and in the case of crime fiction, generic expectations generally fall on either hard-boiled urban locations, or idyllic rural ones: the murky back-alleys of Sherlock Holmes' London or the innocent idyll of Miss Marple's St. Mary Mead. Recent studies, like Fletcher's own *Popular Fiction and Spatiality* (2016), have begun to unpick the relationship between place and genre writing, focusing explicitly on the ways that, as Fletcher puts it, "place and storytelling should be considered together" (2). While geocriticism has reconsidered literary texts in light of their relationship with place, studies such as Fletcher's draw out the significant impact of genre expectation and the consequent "influence the representation of space and place [has] in some of the most widely read books in the world" (4). By considering texts that respond as explicitly to genre as Griffiths', this chapter responds to Fletcher's call to push Robert Tally's concept of the novel as a "cartographic project" towards an understanding of the role of genre – as well as the writer – in this process (13).

The existing connection between British crime fiction and rural landscapes is long documented: John Scaggs refers to the implicit pastoralism of the term 'Golden Age' itself as a desire to "restore or return to a lost order that, in all aspects, is superior to the present world" (50). Likewise, he draws on W. H. Auden who famously suggested that the "more Eden-like [the setting] is, the greater contradiction of murder" (50). Much of the shift towards place in twenty-first century literature similarly links in some ways to a revisiting of the pastoral as a term, in part, as Deborah Lilley suggests, because of "contemporary environmental concerns" that encourage authors to "put [the pastoral] to new uses" (*The New Pastoral* 17). This is true of crime fiction as it is of many other genres; Nathan Ashman mentions the recurrent theme in American hard-boiled crime writing of "environmental exploitation" and the destruction of the "pastoral myth" (45). And yet, in some ways, Griffiths' novels diverge from the wider contemporary literary preoccupation with a sense of ecological crisis despite her coastal setting. In this sense, while there are connections between her crime fiction and the American pastoral trend towards what Ashman refers to as "longing for an idyllic landscape that is always in the process of disappearing" (45) – seen most literally in the very nature of the saltmarsh and its continually shifting topography – in other ways, her work less overtly confronts any sense of crisis.

While recognising the generic preoccupation with location within crime fiction, this chapter also suggests that the deep and detailed interweaving of place and character owes much to the trend towards the narrative retelling of landscape consciousness that is more widely recognised in creative non-fiction. A genre which looks set to have been boosted by pandemic-fuelled interest and growing awareness of the environment, place writing (sometimes referred to as nature writing) saw a revival in the early twenty-first century that shows no sign of abating. Figures such as Robert Macfarlane, Richard Mabey and Mark Cocker have fronted a form that has much older roots, the works of which have been similarly rediscovered more recently. The writing of John Lewis-Stempel, Helen Macdonald, Juliet Blaxland and William Atkins, amongst many others, explore their localities, giving beautifully poetic descriptions of the abundance of life therein alongside deep reflection and introspection on the place of nature within our lives. These works are skilfully interwoven multidisciplinary pieces, bringing together social and cultural histories, the natural sciences, geology, ornithology, autobiography and literature. Prizes such as the Wainwright Prize for UK Nature Writing have garnered market attention and the beauty of many of the cover sketches, paintings and prints have reignited the recognition of books as art.

Equally significant is the abundance of *East Anglian* nature or place writing.[1] Indeed, many of the original nature writers – those whose works are now being reissued in editions that share aesthetic commonalities with new texts – were from the eastern-most edges of Britain. Roger Deakin famously lived in a Suffolk farmstead; Adrian Bell near Beccles in North Suffolk; Ronald Blythe on the Suffolk/Essex border; Mark Cocker in Claxton, rural Norfolk and Richard Mabey in the Waveney Valley. As Joe Moran has suggested, it is "striking that a number of practitioners of the new nature writing (Cocker, Deakin, Mabey and Macfarlane) are based in and often write about a highly distinctive region: the fens" (52). Much of their writing draws on the authors' own processes of becoming and belonging within their familiar landscapes in an autobiographical tradition of telling the story of their relationship with place as an extension of their journey to understanding self. In this "liquid landscape", as Moran calls it, this process challenges the categorisations commonly found in new nature writing: of edgeland, wilderness, periphery – as Jos Smith refers to them in *The New Nature Writing* (2017).

There is also an arguable cross-fertilisation of place writing and fiction that can be identified in works by Melissa Harrison (also based in Suffolk), Sarah Hall (Norwich), Jeremy Page, Sarah Perry and Michelle Paver. This group of authors showcase the role that poetic descriptions of place can play in narrative fiction, but many also demonstrate a distinct anti-pastoralism that challenges the conceptualisation of rural landscapes as unproblematically idyllic. Griffiths plays a part in this deconstruction of picture postcard imagery by drawing out violent and haunting undercurrents in a manner that can be traced back as least as far as Agatha Christie. Unlike St. Mary Mead, however, Ruth's home carries its own undercurrent of threat that posits the saltmarsh as in tension with the wider locale – pretty seaside villages and tourist destinations. It is at the same time a seemingly straightforward rural setting and a complexly anti-pastoral landscape. Equally, the use of Kings Lynn as the key urban focus of the series problematises depictions of Norfolk as a predominately leisured space – and it is itself often seen as a peripheral town, overshadowed by its more powerful counterpart at the centre of county, Norwich. More telling still, the type of crime found in Griffiths' novels, particularly the repeated narratives around violent crime inflicted on children, raises the connection between the isolation of the saltmarsh and the deep fear of the shifting landscape and treacherous waters. Griffiths, like many other crime writers, takes the hidden horror of place and renders it fully into consciousness; childhood anxiety about the tide becomes a raging reality of parental worst nightmares – innocent children snatched, murdered and lost to the sea.

Unlike many of these other authors, Griffiths is not an East Anglian writer – she lives in Brighton – but her novels set in North Norfolk are arguably much more centred on place than those she locates in her hometown. In this, she is carrying on a long tradition of finding inspiration in the East Anglian coastline, drawing on the wide array of cultural texts that have taken locations such as Blakeney, Walsingham and others as their focus. This study of her work will demonstrate not only the cross-correlation between contemporary crime fiction and place writing but the significance of the saltmarsh as a cultural space. In doing, so it will examine the depiction of the saltmarsh as a layered and unknowable landscape and the deep development of character through place.

The Saltmarsh

At the core of these novels is the complex landscape that is the saltmarsh, about which Elly Griffiths herself has written:

> the ancient belief that marshland is in itself somehow sacred; an in-between place, neither land nor sea, neither life nor death. Archaeologists think that prehistoric people saw marshland as a bridge to the afterlife, which is why they sometimes buried their dead there, the so-called bog bodies.
>
> *(Griffiths* Crime Reads*)*

Neither sea nor shore, fully water nor land, the saltmarsh is thus deeply symbolic, as she says, not only as a place that defies definition now, but which was likely equally weighted with a complex sense of spatial identity, even in prehistoric times. While various critics and thinkers have attempted to theorise the spatiality of the sea and the shore, there is a slippage that renders marshland ungraspable. If as Christoph Singer argues, "the shore is transformation spatialized" (11), then the marshland presents an even greater challenge because it often retains its status as either water or land mass in less predictable ways than the tide offers the beach – what Griffiths refers to as "uncertain ground, half-land, half-sea" (*Stone Circle* 7). What is more easily applicable is the sense that, as Singer suggests, the literary personality of the sea remains potent:

> The individual subject also finds her or his liminal experience between the sea and the land: John Banville's *The Sea* and Ian McEwan's *Chesil Beach* present the shore as a site of memory, of sexual awakening, and align the space, almost in a romantic fashion, with the inner torrents and tempests of the protagonist's psyches.
>
> *(Singer 21)*

This certainly speaks to Griffiths' use of characterisation and the tumultuous relationship that the key figures have in the novels with both the saltmarsh and the sea itself. And if the sea is seen as not only a space of liminal experience, but also as a site thronged with what Ursula Klawick and Virginia Richter have called "the desire and fear in the human imagination" (12), then the saltmarsh lives up to this literary expectation. The terrifying crosscurrents and dangerous shifting sands of the marsh add increased peril and unpredictability to a space already recognised within human consciousness as inherently destructive.

Where Griffiths' work challenges some of the assumptions that have come before is in the confrontation of the coast as a male-dominated landscape (as Doreen Massey identified it [6]), and the creation of a female protagonist who has an immovable desire for the isolation and emptiness of the beach. In this aspect, the novels share much in common with the writers of place, many of whom, like Amy Liptrot and Kathleen Jamie, challenge the assumptions around isolation and being part of a wild landscape as being accessible only for what Jamie called the "lone, enraptured male" (Jamie 2008) – a figure equally resonant in the crime genre. One of the key features of the first novel, *The Crossing Places*, is a growing awareness of the flawed behaviours of this type of male figure who, at first, seems to stand on the saltmarsh in a position of mastery, but whose hubris results often in catastrophe. The narrative focus, instead, is on a female intuition of place that, ultimately, outwits the domination of landscape seen in the attitudes of the male characters.

In the first novel of Griffiths' series, the North Norfolk saltmarshes are warranted as much attention as the main characters. Referred to early on as "a liminal zone, between land and water" and as "vast and completely featureless" (*Crossing Places* 122, 124), Griffiths nonetheless unpicks the ambiguity of the empty landscape through the archaeological dissection of the marsh itself, revealing the centuries of human interaction. Similarly, the comparison between Ruth, who is described as dressing in "shapeless" black clothing, in a bid not to draw attention to her body, and the 'featureless' marshes is fairly clear (4). Ralph Crane and Lisa Fletcher refer to this as an integral part of the term 'locale' which suggests a "human relation" and therefore may be useful for stressing the mutual dependency of characterisation and setting in fictional texts (12). The deep connection between a character who prefers to remain on the sidelines and a setting that is frequently overlooked in favour of more picturesque locations drives the empathy between the two. At the same time, this relationship between place and character is fundamental. Anne-Julia Zwierlein talks about fiction of the sea in the turn of twenty-first century texts as "the ever-shifting margin between land and sea [which] functions as symbolic catalyst for existential situations for the probing and unravelling of human relations"

(63). As the marshes reveal their archaeological secrets to Ruth, her role and relationships alter; she is drawn not only towards but also out from the saltmarsh into the complex world of crime as well as the seemingly opposing sphere of love and motherhood.

From the prologue of the first book, the saltmarsh is imbued with personality. Ruth and policeman DCI Harry Nelson meet because Ruth is a specialist in bones, and the police require help in ageing a corpse found buried on the marsh. Fearful that it is the body of a recently missing child, the pair set out on a bleak mission to recover the bones, but the site seems almost reluctant to give up its secrets:

> The wind is whispering through the reeds, and here and there they see glimpses of still, sullen water reflecting the grey sky. At the edge of the marshland Ruth stops, looking for the first sunken post, the twisting shingle path that leads through the treacherous water and out to the mudflats. When she finds it, half-submerged by brackish water, she sets out without looking back.
>
> *(1)*

The suggestion here that the saltmarsh is deliberately hiding something evokes a relationship between it and Ruth, suggesting that she must coax from the marsh the body that they seek. While Nelson is an interloper, Ruth is able to identify and follow the deep-sunken signs that lead safely across the landscape. Nelson later attempts a daring rescue, but his lack of knowledge and experience prevent him from making safe passage; at the same time, those characters capable of 'reading' the marshland are imbued with almost magical and mythical qualities, with the exception of Ruth, whose stubbornness of character and intuitive understanding of the landscape is what protects her. It is her reluctance to leave her saltmarsh home, regardless of the danger this presents, and her determination to find that which she seeks out on the marsh, that see her protected against the elements. Zwierlein refers to this trope more widely in sea fiction as a "Darwinisticist" will to survive that is common in protagonists by the sea (69).

This interrelation of landscape and Ruth's development begins with the ways in which the saltmarsh is immediately destabilised and rendered more compatible to human life through the early description of Ruth's home as its antithesis – a space that is both claustrophobically enclosed but also cosy and comfortable. Her cottage is described in the first novel as "tiny", with "very steep stairs", a kitchen that "barely has room for a fridge and cooker" and what Ruth's mother refers to as an "unsanitary bathroom" (*Crossing Places* 4–5). Yet it also contains Ruth's prized books, "a comfortable sofa, large flat-screen TV" and "her favourite place" – a front window that looks out onto "nothingness" (5). Or rather, what Ruth describes as: "just miles and miles of marshland, spotted with stunted gorse bushes and criss-crossed with small, treacherous streams" (5). This view becomes instrumental to Ruth's sense of home and well-being, despite there being various events over the course of the series that see the character imperilled both out on the marsh itself and in her cottage at its edge, hinted at in the description of the 'stunted' growth and 'treacherous' water. Inhospitable as it seems, this is mirrored by Ruth's general attitude to, as she sees them, invasions of privacy, preferring her own company to that of almost anyone else's.

In her pride at the emptiness of her landscape, there is also a statement about the sheer independence of Griffiths' character. Ruth's desire to be alone on the saltmarsh is a contributing factor to the way Griffiths writes her as both a figure subsumed by romance and its antithesis. Indeed, when Ruth states: "sometimes the saltmarsh's sheer loneliness and splendour gives her a thrill of almost sexual pleasure" (*Janus Stone*), there is evidence of the ways that Griffiths is anthropomorphising place as a romantic partner. This connection between Ruth and the marsh equally highlights her independence – the landscape provides all that Ruth's mother, with her obsession with Ruth marrying, and society more widely, might expect to be missed. While crime fiction so often focuses on the bachelor detective (Holmes, Hercule Poirot, Inspector Morse, Inspector Rebus) and, as Tim Hannigan points out, the "most common practitioner" of new nature writing is "the strongly gendered (male) figure, foregrounding intellectual and physical mastery", Griffiths offers a refreshing perspective towards both (15). Ruth is at the top of her field; she is an "expert on bone preservation" and is happy to reinforce her position when confronted by male characters who assume greater knowledge. While there is a human entanglement with DI Nelson, it is frequently and negatively contrasted to the ownership and freedom of life in the cottage on the edges of the marshland (*Crossing Places* 13). Indeed, each time something happens between Ruth and Nelson, Griffiths uses this as an opportunity for Ruth to reflect on her relationship with the saltmarsh as in perfect synchrony. Nelson is posited as a threat to the order of her life with her cat and later her daughter in their tiny cottage.

Indeed, part of the saltmarsh's appeal is the fact that it is not valued by others – her colleagues question why she lives in such an "inconvenient" place, while her mother refers to it as "that awful place where you live" (*Crossing Places* 170). These comments echo those from P. D. James's *Devices and Desires* (1989), where the marshes are referred to as a "funny place to choose to live" (428). Ruth's response is that even she cannot explain why "a girl born and brought up in South London can feel such a pull to these inhospitable marshlands, these desolate mudflats, this lonely, unrelenting view" (*Crossing Places* 7). Ruth becomes one with this landscape in a manner similar to that often seen as a process of complete immersion with place in nature writing – see for instance, Rob Cowen's *Common Ground* and the author's relationship with the edgeland near his new home. Despite the cosiness of the cottage, the emphasis in this first novel is all about the uncompromising nature of the saltmarsh, a view that is reinforced in wider cultural depictions of the landscape. It is, however, aligned with Ruth's own unwillingness to compromise on her freedom. As Jeremy Page puts it,

> Between the sea and the land I found an imaginative territory that was free and exotic. For others, the saltmarsh is a place of infinite isolation and, with its dangerous flows of creeks and tides, a place of psychic peril.
>
> *(Page "East Anglian Literature")*

In his own novel, *Salt* (2007), Page describes the saltmarsh through the eyes of a character, Lil' Mardler, who is, unlike Ruth, of the landscape but who is similarly awestruck by its seamlessness: "She inhabits a landscape that is so big and flat is seems the edges slope up into the sky all round" (50). The novels are from different genres – *Salt* is described as a "mystic family saga" (Hagestadt 2011) – yet there are commonalities to their engagement with this landscape. Ruth's attitude certainly mirrors that of Page, in the recognition of both the freedom and the dangerous isolation of the marsh.

In this, even the road that takes Ruth to her home signals a shift from the relative safety of the better-known and more densely inhabited areas. When she travels from the vast, unpopulated marshes to Kings Lynn and the university, the reader gets a sense of the contrasting landscapes in which she exists:

> The road is clear, with only the ever-present wind blowing a thin line of salt onto her windscreen. She squirts water without noticing it, bumps slowly over the cattle grid and negotiates the twisting road that leads to the village. In summer the trees meet overhead make this a mysterious green tunnel. But today the trees are mere skeletons, their bare arms stretching up to the sky.

Brosseau and Le Bel refer to crime fiction as "very much an urban genre", and yet the marsh captures this atmospheric gothic-ness whilst imbuing an additional sense of total loneliness and isolation (45). The terror created in the simple rustic description of the winter trees and the road that is difficult to navigate establish Ruth's vulnerability as well as foregrounding her position as a mediator/translator between those murdered ("mere skeletons, their bare arms stretching up") and those responsible for justice as together they "negotiate the twisting road". In *The House at Sea's End*, the third novel in the series, Ruth describes driving home to her cottage at night as "driving into nothingness, isolated, vulnerable", and the high point of the narrative drama centres on a road closure caused by snow, trapping Ruth and the killer together in a remote coastal setting (59).

Equally, as Klawick and Richard make clear, there is a potent atmosphere connected to the beach itself. As they suggest, "the encounters here are not always peaceful but, more often than not, are complicated and dangerous" (12). Even the act of bird watching, associated with quiet reflection on the natural world, is given violent undertones on the beach in the first novel. Ruth's cat is killed by a murderous reserve warden because it is seen as a menace to the wildlife on the marsh. This is a feature too of J. M. Dalgliesh's *The Dead Call* (2020), where a prominent local bird watcher is found brutally murdered at Blakeney Point after a campaign of macabre bird killings. Finally, the ending of *The Crossing Places* involves a wild chase in the dark across the marshes towards a bird hide, with the consequent drowning of two characters as the tide races in.

The saltmarsh is thus a potentially dangerous landscape, not only because of complex human interactions but due to the ways that rip currents and tidal channels sweep in unnerving ways and with terrifying speed. In *The Crossing Places*, it is a combination of human and littoral activity that results in death. Folkloric tales draw upon the fact that saltmarsh is also a disorientating place in that the sky gives little by way of navigational features when the water appears in multiple directions around you. And yet, potentially the more terrifying threat of the marsh is not physical but psychological, caused by the isolation Ruth identifies. In *Salt*, Page writes that the family at the novel's core are "made by it and made a little crazy by it" with disastrous consequences when the two worlds – of the saltmarsh and its nearby village – "rub too closely together" (171). There are similarities between the saltmarsh and the cave "anti-places" that Crane and Fletcher examine; like a cave, the saltmarsh is "a function of both the ideas [that] prevail in Western culture and the expectations of their genre" (12). The marsh is culturally embedded as a dangerously liminal landscape and it functions both as a site of madness in a novel that takes a deeply introspective exploration of self and place and as a layered necropolis in crime fiction.

As part of this, the saltmarsh is inescapably bound up in the type of narrative events that take place within the stories. As an archaeologist, Ruth specialises in digging beneath the surface of the land to find clues to the past, and yet the saltmarsh provides an interruption to the cartographies of time. For instance, each of the novels has a different temporal emphasis that comes not from the situation of the main characters but the crime that is being investigated, often involving distortions caused by the complex preservation of bones. In *The Crossing Places*, a prehistoric child's body is found preserved in a state that mirrors those of a much more recent child murder victim, disrupting the concept of grief as something that passes with time. The desolation of the marsh is also powerful enough to provide the perfect hiding place for a kidnapped third child, while the murderer is lured to his death by the avenging sea when attempting to cross the marsh at night. In this instance, the marsh is reinforced as a dangerous and unpredictable space, that can both be held responsible for taking life and avenging wrongdoing. As Ruth puts it: "What the Sand gets, the Sand keeps forever" (*Crossing Places* 124).

The establishment of the Sand as an autonomous being evokes also the haunting that is commonly associated with the saltmarsh itself, as well as East Anglian landscapes. Page's novel, *Salt*, is a complex tale of a family that struggles to escape from the generational pull of deep history embedded in the sand. Likewise, Ruth's various discoveries within the shifting sands continue to haunt her across the series with the memory of the first dig and the finding of an early prehistoric henge continuing to influence the narrative arc of the stories beyond the first novel. *The Stone Circle*, the eleventh book in the series, makes this clearest: "the whole thing – the dig on the Saltmarsh, Erik's son materialising – feels uncomfortably like one of those dreams where the past replays itself but with certain details subtly altered" (38). This paralleling of the setting and the narrative concern with haunting is clear in the cyclical nature of the events. Just like the reiterative tide, Ruth's mentor, Erik, and his death on the marshland are repeatedly referenced as key moments in Ruth's understanding of the dualism of love and betrayal; his character haunts her decisions as well as her identity as an archaeologist.

More commonly personified as a ghostly presence, the sea is a wider cultural reference within the novel. As Lucie Armitt suggests, in her discussion of "ghost-al" erosion in M. R. James's East Anglian writing, there is a tradition of "endow[ing] the sea with a predatory form of supernatural agency" (97) which becomes particularly apparent in *The Crossing Places* as tensions rise in the police investigation, and Ruth gets drawn further into the search for a missing child:

> Dark clouds are gathering over the sea, which gives the whole place an ominous feel. Shadows race over the mudflats and the seagulls are flying inland, sure sign of a storm to come.
>
> *(102)*

Ruth's friend, Shona, captures this eerie atmosphere with a casual, "this place gives me the creeps" (102). This connection between the creepy and the ghostly is emphasised in James's reiteration of "nothingness", as Armitt suggests (106), but it also forms the basis of Ruth's relationship with the saltmarsh. Ruth both requires the intense solitude of the marsh but while her scientific rationalism makes sense of the ghostly apparitions, she also renders the landscape anthropologically meaningful, seeing the layers of human interaction. Her work on the saltmarsh forms a process of knowing and, in some ways, owning, what the sea attempts to render unknowable and unpossessable. It is the culmination of Ruth's work as a forensic archaeologist and yet it is also ultimately resistant towards her practice, with the tide filling any dig and washing away her discoveries. In this sense, the saltmarsh (and the sea) defies the resolution sought at the end of a crime novel, yet simultaneously offers tangible evidence in the exploration of the past.

This becomes more apparent in some of the novels – indeed Armitt mentions Griffiths' *House at Sea's End* as an example of a text that deliberately "localise[s] deep time" in a bid to convey both the longevity and the vulnerability of the human coastline (70). As one of the archaeologists in the novel puts it,

> "Tides change . . ." says Trace shortly. "Sand gets moved, parts get filled up, other parts uncovered . . . things that were buried become exposed".
>
> *(House at Sea's End 11)*

Trace's casual profundity captures the essence of the saltmarsh and its impenetrable yet charismatic and revealing nature. In Griffiths' series, this becomes a truism not only of the tide but of time, of archaeology and police work, of human psychology, relationships and the exploration of the self.

In *The House at Sea's End*, Griffiths makes use of the danger and threat of the castle-like home tumbling from the cliffs into the sea below to imbricate a wider atmosphere of fear – focused, in this instance, on wartime coastal communities and the threat of invasion. There is certainly an underlying anxiety about being cut off (both metaphorically as a community and literally by the tide) which speaks to the recurring theme of isolation, but it rests alongside a layering of human vulnerabilities which reinforces this as a commonality of human experience across time. Ruth is outside of society but part of a community that spans nearly a million years; her job as an archaeologist reinforces even this perilously unstable seascape as one that is culturally and geographically layered, within which individual stories can be extracted and former injustices righted.

Thus, the novels articulate the shared purpose of both detective work and archaeology; as Lucy Franklin suggests, the creation of narratives around "archaeological features [is] an attempt to explain their existence and origin when, as we see it, the true nature of their origins was unknown" (146). In Griffiths' books, there are paralleled narrative-building deductions taking place as Nelson solves the crimes and Ruth interprets the past, and while these fit the generic expectations of crime fiction, there is also a strong connection with the type of discovery of place and self that is common in place writing. The development of Ruth as a character owes much to her growing understanding of the saltmarsh itself:

> There is something grand and terrible about the great expanse of sea and sky, something terrifying, yet at the same time exhilarating. We are nothing, Ruth thinks, nothing to this place. Bronze Age man came here and built the henge, Iron Age man left bodies and votive offerings, modern man tries to tame the sea with walls and towers and bridges. Nothing remains. Man dwindles into dust, less than sand; only the sea and sky stay the same. Yet she walks jauntily, with a spring in her step, stepping lightly over mortality.
>
> *(Crossing Places 288)*

Here Griffiths brings together the long vision of the past with an exploration of the awareness of self within the landscape, in a way that chimes with the philosophical leaning of much of contemporary place writing and ecocritical practice. Her work makes the saltmarsh a catalyst for a deeper penetration of anthropological, geological and philosophical concepts that find voice through the articulation of Ruth's character.

A Spiritual Saltmarsh

Saltmarsh as a topography thus has a complex status; one recurrent description from Griffiths' books refers to it as "where the earth meets the sky", and this categorisation, while speaking to East Anglian open landscapes more widely, is particularly evoked in the ever-shifting boundaries of land, sea and sky in the treacherous sands of the marsh. Ruth refers, in the first novel, to the significance of prehistoric saltmarsh as a "kind of symbolic landscape . . . a link between the land and the sea, or between life and death" (*Crossing Places* 16). This speaks to the general humanising impulse of the novels as one which, unusually perhaps for crime fiction, offers a positioning that is deeply responsive to the saltmarsh as a landscape. There is certainly a connection here between the marsh and the "otherworldliness" that Franklin identifies as a key aspect of landscape, archaeology and folklore in a Devon context (147). She argues that wilderness areas like the saltmarsh are often considered "the haunting grounds of ghosts, demons and other spirits", which ties in with the evocation of the crimes that have historically and recently take place near Ruth's home (147).

Across the series, there are a number of ceremonial scenes that take place on the beach or marsh that create a mirroring across key narrative events; for instance, in the original blessing of the stone henge in *The Crossing Places* with Erik and the almost recreation of this event in *The Stone Circle* with Erik's son, Leif:

> The sun rises over the marshes, turning the inland pools red and gold. The sand stretches out in front of them, rippled like a frozen sea. A flock of birds flies from the reed beds, zigzagging into the light. Leif raises his arms: "Goddess of the earth. Bless our endeavours today".
>
> *(66)*

Leif's blessing of the dig on the marsh offers a pagan ceremony where later, when a young girl's body is found nearby, her parents perform a silent Hail Mary in grief. This interlacing of multiple spiritualisms later finds further expression in the memory of Erik's tales about the mystical shape-shifting Nix who "lur[ed] sailors to death on the jagged rocks" (*Stone Circle* 211). These connections to a wide and multifaceted spirituality are all symbolic steps in Ruth's understanding of, and journey to, becoming part of the marsh itself. Indeed, it is when her own memories on the marsh, the constant reiteration of the archaeological past (which only she can uncover) and the interrelated pasts of the other characters come together that her journey to the heart of both herself and the saltmarsh can be understood.

Akin to what can be referred to as "ecological pilgrimage", this is a common trope in nature writing too. As Neal Alexander puts it,

> Contemporary landscape writing often seeks to revalue and explore ordinary landscapes and everyday habitats . . . the same texts also tend to employ a common set of quasi- religious tropes that imply a deeper network of associations linking ideas of landscape to the sacred, the mystical, and the extra-ordinary: theologies of the wild.
>
> *(1)*

This spiritualisation of place comes across in the repeated cultural cross-connections between landscape and individual in these novels; in the Christian (and particularly Catholic) inferences drawn, for instance, when a female figure in blue is discovered in a Walsingham graveyard or in the sacred tropes of the henge itself, found out on the saltmarsh.

These instances fit in with wider symbolic references throughout the novels, most of which are centred on the character of Cathbad, described as a druid but equally as someone who has many varied spiritual interests and influences – the pagan god Hecate, the moment of Imbolc, his telepathy and the drug-fuelled trance into which he goes to save Nelson's life. Clare Carson's spy/crime thriller, *The Salt Marsh* (2016), interestingly features many of the same mystical tropes – including a character who is described as a Magus, a collection of bones and other beach ephemera as runes, and references to 'will-o-the-wisps'. There is something culturally significant in the interconnection between the saltmarsh and a long-standing association with mysticism. Mark Gatiss put it this way, when talking about M. R. James' coastal settings: "whenever you get a margin between different landscapes you often get a greater than usual awareness of the supernatural" (Arnitt 98). This liminality presents, at times, a dangerous spirituality that is captured in the essence of the 'will-o-the-wisps' themselves that, as Ruth describes to Nelson, "lead travellers onto dangerous ground and so to their deaths" (*Crossing Places* 74). Ruth's explanation of the wisps as phosphorus reminds the reader of her position as a scientist, but her rationalism is happy to coexist with the various spiritualities in the series, and a balance is somehow created between the mysteries solved by archaeological exploration and those which are more miraculously explained.

In part, this is a result of the connections between what Kim Wilkins calls "the deep past" and paganism which are deeply entrenched in the novels through the almost tangible metaphysical links between archaeology and the "sacred sites" that are dug (111). As Willis and Blain argue, "specific narratives are forming [around] general pagan relationships with landscape" and Griffiths' novels speak to this more open interpretation between archaeological sites and spirituality that hangs quite specifically on the construct of Cathbad as a blend of pagan, Druid and lapsed Catholic, but also as a scientist and amateur archaeologist (310). Ruth's reverence at the discovery and retrieval of fragments of the past are often intertwined with Cathbad's ceremonial devotions which ties in with Wilkin's description of the "sacralizing practices in evidence at [pagan] sites" (111). Wilkins talks about fantasy fiction as a "kind of pagan tourism" (112) but there is an equally clear sense of Griffiths' novels as part of a "religiogeography" of East Anglian paganism, as well as, at times, an exploration of the intersection between paganism and Catholicism. Many of the characters identify as Catholic, or were raised within Catholicism, and references to the rituals and belief of the faith are combined with recurring priest characters and even one narrative set on holy pilgrimage in Walsingham (in a nod to older East Anglian crime such as S. T. Haymon's *Death and the Pregnant Virgin* [1980]).

Franklin discusses the interconnectedness between archaeology, landscape and folklore in reference to medieval Devon and Cornwall, but many of the points she makes tie in equally with Griffiths' depiction of Norfolk. She refers to the ways that "visual archaeological features served as powerful mnemonic [in] giv[ing] an area a strong spatial identity" (112). While the henge depicted on the saltmarsh has not the status of Stone Henge in the West Country, there is no doubt that it defines the landscape both physically and spiritually in the novels, becoming a symbolic marker in Ruth's own life, aligned with the key relationship with her mentor, Erik; her friend, Shona; her ex-boyfriend, Peter and Harry Nelson. Its Neolithic identity places it firmly within the period that "many of the best known pagan places of Britain were constructed", giving it an authenticity that places the saltmarsh on an archaeological parallel with Stone Henge (Franklin 116). It is also, as Griffiths mentions in the acknowledgements to *The Crossing Places*, inspired by the seahenge found at Holme-next-the-Sea, as evocatively described by the pre-eminent archaeological expert on fenland, Francis Pryor, in *Seahenge* (2001). Ruth refers to the discovery of the henge and the controversy surrounding its removal from the sands and preservation in a museum, as a continued point of conflict, drawing upon this metaphor in the depiction of her relationship with a wider entity summarised as 'nature' but also with archaeology itself.

Part of Ruth's consciousness rests on a presumption that the saltmarsh owns the henge and has retained it against the efforts of the tide to take it away. This is connected in many ways to the saltmarsh as a mythologised space. While not as widely discussed as the sea itself, there are echoes here of the relationship between the sea and folklore that are summarised by Franklin as being an "intertwining" of "landscape/seascape, magic, ritual and fate" (145). Indeed, saltmarsh is at the very crux of "landscape/seascape" in its seamless and rhythmic tidal slippage between each. References to local myths like Jack O' Lantern and Snatch Valentine situate the Norfolk saltmarsh at the crux of rural localism, while the wider pagan and Catholic symbolism and inferences of Cathbad's almost magical powers draw the characters into a narrative that draws them towards their fate, both collectively as a group of friends and individually. In this sense, the landscape becomes a "socially constructed entity" that gains meaning through the interactions of the lead characters and their experiences – both ritual, spiritual, and near fatal – of it (Franklin 145). By the resolution of each novel, however, the generic conventions re-emerge. Unlike Sir Arthur Conan Doyle's belief in spiritualism, which seemed to contradict his famous detective's rationalism, the saltmarsh and its status in folklore and mythology allows for an evocation of mysticism within a rational discourse. In this, the characters are also key: Ruth, as an archaeologist, connects the worlds of ritual, sacrifice and symbolism with Nelson's evidence-based approach to the 'truth'; a contrast articulated by Ruth herself at the close of the first novel, when she returns to the saltmarsh to meet Nelson, saying: "The questions are more important than the answers" (295).

Conclusion

In Griffiths' novels the saltmarsh is constructed as a site of deep and terrifying haunting, with its generational layering of buried bodies that date from the prehistoric child to events that take place within the story itself. In this sense, Griffiths picks up on the long-standing signification of the Gothic East Anglian landscape evoked by M. R. James, whilst channelling, in a geographical sense, the erosion that simultaneously covers and reveals the history of the sands. Throughout these evocations, the novels construct Ruth as a character who in knowing the saltmarsh gains deeper understanding of herself. While individually, the novels conform to generic expectations (each novel features its own set of murders that are introduced, solved, explained and thus contained within the individual narrative), the wider overarching story, woven across the series, is dominated by two relationships. The first is a conventionally complex long-term romance that sees Ruth and Nelson navigate a series of trials, disputes and narrative twists. The second, and in many ways, more prominent, is the development of Ruth's connection with the landscape of the marsh which parallels Ruth's understanding of self. In this, Griffiths' fiction draws on many of the same methods and tropes of contemporary place writing; by creating a character who draws out the inherent complexities of the saltmarsh while using these to reflect upon life's meaning, Griffiths demonstrates the intrinsic relationship between people and place.

At the same time, as Jamie says of Robert Macfarlane's *The Wild Places* (2008), there is a "sense of beguiling solitude" ("Lone Enraptured Male") in Ruth's saltmarsh that is both powerful and potent yet restorative and freeing. This desire for isolation and wilderness that features so prominently in many of the works of new nature writing, is also a key aspect of Griffiths' novels, and the characterisation of a female protagonist who seeks life on the periphery of society complicates the ideas often associated with solitude and gender (as Jamie's 'lone, enraptured male' implies). For Ruth, alone on the saltmarsh, there is a spiritual harmony in the muddy merging of the past, present and even future, as she is enraptured not only by the beauty of the moment but this deep sense of anthropological legacy within the landscape. The saltmarsh both disrupts and preserves, gives rhythmic order and unpredictable behaviours and defies cultural categorisation, rendering it a natural setting for a fusion between the crime and place writing genres.

Note

1. The long-established creative writing programme at the University of East Anglia has fostered the connection between place and writing, with many of its alumni – writers such as Edward Parnell – continuing to emphasis the significance of East Anglia on the literary map. See also, the very recently published *Fieldwork*, an anthology of East Anglia new nature writing developed across a series of public workshops and open submissions from people across the region and published by UEA.

Bibliography

Alexander, Neal. "Theologies of the Wild: Contemporary Landscape Writing." *Journal of Modern Literature*, vol. 38, no. 4, 2015, p. 1, doi:10.2979/jmodelite.38.4.1.

Arnitt, Lucie. "Ghost-al Erosion: Beaches and the Supernatural in Two Stories by M. R. James." *Popular Fiction and Spatiality: Reading Genre Settings*, edited by Lisa Fletcher, Palgrave Macmillan, 2016, pp. 95–108.

Ashman, Nathan. "Hard-Boiled Ecologies: Ross Macdonald's Environmental Crime Fiction." *Green Letters*, vol. 22, no. 1, 2018, pp. 43–54, doi:10.1080/14688417.2018.1431139.

Brosseau, Marc, and Pierre-Mathieu Le Bel. "Chronotopic Reading of Crime Fiction: Montreal in *La Trace de L'Escargot*." *Popular Fiction and Spatiality: Reading Genre Settings*, edited by Lisa Fletcher, Palgrave Macmillan, 2016, pp. 45–62.

Carson, Clare. *The Salt Marsh*. Head of Zeus, 2016.

Fletcher, Lisa. "Introduction: Space, Place and Popular Fiction." *Popular Fiction and Spatiality: Reading Genre Settings*, edited by Lisa Fletcher, Palgrave Macmillan, 2016, pp. 1–8.

Fletcher, Lisa, and Ralph Crane. "Cave Genres/Genre Caves: Reading the Subterranean Thriller." *Popular Fiction and Spatiality: Reading Genre Settings*, edited by Lisa Fletcher, Palgrave Macmillan, 2016, pp. 9–24.

Franklin, Lucy. "Imagined Landscapes: Archaeology, Perception and Folklore in the Study of Mid Devon." *Medieval Devon and Cornwall: Shaping an Ancient Countryside*, edited by Sam Turner, Windgather Press, 2005, pp. 144–161.

Griffiths, Elly. *The Crossing Places*. Quercus, 2009.

———. *The Janus Stone*. Quercus, 2010.

———. *The House at Sea's End*. Quercus, 2012.

———. *The Stone Circle*. Quercus, 2019.

———. "Why Marshes Capture Our Imaginations-and Inspire Some of Our Most Unsettling Folklore." *CrimeReads*, 13 July 2020, crimereads.com/why-marshes-capture-our-imaginations-and-inspire-some-of-our-most-unsettling-folklore/.

Hagestadt, Emma. "Salt: Review." *The Independent*, Independent Digital News and Media, 5 June 2008, www.independent.co.uk/arts-entertainment/books/reviews/paperback-salt-by-jeremy-page-841042.htmlmost-unsettling-folklore/.

Hall, Sarah. *The Carhullan Army*. Faber & Faber, 2007.

Hannigan, Tim. "Foreign Country: Lone Enraptured Males, Healing Females and the Othering of Rural Britain in the 'New Nature Writing'." *Encountering Difference: New Perspectives on Genre, Travel and Gender*, edited by Gigi Adair, Vernon Press, 2020, pp. 15–28.

Hausladen, Gary. *Places for Dead Bodies*. U Texas P, 2000.

Ishigaru, Kazuo. *Never Let Me Go*. Faber & Faber, 2005.

James, P. D. *Devices and Desires*. Faber and Faber, 1989.

Jamie, Kathleen. "Kathleen Jamie a Lone Enraptured Male: The Cult of the Wild LRB 6 March 2008." *London Review of Books*, 21 May 2020, www.lrb.co.uk/the-paper/v30/n05/kathleen-jamie/a-lone-enraptured-male.

Kluwick, Ursula, and Virginia Richter. "Introduction: Twixt Land and Sea." *The Beach in Anglophone Literatures and Cultures: Reading Littoral Spaces*, edited by Ursula Kluwick and Virginia Richter, Routledge, 2015.

Lilley, Deborah. "Unsettling Environments: New Pastorals in Kazuo Ishiguro's *Never Let Me Go* and Sarah Hall's *the Carhullan Army*." *Green Letters*, vol. 20, no. 1, 2016, pp. 60–71, doi:10.1080/14688417.2015.1123103.

———. *The New Pastoral in Contemporary Writing*. Routledge, 2020.

Massey, Doreen. *Space, Place and Gender*. Polity Press, 1994.

Moran, Joe. "A Cultural History of the New Nature Writing." *Literature & History*, vol. 23, no. 1, 2014, pp. 49–63, doi:10.7227/lh.23.1.4.

Page, Jeremy. *Salt*. Penguin, 2007.

———. "Is There an East Anglian Place Writing." http://jeremypage.co.uk/Site/inside%20the%20shed . . . /5481C8CE-9791-40A8-B110-3C3B82ECBE2D.html.

Scaggs, John. *Crime Fiction*. Routledge, 2005.

Singer, Christoph. *Sea Change: The Shore from Shakespeare to Banville*. Brill, 2014.

Smith, Jos. *The New Nature Writing: Rethinking the Literature of Place*. Bloomsbury, 2017.

Wallis, Robert J., and Jenny Blain. "Sites, Sacredness, and Stories: Interactions of Archaeology and Contemporary Paganism." *Folklore*, vol. 114, no. 3, 2003, pp. 307–321, doi:10.1080/0015587032000145351.

Wilkins, Kim. "Pagan Places: Contemporary Paganism, British Fantasy Fiction, and the Case of Ryhope Wood." *Popular Fiction and Spatiality: Reading Genre Settings*, edited by Lisa Fletcher, Palgrave Macmillan, 2016, pp. 109–124.

Zwierlein, Anne-Julia. "'Gripping to a Wet Rock': Coastal Erosion and the Land-Sea Divide as Existential/Ecocritical Tropes in British and Irish Fiction." *The Beach in Anglophone Literature and Culture: Reading Littoral Space*, edited by Ursula Klawick and Virginia Richter, Routledge, 2015, pp. 59–75.

4

THE BIG DEEP

The Ecological Turn in Nordic Noir

Michael Hinds and Tomas Buitendijk

The Danish crime drama *DNA* (2019) features the ominous image of a storm-stricken car ferry as a leitmotif throughout, bobbing on the waves and at the mercy of the sea. This corresponds with the recurring image of the rocking cradle in D. W. Griffith's *Intolerance* (1916), a portentous object that the director uses to explain the ineradicable violence of the human condition. Suffering is measured there in rock-a-byes; in *DNA*, the ferry performs a similarly symbolic function, but also provides the key plot event in the narrative. A Danish policeman is travelling on a ship heading to Poland as part of a child abduction case. Incredibly, the policeman has brought his baby daughter along in a pram, since his wife, a flight attendant, is away at work. The violence of the storm has the policeman go outside on deck for air; once there, he is seized by the need to vomit. Rather than taking the obvious route of throwing up over the side, or indeed leaving the baby safely indoors, he locks her pram in place and runs to the nearest bathroom. When he comes back out, the pram has tumbled down a set of stairs, and the child has vanished. Apart from the grief-stricken policeman, everyone assumes that the sea has taken her; it takes years for him to discover the evidence that proves otherwise.

Despite the ultimate explanation – that someone abducted the baby while the policeman was vomiting – it is interesting to reflect on the way in which the sea setting is used to license all of this absurdity. The policeman acted consistently in a way that endangered his child, almost as if being at sea made that inevitable. Or rather, only at sea could such licence plausibly be afforded. This trepidation towards the sea as a metaphorically cruel or hostile place is a classic theme in crime and detective fiction, but contemporary Nordic dramas have moved it into a new zone of significance. Here, the sea is no longer imagined comfortably within metaphor or allegory, but rather within the terms of its own agency. Put more radically, it can no longer be imagined, rather only acknowledged. If crime fictions are traditionally structured around the idea of pursuing a mystery through to its resolution, these dramas contrarily emphasise the experience of getting lost. In the oldest sense of the term, they show the world "at sea".

Acts of violent carelessness towards children might speak of classically Freudian anxieties, but in a Nordic context they take on a specific aspect. In her study of Scandinavian crime fictions, the American critic Wendy Lesser queries continuously why it is that "[k]ids evidently have a hard time of it in Scandinavia" (64), going by the multiple infanticides and forms of abuse that appear in its thrillers. Both killers and victims are represented as sufferers of abuse. Childhood is a unifying trauma, the dire scenario that everyone understands: "It's as if the Scandinavian writers feel they can't generate enough concern on behalf of the cops alone; they need to put a child at risk if we are to

DOI: 10.4324/9781003091912-6

feel sufficiently anxious" (Lesser 66). Lesser reiterates the question of why this preoccupation exists throughout her study, and reaches for an answer as she watches a multicultural and diverse group of Swedes dancing at a midsummer festival:

> It is not just their own children's innocence the adults are guarding from imaginary cries and violations, but the innocence of the children they once were. Childhood is evidently a period these Swedes recall with tremendous fondness- a time, as they truly or falsely remember it, when everything was joyful and fun. (Perhaps this feeling applies even more strongly to those who did not have a happy childhood, but who feel, for various cultural and conventional reasons, that they *ought* to have had one).
>
> *(240–241)*

The problem with this explanation is that it remains locked within the self-concern of a baby boomer complacency, and that it does not account for the tensions between generations that necessarily emerge out of material concerns. Lesser centres her explanation on an idea of happiness, as if the spectre of violence against children was only mustered as a fantasy necessary to the continuing enchantment of childhood. Yet it can also be argued that this fantasy is in fact deeply implicating and based on an apprehension of a more thorough and inevitable violence. The evidence for this also comes from Sweden, in Greta Thunberg's 2019 address to the UN climate action congress in New York:

> This is all wrong. I shouldn't be up here. I should be back in school on the other side of the ocean. Yet you all come to us young people for hope. How dare you. You have stolen my dreams and my childhood with your empty words.
>
> *(Vaughan)*

Thunberg's accusation has a double implication: the crime of responding inadequately to climate change is also a crime against children.[1] In such a context, the serial crimes against children that occur in Nordic crime fiction on screen and page can be seen as eco-metaphorical. The father from *DNA* may as well throw his baby into the sea, because the waves will claim her eventually. By referring to her life on "the other side of the ocean", Thunberg seems to acknowledge that the sea is something that should not be crossed lightly. With that, a further implication emerges: perhaps the ultimate terminus of global warming will become clear when there is nothing left but water. This is the truth that the world works assiduously to avoid or deny, and yet the sea just keeps on rising.

The sea in *DNA* is seen as profoundly antisocial and corrupting, a place where there is little limit as to what might be imagined or possible, a place where criminality might thrive. Most crime stories are grounded in land-bound plausibility, yet putting people on a boat seems to make anything possible. This is an old idea in fiction; the further the crew of the Pequod get from Nantucket, the more susceptible to madness they become. Yet it is also curious that sea-fearing and seafaring Nordic nations should regard the sea in this way, as something where realism might be suspended. The sea is a constant factor of life there; Norway boasts over 25,000 km of coastline, from which one is never too far. Wendy Lesser stresses that "criminals in these mysteries seem able to flee relatively easily from one place to another, in part because of the prevalence of watery modes of transport" (21). Yet although Nordic nations coexist with the sea, they do not do so in complacency; the sea acquires metaphorical significance according to shifting circumstances. The transference of criminality to the sea is symptomatic of the profound ambivalence that attends the prosperity of the Nordic nations in modernity, which Oxfeldt, Nestingen, and Simonsen identify as the problem of having to reconcile the evaluative idea of happiness with the affective experience of it. Evaluatively, Nordic people ought to be happy, according to the United Nations' World Happiness Report (Helliwell et al.); they have multiple social protections and relatively robust economies (Oxfeldt et al.). Yet disquiet persists and not just in the mind of Thunberg. There is a feeling that such "happiness" has been bought at the expense of future generations. On the one hand, the wealth of Norway is stupefying, an astounding stroke of luck for its inhabitants; on the other, the banal and even dirty explanation is that Norway hit the jackpot through petrodollars that spumed out of the sea, and this generates thorough anxiety as a counterpoint to extraordinary prosperity. The 2017 update on the World Happiness Report placed Norway first in the world, with Denmark and Iceland next, Finland in fifth, and Sweden in tenth place (Helliwell, Layard, and Sachs); yet the more evaluatively happy these countries were by socio-economic measures, the more they produced traumatic dramas about crimes for mass consumption on television, broadcasting both to themselves and the world that there was trouble in so-called paradise.

Two recent series, *Twin* (Norway, 2019) and *Trapped* (Iceland, 2015–) are of particular interest for the recurrence of deep-seated ecological, social and economical anxieties, articulating the uneasy relationship between modern Nordic societies and the natural environment. They also manifest the independent agency displayed by a wide variety of non- and more-than-human agents, many of which stage decisive interventions in the plots of the two series. This takes on particular significance with the representation of the sea and shows the contemporary shift in Nordic noir towards a pluriagential, rather than a singularly human, configuration of drama and mystery.

'Do You Want to Live in a Place That's Full of Shit?': Ecological Anxieties in *Twin*

The second most important factor in the Norwegian economy is coastal tourism. This tourism is unthinkable without the sea, and yet the country's primary source of wealth is something that actively menaces the marine environment with potential contamination.[2] If the fjords are sold to tourists as sites of eternal and transhistorical purity, they are nevertheless always imperilled by the petro-wealth that dare not speak its name. In other words: repression rules.

Ecotourism features heavily in the Norwegian crime drama, *Twin*, but what really predominates is the theme of repression and its consequences. The plot features twin brothers, Adam and Erik (both played by Kristofer Hivju), both of whom loved the same woman, Ingrid (Rebekka Nystabakk), and who also shared an adolescent love of surfing. The brothers became estranged from one another after Adam won Ingrid's affections while Erik was away travelling (Erik and Ingrid had been together prior to his departure). Adam and Ingrid got married and ran her family's hotel business, offering boat-rides to whale watchers. Erik remains a cash-poor but apparently self-contained beach bum, running a surf school but without any interest in "growing" his business. The brothers collide after a long period, with Erik seeking financial help and a bed shortly after crashing his campervan into the sea. They end up fighting on a boat belonging to the hotel. Ingrid intervenes and hits Adam on the head. Erik determines to bring Adam to the nearest hospital, and heads off into the night sea, with Ingrid's father (John Sigurd Kristensen) calling out that the boat is unsafe. At that point, the action cuts to a police search in the fjord in total darkness, and a body is removed from the water. What happens between Erik heading out to sea and the discovery of the body represents an abyss in the plot that is fundamental, shifting the emphasis from the traditional idea of crime (as an essential human activity) to energies and actions that are uncodifiable and more thoroughly mysterious.

It is never revealed what actually transpired that night; Ingrid finds Erik the next morning on the shore, nearly catatonic with hypothermia. He reports that Adam "just fell in . . . there was a net or something . . . I lost the boat, it disappeared, and . . .". Erik's mind has been obliterated by what has occurred, and the audience is never given any more information; as such, this becomes a crisis of agency. The uncertainty over what happened also translates into a crisis of genre: perhaps curiously so, *Twin* operates most productively as a comedy of mistaken identity rather than a tragedy of fratricide or mariticide. This is a crime drama that cannot act upon the logic of a whodunnit, not in the least because nobody (including the viewer) knows what exactly has happened, and what role the sea itself may have played in the course of events. Adam's body might have been caught up in a fishing net and pulled under the water; if so, he died by drowning, almost like a reverse fishing catch. At base, Erik might know what happened to Adam and is lying, but it also seems that his stupefaction is a way of telling the whole truth. Erik is especially well qualified to attest to the unquestionable remorselessness of the sea, as he has spent most of his adult life on the beach. In this wealthiest of countries, where nobody should be poor, Erik seems a peculiar outlier in Norwegian society, someone who will not or cannot pay his debts. More than being a spendthrift, he simply refuses the logic of debt and payback that the human world depends upon. His girlfriend, thinking him to be dead, describes him in simultaneously watery and financial terms: "He lived in his own bubble". As such, Erik corresponds to Jon Anderson's argument that surfers should be defined as much "through their relations to fluid, watery spaces as they are to terrestrial locations" (237).

The bubble has to break; to protect Ingrid, and with her encouragement, the beach bum twin assumes his dead sibling's identity, which means having to simulate his own death. It also turns out that Erik is the real father of Adam's and Ingrid's daughter (which further explains the fraternal enmity). This would be a dark gothic tale in many tellings, but not here. Nobody is seen as entirely villainous. Rather, the characters are hapless, contending to keep pace with the absurdity of their situation, and this is especially acute when Erik tries to impersonate Adam. *Twin* asks the viewer to forgive its contemporary Cain for killing Abel (even though he did not kill him) and then empathise with his struggles to impersonate a successful businessman. Adam had boasted to Erik how he had figured out how to farm tourists and extract profit from the sea: "This place is boiling over . . . We've got Chinese, Germans, the lot". Erik's business, on the other hand, was much more about the pleasure of surfing than any kind of active schooling. Now Erik is required to take part in an interview with a travel correspondent from *The Guardian*, and to pose hopelessly as a successful eco-hotelier. It is not guilt over the death of his brother that makes this impersonation difficult, but more simply that these acts of self-presentation require considerable skill. The art of bullshitting is difficult to master; Erik would rather flee to the sea.

Erik's friend Frank (Gunnar Eiriksson), a policeman who suspects that there has been foul play, reveres his memory, proclaiming him to be a "steady rock", "with no earthly chains holding him to all kinds of nonsense". He also avers that drowning represents a kind of happy ending for Erik: "In a way, you could say he's finally reached his Utopia". Meanwhile, Erik is living a nightmare. He discovers that Adam was a closeted homosexual, which introduces yet another repressed truth he has to live out on his dead brother's behalf. All of this occurs in a curious context in which the police are not particularly central to the action, which is unusual for a crime drama. For example, Frank's analysis of what has happened (in many regards correct, apart from his understandable conviction that it really is Erik who has died) is dismissed out of hand by the senior detectives handling the case.

This kind of denial might be a national theme. The green tourist bears a heavy carbon footprint, and the pristine fjord might swim in oil at any moment; doubleness is everything. It can also be inferred that the thrill of ecotourism is only heightened by the sense that the environment is imperilled. This guilty pleasure is contaminating; voices emerge to denounce the tourism that is nevertheless proclaimed as bringing life to the area: "The tourists have been shitting there, shitting all over the place. Do you want to live in a place that's full of shit?"

There are many ways to be "full of shit". Erik knows that he is an impostor, but also recognises that this is a betrayal of the kind of life that he had been trying to live, sleeping on the beach in a repurposed shipping container. Shipping containers are key objects in transnational crime dramas (recall the second season of *The Wire*, set on the Baltimore docks, USA); they are the modular blocks out of which all global business is made, whether the economy is black or whiter than white. Ordinarily, their significance in crime plots is dependent upon their capacity for concealment; one can smuggle anything within something so practically impenetrable to sight. Yet Erik's container is open, facing the sea, practically welcoming the elements, revealing to all that he possessed nothing. This makes him a curious protagonist for noir, which tends to install world-weariness into its characters. Erik is guileless and apparently without cynicism; he has refused to function in the "real" world of business and its attendant corruptions. At the end of *Twin*, it remains unclear whether he can really be considered a criminal. He alone knows what happened at sea, yet he also has no words to translate that experience into a form understandable to more conventional people, including the audience. Every other character exists within the domain of necessity and remains oblivious to the peculiar terror (and pleasure) of the sea that only Erik understands. He places the familiar noirishness of the plot and the rest of the cast in a new context of radical uncertainty, where only he has the imagination to understand what the natural world is really capable of. Yet that knowledge is also too vast and powerful to be communicated. Erik is simultaneously absurd and liberated, the only character in the series that does not behave as if he is in a crime drama.

"Thing-Power" and the Abyss

Twin appears to surrender its connection to many conventions of crime fiction, and yet its subject material does correspond thoroughly with that of a "classic" mystery. Both *Twin* and Edgar Allan Poe's "Descent into the Maelstrom" are set in the Lofoten archipelago of Norway, with the latter story also featuring the loss of a brother in a mystery at sea. Whatever happened to Erik in the storm is unsayable: this might be because of his shock and awe at what has happened, but it also seems to be a matter of the experience taking him beyond language, even beyond the human. Poe's narrator is similarly afflicted after he emerges from the maelstrom: "[A] boat picked me up – exhausted from fatigue – and (now that the danger was removed) speechless from the memory of its horror" (448). Recounting the events years later, the historical distance between him and the drama means he now almost has *too much* to say. Fundamental to his recollection is the fact that the maelstrom represented something, a kind of "Thing-Power" (Bennett 6), that made him re-coordinate his understanding of reality. He found peace in the non-humanity of the drama:

> It may appear strange, but now, when we were in the very jaws of the gulf, I felt more composed than when we were only approaching it . . . how magnificent it was to die in such a manner, and how foolish it was in me to think of so paltry a consideration as my own individual life, in view of so wonderful a manifestation of God's power.
>
> *(443)*

Compared to Erik's experiences surrounding the death of Adam, one critical distinction must be acknowledged: Poe's narrator has glimpsed the conventionally divine, but Erik remains unable to qualify what happened to him in the storm. Yet Erik's speechlessness represents its own form of eloquence, one that closely aligns with the non-human forces encountered by Poe's narrator. Both point towards the limits of language, the power of the other-than-human, and the possible loss of meaning in an abyss that defies (human) understanding.

Poe's narrator recounts how the maelstrom reduces ships, people, and trees to a new unity of thingliness, a "wide waste of liquid ebony on which we were thus borne . . . I perceived that our boat was not the only object in the embrace of the whirl" (445). He continues:

> Both above and below us were visible fragments of vessels, large masses of building timber and trunks of trees, with many smaller articles, such as pieces of house furniture, broken boxes,

> barrels and staves. I have already described the unnatural curiosity which had taken the place of my original terrors. It appeared to grow upon me as I drew nearer and nearer to my dreadful doom. I now began to watch, with a strange interest, the numerous things that floated in our company. I *must* have been delirious – for I even sought *amusement* in speculating upon the relative velocities of their several descents toward the foam below. 'This fir tree', I found myself at one time saying, 'will certainly be the next thing that takes the awful plunge and disappears', – and then I was disappointed to find that the wreck of a Dutch merchant ship overtook it and went down before.
>
> *(445–446)*

Poe's maelstrom reorganises and reorientates things as it chooses, not respecting any force other than its own. By extension, this also explains why *Twin* is so peculiarly disinterested in the workings of the police, as they cannot comprehend how to bring what happened at sea under their jurisdiction.[3] Even the bottom-line method of criminal detection in modernity, the DNA test, is of no use to them; they are unable to figure out which twin is dead. The impossibility of knowing what happened to Adam at sea, which most clearly is demonstrated by Erik's inability to find words for it, is a *mise-en-abîme* that does not allow for the confident understanding that a crime has occurred. The mystery is genuinely unfathomable, and the power of explanation remains with the water. As the sea also holds all relevant physical evidence, the police are entirely subject to its material authority. With regard to the latter, Poe's narrator remarks how the sea makes its power of matter visible even outside of the maelstrom:

> I called to mind the great variety of buoyant matter that strewed the coast of Lofoden, having been absorbed and then thrown forth by the Moskoe-ström. By far the greater number of the articles were shattered in the most extraordinary way – so chafed and roughened as to have the appearance of being stuck full of splinters – but then I distinctly recollected that there were some of them which were not disfigured at all.
>
> *(446)*

This capacity of the sea to either radically alter or preserve matter finds parallel expression in *Twin*: the body of Adam is yielded up, practically in perfect condition, but at the same time a fragment of Erik's camper van is seen by Frank, bobbing near the shore. The sea makes compost out of the material things of the land, shredding them to a pulp; but it can just as well preserve objects in their original state, and offer them back to society. This means that the body of Adam is not so much discovered by the police as given up by the water. In the same way, the sea may well have spared Erik in the storm while dragging his brother to his death. The distinct dehumanisation of the plot in *Twin* is in many ways a logical culmination of the ecological turn at work in Nordic noir. Human (or viewer) omniscience is unsettled, which voids the epistemological contract of the genre. In *Twin*, one cannot 'find out' because things are never neatly resolved. The sea has become a law unto itself, meaning and intention have collapsed, and human frames of reference have lost their sense of priority. In other words: we are looking into the abyss.

'This Really Is Disgusting': *Trapped*'s Two Ecosystems

The first season of the Icelandic series, *Trapped*, also begins with a body being yielded up by the sea. More precisely, a small fishing trawler hauls up a torso, missing its head and all of its limbs. This discovery is made as a car ferry from Denmark simultaneously approaches its destination at a nearby port. The town is unnamed in the show, but reference is made throughout to it being in the north and a far distance from Reykjavik on the other side of the country.[4] This becomes significant when a blizzard (and subsequent avalanche) isolates the town and makes any land travel impossible. The local police chief, Andri (Ólafur Darri Ólafsson) and his colleagues, Hinrika (Ilmur Kristjánsdóttir) and Ásgeir (Ingvar Eggert Sigurðsson), are tasked with keeping things under control until the Reykjavik police arrive. They instruct the ship to stay in port, enforcing this with a court order, and so total immobilisation is achieved: there is no way out, by any means of transport. This immobility conforms to the classic mystery device of the closed circle of suspects but also has a kind of viral impact in terms of revealing crimes, which proliferate to a remarkable degree (even for the noir genre).

Suspicion falls immediately upon the ship. A known member of the Lithuanian Mafia attempts to escape in a camper van, along with two young African girls that are being trafficked for the sex trade. A member of the crew turns out to be a young local man, once imprisoned for involvement in the death of his girlfriend in an arson attack. Mutually discomforting looks are exchanged between the captain and a man who appears to be an engineer. *Trapped* initially appears to be set on the same course as *DNA*, using the trepidation of both ship and sea to license its plot. However, while the ship was being used for sex trafficking under the control of the sociopathic faux-engineer, this is seen as much less significant than the steadily unveiling criminality of the town itself. The police chief's daughters are shown cruelly teasing a small boy in their school where a teacher is having an affair with a teenager. An old man in a wheelchair spies on the town's inhabitants through a telescope. The mayor beats his wife and has violent sex with her; she, in turn, manipulates prominent citizens in the town to secure a lucrative but ethically suspect business deal for the construction of a Chinese megaport by buying out the properties of local residents. There are so many further expressions of depravity in the town that the extent of criminality seems unfathomable; as one character says to Andri, "You're not native to this town. You can't see into the abyss". A strange chorus to all of this is provided by the stranded passengers from the ship, who are shown corralled unhappily in either the school gymnasium, the ferry terminal, or the hotel bar. They are not allowed much of a voice, as if what is going on has nothing to do with them.

Arguably, they are right. *Trapped* is in one sense heavily allegorical, or more precisely metonymic, suggesting that this isolated community is a concentrated version of the entirety of Iceland itself. This is especially relevant in the aftermath of the financial crisis of 2008 and subsequent revelations about widespread corruption in the nation's elite. The headless corpse that started everything is therefore readable as emblematic of the Icelandic body politic. When stolen from its storage in the factory, it presumably begins to putrefy, just as the town reveals its many corruptions. In one of several moments of mordant comedy, Andri first inspects the body and says: "Not in the water for long. It would be more swollen, and the shrimps would get to it". Whether nipped by bottom-feeders or hacked by criminal conspirators, bodies can only be read cynically in such a context: "This really is disgusting". If *Twin* is symptomatic of the inevitable collapse of the petro-economy and guilt about the riches it has generated for Norway and Norwegians in the meantime, *Trapped* indicts Iceland for not feeling guilty enough for its shortcomings, including the failure to excoriate itself sufficiently both before and now.

As this carnivalesque nefariousness reveals itself, the show uses establishing shots of the weather and the waves to remind us that they are the only source of control in this otherwise sour environment. As the blizzard arrives, the snow seems illuminatory and balletic, while the sea is shown as self-possessed and largely unexpressive, refusing to be drawn into facile metaphor. Buildings and even projected Chinese megaports, cars, and ships all seem entirely inadequate structures to protect the population from the elements that engulf them. Yet the violence of the elements is seemingly harmless in comparison to what people perpetuate; the killings and cycles of violence in the town are only symptomatic of a broader tendency towards self-destruction that is shown as essentially human. Even as the devastation appears to be over, at the end of the series the mayor's widow Kolbrun (Sigrún Edda Björnsdóttir) is shown reviving the negotiations for the Chinese deal. *Trapped* returned for a second season, and this time looked at the contamination of the water table by an

American aluminium plant and the corrupt political interests that had invited them there. A prominent role is played by Kolbrun, now a local politician herself.

In this sense, *Trapped* shows two ecosystems. One is socio-political and limited to humans, in which they end up cannibalising or asset-stripping one another; and then there is a broader system that has a completely different relationship to time and space. *Trapped* generates a sense that the sea and the mountains are biding their time, awaiting the conclusion of the human death spiral. These other-than-human factors play at such vast timescales that the human ones barely matter anymore. The mayor's office in the series hints at this discrepancy in temporal significance, as well as the inability of the townspeople to accept their own relative irrelevance: the room contains a giant framed picture of the wrecked ship in the village fjord, visible to all yet never mentioned in the dialogue. The picture seems to acknowledge the inevitability of structural ruination, not metaphorically, but as fact.

Beyond this Ozymandian scenario, the marine environment can further be seen as the prime witness in the entire drama. The sea initially yields up the torso, and another body part later on at a vital juncture, enabling the investigation to progress; every time it performs this kind of intervention, it generates paradox. The plot moves further towards a kind of exposition, but it also gets more complicated. The ship's captain shows some insight here. He refuses to turn on the heating on his ship, saying: "I can't do that, I'll use all my fuel, and I'll be stuck here". Resisting the immediate claims of human comfort, he realises that this is not a place to be stuck. At least there are fewer people out at sea. The human domain is a moral abyss, dominated by cyclical violence, intergenerational conflict, and unrepentant profiteering. The actual marine abyss is chucking up body parts, not allowing the human one to deny responsibility. As in *Twin*, the sea composts, but it also offers up unignorable truths, the most fundamental of which is that this space exists outside of human control, and that it will still exist after human auto-destruction is complete.

At the conclusion of *The Big Sleep*, Raymond Chandler's Philip Marlowe offers the limited consolation that when you die, you die. He envies the dead Rusty Regan:

> You were sleeping the big sleep, you were not bothered by things like that, oil and water were the same as wind and air to you. You just slept the big sleep, not caring about the nastiness of how you died or where you fell. Me, I was part of the nastiness now. Far more a part of it than Rusty Regan was.
>
> *(213–214)*

Ultimately, these fictions all seem to seek the peace that might lie beyond the frenzy of being human, a place where wind, air, water and even oil can coexist without the contaminations characteristic of the Anthropocene. Marlowe calls it nastiness, as if it is a moral defect, but this also suggests that life is ultimately within human control. In the Nordic imagination, human nastiness can turn to the horrific or the quasi-comedic, but it finds particular exposure when set against the phenomenal indifference of the sea towards human agency: one might call it the Big Deep.

Towards a Pluriagential Nordic Noir

Far beyond their intricate, sometimes ridiculously contrived dramas, both *Trapped* and *Twin* emphasise the ultimate insignificance of human activity when measured against the enormity of a more-than-human reality. In the same context of pluriagentiality, they explore aspects of chaos, violence and the staking of individual and group territories during the interaction between different (kinds of) beings. Featured conflicts between humans become symptomatic of a larger process, in which all participants in the complex realities of Seyðisfjörður and the Lofoten islands carve out their respective spaces, forge alliances, and settle debts. All this is reminiscent of French sociologist Bruno Latour's recent (re)appropriation of James Lovelock's Gaia hypothesis, and in particular Latour's focus on the inevitability of territorial encroachment and violence (*Facing Gaia*). Close reading of *Trapped* and *Twin* reveals that, exactly as Latour argues, there is a significant assumption of independent agency for a wide range of animals, objects, and phenomena in both series, as well as frequent intervention by these other-than-human actors in human affairs (*Facing Gaia* 98–99). Arguably, either plot is only able to accelerate because of intermittent non-human interventions. There is the collusion between sea, fishing net, and other 'incredible forces'[5] that allegedly make Adam's body disappear in *Twin*, only for the "undertow" to make it reappear elsewhere; but there is also the closed circle of suspects scenario enforced by the confluence of weather events in *Trapped*, and the (delayed) revelations of clues in the same series due to snow cover or whimsical sea currents.

In Latourian terms, the full array of other-than-human agents in *Trapped* and *Twin* are effectively shown to evolve from reliable intermediaries to unreliable mediators. Latour observes that while the former "[add] . . . predictability to the setting, [the latter] might suddenly make it bifurcate in unexpected ways", which means that "[i]f any of the intermediaries mutates into a mediator, then the whole set up, no matter how solemn or controlled, may become unpredictable" (*Reassembling the Social* 202). The sea may long have been considered a more or less immobile backdrop for resource extraction, ecotourism, and transport, but now reveals itself to be an agent in its own right, intervening indifferently yet decisively in the making of human and non-human lifeworlds. In a similar way, everyday objects (e.g., the net in *Twin*), environmental factors (snow and the storm in the first season of *Trapped*), and even the earth itself come to reshape the course of action, albeit without clear motive (normally so important in film noir) or due attention to timing. This unfolding autonomy of expression (what we referred to as "Thing-Power" earlier in this chapter) is particularly clear in the shift of pace between seasons one and two of *Trapped*: while the former "merely" figures non-human agents as enforcers of classical noir tropes, the latter casts the more-than-human environment as a rogue agent that acts on and reacts directly to human interference with the Icelandic environment. Earthquakes and the poisoning of a host of individuals occur seemingly at random, making a mockery of the human drama of the series. Even if it later transpires that the source of the poisoning was illegally dumped toxic waste, the acceleration of the plot is explicitly affected by run-amok non-humans: the poisoned water has gone "feral" (Tsing et al.), making its independent way through the landscape to wreak further havoc in human communities. The dominance of this non-human agency over human involvement becomes strikingly clear when all rivalries are temporarily placed on hold after the poisoning of Skuli, a naïve young "outlaw" who ends up in critical care after drinking the seemingly pristine water.

Skuli's poisoning reiterates the Latourian themes of conflict and violence in the meeting between human and non-human individuals and communities. The second season of *Trapped* partly focuses on a right-wing environmentalist group, "Thor's Hammer", but the action goes against the finding of common ground between (native) humans and non-humans; there are no attempts at (positive) recuperation nor at 'getting on together' in order to arrive at a more balanced relationship between the human and the other-than-human (Haraway 10). One might question the motives of Thor's Hammer: is their "environmentalism" anything more than a clash over available resources, or the ability to make a living by drawing from the natural environment, just like everyone else (think of ecohoteliers, surfers, megaport owners, and aluminium smelters)? Deeper human sensitivity to the agency of the other-than-human clearly has no place in the contemporary Nordic imaginary. As demonstrated, the latter is still caught up in appropriations of the environment as an indifferent adversary, cornucopia, or sightseeing destination, and already fails to conjure up an appropriate response to either of these projections. There is no cross-species confluence of experience; instead, both series emphasise time and again the violence of Gaia and drive to extremes the ferality of the non-human other: occasionally (and at other times questionably, in case of the weather) anthropogenically enabled, but always out of human control. The result is the same as well: the main plot drivers are now of a non-human character.

Many theorists point at the pluriform connections between human and non-human bodies (Alaimo; Bennett; Haraway), which effectively blur the boundary between the (human) self and the (non- or more-than-human) other. This plays some role in the two series as well. Erik's surfing in *Twin* suggests a deep connection between man and sea, a place where he can be himself, and conversely a place that seems to understand him. Negative instances of such connections are even more abundant: the avalanche in the first season of *Trapped* emphasises the dependence of Icelandic society on beneficial weather, and the poison in its second season literally makes its way into human and non-human bodies. At the same time, these incursions into the convergent experience of (human, non-human, and more-than-human) life are continually off-balance or turn out to be a cog in the machinery of hidden agendas. Does Erik really feel a meaningful connection with the sea, or is he simply running away from societal responsibilities? Is he a physical or a social "fish out of water" when on terra firma?

Trapped and *Twin* can be said to constitute (further) proof of a contemporary shift in attention towards the inevitability of more-than-human agency. This development has been signalled before by Helen Mäntymäki, among others, in her discussion of Swedish crime series *Jordskott* in a 2018 special edition of the journal *Green Letters* on ecological crime fiction.[6] Mäntymäki suggests that "[t] he crime genre has, during the past few decades, evolved into a many-sided and flexible means of responding to social concerns and developments" (90). Series like *Jordskott* (and *Trapped* and *Twin*) have shifted the critical perspective beyond the human and towards a "wide network of acts and subjects including human and non-human as well as alive and non-alive ones" (Mäntymäki 98). In other words: Nordic noir is moving in a plurispecies, pluriagential direction. Possibly owing to the nature of the genre itself, the main characteristics of this more-than-human approach to drama are encroachment, negotiation, and violence; but we also find a decentralisation of human interests, expansion of timescales from the human here and now to weather-affected, seasonal, or even deep/geological time, and an intentional confusion over questions of agency, intention, and motive. The shift remains incomplete; this, again, is evident from the change of pace between the first and second seasons of *Trapped* (which can be read as proof of a growing eco-awareness, even in its own makers) and the almost comedic vacillation of agency, intention, motive, and plot resolution in *Twin*. Even as the new ecological narrative emerges, old tensions in the human relationship to the non-human environment continue to be worked through.

Characteristics of the New Eco-Noir

One question at this stage is whether the general public is ready for a more ecologically attuned noir genre. Both *Trapped* and *Twin* were generally well received, but most of the praise was related to cinematic quality and the actors' performance, with relatively little attention paid by reviewers to the shift in agency underpinning their respective plots. *The Guardian*, for example, hails the second season of *Trapped* for "[tackling] the far right, economic anxiety and environmental doom . . . and woolly jumper porn too", but limits its commentary on the role of non-human agents to the observation that "[n]ature lies like a dank, heavy flannel over the not-so-sleepy town", and that "the landscape [pins] the town in, and its residents down" (Bramley). Such remarks are accurate yet leave much unsaid about the very active role played by non-human others in plot development and resolution. Conversely, it is possible to criticise *Trapped* for trying to "do it all", tackling too many different themes in the space of only a few episodes. In this view, the "eco" character of the series becomes a nuisance; a sign that the spectre of political correctness is invading even that grim bastion of (fictional) crime: Nordic detective noir. It can be countered that such complaints are familiar in the context of debate about climate change and that its hard truths require actions that are apt for the complexity of the situation. Furthermore, Nordic noir fiction has always been politically engaged, from the declaredly Marxist Maj Sjöwall and Per Wahlöö onwards.

A possible concern regarding the transition from a "classic" to a more ecologically attuned noir is the impact such a shift can have on key traits of the genre. However, changes to the character of Nordic noir that occur as a result of its recent or pending ecological turn do not necessarily drive the genre to destruction. In fact, the befuddlement of actions and reactions that characterises the expanded host of characters – humans, non-humans, and more-than-humans vying for the attention of the viewer, and each other – is far from foreign to the machinations of the genre: both "Gaia" and detective noir are typified by the disappearance of former allegiances and individual interests under the pressures of a new kind of plot. This means that the described introduction of non-human interests to what has long been a stronghold of exclusively human drama might well lead to a revitalised genre, an eco-noir that builds on novel interpretations of its predecessor's ("normal" noir) signature confusion of dichotomies, such as good/evil, innocent/guilty, victim/perpetrator, and more. It is only to be expected that forthcoming noir series will feature crimes and mysteries in which no human instigator can be found, or in which (much like in *Twin*) the police take the back seat as non-human forces conspire to resolve or thicken the plot. An example of the former is already found in recent Netflix series *Katla* (2021), where long-dead and missing Icelanders return to their families amidst a violent, drawn-out volcanic eruption.[7]

Furthermore, whereas "classic" noir drama is often driven by greed, lust, revenge, or other (human) agendas and emotions, the sheer indifference of non-human protagonists can be expected to have a significant impact on the characterisation of their interventions and the plots they accelerate or complicate. For example, one might expect new explorations of the abandoned legal device of the "deodand", used until the mid-nineteenth century to give meaning to and mete out justice in those situations where a non-human agent was the sole cause of harm to a (human) citizen (cf. Bennett 9). Meanwhile, it is not guaranteed that forthcoming noir dramas revolve exclusively around characteristically "ecological" themes. *Trapped* already suggests that non- and more-than-human agents do not pursue by default an agenda of nature preservation: there, the locally dumped poison is carried away from the original source of contamination by means of non-human actions (animals, water, the wind), effectively driving other parts of the ecosystem to destruction. In many senses, the ecological turn in Nordic noir directs the public's attention towards issues that are as pressing off-screen as they are on it, giving new complexity and impetus to a genre that has a tendency to rely on familiar tropes and plot devices.

The varying, at times conflicting approaches to narrating the climate catastrophe that characterise works of eco-noir like *Trapped* and *Twin* highlight the capacity of the genre to practise both responsibility and 'response-ability', as Haraway calls it: "to venture off the beaten path to meet unexpected, non-natal kin, and to strike up conversations, to pose and respond to interesting questions, to propose together something unanticipated, to take up the unasked-for obligations of having met" (130). *Trapped* amounts micro-aggressions and violations against the environment as part of a thorough (but familiar) portrayal of human criminality; *Twin* complements this on a phenomenally vast scale, where humanity is unable to comprehend the enormity of the problems it must face. *Trapped* demonstrates the need for unending local actions and remedies, given the recidivist nature of human-on-environment crime. At the other end of the Nordic spectrum, *Twin* worries about the inevitable insufficiency of such measures. These demonstrate the imaginative limits of human agency, rather than suggesting the simultaneously visionary and pragmatic capacity that is required to avert outright disaster. New terrors generate new forms of expression and in turn demand a radical response. Our examples of contemporary Nordic eco-noir combine to suggest the extraordinary dimensions and ramifications of environmental responsibility that are necessary to the direful scenarios of climate change.

Notes

1. The Australian political commentator Andrew Bolt, a conservative and climate change sceptic, countered by claiming that Thunberg was rather the victim of child abuse by the environmentalists (including her family) who had encouraged her activism.

2. In a 2016 feature essay for *The New Yorker*, Jonathan Franzen describes going on a diesel-spitting luxury cruise to Antarctica with the uneasy motivation of "seeing it before it melts". This very same tension underpins much of modern ecotourism, with the irony particularly visible in left-leaning Nordic petro-culture.
3. This can be seen as metonymic of the struggle to bring the natural environment into crime fiction. The sea and other landscape elements do not fit neatly within existing legislative boundaries: they cannot be held accountable for their transgressions in the same way as human perpetrators. This is further complicated by the fact that human industrial activity is the root cause of climate change; in many ways, society has brought different forms of non-human violence to bear on itself.
4. The show was shot on location in the eastern fishing town of Seyðisfjörður, which is nearly as far away as one can get from the capital, which lies in the southwest of the country.
5. Ingrid's father, who forewarned that the boat on which the twins tussled was not safe, continues this train of thought after the incident by suggesting that 'it came loose in the storm' and that 'there were incredible forces at work here'. This essentially reiterates the possibility that everything that transpired during the night of Adam's death was beyond human control.
6. Much like *DNA* (see introduction to this chapter), *Jordskott* connects the threat of violence against children with that against the natural environment, in this case with relation to a destructive logging and mining operation somewhere in rural Sweden.
7. The reduced importance of the human in this more-than-human drama is aptly captured in a remark by one of the protagonists: "The sun goes up, the sun goes down. We are born, we die. Who am I to question the nature of things?" It might come as no surprise that the series was co-created by Baltasar Kormákur, who is also the writer and executive producer of *Trapped*.

Bibliography

Alaimo, Stacy. *Exposed. Environmental Politics & Pleasures in Posthuman Times*. U Minnesota P, 2016.

Anderson, Jon. "Surfing between the Local and the Global: Identifying Spatial Divisions in Surfing Practice." *Transactions of the Institute of British Geographers*, vol. 39, no. 2, 2013, pp. 237–249, doi:10.1111/tran.12018.

Bennett, Jane. *Vibrant Matter. A Political Ecology of Things*. Duke UP, 2010.

Bolt, Andrew. "'Filthy Climate Lies' Are Simply Child Abuse." *Sky News Australia*, 25 Sept. 2019, www.skynews.com.au/details/filthy-climate-lies-are-child-abuse-to-greta-turned-martyr. Accessed 24 June 2021.

Bramley, Ellie Violet. "Make Iceland Great Again! The Return of Trapped, 2019's Most Timely Show." *The Guardian*, 21 Feb. 2019, www.theguardian.com/tv-and-radio/2019/feb/21/the-return-of-trapped-2019s-most-timely-show-make-iceland-great-again. Accessed 27 May 2021.

Chandler, Raymond. *The Big Sleep*. Pocket Books, 1950.

DNA. Created by Torleif Hoppe, ARTE, Nordisk Film Production, and TV2, 2019.

Franzen, Jonathan. "Jonathan Franzen: A Voyage to the End of the World." *The New Yorker*, 16 May 2016, www.newyorker.com/magazine/2016/05/23/jonathan-franzen-goes-to-antarctica.

Haraway, Donna. *Staying with the Trouble. Making Kin in the Chthulucene*. Duke UP, 2016.

Helliwell, John F., et al. (Eds). *World Happiness Report 2021*. Sustainable Development Solutions Network, 2021.

Intolerance. Directed by D.W. Griffith, Triangle Distributing Corporation, 1916.

Katla. Created by Sigurjón Kjartansson and Baltasar Kormákur, RVK Studios, 2021.

Latour, Bruno. *Reassembling the Social. An Introduction to Actor-Network-Theory*. Translated by Catherine Porter, Oxford UP, 2007.

———. *Facing Gaia. Eight Lectures on the New Climatic Regime*. Translated by Catherine Porter, Polity Press, 2017.

Lesser, Wendy. *Scandinavian Noir*. Farrar, Strauss, and Giroux, 2018.

Mäntymäki, Helen. "Epistemologies of (Un)Sustainability in Swedish Crime Series *Jordskott*." *Green Letters*, vol. 22, no. 1, 2017, pp. 89–100, doi:10.1080/14688417.2017.1415159.

Oxfeldt, Elisabeth, et al. "The Happiest People on Earth? Scandinavian Narratives of Guilt and Discontent." *Scandinavian Studies*, vol. 89, no. 4, 2017, pp. 429–446, doi:10.5406/scanstud.89.4.0429.

Poe, Edgar Allan. "A Descent into the Maelstrom." *Poetry and Tales*, edited by Patrick Francis Quinn, Library of America, 1984, pp. 432–448.

Trapped. Created by Baltasar Kormákur, RVK Studios, 2015.

Tsing, Anna L., et al. *Feral Atlas. The More-Than-Human Anthropocene*. Stanford UP, 2020.

Twin. Created by Kristoffer Metcalfe, Nordisk Film, 2019.

Vaughan, Adam. "Greta Thunberg: You Have Stolen my Childhood with Your Empty Words." *New Scientist*, 23 Sept. 2019, www.newscientist.com/article/2217418-greta-thunberg-you-have-stolen-my-childhood-with-your-empty-words/. Accessed 27 May 2021.

5

AESTHETIC IMAGINARIES OF NATURE AND NATIONHOOD IN THE WORKS OF ARNALDUR INDRIÐASON

Priscilla Jolly

Over the years, Scandinavian crime fiction has become popular for its bleak atmospheric offerings. The beginnings of this trend, which stretches far beyond its origins in Scandinavia, can be located in the publication of Maj Sjöwall and Per Wahlöö's ten volume Martin Beck series, published between 1965 and 1975. The series established many of the tropes frequently associated with Scandinavian crime fiction, including the broken welfare state, the threatening natural environment and the traumatised or flawed detective (Stougaard-Nielsen 39). With regards to the adjective Scandinavian, some authors use the term to designate a geopolitical region that includes Denmark, the Faroe Islands, Iceland, Norway and Sweden, a convention that will be followed here (Arvas and Nestingen 6). While Scandinavian crime fiction has been associated with several tropes, this chapter will focus on one of them, that of the setting. Drawing from authors who have argued that Scandinavian genre fiction features "hyperlocations" that provide a critical entryway into the texts, this chapter will analyse how setting is deployed in the work of Icelandic crime writer Arnaldur Indriðason's Erlendur novels, reading his fictional representations of Iceland in conjunction with the nationalistic imagination of an 'Icelandic landscape' (Stougaard-Nielsen 114).

Detective Erlendur shares many of his traits with another detective in the Scandinavian tradition, that of Henning Mankell's Kurt Wallander. Similar to Wallander, he has a strained relationship with his family, is old fashioned and feels out of place in a rapidly changing society. Despite these clear historical lineages, Erlendur differs from the conventional Nordic detective through his specific relationship to landscape. The argument presented here is twofold. First, this chapter studies how landscape has been historically perceived in Iceland and how it has been instrumental in framing Iceland as a nation. Accordingly, it examines the history of landscape representation in Iceland, Icelandic art history and the subsequent development of the Nordic sublime. Secondly, it will investigate the links between these representations and tourism, with a specific emphasis on touristic practices produced by crime fiction. By situating these niche practices within the larger ambit of state initiatives to bolster tourism in Iceland, this chapter will reveal how crime fiction participates in building a national imaginary in Iceland.

Within this framework, representations of landscape in Indriðason's work function on two levels. First, they respond to state-sponsored attempts to cultivate a mode of landscape perception and appreciation through landscape art and the national tourism website. To capture this process, by which the state foregrounds a specific aesthetic, I employ the term 'directed gaze'. The term refers to practices

DOI: 10.4324/9781003091912-7

which valorise certain modes of framing and perceiving landscape, and how these practices condition the viewers to look upon landscape in specific, directed ways. On one hand, Indriðason's textual work and book covers respond to the directed gaze by foregrounding a landscape aesthetic similar to the state-sponsored 'touristic' discourse. On the other hand, the novel complicates this picture of landscape by building on a long Icelandic tradition of storytelling, which emphasises a supernatural element that exceeds the gaze. Studying Indriðason's treatment of the supernatural and Erlendur's relationship to these phenomena, the second part of this chapter will situate the supernatural in Indriðason's writing in relation to Icelandic legends and sagas. Drawing from the connections made between medieval Icelandic sagas and crime writing (Tulinius), this chapter attempts to unravel the linkages between historical representations of Icelandic landscape, touristic practices that capitalise on these representations and the construction of a national imaginary. These allusions to a historical storytelling tradition in Iceland, and the representation of nature in this tradition, constitute yet another opportunity to construct a national imaginary. By a national imaginary, I refer to the techniques and processes by which the idea of a nation is imagined. To account for this representation of landscape that exceeds the directed gaze, I turn to the sublime, but a sublime that is closely related to the apocalyptic. Drawing from the Levinasian notion of the *il y a*, I will argue that Indriðason's work foregrounds a representation that is more allied with terror and strangeness, as opposed to human familiarity. Thus, Indriðason's work capitalises on two impulses – the first that relies on a state-sponsored 'directed gaze' and the second that relies on an older Icelandic tradition that exceeds the gaze – to construct a national imaginary. These connections will be studied through Indriðason's Erlendur novels, with an emphasis on the texts *Hypothermia* (2007) and *Strange Shores* (2010).

Through the Lens: Landscape and Representation

The North Atlantic part of the world, which includes Iceland, and which, being located on the edges of the 'known' world, was "a marginalised region of the Danish-Norwegian kingdom" in the eighteenth century and generally considered an exotic wilderness by travellers from Europe (Oslund 7). In the early days of the sagas, the distance from the Christian centre of the world, Jerusalem, was instrumental in structuring the binary between civilisation and wilderness. The assumption being that the further one was removed from this centre, the more extreme the climate and the more monstrous the people residing in those lands. This imagination is at the heart of Snorri Sturluson's *Edda* (1300), where monstrous creatures inhabit the outermost realms of the earth, awaiting a war that will bring about the apocalypse (Vídalín 147–148). As the North Atlantic was settled in the period ranging from mid ninth-century to 1000, the territories in this geographical unit became environmental and civilisational "outposts" (Oslund 14). Eighteenth and nineteenth-century travel writing about Iceland describes the country as different and exotic, in continuation with its conception as an environmental outpost. With the rise of the romantic sublime in the nineteenth century, the Icelandic landscape, made of lava flows and rock formations, was re-evaluated by Icelanders. As a result, landscape became a part of imagining Iceland as a nation, evident in the Icelandic struggle against the Danish crown. Despite remaining the subjects of the Danish crown until the twentieth century, Icelanders "retained a distinct culture shaped by their harsh and isolated island environment" (Magnússon 19).

The Laki volcanic eruptions of 1783 became a landmark event in defining the relationship between Iceland and its landscape. It became evidence of nature's forces at work, as well as "the failures of the Danish state and its administration of the island" (Oslund 45). In this tradition, the "motifs of Icelandic nature were commonly viewed as literal signifiers of a folk history and a national history. Volcanoes and glaciers became suggestive of Icelandic spirit of independence and stoicism" (Ibid 47). The rise of an Icelandic tradition in landscape art was tied to the nationalistic movement which advocated for independence from Danish rule. Modern landscape art in Iceland began with an exhibition by Thórarinn B. Thorláksson in December 1900. This exhibition included summer scenes and made landscape painting "the foundation of Icelandic visual art tradition" (Ólafsdottir 23). These early artists were trained in Denmark and their work reflected an idealised landscape characterised by bluish hues and wide horizons (Ibid 24–25). Ásgrimur Jónsson's (1876–1958) *Thingvöllum* (1905) is an example of this approach. The beginnings of what has been called the Nordic sublime can be further located in the work of Thorláksson and Jónsson. As Martia Sawin suggests:

> Using a Northern transcendental approach to landscape, sometimes called the "Nordic Sublime", Thórarinn B. Thorláksson and Ásgrimur Jónsson idealised the surroundings . . . their views of a serene and peaceful land can also be linked to the stirrings of Icelandic nationalism.
>
> *(16)*

Thórarinn B. Thorláksson's *Langisjór Lake and Vatnajökull Glacier* (1921) also captures the elements discussed here.

These tranquil images of nature have latterly been appropriated by the official Icelandic tourism website (visiticeland.com), which features landscapes viewed from a sweeping aerial, panoramic gaze, presenting a romanticised snapshot of different variegated terrains. The tradition of idealising landscape thus finds expression in state discourse surrounding the natural environment. The vision of an unspoilt pristine realm informs these attempts at idealisation, a tendency that also merits mention in Indriðason's *Arctic Chill* (2005). In the novel, Erlendur meets a woman who takes advantages of the economic possibilities offered by a pristine landscape by organising safaris through the Icelandic wilderness (83). Foreign visitors being attracted to Iceland for its sublime landscapes is certainly not a new phenomenon. The historian Guðmundur Hálfdanarson writes about nineteenth-century European travellers who were interested in folk cultures of remote European societies:

> Drawn to this northern country by their hunger for exploring the exotic, the upper-class travellers certainly got what they were looking for. The landscape of the desolate island was unlike any they had experienced; scars of an unceasing struggle between the natural elements abounded and made large parts of the country a wasteland . . . the harsh climatic conditions set their distinctive mark on the cultural landscape.
>
> *(quoted in Magnússon 19)*

A google search with the key words 'Iceland' and 'wilderness' offers multiple websites that offer touristic experiences. A report by the Icelandic Tourism Board provides a breakdown of the reasons why travellers choose to come to Iceland, the most popular of which was Icelandic nature (Óladóttir 18). Fifteen percent of the respondents also mention Icelandic literature or a book featuring Iceland as a reason for their visit. Through the official website and other state policies surrounding tourism, Iceland actively promotes the natural landscape and the potential methods for a tourist to enjoy them. To this end, the official tourism website has a section that lists activities that a tourist can undertake in Iceland. The website thus reflects a marketing strategy that foregrounds the development of an 'Icelandic' brand by portraying Iceland as a peaceful destination where one can relax in the lap of nature. Similar to the section that introduces different Icelandic regions to a potential tourist, the section on 'Things to Do' bookmarks each activity with an array of photographs featuring Iceland's natural topographies. The activities listed include seeing the northern lights, birdwatching, whale watching, touring on helicopters, riding Icelandic horses, hiking, cycling, running, caving and ice climbing, to list a few. With the aesthetic gaze that evolved from landscape painting, the website combines the nationalistic gaze of the state with suggestions about how to position oneself as potential tourist in the Icelandic landscape.

Though not state sponsored, crime fiction has also become part of this touristic imaginary. A 2016 entry on the website Crime Fiction Lover (crimefictionlover.com) showcases Icelandic noir and a trail of locations featured in Icelandic writing under the moniker of 'Magical Murder Tour' (*Iceland Noir*). The various stops on the tour include locations featured in Indriðason's work. Among other aspects highlighted as part of the tour, the account draws attention to the "lava formations" and "the other-worldly landscapes" (Ibid). Referring to the history and folk tales that are associated with Icelandic locations, the write-up claims that Icelandic crime writing "seems like the layer of greenish-grey moss that grows on the lava in Reykjanes" (Ibid), suggesting that crime writing draws from older Icelandic folklore and place-based stories. While the website Crime Fiction Lover is not associated with state-sponsored initiatives to bolster tourism, it is worth noting that the website and Indriðason's *Silence of the Grave* (2010) reproduce some of the same techniques employed by the state website for tourism.

The home page on visiticeland.com features a link to yet another page titled "The Regions", which introduces seven different territories of Iceland to a potential tourist. The top of this page is marked by the presence of a map of Iceland. In a similar vein, the tourist account on Crime Fiction Lover also begins with two maps, as does the opening page of *Silence of the Grave*. The previously cited example about Icelandic safaris from *Arctic Chill* is also relevant here. As a visual object, the map functions as a sign of territoriality. It has been argued that the map, in general, is an important artefact in creating what has been termed as a "geo-body" of nation. Geo-body describes "the operations of the technology of territoriality which created nationhood spatially" and is an "effect of modern geographical discourse whose prime technology is a map" (Winichakul 16–17). These maps of Iceland thus have "meanings and values and can send messages because they refer to the map of . . . a nation, which has been loaded with meanings and values of nationhood" (Ibid 138). When placed in the context of Benedict Anderson's proposition that the novel is an avenue through which nation as "imagined community" manifests (Anderson 24–25), maps and crime fiction serve as avenues for a state-directed national imaginary to flourish.

While crime fiction is not state endorsed, the development of a tourism industry around Icelandic crime fiction and its inclusion of 'Icelandic' landscapes as a part of the tour speak to the presence of a national imaginary at work. This national imaginary comes into being as a result of certain techniques of seeing, which in turn foreground an aesthetic which is imbued with national values. Hinging on the landscape aesthetic, the official state discourse surrounding tourism constitutes an opportunity for the state to define what constitutes 'Icelandic', both in terms of landscape and a national identity. Touristic practices inspired by this definition, trafficking on 'Icelandic landscape', as is the case with the 'Magical Murder Tour', reproduce the same aesthetic. The reproduction of certain idealised landscapes, both on the tourism website and via Indriðason's book covers, as shown in Figure 5.1, thus renders them hypervisible.

Such imagery is not distinct from the idealised landscapes painted by pioneering Icelandic artists discussed earlier. In the twentieth century, a few artists rebelled against the tradition of painting idealised soft landscapes. This group of painters, including Guðmundur Einarsson of Miðdal and Finnur Jónsson, utilised subjects that were previously thought to be unattractive. Finnur Jónsson's paintings about the Laki craters fall into this category (Oslund 52). This shift towards a darker aesthetic, wherein the landscape harbours violence, is also evident in the covers for Indriðason's work. As opposed to the well-lit idealised landscapes characteristic of the older perceptions, the book covers emphasise bleakness, darkness and isolation, often portraying a lone figure in desolate snowscapes. Nonetheless, there is a recognisable aesthetic at work in these covers, with their vast, panoramic, seemingly untouched landscapes. For instance, the Vintage Digital editions of Indriðason's books in the United Kingdom, such as *Voices* (2008), *Hypothermia* (2009), *The Draining Lake* (2009), *Silence of the Grave* (2010) and *Oblivion* (2015), feature images of landscapes that are similar to the ones on the tourism website,

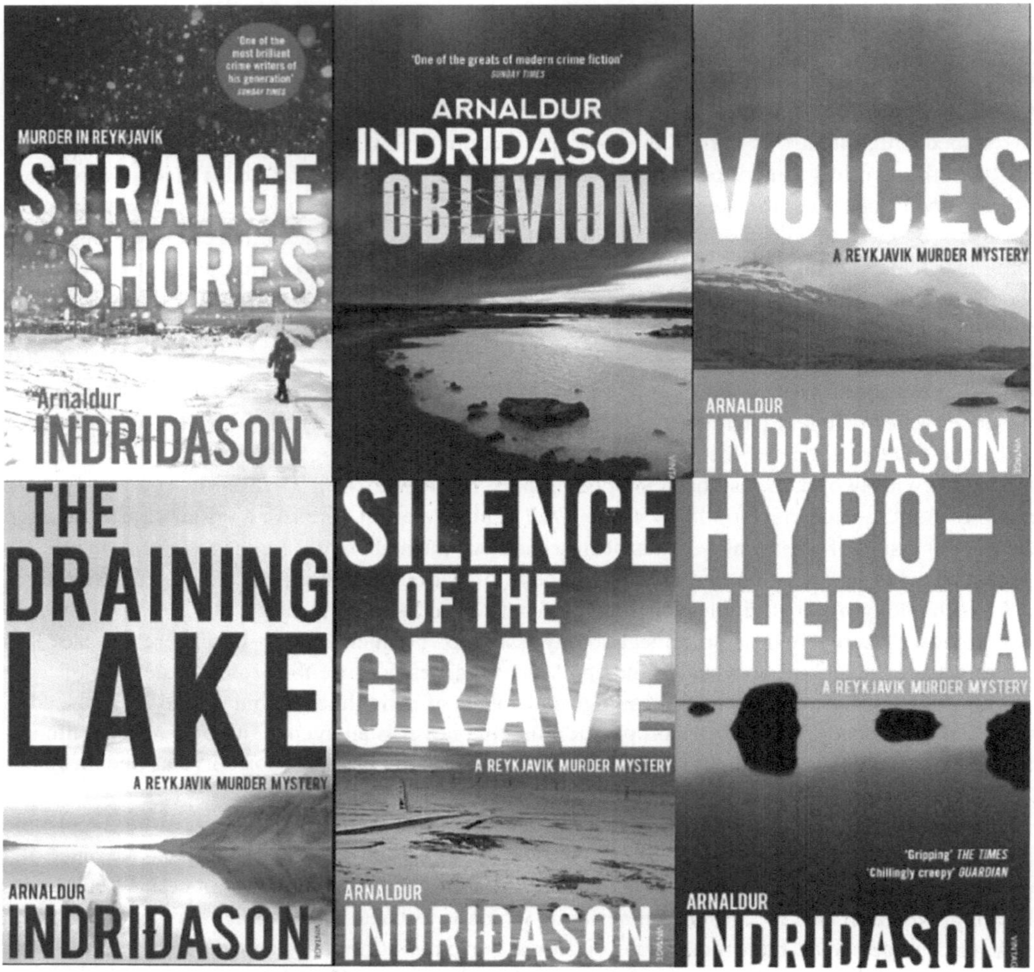

Figure 5.1 A selection of Arnaldur Indriðason's book covers.
Source: Courtesy of Penguin Random House UK.

showcasing desolate, yet beautiful vistas. As exemplified by the website Crime Fiction Lover, Indriðason's *Strange Shores* traffics in both conceptions of landscape: the one that is idealised and the one that harbours violence.

Set in the East Fjords of Iceland, *Strange Shores* features descriptions that emphasise the picturesque nature of the setting. At one point in the novel, Erlendur's driving route is described as something that could be in a tourist brochure: "The road followed the shoreline, threading in and out of one picturesque fjord after another, past endless ranks of mountains characterised by the distinctive sloping strata of the East Fjords" (Indriðason 195). One of the tensions that the novel presents is the tussle between the old and the new. Erlendur, who belongs to the older generation, is unsettled by the landscape's transformation at the hands of capitalist redevelopment projects: "He couldn't understand how on earth an unaccountable multinational, based far away in America, had been permitted to put its heavy industrial stamp on a *tranquil fjord* and *tract of untouched wilderness* here in the remote east of Iceland" (Ibid 8, emphasis added). The older generation in the novel, also represented by Bóas the hunter, sees the signs of progress as being at the cost of the natural landscape. Bóas calls the new factories that have been set up in Hvalfjördur "monstrosities" that spew poison (Indriðason 10). The text contrasts these markers of "relentless progress" with the "tranquillity" of the fjord and the "snow-capped mountains" on several occasions, indicating a tension between the affordances of modernity and the need to preserve the past (Ibid 62).[1] While the tension between progress, modernity and the tranquillity offered by natural landscape is not uncommon in novelistic representations of Iceland, Indriðason adds an element of the supernatural. Erlendur and other characters in Indriðason's universe, such as Hrund from *Strange Shores*, inhabit a landscape haunted by ghosts. They tell stories about people disappearing in Icelandic landscapes, and how some of them linger on as spectral presences. Indriðason creates an older generation conversant in these tales associated with landscapes, aware of the connections stretching back into the past. This framing of character then raises questions about the quality of being Icelandic, both in terms of landscape and character: what counts as 'Icelandic' landscape and who gets to claim a connection to it?

Strange Shores frames markers of 'progress' as antagonistic to the natural world, while veering towards a darkening aesthetic of landscape perception as typified by the death of Erlendur's younger brother, Bergur, in a violent blizzard when he was a child. This aspect of the landscape is something that the novel frames as characteristic of the older generation. For instance, when Erlendur mentions his brother who died on the moors, Hrund – who provides Erlendur information about Matthildur, another victim of the moor – says that Iceland is "an unforgiving country" (Indriðason 26). Through the story of Bergur, Indriðason deploys landscape as a device for the development of the protagonist, as Erlendur will continually return to the moors in the hope of discovering his lost brother's remains. The novel subsequently situates itself in a tradition of Icelandic art that features hostile encounters with landscape and views this as constitutive of an authentic national experience. While the images from the website, crime fictions tours and Indriðason's book covers thus signify a shared visual aesthetic of being Icelandic, the novelistic text associates this quality with a certain generation who have been shaped by their struggles with the natural world.

Liminal Landscapes and the Supernatural

As stated, although the discourse around tourism frames Iceland as a destination of natural beauty for a potential tourist to escape into, the Erlendur novels depart from this imagination. This departure is primarily accomplished with reference to the blizzard in the picturesque East Fjords, which resulted in the death of Erlendur's brother, Bergur. The blizzard is mentioned in multiple novels such as *Voices*, *Silence of the Grave* and *Strange Shores*. The reason for Erlendur's guilt is revealed in *Strange Shores*. His father had originally planned to take only Erlendur up into the moors, but as they are about to leave, Erlendur demands that "Bergur must come too" (Indriðason 180). Landscape is not a refuge from modernity and city life; for Erlendur, it is a constant reminder of loss and violence. As a policeman, Erlendur is therefore deeply interested in missing persons cases and stories of ordeals in the wilderness. Though the Fjords appear both on the official tourism website and in the novel *Strange Shores* as picturesque locations, revealing the landscape aesthetic at work, the incidents described by Indriðason dispel the associations that the website painstakingly cultivates.

While the characters presented in *Strange Shores* illustrate the tussle between the old and the new, often marked by a nostalgia for the good old days, Erlendur is not seeking such a return. Bóas, a fox hunter whom Erlendur meets at the beginning of the novel, illustrates the tension between modernity and tradition. The hunter is dissatisfied with the new industrial projects and reminisces about the past, questioning whose interests are served by the new factories in the region: "A bunch of insanely rich foreigners who couldn't even find Iceland on a map. Is that our fate?" (10). As opposed to Bóas who likes to reminisce about the past (11), Erlendur finds no comfort in the past or in the familiar landscape of his childhood:

> Deep down he knew that he had been avoiding this place, not just physically but in his mind. The light Arctic night offered no comfort. On the contrary it illuminated with painful clarity all that was most difficult and distressing about this homecoming. He was convinced there and then that he would never be a happy man.
>
> *(80)*

Though he is younger than Bóas, Erlendur is presented as part of the older generation. He sees a young boy wearing his trousers low to reveal his boxer shorts and thinks "What is the world coming to?" (56). Bóas thinks that the countryside should remain uninhabited and not be subject to the vagaries of progress (Indriðason 10), whereas Erlendur understands the inexorable nature of progress and its ramifications (Indriðason 10). While Erlendur empathises with Bóas, he is also able to see the consequences of innovation.

Landscape is approached in *Strange Shores* in two ways. In the first instance, even though not foregrounded, the novel works with the aesthetic that drives the official tourism website, with references to picturesque fjords, (195) tranquil inlets and untouched wilderness (8). In this way, the novel lends itself to being read as a critical reflection on the unstoppable and homogenising power of modernity, particularly its role in creating fraudulent and fragile national myths. As Timothy Morton puts it: "Our notions of place are retroactive fantasy constructs determined precisely by the corrosive effects of modernity" (11). These flights of nostalgia are driven by ideas that took form in the Romantic period. On the connections between Romanticism and how nature is imagined, Morton continues:

> The "thing" we call nature becomes, in the Romantic period and afterward, a way of healing what modern society has damaged. Nature is like that other Romantic-period invention, the aesthetic . . . Contact with nature, and with aesthetic, will mend the bridge between subject and object . . . Subject and object require a certain environment in which they can join up together. Thus is born the special realms of art and nature, the new secular churches in which subject and object can be remarried.
>
> *(Ibid 22–23)*

Both the tourism website and the novel offer this aesthetic where the subject can freely reconvene with nature to feel rejuvenated. For instance, the webpage on West Fjords promises the tourist "relatively unspoilt wilderness" (*Westfjords*), while horse riding listed under the activities section of the website lists it as "a great way to explore unspoilt nature" (*Horses of Iceland*). Besides, Iceland's topography, one that combines snowscapes and lava fields, also lends itself to being read alongside yet another enduring legacy from the Romantic period: that of the Sublime. Travellers arriving in Iceland in the eighteenth and nineteenth centuries were fascinated with the romantic sublime and Iceland was perceived by these travellers as a space where subliminal encounters were possible (Oslund X). Synthesising the imagination of sublime in different theorists, William Cronon points out how the sublime became a keyword to indicate experiences that revealed human insignificance.

> In the theories of Edmund Burke, Immanuel Kant, William Gilpin, and others, sublime landscapes were those rare places on earth where one has more chance than elsewhere to glimpse the face of God. Romantics had a clear notion of where one could be most sure of having this experience. Although, God might, of course, choose to show Himself anywhere, He would

most be found in those vast, powerful landscapes where one could not help feeling insignificant and being reminded of one's own mortality.

(*Cronon 10*)

In this sense, the sublime eventually becomes an edifying experience, preserving the integrity of the subject in question. This is the first way of reading landscape in the novel, its seeming tranquillity affirming the subject's need to feel at one with nature. Thus, nostalgia for the "uninhabited countryside", which Bóas expresses, could be read as sign of the anxieties of a subject that feels out of place in a new rapidly progressing industrial environment that is changing the Fjords (Indriðason 10). The tourism website also capitalises on this impulse to have an edifying experience in nature by selling a vision of untouched nature to the potential tourist.

Through characters such as Bóas and Hrund, Indriðason touches on a prevalent theme in Scandinavian crime fiction: nostalgia as a method to deal with the anxieties that are borne out by the diminishing role of the nation-state in a globalised world. The emphasis on tourism and the promotion of an 'Icelandic nature' brand is also a testament to the forces of capital that circulate across national borders. In fact, Jason Moore argues that capitalism is a way of organising nature (Moore 2). The touristic imaginary that the government website capitalises on thus employs the natural landscape as capital to be exploited. While a part of this imaginary derives its sustenance from the 'pristine' landscapes, the imaginary also functions by alluding to national histories embedded in those very landscapes. Hence, modern crime novels reveal an obsession with "the past, with national histories, personal and familial traumas and even the deep past of myth and primordial nature" (Stougaard-Nielsen 117). In fact, it has been argued that Indriðason's novel *Voices* serves as an articulation of an Icelandic nationality on the wane. Katrín Jakobsdóttir argues that "Icelandic nationality has become a meaningless tag which nobody understands anymore; a hollow sign, loosely connected to a place where pathetic and squalid murders are committed" (57).

Contrary to these assertions, I argue that it is the intersection of tourism discourse and crime writing that works to produce a national imaginary, the quality of being Icelandic. While the government tourism website actively promotes a state-sanctioned aestheticised gaze through the framing of certain landscapes, the crime fiction discussed in this chapter approaches the production of the nationalistic imaginary in a different way. The 2018 report by the Icelandic Tourist Board, which was cited earlier, not only reveals the state interest in the movement of foreign bodies in Iceland, but also sheds light on the movements of Icelandic citizens. Echoing the earlier attempts made through landscape art to create an interest in a national landscape, the 2018 report also includes a section on the effects of tourism and foreign travellers on Icelandic citizens. The report notes that nearly half of the respondents of the survey believed that the tourists had become a catalyst for Icelandic citizens displaying an increased interest in the natural landscape of the country (Óladóttir 28). The state, in this situation, actively creates a category of subjects through what I refer to as 'directed gaze', which captures the state-sponsored attempts at shaping the landscape. Hence, landscape painting which was intended to create a national idea of Icelandic landscape forms part of the directed gaze. The tourism website and Indriðason's book covers further reproduce the directed gaze by soliciting a potential tourist or a citizen to perceive 'Iceland' in specific ways. For this reason, the tourism discourse and crime writing can be considered as a part of a part of larger network that produces specific subjects who participate in constituting 'Icelandic' landscapes by responding to the directed gaze. The 2018 report shows that marketing strategies, which frame Iceland as a place of great natural beauty, aimed at tourists, come to bear upon practices that produce citizen-subjects of Iceland, who are, in turn, called to gaze at the 'Icelandic' landscape in prescribed ways. Crime fiction participates in this exercise by reproducing manifestations of the landscape aesthetic reflected in the book covers and the setting.

In addition to the visual landscape aesthetic that is foregrounded by the book covers and on the tourism website, the national imaginary is also reflected in the liminality of Erlendur novels which straddle a terrain between the real and the supernatural. This aspect of Indriðason's writing comes to the forefront in *Strange Shores* and *Hypothermia*. In the former, the readers are given an extensive description of Bergur's death and how Erlendur is haunted by these events. In the latter, Erlendur investigates a suicide and uncovers a group of medical students who have been experimenting with 'flatlining': slowing someone's heart down until they are medically dead and then bringing them back to life. This emphasis on liminal landscapes and the spectral presences contained in it can be situated within a long tradition of Icelandic story telling. Old Nordic landscapes, in particular Icelandic farms, used to contain "untouchable sites" or "*álagablettir*" which stands for "enchanted sites" since they were associated with the dead (Gunnell 31). These spaces were associated with powerful liminality and were "portals of a kind in which one could communicate with the other world, and in which the other world had access to ours and could wield influence over the way in which things worked" (Ibid 33).

In *Strange Shores*, Erlendur goes back to his childhood farm at Bakkasel and is subject to haunting experiences. The opening pages of the novel establish the liminal space that Erlendur inhabits by referring to the "borderline between sleep and waking", strange tricks of the mind that shuffle between the past and present and through time and space (Indriðason1). The novel acknowledges the spectral presences by drawing attention to the Icelandic story telling tradition. In response to a story about a woman who goes missing on the moors causing a shipwreck, Erlendur reflects on these ghost stories:

> They were part of the old Icelandic story telling tradition that had peopled the landscape with ghosts, elves, trolls, magic stones and unseen beings, linking man to his environment with invisible bonds. In the past people had lived more closely with nature and their lives had depended on it. Respect for the land and the forces latent within it was the theme of many a folk tale, and implicit in them was the warning that no one should underestimate the power of nature.
>
> *(Indriðason 38)*

Thus, the novel traffics in a dual presentation of landscape; on the one hand it portrays landscape in accordance with the directed gaze employed by the state, which produces a certain citizen subject or a tourist subject. The second mode of landscape representation is captured by Erlendur's experiences on the moor. These reveal how landscapes that are rooted in the old traditions that cannot be measured by the gaze since they exceed methods of representation and the limits of human experience. This is exemplified by Erlendur's experience on the moors and his subsequent development as a character, where he is haunted by his brother's disappearance. His interest in tales of survival in Icelandic wilderness and missing persons cases colour the landscape with a spectral presence, which exceeds the ambit of the directed gaze. Despite not being confined by the directed gaze, the allusion to the long history of an Icelandic tradition and the production of a protagonist who embodies liminality is yet another opportunity for a national imaginary to function. As stated earlier, by imbuing the landscape with a supernatural element and by framing Erlendur as someone who is in tune with Icelandic tales and traditions, the novels allow the national imaginary of being Icelandic – both in terms of personhood and landscape – to function in a different way. In addition to the picturesque landscape aesthetic employed in the tourism discourse and the book covers, allusions to a long-standing Icelandic tradition constitute the second aspect of landscape representation that is at work in Indriðason's writing. Through the character of Bergur, Indriðason turns the moors of Erlendur's childhood into a phantom landscape that straddles the worlds of the living and the dead. Not only does Indriðason establish a supernatural element in *Strange Shores* and *Hypothermia*, but he actively incorporates motifs from Icelandic legends into Erlendur's narrative.

By integrating the elements from old Icelandic narratives into crime fiction, Indriðason brings back the gothic influences that shaped early crime fiction. At the end of nineteenth century, novelists and critics formed detective fiction into a genre by "denying its sensational heritage . . . in order to emphasise its rational character" (Ascari 1). On the one hand, the emphasis on rationality prompted an abandonment of the gothicism and supernatural in detective fiction; on the other hand, writers were unwilling to part with a tool that held public attention. As a compromise, writers invented the technique of explaining their plots.

> In order to assert a new standard of verisimilitude, the supernatural was marginalised both as an instrument of detection and as a criminal tool, but it re-entered the genre through the back door . . . Re-enacting the transition from the gothic proper to the rationalised gothic, crime writers resorted to "staging" the supernatural and deconstructing their own "plots".
>
> *(Ibid 62)*

In Indriðason's work, the supernatural is not rationalised away. In fact, *Strange Shores* presents a narrative structure wherein the focalisation of the novel shifts from Erlendur's investigations to his own spectral experiences in his ruined childhood home at Bakkasel.

The novel portrays Erlendur as someone who is continually subject to visions and hallucinations, often relating to his brother, Bergur. Erlendur's encounter with the blizzard changes him in irrevocable ways, as he directly acknowledges in *Voices*: "I was found and rescued but, I died too. Something inside me. Something I had before. I don't know what exactly it was. My brother died and I think something inside me died too" (Indriðason 252). For Erlendur, being in the blizzard thus brings about a fissure between the world of the living and the dead. The motif of a protagonist undergoing suffering to uncover a different part of their identity is also a part of the gothic tradition in Scandinavian writing. As Yvonne Leffler suggests:

> In Scandinavian horror the underworld is the Nordic landscape where the protagonists undergo a series of transformations or ritual sufferings that lead to the painful discovery or recovery of their lost or hidden identities, as well as their connection to a pagan past.
>
> *(144)*

Though Erlendur inhabits modern Iceland, his experience on the moors makes him a conduit for spectral presences.

In *Hypothermia*, for instance, Erlendur encounters a medium during his investigation, one who claims that Erlendur is surrounded by multiple spectral presences:

> "There is nothing malign about the spirit world", Andersen said. "We all have our ghosts. You not least." "Me?" Erlendur said. Andersen nodded.
>
> "A whole crowd", he said. "But don't worry. Keep looking. You'll find them." "You mean him", Erlendur said. "No", Andersen said, contradicting him and standing up. "I mean them."
>
> *(Indriðason 189)*

The snowstorm in which Bergur dies transforms Erlendur into someone who is in tune with the supernatural world, one inhabited by forces that are beyond human comprehension. Erlendur's preoccupation with missing persons cases provides an opportunity to look for these ghosts, such as the case of Matthildur in *Strange Shores*, the woman who goes missing on the moors. Indriðason draws from Icelandic legends when he writes about the blizzard. In *Strange Shores*, Erlendur's father decides to go up to the moors because he fears that the ewes the family have been raising might be lost. The motif of going into a different terrain, such as moors or mountains, and then being struck by weather which reduces visibility, is a trope that recurs in Icelandic legends. In Icelandic legends about outlaws, a common feature of such narratives is a male character who goes in search of sheep into the wilderness before being surrounded by fog which affects their perception. This development can be found in tales such as Úlfsvatn, The Herdsman, among others (Árnason 101–110). These legends are also tales of transgressed borders:

> Icelandic outlaw legends often underline this concept of the wider meaning of the geographical borderline by having their law-abiding heroes suddenly find themselves engulfed in fog the moment they leave the organised cosmos of rural society and enter the chaotic wilderness inhabited by outlaws, trolls, ghosts and elves.
>
> *(Gunnell, "Legends and Landscape" 314)*

When Erlendur, his father and his brother go into the moors to look for their lost ewes, the blizzard strikes suddenly, "reducing the visibility to zero", making the snow "blinding" (Indriðason 269).

Similar to the outlaws who undergo transformative events in dangerous terrains while looking for lost sheep, Erlendur's experience in the blizzard grants him a connection to a non-human, phantom world which cannot be rationalised away. Thus, the Erlendur novels create a strange, localised ecology wherein the rational world coexists with a spectral world imbued with magic. As it can be seen from the older Icelandic legends and Indriðason's account of the blizzard, this world defies the regime of visuality. Hence the directed gaze of the state, discussed earlier and at work in the tourism website and the book covers, is not adequate to account for this turn in writing. However, similar to the directed gaze which works to build a national imaginary, Indriðason's protagonist, who is fashioned after Icelandic heroes, can also be placed in a history of national imagination, since Erlendur follows in the wake of Icelandic heroes from outlaw stories. Following the first presentation of landscape in terms of the directed gaze, the second presentation of landscape in Indriðason's writing creates a national imagery by alluding to old Icelandic legends. However, as shown earlier, this aspect of the text cannot be encompassed in a visual regime. For this reason, I return to the notion of sublime, but not the sublime as it has been imagined traditionally, but rather a sublime that alludes to the apocalyptic. In contrast to the sublime that became "domesticated" by the end of the nineteenth century, the landscapes presented in Indriðason's writing, inspired by older traditions in Icelandic writing, continue to inspire terror (Cronon 12). Erlendur's apocalyptic encounter with the blizzard on the moors is an example, an event which opens up a pathway into another plane of being.

Alongside Erlendur, yet another figure in the novel *Hypothermia* is shown to inhabit the space between life and death. Trygvvi was part of an experiment conducted by a group of medical students in which his heartbeat was slowed down by inducing hypothermia so that he was medically dead. Afterwards, Tryggvi was brought back to life using a defibrillator. After he is brought back to life, he ruminates on how "nothing mattered anymore; not other people, not my studies, not my surroundings. Somehow life stopped mattering. I didn't feel connected to it any longer" (Indriðason 149). The experience estranges Tryggvi from his own being; his subjectivity erodes away to such a point that Erlendur looks at him and thinks that "this was probably the closest he would ever come to meeting a ghost" (150). Erlendur also believes that something inside him died the night his brother disappeared. Erlendur and Tryggvi thus inhabit a liminal position, simultaneously human and not human. A protagonist that belongs to two separate spaces, a human and a non-human one, is also a feature that can be found in Icelandic sagas such as the Grettir Saga. Similar to Erlendur, who is shown to belong to both the human and a spectral world, Grettir also belongs to

> two separate spaces, one of which is metonymically related to society by being part of the social space, and the other being metaphorically related to society in its separate, largely non-social space. In that sense Grettir is both "of this word" and "of the world of non-humans".
>
> *(Hastrup 163)*

Occupying the juncture between these two worlds, Erlendur and Tryggvi can perceive the landscape as something that exceeds the gaze, as something primordial and strange.

This creates a different sublime rooted in terror, one in which the very nature of subjectivity is threatened or even erased. Relying on the Levinasian notion of the *il y a* or *there is*, John Sallis writes about the return of nature:

> In its return nature will forsake its immediacy and familiarity. As it returns it will appear strange, as if belonging to a region distant from and alien to the human world. In a sense it will have cast off its disguise: it will no longer be the nature that is shaped and formed within the human world and in accord with the measures of that world, but rather a nature capable, in its excess, of evoking feelings of both sublimity and terror.
>
> *(Sallis 87)*

Not only is the *il y a* characterised by sublimity and terror, but also by a horror that chips away at subjectivity itself, as happens with Erlendur and Tryggvi. For Levinas, the quintessential image of horror is encapsulated by the "haunting spectre, the phantom" (Levinas 33). Levinas further reinforces the connections between the *il y a* and the spectral by referring to Shakespeare's plays and how the presence of ghosts allows Shakespeare to create a space that is close to the *il y a*. About Shakespeare, Levinas writes: "Spectres, ghosts, sorceresses . . . allow him to move constantly towards this limit between being and nothingness" (Ibid). Tryggvi who is brought back to life from the point of death thus embodies the boundary between being and non-being. He does not feel connected to his being anymore and is constantly waiting for something, though he is not sure what. The spectral presences and figures such as Erlendur and Tryggvi capture the terror in the landscape that is characterised by the uncanny.

Contrary to the directed gaze wielded by the tourism website and the book covers, which work to project a sublime image, the spectral aspect of the landscape in Indriðason's work exceeds the gaze. Yet even though this landscape outstrips the regimes of visuality that structure the directed gaze, it still situates itself in a national imaginary by drawing from Icelandic legends and folktales. As already shown, the topography in Indriðason's novels derives from the long tradition of Scandinavian horror in which characters encounter wilderness. Because of the childhood incident on the moors, Erlendur is marked by a lifelong preoccupation with encounters of people in the wilderness and uncanny survival stories. The element of the supernatural, a defining characteristic of old Nordic landscapes, is brought to the forefront in Indriðason, without the supernatural being rationalised away. *Strange Shores* builds on this tradition, with references to how Icelanders love to make up ghost stories (26). The text also features remarks about the old conception of Icelandic landscape being imbued with magic and magical beings (38). Thus, even though this conception of landscape cannot be apprehended through visual cues, since it builds on a long tradition of Icelandic storytelling, it is still capable of mobilising a national imagery.

Conclusion

This chapter has suggested that Indriðason's work deploys two elements that can constitute a national imaginary in Iceland. The first is a direct result of the state-sponsored directed gaze, which has its origins in the initial days of Icelandic landscape art and attempts to construct a national landscape. This directed gaze is also at work in the state-sponsored website for Icelandic tourism, employing an idealised representation of 'Icelandic' landscape. Indriðason's book covers and crime plots participate in reproducing the directed gaze. However, Indriðason complicates this representation of landscape by bringing in a second aspect with regard to landscape production in his work. Drawing from the older Icelandic tradition of outlaw legends and sags, Indriðason creates a spectral topography which exceeds the human attempts to contain it via the gaze. Yet, by situating Erlendur in a tradition of distinctly 'Icelandic' protagonists, Indriðason builds yet another national imaginary connected to the representation of landscape and Erlendur's relationship with it.

Hinging on the landscape aesthetic, the nexus between the official state discourse surrounding both tourism and crime fiction offers an opportunity for the state to define what constitutes 'Icelandic'. The seeming contradiction of Iceland being a peaceful country which produces crime writing also serves as an opportunity for the state to highlight its territorial security. In the 2019 "Travel and Tourism Competitiveness" report released by the World Economic Forum, Iceland ranks second in the 'safety and security' parameter of the report (Calderwood and Soshkin 71). On the official tourism website, Icelandic is defined through a curation of activities that employ the state-directed gaze. These activities direct a potential tourist to engage with the landscape that has already been shaped through a landscape aesthetic. Attempts by the state to build a national identity surrounding 'Icelandic' is not just restricted to its official website, but also extends to the production of Icelandic crime, despite it not being formally endorsed by the government. The idea of an Icelandic landscape is built up through cover design and the extensive descriptive prose, aided by a distinctive cultural tradition that values literacy. Historian Sigurður Gylfi Magnússon writes about how Icelandic culture attempts to impose order on a chaotic landscape through literature.

> Iceland is in some senses a wasteland. One can, if one wishes, see some kind of reflection of the physical surroundings in the Icelander's cultural obsession with literacy, an urge to impose order on the desolation of Icelandic landscape, to build a wasteland with words.
>
> *(Magnússon 267)*

Indriðason's writing, while responding to the directed gaze, creates an Icelandic national imaginary by reimagining the sublime with an emphasis on the older spectral landscapes of Icelandic legends and sagas. Thus, the books themselves, while engaging with an idea of aestheticised landscape, also destabilise it through a series of spectral encounters, and subsequently point to the multiple influences at work in the production of a national imaginary. The landscape, and the associated environmental concerns surrounding it, in Indriðason's crime fiction become a focal point for articulating who or what constitutes 'Icelandic'.

Note

1. In response to a survey conducted amidst Icelanders, where "three out of every four respondents in the survey in 2018 agreed with the statement that the burden of tourists on Icelandic nature was too great" (Óladóttir 28). On one hand, the official tourism website encourages people to visit Iceland; on the other hand, this global mobility threatens 'Icelandic' nature.

Bibliography

Anderson, Benedict. *Imagined Communities: Reflections on the Origin and Spread of Nationalism*. Revised ed., Verso, 2006.

Árnason, Jón. *Icelandic Legends*. Translated by George E. J. Powell and Eiríkr Magnússon, Longmans, Green, and Co., 1866.

Arvas, Paula, and Andrew Nestingen. "Introduction: Contemporary Scandinavian Crime Fiction." *Scandinavian Crime Fiction*, edited by Paula Arvas and Andrew Nestingen, U Wales P, 2011, pp. 1–17.

Ascari, Maurizio. *A Counter-History of Crime Fiction: Supernatural, Gothic, Sensationa*. Palgrave Macmillan, 2007.

Calderwood, Lauren Uppink, and Maksim Soshkin. "The Travel & Tourism Competitiveness Report 2019.Pdf." *World Economic Forum*, 2019, www3.weforum.org/docs/WEF_TTCR_2019.pdf.

Cronon, William. "The Trouble with Wilderness: Or, Getting Back to the Wrong Nature." *Environmental History*, vol. 1, no. 1, Jan. 1996, pp. 7–28.
Gunnell, Terry. "Legends and Landscape in the Nordic Countries." *Cultural and Social History*, vol. 6, no. 3, Sept. 2009, pp. 305–322.
———. "Spaces, Places, and Liminality: Marking Out and Meeting the Dead and the Supernatural in Old Nordic Landscapes." *Landscape and Myth in North-Western Europe*, edited by Matthias Egeler, Brepols Publishers, 2018, pp. 25–44.
Hastrup, Kirsten. *Island of Anthropology: Studies in Past and Present Iceland*. Odense UP, 1990.
Horses of Iceland. https://visiticeland.com/article/the-icelandic-horse. Accessed 18 May 2021.
Iceland Noir: Magical Murder Tour | Crime Fiction Lover, 21 Nov. 2016, https://crimefictionlover.com/2016/11/iceland-noir-magical-murder-tour/.
Indriðason, Arnaldur. *Arctic Chill*. Translated by Bernard Scudder and Victoria Cribb, Victoria. Minotaur Books, 2005.
———. *Hypothermia*. Translated by Victoria Cribb, Minotaur Books, 2009.
———. *Strange Shores*. Translated by Victoria Cribb, Vintage, 2014.
Jakobsdóttir, Katrín. "Meaningless Icelanders: Icelandic Crime Fiction and Nationality." *Scandinavian Crime Fiction*, edited by Andrew Nestingen and Paula Arvas, U Wales P, 2011, pp. 46–61.
Leffler, Yvonne. "The Devious Landscape in Contemporary Scandinavian Horror." *Gothic Topographies: Language, Nation Building and "Race"*, edited by P. M. Mehtonen and Matti Savolainen, Routledge, 2016, pp. 141–152.
Levinas, Emmanuel. "There Is: Existence without Existents." *The Levinas Reader*, edited by Sean Hand, translated by Alphonson Lingis, Basil Blackwell, 1989, pp. 29–36.
Magnússon, Sigurður Gylfi. *Wasteland with Words: A Social History of Iceland*. Reaktion Books, 2010.
Moore, Jason. *Capitalism in the Web of Life*. Verso, 2015.
Morton, Timothy. *Ecology without Nature: Rethinking Environmental Aesthetics*. Harvard UP, 2007.
Óladóttir, Oddny Þóra. *Tourism in Iceland in Figures 2018*. Icelandic Tourist Board, 2018, p. 28, www.ferdamalastofa.is/static/files/ferdamalastofa/talnaefni/tourism-in-iceland-2018_2.pdf.
Ólafsdottir, Audur. "Visions of Nature in Icelandic Art." *Confronting Nature: Icelandic Art of the 20th Century*, edited by Ólafur Kvaran and Karla Kristjándóttir, National Gallery of Iceland, 2001, pp. 23–38.
Oslund, Karen. *Iceland Imagined: Nature, Culture and Storytelling in the North Atlantic*. U Washington P, 2011.
Sallis, John. "Levinas and the Elemental." *Radicalizing Levinas*, edited by Peter Atterton and Matthew Calarco, State Univeristy of New York Press, 2010, pp. 87–94.
Sawin, Martia. "Updating the Nordic Sublime." *Confronting Nature: Icelandic Art of the 20th Century*, edited by Ólafur Kvaran and Karla Kristjándóttir, National Gallery of Iceland, 2001, pp. 15–20.
Stougaard-Nielsen, Jakob. *Scandinavian Crime Fiction*. Bloomsbury Academic, 2017.
Tulinius, Torfi H. "D'origine étrangère ? Arnaldur Indridason et le roman policier en Islande." *Etudes Germaniques*, vol. 260, no. 4, 2010, pp. 893–908.
Vídalín, Arngrímur. "From the Inside Out: Chronicles, Genealogies, Monsters, and the Makings of an Icelandic World View." *Supernatural Encounters in Old Norse Literature and Tradition*, edited by Daniel Sävborg and Karen Bek-Pedersen, Brepols, 2018.
Westfjords. https://visiticeland.com/article/the-westfjords. Accessed 18 May 2021.
Winichakul, Thongchai. *Siam Mapped: A History of the Geo-Body of a Nation*. U Hawaii P, 1994.

6

UNSETTLEMENT, CLIMATE AND RURAL/URBAN PLACE-MAKING IN AUSTRALIAN CRIME FICTION

Rachel Fetherston

Ranging from Kenneth Cook's *Wake in Fright* (1961) and Greg McLean's *Wolf Creek* (2005) to Jane Harper's *The Dry* (2016) and Garry Disher's *Bitter Wash Road* (2013), the Australian crime and thriller genre is known for its emphasis on the supposed violence and terror of the outback. Much of the scholarship on the subgenre also highlights its concern with the proximity of rural locales to Australia's "unknown" and "untameable" bush. It is therefore surprising that many Australian literature researchers overlook how such texts engage with the ecological anxieties of these places, particularly in terms of the city/country binary that is often implied. In this chapter, I will explore how the representation of environmental concerns in both Indigenous and non-Indigenous Australian crime fiction not only reflects human/non-human relations in the age of environmental crisis, but calls settler-colonial belonging into question, particularly through an erosion of the problematic dichotomy of the urban and the rural and, in the case of Indigenous crime fiction, a strong engagement with Indigenous ontologies and epistemologies.

The notion of "unsettlement" is key to this discussion and is something which has been explored by many scholars within the framework of postcolonialism. In the context of Australian literary studies, unsettlement can be understood as referring to "the text's relation to settlement . . . and its material negation and resistance" (Farrell 7). Lisa Slater also describes "unsettlement" as a "general white response to encountering the materiality of Indigenous people and life: the density of people's lives rather than representations" (1). Related to this understanding of unsettlement is "material location" (Farrell 7), and so a crime text's engagement with setting and place-making is key to how said text explores notions of unsettlement. Approaching Emma Viskic's Caleb Zelic series (2017–2022) as an example of non-Indigenous, settler-colonial Australian crime fiction and Julie Janson's *Madukka: The River Serpent* (2022) as an example of Indigenous Australian crime fiction, I will investigate how unsettlement, colonial crimes, Indigenous sovereignty and ecological considerations are represented across urban and rural places in these texts.

Ecocritical studies of Australian literature have yielded some significant findings on how place is configured in relation to ecology, including work by scholars Emily Potter, Brigid Magner, Tony Hughes-d'Aeth, and Graham Huggan, amongst others. However, there has been little explicit investigation into how notions of place and ecology interact in Australian genre fiction, particularly crime fiction. Existing scholarship includes Nicole Watson's work on the representation of place in Indigenous crime fiction and Sue Turnbull's discussion of the power of crime fiction to evoke empathy in its portrayal of domestic abuse and ecological crisis. Stephen Knight has also interrogated colonial Australian

DOI: 10.4324/9781003091912-8

crime fiction's preoccupation with the land as "an avenger", with those settlers who visit or lay claim to it "becoming[ing] disturbingly involved in a power greater and more threatening than they or the inhabitants can sometimes comprehend" (*Continent of Mystery* 144). The amount of scholarship still seems fairly limited, though, when one considers the aforementioned emphasis on place in crime fiction generally and recent trends in Australian crime publishing which have seen works that engage strongly with Australian environmental concerns gaining critical (at times international) acclaim, like Jane Harper's *The Dry*.

Australian settler-colonial crime fiction is a generative starting point for investigating relationships between settler belonging, place and ecology. This is because the genre has long been concerned with human/land relations and the supposed lawlessness of areas that are closer to the isolated bush and outback, far from more densely populated urban centres. While literary links between place and ecology are of course not restricted to Australian settler-colonial crime fiction, nor to crime fiction generally (all fiction arguably deals in place), this genre is uniquely positioned to engage readers and scholars with ideas concerning postcolonial place, including problematic settler views of nature and the non-human and the violence associated with such classifications.

It seems remiss not to discuss and problematise how ecological concerns are being represented in non-Indigenous Australian crime fiction, given that so many settler-colonial crime authors still write into a tradition that locates the settler human in opposition to nature and the non-human, with crimes and violence often linked to extreme weather events and the "wildness" of the bush and outback. As mentioned, the ecocritical possibilities of such texts have been largely neglected by scholars, but other scholarship on settler-Australian crime fiction highlights the genre's engagement with "the rural". One example is Ken Gelder's and Rachael Weaver's exploration of the bush in early colonial Australian crime fiction and how its representation relates to settler anxieties about belonging. Knight is also concerned with the genre's evocation of Australian country landscapes, particularly those affected by fire and flood ("Crimes Domestic and Crime Colonial"). It is more than this, though, with such texts also conveying the notion of unsettlement and hinting at the other, historical crimes which lie not far below the surface of more contemporary crimes – that is, the dispossession and genocide of Indigenous Australian peoples that have facilitated settler ownership and belonging.

My work places a particular emphasis on the city/country binary of Australian crime fiction, suggesting that the dismantling of this binary by contemporary settler-Australian crime authors evokes important understandings of place and ecology. It is not enough, though, to only consider how settler-Australian crime authors may be shifting their perspectives of place and ecology. In comparison to settler-colonial Australian crime fiction, Indigenous Australian crime fiction is even more concerned with such issues, as "the reader is exposed to Aboriginal perspectives of history, and the complexities of being an Aboriginal person in a nation that is yet to confront its colonial past" (Watson, "Deadly Detectives" 78). Unsettlement is therefore a key theme of Indigenous crime fiction, too, as this genre explicitly addresses connections between "everyday" crimes and colonial crimes, challenging settler authority and belonging through an assertion of Indigenous sovereignty and knowledge systems. It is vital that scholars also explore how Indigenous crime fiction connects colonial crime, modern crime and environmental crime through a characterisation of place, lands and waters that stems strongly from Aboriginal cultures and traditions. This is why, from my position as a non-Indigenous Australian scholar, I will also investigate how Indigenous Australian crime fiction compares to settler evocations of place and is, as Iva Pollack argues, "an important tool for dismantling the fallacies of the postcolonial space in which settler-colonial past and present remain deadlocked" (1).

In my analysis, I implement a postcolonial, ecocritical framework informed by the work of Ross Gibson and Emily Potter. I also draw on scholarship that investigates how place and belonging are configured in non-Indigenous Australian crime fiction, with a comparative discussion of how Indigenous Australian crime fiction interrogates notions of place-making and the non-human in more relational, less dualistic ways in comparison to white, settler-colonial Australian crime writers. Place-making is of course more than just the naming and ownership of places – it represents the formation of one's sense of place in the world, which is "contingent upon both having and knowing a solid and permanent place that one owns or can lay claim to" (Potter, *Writing Belonging at the Millennium* 15). Here, I take place to be what Tim Cresswell, drawing on John Agnew's definition, describes as a "combination of location, locale, and sense of place" (Agnew, cited in Cresswell 29); that is, "as well as being a location . . . [place] has a physical landscape . . . and a 'sense of place' – meanings, both personal and shared, that are associated with a particular locale" (Cresswell 31). In 1984, Leonard Lutwack argued that literary scholars do not want to just investigate place as "a formal element" of literature, but also as a means of understanding environmental concerns. This is because "earth as a place . . . is being radically changed", and it is through an "increased sensitivity to place" that one might begin to better understand anthropogenic impacts on the environment (2). This idea still stands today – if we cannot comprehend our own relationship to place, then how can we comprehend human-nature relations?

Included alongside place and place-making, my use of "ecology" covers the interconnections between the human and non-human, the human and human, and the non-human and non-human. I am therefore not using it to replace the term "nature" or "non-human", but to encapsulate the connections between and within these concepts. Climate is of course just one facet of ecology. I will be focusing on climate in parts of this chapter, not only due to its obvious relevance in any ecocritical discussion of the twenty-first century, but because, as I will go on to describe, climate is key to so much Australian crime fiction in terms of plot, characterisation and setting. Interrogating how climate is represented in Australian crime fiction is, I believe, a productive means through which to discuss both the problematic and more ecologically attuned ways of thinking about nature and the non-human in a postcolonial Australian context.

Modern Australian crime fiction authored by white, settler-colonial writers is inextricably entangled with settler-colonial perspectives of city, country and the non-human, while Indigenous authors of crime fiction emphasise Indigenous understandings of place and human/non-human relations, in turn challenging the authority of Western, settler dualisms of culture/nature, human/non-human and civilisation/wilderness. Settler-colonial writers are often focused on highlighting the links between crime and elements of the Australian environment, particularly heat, drought, bushfire and flooding, as well as animals traditionally perceived as threats to settler expansion into the bush, such as snakes, dingoes and non-native species such as rabbits and foxes. Place-making, particularly settler place-making, and its relationship with ecology is therefore of particular importance here. Understanding climate in relation to non-Indigenous place and belonging is especially pertinent given that, as Potter describes, climate change is an event that "tap[s] into longstanding non-Indigenous anxieties concerning their legitimate place and sovereign claim over the Australian environment" ("Climate Change and Non-Indigenous Belonging" 30–31). How settler-colonisers have engaged with place-making over the years since Australia's colonisation also reveals how such a practice is directly connected with the ecological crises that we are seeing today – European-style farming, extractivist practices such as logging and mining, and urban expansion can all be linked to a variety of environmental issues in modern Australia.

Like many countries in the world, the majority of Australians reside in cities, with 66.9% living in capital cities and 33.1% in regional areas (ABS). In fact, *The Mystery of a Hansom Cab* (1886) – one of Australia's earliest, most commercially successful works of colonial crime fiction – is set in the city of Melbourne and not in a rural locale at all. Since the publication of this text, though, the majority of colonial Australian crime fiction and much contemporary Australian crime fiction is set in rural and/or "wilderness" locations, far from the bustling cities; a setting that is used by many authors to convey a sense of isolation, distance and fear to the majority of Australian readers residing in cities.

Of course, this focus on rural settings also speaks to postcolonial anxieties, revealing more viscerally than an urban setting what Knight refers to as "the relationships between crime domestic and crimes colonial" ("Crimes domestic and crimes colonial" 33). Crime fiction, he argues, is uniquely capable of exploring identity and belonging in a postcolonial context because of the way it presents both domestic and colonial crimes "in morally and emotively legible personal terms" (Knight, "Crimes domestic and crimes colonial" 33). As I will go in to describe in the context of this chapter's texts, I certainly agree that postcolonial Australian crime fiction – both Indigenous and non-Indigenous – is well-positioned to explore ideas about settler belonging and anxieties in the modern age, but I would go another step further to say that it in fact *unsettles* such notions. Climate change is particularly key to this, providing what Potter describes as "a conceptual site through which belonging anxieties [are] further played out" (*Writing Belonging* 6).

Relocating the Badlands in Non-Indigenous Australian Crime Writing

Non-Indigenous Australian crime writer Emma Viskic speaks to this uncertainty of settler identity and belonging in her exploration of the links between rural and urban crime in Australia and how such crime affects and is affected by the colonial systems of dominance that uphold culture/nature and human/non-human binaries in the present day. Her work speaks both to and against an idea that is often implicit in settler-Australian stories about living on the land: that "the White Australian dreams tried to believe that a past so recently tumultuous and polyglot could be gone simply by disregarding it" (Gibson 158). Viskic's crime fiction engages with the tensions characteristic of what is known in Australia as the post-Mabo period. The Mabo decision – the recognition of Native Title passed into Australian law in 1992 – resulted in what Kieran Dolin summarises as an "uneasy or doubled sense of being at home" (4), also described by Ken Gelder and Jane Jacobs as "the postcolonial uncanny" (cited in Dolin, 4). Understanding the "postcolonial uncanny" means understanding the idea of "reconciliation" between Indigenous and non-Indigenous Australians: "It is not simply that Australians will either be reconciled with each other or they will not; rather, these two possibilities . . . coexist and flow through each other in what is often . . . a productively unstable dynamic" (Gelder and Jacobs 34). The uncanniness of reconciliation and decolonisation lies in the settler's perspective that "what is 'ours' is also potentially, or even always already, 'theirs': the one is becoming the other, the familiar is becoming strange" (Gelder and Jacobs 33). Viskic's protagonist, Caleb Zelic, navigates both the possibilities and perceived threats that the post-Mabo unsettlement made real for non-Indigenous Australians.

Caleb is in some ways far from the traditional crime solver of classic colonial Australian crime fiction, being profoundly deaf and with close familial connections to the Indigenous community, but he is still white and often presumptuous about his sense of belonging in the world – a sense that is then upset through his own confrontation, as a white Australian, with the traumatic events, both past and present, that Indigenous people in Australia have been experiencing since the continent's colonisation.

Ross Gibson's exploration of "Australian badlands" bears particular relevance here. The way that some rural locations are treated in Australian settler culture match what Gibson describes as "a place where evil can be banished so that goodness can be credited, by contrast, in the regions all around" (17). The fictional Resurrection Bay of Viskic's novels is indeed such a place. It is far from the increasingly cosmopolitan country towns closer to Australian cities that are regularly frequented by tourists. Rather, Resurrection Bay represents the more common, but oft-forgot, isolated rural location that is far enough away from the city for crime to potentially thrive unnoticed. Such places are viewed by some as retaining a sense of the colonial frontier – an environment where earning a living is still hard-fought in the face of harsh bushland and climate and there is no easy country life to be had, even in contemporary Australia. I understand Gibson's concept of the Australian badlands as a vital interface between the human and non-human, historical colonial crime and everyday crime, and Indigenous and non-Indigenous belonging. The badlands are places that evoke a strong sense of both past and present crimes, emphasising the links between Australia's colonial past and postcolonial present, including settler exploitation, the attempted annihilation of Indigenous peoples and cultures, and the Australian environment at large. A confrontation with such places also has the potential to unsettle white, non-Indigenous Australians who must recognise the dark colonial history of these badlands – a history that made their own belonging (or at least some sense of belonging) possible.

Viskic's Caleb Zelic series attempts to address this unsettlement, with a particular focus on how crime, both present and historical, is perceived across rural and urban places. Viskic's award-winning series is four books in length; this chapter will specifically examine the first two in the series – *Resurrection Bay* (2015) and *And Fire Came Down* (2017). Viskic is a non-Indigenous Australian with Dalmation and Irish ancestry. Her husband was raised in an Indigenous family (see Cosic) – a relationship dynamic which perhaps informed her character of Caleb Zelic, who is a non-Indigenous Australian married to an Indigenous woman.

Caleb Zelic is an ex-insurance investigator now applying his skills to his own private investigation business. The opening of *Resurrection Bay* describes the death of one of Caleb's closest childhood friends, Gary, a police officer. Caleb sets out to solve Gary's murder with the help of his friend Frankie, an ex-police officer. Caleb finds himself returning to his small coastal hometown of Resurrection Bay to recover from a stab wound inflicted during his investigation in the city but is followed there by the men he suspects are involved in Gary's death. As mentioned, Caleb has been deaf since childhood; he is adept at lip-reading and can sign with others who also know Auslan (Australian Sign Language).

The second book in the series, *And Fire Came Down*, begins seven months after the events of the first novel. Caleb is dealing with the fallout of these events, in particular his complicated relationship with his ex-wife Kat and some serious PTSD symptoms. In the opening chapter, Caleb is walking the streets of Fitzroy – the inner-city Melbourne suburb where he lives – and is approached by Portia, a terrified woman who tries in vain to communicate an important message to him before she runs from an unknown man and is killed after falling in front of oncoming traffic. The remainder of the novel sees Caleb investigating Portia's motivations for seeking his help, leading him back to his hometown of Resurrection Bay where he discovers links between the woman's death and the local drug trade. Tied to this are acts of serious vandalism and arson committed against local businesses, particularly those owned by Indigenous people, and Caleb finds himself trying to re-establish connections with his in-laws in order to solve the crime – this is Kat's family, who identify as Koorie[1] and are particularly affected by these crimes.

The crime-solving in these novels takes place across city and countryside, with Caleb often travelling between the city of Melbourne and Resurrection Bay, which provides an apt opportunity to interrogate Gibson's conceptualisation of the Australian badlands. I wish here to partly complicate Gibson's theory in terms of what the badlands are, or can be, particularly when one considers the divide between city and country that characterises so much of Australian, specifically settler-colonial, perspectives on the environment. As in many places in the world, Australian cultural norms assume that city people are removed from nature, and rural residents are deeply entrenched in it (even if in a way that is about exploiting or controlling it, such as through farming or mining). Viskic's crime fiction both challenges and endorses what is a very Anglo-settler perspective on the city/country and culture/nature dualisms that still pervade mainstream Australian culture, affecting political decisions around climate change and other ecological crises. For example, such dualisms strongly influenced the previous, conservative Australian Federal Government led by Prime Minister Scott Morrison to prioritise coal power and mining over policies regarding renewable energy and climate action (see Murphy; Slezak). It is significant that, in Viskic's work, there are no easy answers when it comes to where the "bad places" really are. There is often no clear safe space for Caleb to turn to – his hometown of Resurrection Bay is just as entrenched with dangerous crimes as the city. Both locations are not only breeding grounds for violent crimes but are settings that demonstrate the intimate links between such crimes and a vast and complex history of colonial power and violence.

First, though, let us consider the country setting of Viskic's work as the true badlands. As per Gibson's description, Resurrection Bay – the small coastal town where Caleb grew up – is characterised as a place where "old passions and violent scenes are lying around in a million clues and traces" (Gibson 2). These sorts of clues and traces are largely absent from the first book, but they are made more explicit in the second novel, *And Fire Came Down*, with several references to the dispossession and murder of Indigenous people (referred to as Koorie people throughout the text) in this area during colonial times, such as the following:

> Caleb parked in front of the police station, bunker-grey with a neat garden of palms and pine bark. Its windows were too high to give occupants a view of Red Water Creek opposite. Then again, maybe the cops didn't want a daily reminder of how the creek got its name back in the 1840s. Twenty-three Koori women and children chased into the water by officers of the law; guns used, machetes, axes. A small massacre by local standards.
>
> *(AFCD 42)*

The description of the police station, particularly the mention of the "neat" palms and the military-eque "bunker-grey", stand in contrast to the visceral brutality of the massacre described. There is a clear implication that the police still maintain control of this area, preventing people in the station from viewing the site of the massacre committed by past "officers of the law". Retaining the brutal name of the creek – which we might understand to be one of Gibson's "clues and traces" – also shows that such history can be simultaneously obscured (uncomfortable for the modern police working in the station who do not wish to view the creek from their desks) and recalled (through the name of the creek that reveals the dark historical crimes of the location).

While Viskic is evidently tying the rural location of Resurrection Bay very directly to colonial crimes in such as a way as to clearly make them "badlands", the city reveals its violent colonial past in other ways throughout the narrative, particularly in the conclusion to *Resurrection Bay* which sees Kat, an Indigenous woman and Caleb's wife, brutally tortured under the orders of a corrupt cop, Hamish McFarlane. The power dynamic implied here between Indigenous and non-Indigenous persons is palpable, and the setting of this scene – a large warehouse set amongst "a cluster of tin sheds" (259) – reflects the ominous residue of colonial industrialisation and authority in this location.

Notably, Viskic's badlands are often shown to be intimately linked to environmental features and phenomena. The badlands in the city are characteristically described with odour and air quality in mind, such as the "stale-piss smell of the city" (*AFCD* 194). In fact, Gibson suggests that the unsettling quality of the badlands can be linked to a loss of sight (particularly in dense scrub or bushland), which forces settler-colonisers to rely on "noise and smell for their ways of orienting themselves" (97). While Gibson refers here to early colonial explorers of the rural badlands in Queensland, we might take this idea and apply it to Caleb's experiences in the "badlands" of Melbourne. Why does this matter, though? What is significant about denoting, say, an urban location as a "badland" rather than a rural one, as Gibson does? And what does this say about the human relationship with climate and the more-than-human? Gibson states that "Australia's most famous badland has long been the 'dead centre' of the outback" (15). The isolated highways of Queensland in Australia north-east – known as the "Horror Stretch" – are the primary focus of Gibson's work on badlands and are similarly isolated. Such isolation is what defines these badlands, Gibson suggests: "The isolation . . . its eeriness, its narratives of violence" (17) is what make such places significant in understanding the geographies of colonial crimes. So how does the city fit into this framework?

I argue that we need to bring this understanding of badlands to the city in a more concrete way – the city is, after all, where most Australians live and where colonial crime is arguably made the most invisible. The deserts, scrub and bush of Gibson's typical badlands are largely absent in these places but there are sections within cities that evoke a similar isolation and eeriness – industrial areas built on swampland, for example. Viskic's focus on industrial warehouses in *Resurrection Bay* speaks to this, and there is a link made in her work between these sites and the more geographically isolated Resurrection Bay. Rural badlands are of course significant too, though. Viskic shows how colonial crime might appear buried but, as the example at the police station shows, it cannot remain that way given how visible it is across place. Choosing one place as the badlands and the other as the "goodlands" is therefore not what is important here, but rather acknowledging that places other than the typified "frontier" regions of Australia can be these badlands, and that crime – both colonial and the more contemporary "everyday" – occurs in both. In fact, as Viskic's work demonstrates, a crime in one place cannot occur without a crime in the other, with the ice trade in Resurrection Bay directly linked to that in Melbourne. This understanding can then be extended to ecological considerations, not just in terms of environmental crimes – like the link between settler colonialism and climate crisis – but in terms of how attuning to place means attuning to ecology, and vice versa. This point is perhaps not unique to Australian crime fiction; other postcolonial crime fiction, like that written by US authors, makes similar connections between place and ecology. What is novel in this context, though, is the extreme dichotomy that is set up between the Australian city and the Australian country – the latter often homogenised as the isolated and barren outback despite the many populated and large regional centres that exist across Australian states. Viskic's work challenges this separation of Australian city and country, particularly when it comes to issues linked with colonisation, in turn unsettling the notion of non-Indigenous belonging across such places.

Connecting Climate With Settler-Colonial Australian Place-making

While it has become commonplace amongst academics to blame settler-colonial systems of domination for parts or indeed much of current environmental crises, settler-Australian authors are perhaps taking their time to make these links explicit. One might read Jane Harper's *The Dry* and see an *implicit* connection between an unsustainable settler farming lifestyle and climate crisis, but Harper does not explicitly mention nor interrogate this notion (see Fetherston). While Viskic does not necessarily present the opposite of this – that is, a clear and consistent engagement with ecological concerns in connection with Australian social and political concerns – she does approach a more complex exploration of how non-Indigenous belonging in Australia is being unsettled by climate, crime and the city/country divide.

In the town of Resurrection Bay, the impacts of climate and associated environmental phenomena are often felt more viscerally by the characters than in the city, with more extreme weather events characterising the rural locations. As the title implies, *And Fire Came Down* features the constant threat of bushfire in this coastal area: "the blazing sunlight of the firebreak" (59), references to the Country Fire Authority (CFA) "taking advantage of the still weather to do a late-season burn-off" (56), and the "slam of heat" that pushes Caleb off his feet as a raging bushfire descends on him and Frankie in the climax of the novel (311). Viskic draws a striking link here between arson – one of the recurring crimes in this second novel – and climate change. Arson has also been at the centre of several bushfire disasters in Australia, the most well-known being Brendan Sokaluk whose arson resulted in one of the devastating Black Saturday bushfires.

Significantly, bushfire is responsible for revealing the contents of one of the shipping containers which Caleb suspects is relevant to Portia's death and the local drug trade: "Snowflakes swirled from it, each one flaming as it fell, the labels turning as black as the printed names" (311). This is a common trope in Australian crime novels, with environmental phenomena sometimes playing a role in revealing a vital clue in the case. Such phenomena, particularly those associated with climate such as bushfires and floods, are often "agents of justice" in Australian settler-colonial crime narratives (Knight, "Crimes Domestic and Crimes Colonial" 20). This is a potentially limiting factor of Viskic's work, her use of this trope emphasising a settler perspective on the Australian environment – one that portrays something like climate as vengeful and dangerous rather than an opportunity to seek out more complex understandings of human/non-human relations.

And Fire Came Down paints a frightening picture of a town under the impacts of climate change, primarily because the brutally hot summer seems to be exacerbating crime, with arson a particular threat, especially given the possibility of such crimes resulting in widespread bushfires. Caleb notes that he senses "a feeling of relief at the sight of [the town]" (14) when he returns at the beginning of the book – not because of any reassuring familiarity in his hometown but because the air-conditioner in his car "had given out a few kilometres into the three-and-a-half-hour drive" (14). His relief is due simply to the dry heat of summer and his final arrival in a place where, presumably, he can access working air conditioning. The remainder of the novel depicts the town as under constant threat of both bushfire and the aforementioned arson, both of which are a larger-than-normal threat because "the fuel load was heavier than usual this year, the undergrowth flourishing after a long, wet winter and then baking dry in the endless summer" (56). Here, Viskic recognises some important links between the Australian summer, crime and experience of place, particularly in the country, but there are similar allusions to the effects of heat in the urban setting of the books.

In the city, there is a general feeling that climate is defined primarily by the human and is represented less by typical references to weather and more by smells and sounds, which are linked to the human and read as far removed from the natural environment. For example, there is the "scent of stale cigarettes and cheap deodorant" (*RB* 47) in Caleb's apartment, "diesel fumes" on the roads (*RB* 256) and the "stink of unwashed bodies" in a squatter's house (*AFCD* 107). However, climate is made most evident in the scenes in the country, not the city, with the smells relating more so to the non-human, such as heat and fire: "the wind was coming from the north, carrying with it the smell of burnt eucalypt and the promise of dust storms" (260). Viskic is doing what many other settler-colonial Australian authors have done when writing crime – connecting violence with a harsh and unforgiving rural climate. There is a suggestion that an excruciatingly hot summer exacerbates crime in rural areas and that this is how Australians in these towns live – constantly up against the "unforgiving" climate of the Australian bush, and it is only the toughest locals who can stomach it. But there is also some beauty here too, with Viskic shifting between admiring descriptions of the Victorian natural environment and its climate (particularly the ocean, described at one point as "silver in the midday sun, its salt smell lifting on the breeze" [249]) and more horrifying portrayals of it. This speaks to what Knight argues is a settler-colonial tendency that still pervades Australian cultural discourse about land: it is "both dramatically beautiful and radically life-threatening, as if the merits and risks of imperial possession [the declaration of 'Terra Nullius'] remain a part of the national consciousness, or subconsciousness" (17).

While this is problematic, Viskic still presents some nuance in her engagement with climate, and this is achieved primarily through a focus on unsettlement. What she arguably does differently to other contemporary settler-Australian crime authors is reflect the complexities of both city and country in terms of their ability to be places of safety and familiarity while being concurrently imbued with a sense of danger that threatens to unsettle feelings of belonging for non-Indigenous characters. Caleb is consistently portrayed as being without a steady sense of belonging – when he finally feels in reach of it (such as the various times he believes he is about to reunite with his ex-wife and feel at home again in Resurrection Bay), this feeling is shattered. This is made particularly evident in *And Fire Came Down* when Kat tells him that she "can't do this" and Caleb feels "the ground slid[e] away beneath him" (251). Caleb's inability to be open about his investment in an emotionally mature relationship and properly treat his PTSD are key factors in exacerbating his sense of unsettlement, reflecting the inadequacy of the stereotype of the toxic, white male settler. This stereotype is aptly described by Jeanine Leane, who posits that Australian settler-colonial culture has long been (and is still) concerned with celebrating "a handsome, care free, anti-intellectual bushman, apathetic and unaware of the wider world around him and who embraced and practiced the art of mateship" (32). Rather than establishing his belonging through this kind of identity – as is seen in more traditional Australian literary works (such as A.B. Paterson's acclaimed poem "The Man from Snowy River" and Patrick White's *Voss*) where the settler-coloniser establishes dominance through his whiteness and traditional masculinity – Caleb is forced further away from any stable sense of belonging when Kat calls him out on these traits. Matching the warmth of the Australian summer, Caleb feels "a wave of heat" (251) along his neck as he is confronted by Kat, suggesting the inability of the white settler to establish belonging in this kind of climate unless they confront the reality of unsettlement.

Sovereignty and Water Rights in Indigenous Crime Fiction

There is an irony to Viskic utilising a tried-and-true settler-colonial crime narrative to reveal the ineptitude of such narratives to fully confront the exclusionary nature of this storytelling when it comes to Indigenous people, people of colour and more-than-human considerations. Gibson describes the writing of colonial officers who needed to list their achievements and progress on paper as "writing [that] worked on the country, rendering it a known quantity" (71). Such writing is "a performance that showed whites and blacks who was who on the great chain of being" (72). These thoughts may similarly be applied to Australian settler-colonial crime fiction in the sense that it often still endorses a chain of being through a depiction of climate and the non-human as inevitably dangerous, uncontrollable and lacking in agency, subsequently displacing Indigenous understandings of the land as agential and complex.

While non-Indigenous crimes writers – including those like Viskic who explore Indigenous/non-Indigenous relations in their fiction – depict place in a way that unsettles non-Indigenous belonging, Indigenous crime authors write about place to "assert a continuing sovereignty" which is achieved through "the depiction of land as a sentient being" (Watson, "The Role of Place" 226). Place and place-making, in Indigenous crime fiction, is therefore integrative in its inclusion of both human and non-human agency. While a work like Viskic's might approach some nuance in terms of its engagement with the complexities of human-climate relations in a postcolonial context, any sense of more-than-human agency is restricted to traditional settler conceptualisations of the unforgiving bush or outback enacting justice or vengeance. This is indeed a problem if non-Indigenous Australian crime fiction is to properly address ecological crisis and the coloniser's role in it.

We might turn to Indigenous Australian playwright and novelist Julie Janson's recent foray into crime fiction to better understand how Indigenous crime fiction is engaging differently with the notion of place. Janson, a Burruberongal woman, conveys the concerning links between water rights, colonisation and intra-human violence in *Madukka: The River Serpent* (2022). The narrative is set primarily in rural Australia and, while it does not shift between city and country in the way that Viskic's work does, it is still deeply concerned with changing ideas of place and place-making in Australia, with an emphasis on the way that big corporations and city folk benefit from the exploitation of the bush. Additionally, Janson's protagonist, the feisty, unrelenting Aunty June, is a private investigator, like Caleb Zelic, rather than the more traditional police detective. The primary difference is that June is Indigenous – a "Gamilaraay Aboriginal woman" (2) – while Caleb is a settler, and so June's perspectives on justice, colonisation and ecological catastrophe come from her position as an Aboriginal person whose community has been violently oppressed via Australian law enforcement. *Madukka* therefore addresses the extreme ineptitude of the Australian police to enact justice for Indigenous people and the embedded racism that facilitates prejudicial incarceration practices.

Like Viskic's series, Janson's work aptly encapsulates the atmosphere of a declining Australian country town – the fictional Wilga in New South Wales – in the wake of an economic downturn: "Many other businesses stood empty along the road . . . The failing economy had gutted the biggest hotel; it sat there now with blackened walls, graffiti and glassless windows" (2). This rural location is the setting of the violent injustices committed against Indigenous characters, but there is a suggestion that it is not simply the "backwards" and declining rural town that allows this, but rather the broader systems of governance that make their way from the city to the country.

Related to this point is the fact that the Darling River – Wilga's main water source and one of eastern Australia's major rivers – is drying up (both in the novel and in reality) due to climate change and water extraction for use in agriculture and other industries, which has stopped the Darling from flowing into the Murray, another major river. This has had catastrophic impacts on ecosystems and rural communities, not to mention affecting Indigenous cultures and traditions that are deeply tied to the river system. The Australian Federal Government's failure to address the problem, despite significant financial investment, might be viewed as strong evidence of the failure of colonial systems of governance to protect Indigenous cultural values and the environment.[2] Water theft by irrigators has been a compounding factor in all of this and is central to Janson's narrative. The story begins with the disappearance of June's nephew, Thommo, an Indigenous man and environmental activist who she suspects has been meddling with water pumps in an effort to prevent further water theft.

Where Janson's work differs from Viskic's is in its clearer, at times blatant, commentary on the connections between intra-human crime, colonisation and environmental crime. Watson makes the point that, in Indigenous crime fiction, the land itself has a "profound bond with the human characters, and as a consequence, it reflects their pain" ("The Role of Place" 226). In Janson's work, the fate of the land is as much at stake as the fate of the protagonist and her family – Thommo's murder is directly connected to the water theft from the river, and the investigation of his murder leads to further violence committed against the Indigenous characters. Janson emphasises the agency of the river, describing it as full of "life" (107), which is taken from it as the water is stolen by the colonisers. The "great ancient red gum trees" that require the river to live are described as "sentinels", alluding to their guardianship of the river and Country (107).[3] It is only through reconnecting with Country, Watson says, that the protagonists of Indigenous crime texts "both achieve resolution of the crime, and begin to reclaim their identity" ("The Role of Place" 226).

This cements the importance of connection to place in a different way to Viskic's work, with the land and the non-human depicted as agential beings in a very real and not simply symbolic sense. One significant example of this is Janson's depiction of the Darling River as a serpent – "Waarway the Madukka serpent, a being from the night sky of the Milky Way" (5). At the end of the novel, it is this serpent who provides justice for the murder of Steve, an Indigenous man killed in custody by police officer Blackett, when no justice is forthcoming from the settler-colonial authorities:

> The serpent took Blackett away as it tore through Country . . . drowned him in its rage and tumult. It mapped the path of the Old Man Crow who chased the two sisters, and in their footsteps magic places made the land forever Aboriginal land.
>
> *(296)*

Janson's novel suggests that a significant part of reclaiming this identity is by undermining the coloniser's legal systems and processes using Indigenous cultural knowledge, as demonstrated by the river serpent. It is not simply about solving one crime but seeking justice for all of the interconnected crimes of colonisation. As June tells police officer Wayne Johnson, "It isn't about one person. It is much more, who we are" (282).

These interconnected crimes are made particularly evident when June wakes from a night of fitful dreaming about a "river full of dead stinking fish, or signs that read 'Massacre Downs' . . . or 'Treachery

Creek'" (51) – a telling combination of dreams about a ruined ecosystem and the names typically given to places where colonial massacres of Indigenous people have been committed. Similarly, Old Mary – a relative of June's – at one point tells the story of the Gwydir River massacre where her people "were shot down and burnt" (68), their lands stolen. This same river system has now been ecologically devastated after water extraction and ecological mismanagement by the colonial powers. And so, the connections between colonisation, Indigenous dispossession and genocide, and ecological catastrophe are made clear.

This is not to say that settler-Australian crime authors do not consider such connections, but if we take Janson's work as an example of Indigenous crime fiction more broadly, then we might say that settler crime authors obliquely reflect such connections whereas Indigenous authors make them explicit. As aforementioned, I have argued something similar in an analysis of Harper's *The Dry* – there are allusions to the concerning links between drought, climate crisis and domestic violence, but Harper does not undertake an explicit discussion of ecological crises in her work (Fetherston). Janson, on the other hand, makes such concerns explicit, demonstrating an urgent need for colonisers to better understand the intimate connections between crime, violence, settler culture and climate crisis.

The sense of Aboriginal place in Janson's novel is also solidified by the dead, who play a role in the crime-solving. June learns that Thommo has been murdered and is not just missing when his "goonge" – a kind of ghost – visits her at night, revealing that "he was dead and she was responsible for finding the truth" (82). Similarly, Steve, an Indigenous man from Sydney who, while visiting Wilga, is murdered by police in a jail cell, is described as "float[ing] above the room" (240), seen by June as an "eaglehawk swooping and diving on hot air currents" as "he scratched at the windows. Wanting always to come in" (244). The presence of the goonge emphasises the long-standing cultural and spiritual connection to place and thereby unsettles the notion of *terra nullius* – what is often deemed to be the first colonial law exercised in Australia and used in the oppression of Australia's First Nations peoples. This connection is made especially evident when, on his death, Steve travels to the grave of his grandmother – a grave that is "two thousand years old. As old as Christ", existing on the Australian continent for far longer than settler Australians (244). This unsettlement of *terra nullius* and thereby settler belonging casts serious doubt on the system of colonial law that Janson critiques in her novel – one in which people like Steve are arrested on no discernible charge and then murdered in police custody. For June and her family, seeing a goonge is such a common occurrence that it's "like getting on the phone", emphasising that, like the Madukka serpent, the goonge is very real – not just symbolic of violence and crime in the novel, but a tangible being (276).

Here, then, is another key difference between the work of Janson and Viskic. While Viskic certainly critiques the way in which some members of the Australian police exercise the rule of law, the police officers in *Madukka* are entirely inept; so much so that June tells them, "You lot are the amateurs . . . I will get you the proof" when it comes to solving the murder of Thommo (282). In this way, Janson more seriously undermines the entire premise of postcolonial Australian systems of law by drawing attention to the racism and ineptitude of the police and the importance of deeply entrenched cultural practices of Indigenous people, whose spiritual connection to the lands, waters and skies has survived these oppressive systems of colonial law.

Conclusion

The links between place and ecology are no doubt complex, and the entanglement of both in Viskic's work demonstrates how settler-colonial perspectives on these concepts are still being problematised and debated in postcolonial Australia. Viskic's Caleb Zelic series is a crime work that addresses the ghosts of colonial crime, presenting a protagonist who is both literally and figuratively torn between worlds: city and country, Indigenous and non-Indigenous, criminal and authority. Viskic approaches these dualisms in a way that endorses certain settler-colonial assumptions about climate and nature, whilst also subverting some of these assumptions through unsettlement.

However, the inclusion of extreme climate events and natural settings in crime fiction as mere facilitators of intra-human crimes – which is suggested in Viskic's work – risks what Nicholas Birns describes as the depiction of "the land as an inanimate backbone underneath merely temporary feeling" (128). In the post-Mabo era, Australian literature has seen a shift away from this sort of engagement with land, Birns argues (128). This shift has perhaps been seen more so in literary Australian texts authored by settler-colonial writers, such as the work of Kate Grenville, Gail Jones and Alex Miller (Birns), and is not as evident in popular Australian genre fiction, such as crime fiction. This presents a problem when it comes to engagement with ecological concerns in this genre – that is, many texts risk not going beyond environment as mere backdrop, reinstating the traditional Western, settler perspective on nature as separate to the human. This is also an issue when it comes to Indigenous cultural representation in crime texts authored by non-Indigenous writers. Ignoring deeper connections to land (particularly the Indigenous Australian ontology of Country) that go beyond the idea of "land as an inanimate backbone" is in turn neglecting what was recognised by the Mabo decision; it may be construed as a refusal to engage with nuanced perspectives on the Australian environment.

On the other hand, Indigenous Australian crime fiction is doing just this, engaging deeply with Indigenous cultural understandings of land and waterways in an effort to re-establish Indigenous sovereignty in the face of settler-colonial racism and oppression, unsettling non-Indigenous belonging in the process. As one Aboriginal Elder in Janson's novel points out, the white Australian is "a fella without Dreaming . . . he will have to search for it" (113) in order to make any lasting, positive change for the environment during climate crisis. The Dreaming – an important Indigenous Australian philosophical and cultural concept that includes stories of creation – is described by Irene Watson as "an ever-present place of before, now, and the future, a place that we are constantly returned to" (36). This is represented primarily by Madukka the serpent in Janson's novel, with the Elder's words suggesting that the settler-Australian will be unable to attend to the river without first acknowledging Aboriginal sovereignty and knowledge and in turn re-evaluating their own settler connection to place.

Notes

1. The name Koorie/Koori/Gurri broadly refers to the Aboriginal people and communities located in Victoria and southern New South Wales in south-eastern Australia (Koorie Heritage Trust).
2. For a summary of this, see Richard Kingford, "It's official: the Murray-Darling Basin Plan hasn't met its promise to our precious rivers".
3. The Indigenous concept of Country is completely different to the "country" of the "city/country" dichotomy discussed in this chapter. It is a central part of Indigenous Australian knowledge systems and cultural practices and is described by Laklak Burarrwana et al. as "the feeling of home, the feeling of the seasons that communicate with us. It is all the beings of home. It is everything that we can touch or feel or hear or sense, and it is everything beyond that too. It is everything that belongs in Country, with Country and as Country, including us. And it is the relationships between all those beings too" (23). This is just one of many understandings of Country.

Bibliography

ABS (Australian Bureau of Statistics). "Location: Census, 2021." *Australian Bureau of Statistics*, 28 June 2022, www.abs.gov.au/statistics/people/people-and-communities/location-census/latest-release.

Agnew, John A. *Place and Politics: The Geographical Mediation of State and Society*. Allen & Unwin, 1987.

Birns, Nicholas. *Contemporary Australian Literature: A World Not Yet Dead*. Sydney UP, 2015.

Burarrwana, Laklak, et al. *Song Spirals: Sharing Women's Wisdom of Country through Songlines: Gay'wu Group of Women*. Allen & Unwin, 2019.

Cosic, Miriam. "The Long and Winding Road That Leads to Emma Viskic's Gritty Fiction." *The Sydney Morning Herald*, 30 Nov. 2019, www.smh.com.au/culture/books/the-long-and-winding-road-that-leads-to-emma-viskic-s-gritty-fiction-20191125-p53du4.html.

Cresswell, Tim. *Place: An Introduction*. John Wiley & Sons, 2015.

Dolin, Kieran. "Place and Property in Post-Mabo Fiction by Dorothy Hewett, Alex Miller and Andrew McGahan." *Journal of the Association for the Study of Australian Literature*, vol. 14, no. 3, 2014, pp. 1–12.

Farrell, Michael. *Writing Australian Unsettlement: Modes of Poetic Invention 1796–1945*. Palgrave Macmillan, 2015.

Fetherston, Rachel. "'Little Difference between a Carcass and a Corpse': Ecological Crises, the Nonhuman and Settler-Colonial Culpability in Australian Crime Fiction." *Journal of the Association for the Study of Australian Literature*, vol. 21, no. 2, 2021, pp. 1–17.

Gelder, Ken, and Jane M. Jacobs. *Uncanny Australia: Sacredness and Identity in a Postcolonial Nation*. Melbourne UP, 1998.

Gelder, Ken, and Rachael Weaver. "Colonial Australian Crime Fiction." *The Anthology of Colonial Australian Crime Fiction*, edited by Ken Gelder and Rachael Weaver, Melbourne UP, 2008, pp. 1–10.

Gibson, Ross. *Seven Versions of an Australian Badland*. U Queensland P, 2002.

Hammer, Chris. *Scrublands*. Allen & Unwin, 2018.

Harper, Jane. *The Dry*. Macmillan, 2016.

Hume, Fergus. *The Mystery of a Hansom Cab*. Kemp and Boyce, 1886.

Janson, Julie. *Madukka: The River Serpent*. UWA Publishing, 2022.

Kingsford, Richard. "It's Official: The Murray-Darling Basin Plan Hasn't Met its Promise to Our Precious Rivers. So Where to Now?" *The Conversation*, 3 Nov. 2022, theconversation.com/its-official-the-murray-darling-basin-plan-hasnt-met-its-promise-to-our-precious-rivers-so-where-to-now-188074.

Knight, Stephen. *Continent of Mystery: A Thematic History of Australian Crime Fiction*. Melbourne UP, 1997.

———. "Crimes Domestic and Crimes Colonial: The Role of Crime Fiction in Developing Postcolonial Consciousness." *Postcolonial Postmortems: Crime Fiction from a Transcultural Perspective*, edited by Christine Matzke and Susanne Mühleisen, Rodopi, 2006, pp. 17–33.

Koorie Heritage Trust. "About Us." *Koorie Heritage Trust*, 18 Oct. 2022, koorieheritagetrust.com.au/about-us/.

Leane, Jeanine. "Aboriginal Representation: Conflict *or* Dialogue *in the* Academy." *The Australian Journal of Indigenous Education*, vol. 39, no. S1, 2010, pp. 32–39, doi:10.1375/s1326011100001113.

Lutwack, Leonard. *The Role of Place in Literature*. Syracuse UP, 1984.

Murphy, Katharine. "Australia's Rightwing Government Weaponised Climate Change – Now it Has Faced its Reckoning." *The Guardian*, Guardian News and Media, 22 May 2022, www.theguardian.com/australia-news/2022/may/22/australia-rightwing-government-weaponised-climate-change-reckoning-scott-morrison.

Paterson, Andrew Barton. *The Man from Snowy River and Other Verses*. Angus & Robertson, 1895.

Polak, Iva. "Unpunishable Crimes in Claire G. Coleman's Futuristic Novel *Terra Nullius*." *Humanities*, vol. 11, no. 2, 2022, p. 47, doi:10.3390/h11020047.

Potter, Emily. "Reimagining Place: The Possibilities of Paul Carter's 'Nearamnew'." *Fresh Water: New Perspectives on Water in Australia*, edited by Emily Potter, Alison MacKinnon, Stephen McKenzie, and Jennifer McKay, Melbourne UP, 2007.

———. *Writing Belonging at the Millennium: Notes from the Field on Settler-Colonial Place*. Intellect Books, 2019.

Slezak, Michael. "This Week's Budget Theatrics Occurred as Lismore Flooded Again and on the Ground, They Say We've Hit a 'Low Point'." *Climate Change is One of Australia's Biggest Challenges. But According to the Government and Labor, It's Not so Clear-Cut – ABC News*, ABC News, 2 Apr. 2022, www.abc.net.au/news/2022-04-03/climate-change-challenge-not-priority-in-govt-opposition-budgets/100955218.

Turnbull, Sue. "Monstrous Wounds: Crime, Environmental Catastrophe and Domestic Abuse in Jane Harper's *The Dry*." *Journal of Australian Studies*, 2023, pp. 1–13, doi:10.1080/14443058.2023.2165133.

Viskic, Emma. *Resurrection Bay*. Echo Publishing, 2017.

———. *And Fire Came Down*. Echo Publishing, 2019.

Watson, Irene. "Sovereign Spaces, Caring for Country, and the Homeless Position of Aboriginal Peoples." *South Atlantic Quarterly*, vol. 108, no. 1, 2009, pp. 27–51, doi:10.1215/00382876-2008-021.

Watson, Nicole. "Deadly Detectives: How Aboriginal Australian Writers Are Re-Creating Crime Fiction." *AlterNative: An International Journal of Indigenous Peoples*, vol. 15, no. 1, 2018, pp. 75–81, doi:10.1177/1177180118818187.

———. "The Role of Place in Indigenous Australian Crime Fiction." *Australian Feminist Law Journal*, vol. 45, no. 2, 2019, pp. 225–231, doi:10.1080/13200968.2020.1727663.

White, Patrick. *Voss*. Eyre & Spottiswoode, 1957.

PART II

Bodies and Violence

7

PEST CONTROL

"Wasp Season" in Agatha Christie's "The Blue Geranium"

Alicia Carroll

By acquiescing in an act that can cause such suffering to a living creature, who among us is not diminished as a human being?

Rachel Carson

In Agatha Christie's detective fiction, the cosy cultural work of gardening is often linked to murder. It is no surprise, then, that garden and human pests meet the same fate at the hands of the same weapon, potassium cyanide, in a representative tale such as "The Blue Geranium" (1931). There "wasp season" proves mutually fatal for a wide ontological spectrum. Amongst these, only the death of the annoying human pest, Mrs. Pritchard, is deemed a murder (109). When this crime is solved by Jane Marple, the tale seems to preserve the traditional order of the embowered village, St. Mary Mead. Marple, for example, skillfully identifies a seemingly uncanny perversion of the Victorian language of flowers as "*just camouflage*" meant to deflect attention from Mrs. Pritchard's murderer, the ambitious Nurse Copling (110). Ushering the latter out of the social body, Marple may now direct the long-suffering Mr. Pritchard to a new wife of more "use" than the former (102). But, by the end of "The Blue Geranium", Marple continues to have other victims on her mind: "I know", she says "apologetically", in advance of explaining her solution to the crime, "that I've got wasps on the brain" (110). Worrying about the gardener's casual annual fumigation of wasps, Marple speaks of the "poor things, destroyed in their thousands – and usually on such a beautiful summer's day" (110). They die like Mrs. Pritchard, "poor lady", by inhaling "cyanide of potassium" (110). Marple's awareness of this convergence of death by pesticide reveals that not only are people and insects vulnerable to the same chemicals, but that ranking and taking life is a common human practice, even amongst those who might be considered most "humane": caregivers, gardeners, or otherwise "very good sort[s]" (97). While "The Blue Geranium" and tales like it have been placed in the cosy category – through which the resolutions of popular detective fiction have been thought "conservative", "complacent", and disciplinary – new approaches, as Merja Makinen argues, may reveal the challenges Christie's work poses to concepts of both "law and order" and the "cultural status quo" ("Contradicting" 77). Millennial ecocritics, then, may note that Marple's unresolved qualms over insect extermination at the end of "The Blue Geranium" encourage the reader to see ecological violence *as* violence perpetrated by people against beings deemed of little "use" to them (102).

In fact, Christie's cosy, often embowered detective fiction is relevant to ecological thought today for it generally challenges us to see the extent to which we continue to dole out justice on the basis of

DOI: 10.4324/9781003091912-10

the "artificial nature of taxonomied identit[ies]" we find useful and withhold it from entities we do not (Bernthal 3). While justice may be a matter of life and death determined on the basis of race, class, gender, and sexual orientation in Christie's texts, an instrumentalist "taxonomic" habit of mind also informs our perceptions of which beings receive justice in her work. The presence of instrumentalist language in "The Blue Geranium" is marked in the story through Christie's word play and ironic ontological reversals of human and garden pests. This language remains interwoven, although rarely remarked upon, in popular garden and conservation languages of caring today. If Marple's famous analogic "form" of detection identifies criminality by likening "the same kind of thing" occurring on different scales, her thought process in "The Blue Geranium" demonstrates her "scientific" system of "classification" in action, as she identifies killers as such despite the different species of their victims (Makinen *AG* 59; Christie *MV* 65, *BL* 209). Often turning upon the presence, potency, and agency of animals, plants, insects, and chemicals, the cosiest of Christie's crime stories is an acute and relevant reminder of our ongoing power to take life in a world shared with vibrant entities and elements, the agency of which we often trivialise or discount.

Indeed, in "The Blue Geranium", witnessing "the gardener shaking up the cyanide of potassium" to treat a wasp's nest, Marple thinks "how like smelling salts it looks", comparing the caregiving work of the gardener and the murderous Nurse Copling (110). Throughout the story, this clue maintains its valency for both deaths, lingering in Marple's and the reader's minds even as the murder is resolved. Marple's anxiety over the scale of insect death in St. Mary Mead relocates larger and "graver troubles" to her own garden where wasps, whose nests may contain up to six thousand insects, annually die in great numbers when fumigated (Christie *MV* 65). The questionable morality of fumigating wasps with potassium cyanide began in Christie's "Wasps' Nest", where Poirot aligns extermination with murder and describes the process contemptuously as "another of your English sports!" ("Wasps' Nest" 409). As Christie's fiction displays, according to Bayard, "the difficulty of interpretation . . . deciding what to interpret" for "everything can be interpreted" in a text, so are we forced to consider all deaths in her crime texts as meaningful, suspending the notion of a scale of "trivial" and "graver" deaths until it is too late to forget the former (Bayard 69; Christie *MV* 65). Now witnessing the full impact of global environmental injustice, the terrifying decline in beings we term "pollinators", and the failure of legal systems to include the value of all lives in our justice systems, we might well ask, apparently with Marple, how such injustice begins close to home. What is the "village parallel" to trivialising the life of agentic beings globally (*BiL* 119)? What does ecological violence look like in the cosiest place of all, Miss Marple's garden?

This chapter argues that ecological readings of verdant Golden Age detective fiction may, like popular garden writing, explore how we live with killing on a daily basis and how the representation of such mundane violence, framed as "weed" or "pest control", engages, constructs, or contests the instrumentalisation and trivialisation of life in modernity. To the extent that the paradox of crime in the cosy setting of the garden engages or resists normalisation, such narratives provide an opportunity to go beyond cracking a case; they allow us, as in "The Blue Geranium", to question which takings of life we consider criminal and why. The resulting ecological story, interwoven with ironic plays on words such as "human" and "humane", as well as taxonomic play on human and more-than-human pests, challenges the reader to question how a long-accepted, mundane practice such as extermination, deeply interwoven with patterns of caregiving in the garden even now, impacts our humanity and the lives of other beings on a grand scale, "in their thousands" if not billions (110, 110, 109). We see how the low legal, philosophical, scientific, aesthetic, and cultural position of garden "pests" contributes to human exceptionalism and that the latter is also hierarchical, linked to other forms of elitism and exclusion (Marder "Garden"). As the gardens in *The Thirteen Problems* are then "gently" and expertly dead-headed, weeded, fumigated, and sprayed into order, they both engage and estrange the idea that some beings are "threats" and need to be "neutralized", "subdued", or exterminated (Christie "TMM" 309; Marder "Garden as Form"). This principle of extermination, "The Blue Geranium" and "Wasps' Nest" make clear, is mobile; it spreads from garden to house, applied liberally to people as well as insects and weeds in St. Mary Mead.

As Zygmunt Bauman states in *Modernity and the Holocaust*, "modern culture" itself "is a garden culture" (92). As we design ideal lives and perfect the "arrangement of human conditions", we distrust the diversity of the same in the human and more-than-human world (92). This logic is clear in the make-up of an instrumentalist social world where a chosen design "determines what is a tool, what is a raw material, what is useless, what is irrelevant, what is harmful, what is a weed or a pest. It classifies all elements of the universe by their relation to itself" (92). From this point of view, a being's relation to the garden "is the only meaning it grants them", the only one to be "tolerate[d]" and "all actions are instrumental" (92). Hence, "there are weeds wherever there is a garden. And weeds are to be exterminated. For the sake of the garden, pests must be kept outside society boundaries or killed" (92). Christie's fictional St. Mary Mead is a garden village chemically and socially armed in preparation for its borders and lawns to be breached. "Weed" and "wasp season", then, high summer in "The Blue Geranium", is a killing time that impacts plants, insects, and people whose vitality and animacy fails to contribute to the social plan (109). The tale brings us ever closer to linking human and environmental injustice, understanding how one intersects with another to create mutually supportive instrumentalist networks naturalising the destruction of life on earth.

As Monica Allewaert argues in "Insect Poetics", "although the temptation of our moment is to think ever larger (as the term and concept of the *Anthropocene* suggests) we might instead strive to think" of "the small and the slight", who compose and "recompose this world", offering "far less dominant interests a political parable" in which world making is shared with more-than-human beings, some microscopically small (Allewaert 325). This, Allewaert offers, is "especially necessary as we pass beyond modernity's ways of thinking of agency and struggle to develop new ways of conceptualising animacy, agency, power, organisation, and system" (325). Currently, our legal understanding of the latter discounts all but a few insects' rights to exist with the wasp remaining quite literally on the most wanted list of feared insects (its picture, for example, graces the front page of the British Pest Control Association website). The wasp is still popularly described as only an "accidental pollinator", an "enemy", a "greedy" seeker of "cheap meals", and a potential killer that it is one's "duty" to destroy (Osterloff, Keele "Pest Advice"). The wasp is often represented in popular culture or pest control discourse as a bad bug, who is distinct from the most "beneficial" bug, the bee, nearly always characterised as an "important", "helpful", or useful insect, as of course it is, although it is only one amongst many, including wasps, who pollinate plants and contribute to life on earth.[1] "People often ask", states the nation's most popular gardener, Monty Don, "What use are wasps? What are they for?" (138). The answer of course is that in addition to pollinating, many species of wasps are "carnivores that eat a large number of caterpillars and aphids" (Don 138). Without them, the world would be overrun with insects. However, the decline of the wasp has recently began to outpace the decline of bees in the United Kingdom, as noted in the seminal 2013 *State of Nature* report (11). Yet, the wasp remains feared, a selling point for pest control. The killing of social wasps is still as robust as it was in Christie's time even if the more charismatic bee is gaining ground in the popular imagination. Moreover, if one can no longer sign in with a chemist for an order of potassium cyanide to use as a pesticide, one may now walk in to almost any store in Britain, Europe, or the United States and buy potent pesticides which target wasps but have broad application to other species, such as bees.

Even those useful "pollinators", however, remain almost entirely *legally* unprotected in modernity; that is their right to life remains unprotected and arguments for their existence turn mainly upon their productivity, usefulness, and helpfulness as pollinators and friends of people. Scott points out that "law functions to underestimate the human and environmental costs of" what Rob Nixon terms the "slow violence" wreaked by pesticides, chemicals, and deforestation (Scott 485, Nixon 46). Moreover, law "more generally locates violence outside law", and frequently "legalizes" the deaths of people, plants, insects, and beings determined to be a threat or inessential (Blomley in Scott 489). Justice systems in modernity then do not "contain violence" but are rather complicit with it, trivialising concerns of people and living things who get in the way of neoliberal expansion or its performances (Scott 489). As turning points of cosy, verdant crime fiction then, Miss Marple's narratives perhaps particularly reveal the analogies she draws from the "trivial" issues she notices. These provide us with an opportunity to read the most mundane cultural practices such as pest control in the garden as an ongoing dynamic site effectively shaping life and death on a grand scale. Hence, Marple's insights eradicate the false binary between local and global that contribute to an uneven and exclusive justice system in modernity.

Thinking About Gardens

As plant philosopher Michael Marder notes, it is "extremely difficult to think about gardens" because "our minds are awash with positive, sentimental, and nostalgically inflected cultural associations with these cultivated, carefully manicured green spaces", such as those we associate with St. Mary Mead (Marder "Garden"). Many of us, myself included, have only too recently begun to question our own uses of pesticides in the garden, and the majority of us are aware that we, like Christie's characters, are legally entitled to "ward off external intruders" by "resort[ing] to herbicides, insecticides, repellants, and other poisons, unleashed as part of the global toxic flood we are living or dying through nowadays" (Marder "Garden"). While this toxicity seeps into soil, water, and the cells of all living beings, the designs of our gardens and lawns also cause a reduction of habitat and food supply to support plants, animals, and insects. Human design and human comfort rules in the garden where we often maintain the "rigid confines of what the garden guards, trying to ensure the purity of the species gathered there" in our flower, vegetable plots, borders, and lawns (Marder "Garden"). While "the undifferentiated chemical killers that affect everything and everyone on their path" are problematic, so are our very garden designs, the "stricture of controlled difference, which is the garden" that "relies on lethal indifference and non-differentiation for its consolidation" (Marder "Garden").

Such "lethal indifference" to biodiversity in the garden rarely looks like such. Pleasing colour combinations or a reduction in the garden to a small number of plants for aesthetic impact look like good taste rather than "intolerance" (Marder "Garden"). Many millennial gardeners are happily changing their habits, encouraged by popular organic gardeners such as Monty Don and efforts such as World Bee Day and the work of the Trusts for Wildlife and Bumblebee Conservation. They are opening impenetrable fences and encouraging meadows to grow over lawns. Others, however, are continuing to spray and "treat" plants and insects with chemical applications as a routine part of care giving in the garden. Plant growth, in the meantime, a form of vegetal animacy, is often understood as a force to be formed or eradicated,

> as though plants were mere materials in need of form, as though, that is, vegetal matter had no living forms proper to and co-emergent with it. The garden, then, is the form of form, a meta-order bent on preserving order as such, whether it guards its contents against the trespassing others or serves to consolidate the proprietary power.
>
> *(Marder "Garden")*

"Trespassing others", such as wasps, are sacrificed to this order and the gardener's pleasure which is expressed frequently in Christie scenes where gardeners survey their orderly gardens (Marder "Garden").

The animacy and presence of the wasp in those gardens is perceived as a particular issue in tales such as "Wasps' Nest" or "The Blue Geranium". In both tales, however, the brutality of wasp extermination is remarked upon as a flaw or threat from within the English garden and character. In the former, John Harrison "loved his garden, and it had never looked better than it did on this August evening, summery and languorous. The rambler roses were still beautiful; sweet peas scented the air" (407). The garden epitomises English gentility for him: "This is England", things like violent crime "don't happen like that here. Disappointed suitors don't go about stabbing people in the back and poisoning them" (408). But Poirot baldly asserts "the English are stupid!" (408). Agitated, he points out that lovely gardens "may conceal . . . hate" (409). Although Harrison thinks his fiancé's former suitor "wouldn't hurt a fly", Poirot assures him that is not the case. If the latter "would not take the life of one . . . you forget that he is even now preparing to take the lives of several thousand wasps" (408). Poirot points out that

> There will be destruction, and the [wasps] know it not. There is no one to tell them. They have not, it seems, a Hercule Poirot . . . I tell you, Monsieur Harrison, I am down here on business. Murder is my business.

Characteristically straightforward, the male detective baldly states, and accepts, that killing is a brutal business, an "English sport", despite the species of the victim (409).

Written shortly after "Wasps' Nest", "The Blue Geranium" reveals Christie, like Marple, continued to find wasps intriguing in the late 1920s. Marple's detective work, proceeding seemingly more circuitously than Poirot's, puts wasps on our brain more subtly, revealing the questionable ethics of extermination and its ugly ironies, mass killing, for example, of social summer wasps despite their very short life cycle and usually on "a beautiful summer's day" at the height of their vitality (110). Marple's doubts plant a comparable uncertainty in the reader's mind about human "proprietary power" to weed out or fumigate undesirables (Marder "Garden"). In "The Blue Geranium", Christie, in fact, highlights the Tuesday night dinner club's instrumentalisation and trivialisation of life, stressing each member's callous indifference, drawing a parallel between the gardener's cruelty to wasps and human cruelty towards other humans. Because the business of crime fiction is also, of course, killing, tales featuring garden settings, practices, and discourses, may, as in "Wasps' Nest", reveal how such hierarchical ontological thought is fluid and mobile. It transfers from one practice or institution, pest control and gardening, for example, to another.

In "The Blue Geranium", however, with its dinner club detectives who also act as a jury of sorts and include representative elements of society (law, medicine, military, domesticity, respectability), the juxtaposition of insect and human death reveals the law itself to be instrumentalist, biased against some forms of "life" of no perceptible "use" to the good "sorts" of humans like those in the club (102). The text thus reveals the extent to which what we would now recognise as a hierarchical ecological mindset is deeply intertwined with a raced, classed, and gendered human exceptionalism which informs our conservation practices. At the same time, she exposes the link between our understandings of criminality, justice, and our definition of murder, to hierarchical patterns upheld and normalised as common sense or what "must be" (107). Both environmental action and criminal justice systems in the tale place the value of life on a hierarchical scale in which some lives matter more than others. This garden epistemology seems beyond question when Marple states of extermination, without conviction, "I don't like killing – not even wasps, though I know it has to be, and I'm sure the gardener does it as humanely as possible" (107). This concern about extermination questions the humaneness of human culture at its most *c*ultured, anticipating Rachel Carson's claim that by "acquiescing in an act [extermination] that can cause such suffering to a living creature, who among us is not diminished as a human being?" (58). Certainly, spoken frankly by the Tuesday Night Club, the judgements cast upon Mrs. Pritchard epitomise the "brutal" quality of the social gardeners of St. Mary Mead (102).

Most of all, the ironic intersection between multispecies caring and killing in both the garden and the sickroom introduces a disturbing awareness of the casual ranking of the value of life in "The Blue Geranium". A kind of human pest, surrounded by an anarchic wallpaper herbaceous border in her sickroom, Mrs. Pritchard experiences wasp season alongside insects in "The Blue Geranium". Her fear and pain are dismissed along with her very accurate assessment that "this house is definitely dangerous to me" (99). In the Pritchards' house and garden, the threat to women, weeds, and wasps intersects with their care and is marked on schedules and calendars which record the daily routines of the nurse as well as the gardener's seasonal extermination dates. The killing season in the garden spills over into the house, where along with Nurse Copling, even the virtuous detectives of the Tuesday Night Club are shown to be capable of violence against the annoying Mrs. Pritchard which they go so far as to call justice "with conviction" (96). The resolution of the crime offers more to us, therefore, than the restoration of order. In fact, the tale disturbs the narrative drive of detective fiction towards a conclusion which offers us the satisfaction of knowing who the killer is. When Marple's explication of the crime is prefaced by her admission that she has "wasps on the brain", the reader cannot help but turn back to those insect deaths in garden settings, asking who we are to determine the right of "pests" to live or die, even as Marple drives forward to solve the murder.

"You Might Have Been There Miss Marple"

My recent work on Christie's representations of plants, medicinal herbs in particular, as poison in "The Herb of Death" (1930) and *Five Little Pigs* (1939), argues for reading not just what plants mean in her plots, but what her wildly popular plots mean for plants in an age of rapid deforestation and what Randy Laist terms cultural "defoliation" (Carroll 90–110; Laist 11). In the case of medicinal herbs used as murder weapons in tales like "The Herb of Death" or *Five Little Pigs*, Christie's narratives often keep step with the heightened legal regulation of plants that caused the historical decline of herb farming and herbalism in mid-century England. The discourse of "good" and "bad", wholesome or dangerous, plants permeate those texts, complementing the anthropocentric perspective that naturalises pesticides in "The Blue Geranium". In all, however, the "glamour" of the natural world, its animacy and potency, lingers to challenge garden limits and legal or cultural boundaries (104).

Paradoxically then, Christie's garden ecologies play with the potency and constraints of plant and insect agency, framing both as life forms vulnerable to extermination by modern chemistry. The use of the latter, as in World War I where Christie, as an apothecary's assistant, became familiar with chemicals and former pesticides used in combat and with the discourse of human and insect "pests" as the "enemy", also raise new questions about how "extermination" of this ontologically slippery category of "pests" impacts us as people. As Makinen argues: "Christie's plots are successful precisely because she upheld that murder was not the regime of one aberrant individual who could be excised, because everyone in her novels is a potential murderer" ("Contradicting" 78). Christie's "version of the self in process proves both precarious and perilous, and, as such, anything but cosy", while her plots remind "the reader of the pathos and horrors of life alongside its instability" (Makinen "Contradicting" 79). Often these horrors are distinctly gendered, as when a nurse such as Nurse Copling in "The Blue Geranium" defies her gender role, inverting "her role in modernity, the domestic work of cooking and nurturing" or, indeed, gardening or caring, poisoning her victim with readily available household chemicals (Baučeková 44).

However, the gender-equitable practice of "extermination" Christie explores in "The Blue Geranium" makes human beings in general, rather than women in particular, seem "brutal" by nature, making the very term "humane" ironic (102, 107). "Most men" would have killed Mrs. Pritchard, Mr. Bantry notes. "Any woman on the jury", Mrs. Bantry adds, would acquit Mr. Pritchard of murder and even "reward" him for "braining [his wife] with a hatchet" (96). Nurse Copling, on the other hand, seems paradoxical: "a very good sort" who puts "up with Mrs. Pritchard's tantrums and nerve storms with complete indifference" (97). In one of many of Christie's plays upon words in "The Blue Geranium", a "sensible" nurse lacks sensitivity to the suffering of others (97). The human capacity for violence extends even to Miss Marple: "You might have been there, Miss Marple", Sir Henry notes, as she recounts her analysis of the crime, implying that even she might be the murderous kind (111). Pressing the boundaries of the Golden Age code of fair play, Christie asks us to consider the detective herself as capable of murder. Hence, with Gulddal, I would argue that the most interesting way to read Christie is to "bracket the ending" and read the body of her ironic texts closely to find the larger questions arising out of details of detection which point, in this case, to a discussion of how humane human beings are in general. Tracing Christie's "strong ironical undertone that calls upon readers to question everything, even the authority of the detective" or of human beings themselves is the reward of "The Blue Geranium" (Gulddal). When Miss Marple ponders how she would do it "if I were going to kill anyone", we realise what her mind, the human mind, is capable of, towards "anyone" and anything (107).

Pest Control: Insect and Human

When the Tuesday Night Club gathers to solve the mystery of how the "semi-invalid" Mrs. Pritchard died, they first discuss her character: "capricious, exacting, unreasonable" (96). Pritchard is assumed to be only "semi" invalid, and hence, half hypochondriac. She has an unfortunate "weakness" for the supernatural and spends her days in her embowered room (96). There she is surrounded by a paradisiacal garden: "one of those new wallpapers where you apply clumps of flowers to make a kind of herbaceous border . . . like being in a garden" (101). Mrs. Pritchard's herbaceous border is ironically borderless in many ways. It defies spatial and seasonal limits; flowers bloom out of season amongst companions not found in nature: "bluebells and daffodils and lupins and hollyhocks and Michaelmas daisies all grouped together", bloom simultaneously (101). In this garden setting, the ultimate expression of the gardener's longing for plants to be "mere materials", ultimately malleable and docile to human design, Pritchard feels especially unsafe (Marder). Life for nondesirables is precarious in this garden and its perfection contrasts sharply with Mrs. Pritchard's perceived aggression. She assaults her husband for apparently no good reason, striking out "unreasonably" when she seems unprovoked, curtailing or motivating his movements against his will. "What use is Mrs. Pritchard's life?" asks the woman who will supplant her, Jean Instow (102). While Jean is "good at all games . . . nice-looking, attractive-looking, very fair with a healthy skin and nice steady blue eyes", Mrs. Pritchard, like an errant insect, merely threatens, like an unwelcome wasp, to keep her husband from enjoying his golf game (102). Taking up space at the centre of her garden she refuses the cultural ideal of the good gardening woman which associates gardens with women's health, virtue, wholesomeness, reproductivity, and domesticity (Seaton 18).

Ensconced within her herbaceous border, moreover, Pritchard invites her unwelcome and purportedly useless friends to join her: fortune tellers and spiritualists. Zarida, Psychic Reader of the Future, is distinctly dark and even arachnid like with "black hair in coiled knobs over her ears" and "great black rims" around her eyes (99). She wears a "black veil over her mouth and chin – and she spoke in a kind of singing voice with a marked foreign accent – Spanish, I think" (99). Zarida's handwriting is "big and black", and she claims, correctly, as she is actually the murderer, Nurse Copling, in disguise, to "have seen the future" in which "the blue geranium means death" for Mrs. Pritchard (100). Zarida's dark colouring makes her suspect in the household who view her as an alien, aligning her with the role of the gypsy fortune teller often represented as a charlatan, kidnapper, or thief in British culture. Along with stealing Mrs. Pritchard's peace of mind, Zarida steals or appropriates the nostalgic tradition of "floral fortune telling games" away from a traditional courtship discourse ("he loves me, he loves me not") and towards a series of increasingly dark meanings which pervert the Victorian language of flowers: "the Blue Primrose means Warning; the Blue Hollyhock means Danger; the Blue Geranium means Death" (100). This floral discourse clearly redirects the language of flowers as it appears in such popular

English texts as Kate Greenaway's book of the same name (1884). There, of course, primroses, hollyhocks, and red geraniums signify youth, fecundity, and comfort, respectively (Greenaway 34, 52, 50). Wondering what confidence-game the fortune teller is actually playing, the household is distracted by this usurpation of the cosiest of discourses. Encouraged by Nurse Copling, they seek the "*meaning*" of Zarida's admonitions (100). Hence, they are ever more distracted from the real threat she poses to Mrs. Pritchard's life, along with the alliance she is forming with Mr. Pritchard against his spouse.

Key to the resolution of the plot is Marple's theory that Copling herself has played the part of Zarida the fortune teller; the nurse has cleverly applied chemicals to the wallpaper, pasting litmus paper on it, to change pink flowers to blue with a wave of the ammonia smelling salts Mrs. Pritchard habitually uses. But when these are filled with the pesticide potassium cyanide, and the victim inhales them in her fright upon seeing the uncanny blue geranium, Pritchard dies, ejected from the garden as a being who just seems to be in the way of everyone else's life. Marple points out that "the warnings and the blue flowers were, if I may use a military term", she laughed self-consciously – "*just camouflage*" (110). Theorising that Copling has changed her colouring as she has changed both the colour and meaning of Mrs. Pritchard's flowers, Marple darkens the nurse's character so that the "the real thing", her murderous intent, is revealed (110). Ironically, "wasp season", a season of death, has visited Pritchard's garden. Her room is, it seems, all too much "just like being in a garden" where her distinctly aggressive, disruptive, unbounded behaviour calls for her death. The joke, however, is on Nurse Copling, who is identified as a larger scale pest or threat attempting to remove her patient to woo her husband. In the Tuesday Night Club stories, her "kind" or type, unbound and sexually ambitious, are sometimes in need of "extermination" ("The Companion" 128).[2] The concept of "extermination" evolved in use to apply to "both human and insect enemies" during the inter-war period. At that time, the use of pesticides in chemical weaponry began to allow people to imagine "(and sometimes succeed in) annihilating [human] enemies" as if they were insects (Russell 1509). What set this time in the "twentieth century apart . . . was the scale on which people could plan and carry out killing . . . not just against armies, but against insects and civilians" (1509). As Christie was aware from her service as an apothecary's assistant in the Great War, now one could carry out and could imagine or "fear annihilation on a breath-taking scale across geographic and phylar boundaries" (1509). In Miss Marple's revelation of the crime, she links extermination to murder, apologising for her seemingly trivial concerns. However, the reader of the Miss Marple series knows that it is precisely when she apologises for her brilliant insights that they are the most acute.

Poirot speaks frankly of extermination in the English garden as a macabre kind of sport in a masculine language not accessible to Marple. However, the gentle, nostalgic, and sentimental languages of flowers and gardens, as well as the experience of women "being in a garden", once constituted "a properly edited form of nature for women" that nonetheless historically involved pest control (Seaton 101, 17; Smith 100). The gardening woman showed with "conviction" through her love that she was "pure in her tastes and amiable in her disposition", but always as Smith shows, she had to reconcile that with her destruction of garden interlopers, "garden pests" (Seaton viii; Smith 99). If the "love of flowers is a touchstone of true femininity, an outward expression of woman's finest feelings", it also ensures a garden's seasons of death (Seaton 18). Mrs. Pritchard's presence troubles paradise and must be removed. Nurse Copling/Zarida must go as well. When the language of flowers is restored at the end, it brings with it "a wide range of ideas needed to conduct relationships between the sexes leading to romance and marriage", in Victorian style (Seaton 66). But Mr. Pritchard is revealed to be cruel, "pigheaded", as he "shouts" at his wife and mocks her (Christie 104). Both Mrs. Pritchard's and Nurse Copling's unwelcome presences are edited out of the garden so that another woman, fair, even with "steady blue eyes" to counteract Mrs. Pritchard's seemingly hysterical fears of the colour blue, may enter. Other ambitious nurses are under surveillance by the end of the tale as well. Their kind is suspect within the design of St. Mary Mead itself.

If "all good women love flowers" and the "love of flowers was to purify [a woman's] soul" within the Victorian discourse of flowers (Seaton 18), Copling's modern bastardisation of the same reveals her criminality and makes others like her suspect. Copling's appropriation of the language of flowers startles in its darkness, complemented by her dark hair and eyes, a sign that some "foreign" influence is present in a most English village, the traditional floral iconography of which is being stolen and weaponised by a decidedly modern criminality that seems to play against proper laws of attraction found in nature where the colours of flowers have distinct meanings for the insects which are drawn to them. If, as Beverly Seaton writes, "floral symbolism is quite strongly rooted in the specific culture and its ideals" and "every age writes its own language of flowers in its literature and its customs", then Christie's perversion of the language of flowers in "The Blue Geranium" shifts the code's nostalgic historical associations with Victorian women to intensify the modernity of the sexually and professionally ambitious Nurse Copling (39, 60). Indeed, Nurse Copling is yet another of Christie's suspect professional women who seem to defy established laws of nature in which women were always domestic, neither desirous nor professionally ambitious. As unnatural herself as "blue primroses and blue geraniums", Copling, whose name sounds a great deal like "coupling", thwarts the traditional stereotype of woman as garden queen and becomes an exterminator who is herself exterminated.

By the twentieth century, the language of flowers "was a tired, old-fashioned notion, surviving into the early twentieth century in occasional amateur or commercial applications" (Seaton 84). As in "By the Pricking of My Thumbs", in which a sentimental Victorian gemstone ring spells out "Regard", establishing clues to an older woman's character, in "The Blue Geranium", the usurpation of the Victorian language of flowers relies upon the perversion of a still well-remembered discourse to establish at once the silliness of both Mrs. Pritchard and the gypsy fortune teller. Both are then pushed more deeply into the ontological category of "unreasonable" women through association. When pink primroses and hollyhocks, along with red geraniums, turn blue, they change their homely meanings. As the flowers change colours, the lexicon is destabilised. Plants become agentic and no longer legible to people – in fact they behave as many plants often do, changing colour in relation to seasonal changes or the chemicals they absorb. Miss Marple's identification of the poison and the chemistry which allowed Nurse Copling to terrorise Mrs. Pritchard effectively levels the Tuesday Night Club ontologically, reminding them of uncomfortable intimacies with vegetal bodies. The solution to the crime underscores the realisation that human and plant bodies share chemistries; our body fluids may also be acid or alkaline, our nature is linked to vegetal nature, ultimately transforming into the stuff of the garden, soil itself. This point is driven home by the exhumation of Pritchard's body from the ground. Miss Marple notes this shared chemistry when she points out, from her nursing background, that blue litmus paper turns "red with acids, and red turns blue with alkalis" (111). Reducing human bodies to their chemistry is "not a very pleasant subject" that turns Marple herself "delicately pink", reminding her audience of their own mutable natures (111). The gardens in "The Blue Geranium", real, artificial, and social, are all vulnerable to chemical treatment, disproving what reader and characters think they know about their conception of plants or pests as trivial, fixed, passive beings ultimately different by nature from people. This intimacy with the natural world is not, in Christie fashion, contained to the murderer but liminal.

While characters throughout the tale often seem less than human, brutal, pigheaded or insect-like, Marple flattens such distinctions overall, maintaining that humans are not exceptional creatures, for with "human beings . . . things sometimes happen" (109). Destabilising human subjectivity and lowering people onto a horizontal plane with other beings, she flattens ontological hierarchies themselves when house and garden, people and plants or insects change places. The murder victim's husband, and the likely sympathetic reader, initially see such intimacy as absurd, a joke, coming "out of the blue", and going nowhere (106). Marple, however, spots the shared vulnerability of both humans and insects and what once seemed just a wallpaper garden is by the end of the tale a family portrait of a close ontological relation. "The Blue Geranium" turns upon the shared precarity of multiple life forms. If the resolution of the crime makes this mutual precarity clear, it also encourages us as ecocritics to follow Marple's logic beyond our extant laws and ask if "humane" methods of extermination in the garden may not be just plain murder after all.

"I Dislike Killing"

The restoration of law and order at the end of "The Blue Geranium" is seemingly quite tidy. Larger questions, however, like what it means to be a "human being" or to have power over other living things, remain. Throughout "The Blue Geranium", Christie plays upon the darkly ironic capacity of people to consider themselves "humane" while callously justifying the deaths of people, plants, and insects. This is in keeping with the text's larger thematic critique and exposure of a narrative form of camouflage that obscures or trivialises our shared precarity. Perhaps in that case, "The Blue Geranium" reveals a discussion of justice distributed unevenly in the carefully kept garden that is human society. The tale exposes human brutality and callousness on nearly every page.

As a detective tale begins with a dead body, in this case Mrs. Pritchard's, the reader must pay particular attention to accusations of "unkindness" on the part of the victim's husband (99). When Mrs. Pritchard does die, seemingly of fright at the uncanny appearance of the blue geranium, young Jean Instow "cooly, in so matter a fact tone" asks "Well, that might be all for the best, mighn't it?" In a "brutal and outspoken" fashion the girl explains:

> You don't like my saying that – but it's true. What use is Mrs. Pritchard's life to her? None at all; and it's hell for George Pritchard. To have his wife frightened out of existence would be the best thing that could happen to him.
>
> *(102)*

Although Mrs. Pritchard correctly perceives that the "house isn't safe", her husband refuses to allow her to leave and accuses her of "enjoying herself" (104). He becomes a suspect when a member of the household overhears her saying: "Very well . . . when I am dead, I hope everyone will realise that you have killed me" (106). And

> as ill luck would have it, he had been mixing some weed killer for the garden paths the day before. One of the younger servants had seen him and had afterwards seen him taking up a glass of hot milk for his wife.
>
> *(106)*

If the room Mrs. Pritchard lies trapped in is just "like being in a garden", it makes her vulnerable to being weeded out and this metaphor is not lost on the reader who has come to perceive Mrs. Pritchard as a pest who perhaps deserved to die.

While given multiple insensitive characters' motivation and capacity for murder, Marple joins them by indicating she can think like a killer herself:

> If I were going to kill anyone – which, of course, I wouldn't dream of doing for a minute, because it would be very wicked, and besides I don't like killing – not even wasps, though I know it has to be, and I'm sure the gardener does it as humanely as possible. Let me see, what was I saying?
>
> *(107)*

Marple's distraction from her task proves the extent to which she has the ethics of extermination on her mind while the irony of the "humane" garden exterminator resonates given the tale's focus on human beings' insensitivity, brutality, and callousness. "Prompting" Miss Marple to return to the perceived "graver" death of Mrs. Pritchard, Sir Henry Clithering, Ex-Commissioner of Scotland Yard, commends her: "you frighten me" (108). Clithering knows, as does Marple, "the extraordinary things people" do, hinting at a universal human capacity for violence (95). As Marple puts herself "in the place" of the murderer, she aligns chemical reactions "and the wasp season and everything", resolving the tale so accurately she "might have been there" as the murderer (111). Marple too then seems conflicted, potentially callous, unconcerned about the loss of Mrs. Pritchard's life: "What worries" her is not the victim, but "poor Richard and that nice girl" who wished his wife dead, "Miss Instow. Probably both suspecting each other and keeping apart – and life is so short" (111). Her assessment of human nature at the end is stark after learning of another murder by potassium cyanide: "[it] shows how much wickedness there is in the world, and that if once you give way" anyone may become a killer (111).

Because "The Blue Geranium" allies extermination and murder both as a kind of pest control, we sense that violence simply "has to be" accepted to maintain the English village and more widely, England itself. It comes as no surprise that the language of "extermination" appears again in *The Thirteen Problems* in reference to a host of "certain persons . . . unfortunate females" who don't fit the social design ("The Companion" 128). However, as Marple's concerns and her analogous logic push us as readers towards the resolution of the crime, they also estrange us from human exceptionalism and the normalisation of extermination particularly against those insects we continue to vilify, mark, and destroy with impunity. Grounding us in our own chemical natures and vulnerabilities, "wasp season" in "The Blue Geranium" points out that we share precarity with the smallest beings who build the world we claim to design. If insects do outlive us and the Anthropocene is followed by a Plantocene or Insectocene, those beings will undoubtedly carry the chemical legacies we have left behind.

In the tale, finally, the "*meaning*" of things as signs – flowers, gardens, wasps – is "*just camouflage*", superseded by the larger question of the meaning of the lives and deaths of such things themselves. This shift anticipates ecological thought. As Michael Marder writes, when we consider what "appears to be meaningless and obscure to us", something like a wasp, that being, however small, "becomes meaningful as soon as we try to imagine, at the edge of our imaginative capacity" its perspective as a

> being unconcerned with symbolic meanings. The old questions about the meaning of life should as a result give way to questions about the meanings of lives (both human and non-human) that arise, practically and concretely, from the heterogenous, vivacious action of every single creature including a plant

or an insect (35). Continuously, in "The Blue Geranium", Marple is pulled back from this imaginative edge where she is actively applying her analogic logic to consider beings whose right to exist is not recognised, formally or informally, by the law. Representatives of the law then have to ask her what she is thinking: "'Come now Miss Marple', [Sir Henry Clithering] said persuasively. 'You're lost in a daydream. Won't you tell us about it?'" (108). When Marple returns from her "daydream", she brings with her an expansion of the ontological prejudices of legal protection. From the perspective of ecological crisis, Golden Age fiction, which makes silent, pest-free embowered settings into killing places, can make the cultural status quo of extermination strange, even criminal, reducing the scale of the Anthropocene to something quite local and revealing the value we resist placing on all forms of life. Indeed, the very definition of a life that matters, we see in "The Blue Geranium", is limited by our anthropocentric attitudes deeply culturally embedded in our language and storytelling practices. Only by re-reading these for the meanings of all lives can we hope to create a more just ecological future.

Notes

1. See "Wasp or Bee?" on the 2021 website of the British Pest Control Association: "The way we treat bees and wasps are very different. Bee's [sic] are an important, beneficial species rarely considered a pest. Professional pest controllers only treat bees [sic] nests if it's a significant threat to public health as last resort" ("Wasp or Bee?").
2. In Christie's *The Thirteen Problems*, the original collection of the Tuesday Night Club tales, Christie's "The Blue Geranium" is followed by "The Companion". The tale centers the "extermination" of an "unfortunate female" with a "past" (*Miss Marple* 128).

Bibliography

Allewaert, Monique. "Insect Poetics: James Grainger, Personification, and Enlightenments Not Taken." *Early American Literature*, vol. 52, no. 2, 2017, pp. 299–332, doi:10.1353/eal.2017.0027.

Baučeková, Silvia. "The Flavour of Murder: Food and Crime in the Novels of Agatha Christie." *Prague Journal of English Studies*, vol. 3, no. 1, 2014, pp. 35–46, doi:10.2478/pjes-2014-0016.

Bauman, Zygmunt. *Modernity and the Holocaust*. Cornell UP, 1989.

Bernthal, J. C. *Queering Agatha Christie: Revisiting the Golden Age of Detective Fiction*. Palgrave, 2016.

Carroll, Alicia. *New Woman Ecologies: From Arts and Crafts to the Great War and Beyond*. U Virginia P, 2019.

Carson, Rachel. *Silent Spring*. Houghton Miflin, 2002.

Christie, Agatha. *Murder at the Vicarage*. Harper Collins, 1982.

———. *The Body in the Library*. HarperCollins, 2002.

———. "The Blue Geranium." *Miss Marple: The Complete Short Stories*, pp. 93–111.

———. "The Companion." *Miss Marple: The Complete Short Stories*, pp. 112–133.

———. *Miss Marple: The Complete Short Stories*. William Morrow, 2011.

———. *Hercule Poirot: The Complete Short Stories*. William Morrow, 2013.

Don, Monty. *Down to Earth*. DK, 2019.

Gulddal, Jesper, editor. *Criminal Moves: Modes of Mobility in Crime Fiction Book*. Liverpool UP, 2019.

———. "'That Deep Underground Savage Instinct' Narratives of Sacrifice and Retribution in Agatha Christie's *Appointment with Death*." *Textual Practice*, vol. 34, no. 11, 2020, pp. 1803–1821.

Keele University. "Wasps." *Keele University*, www.keele.ac.uk/arboretum/aboutthearboretum/articles/wasps/.

Laist, Randi, editor. *Plants and Literature: Essays in Critical Plant Studies*. Rodopi, 2013.

Makinen, Merja. *Agatha Christie: Investigating Femininity*. Palgrave Macmillan, 2006.

———. "Contradicting the Golden Age: Reading Agatha Christie in the Twenty-First Century." *Criminal Moves Book Subtitle: Modes of Mobility in Crime Fiction*, edited by Jesper Gulddal et al., Liverpool UP, 2019, pp. 77–92.

Marder, Michael. "Michael Marder the Garden as Form." *The Learned Pig*, 22 Jan. 2019, www.thelearnedpig.org/michael-marder-garden-as-form/5821.

Nixon, Rob. *Slow Violence and the Environmentalism of the Poor*. Harvard UP, 2011.

Osterloff, Emily. "What Do Wasps Do?" *Natural History Museum*, 1 June 2021, 5:00 pm, www.nhm.ac.uk/discover/what-do-wasps-do.html.

"Pest Advice for Controlling Wasps." *British Pest Control Association*, 1 June 2021, 5:00 pm, bpca.org.uk/a-z-of-pest-advice/wasp-control-how-to-get-rid-of-wasps-bpca-a-z-of-pests/188976.

Russell, Edmund P. "'Speaking of Annihilation': Mobilizing for War against Human and Insect Enemies, 1914–1945." *The Journal of American History*, vol. 82, no. 4, 1996, p. 1505, doi:10.2307/2945309.

Smith, Elise L. "Garden Pests and the Inculcation of Virtue in Early Nineteenth-Century England." *Nineteenth Century Studies*, vol. 26, no. 1, 2012, pp. 99–116, doi:10.5325/ninecentstud.26.2012.0099.

"State of Nature." *Royal Society for the Protection of Birds*, 2013, ww2.rspb.org.uk/Images/stateofnature_tcm9-345839.pdf.

"Wasp or Bee?" *British Pest Control Association*, 1 June 2021, 5:00 pm, bpca.org.uk/a-z-of-pest-advice/wasp-control-how-to-get-rid-of-wasps-bpca-a-z-of-pests/188976.

8

GREEN MACHINATIONS

Unknown Poison, Ecology and Female Criminal Agency in L.T. Meade's *The Sorceress of the Strand*

Caitlin Anderson

While plant literacy was a significant field of knowledge for both professional and lay medical practitioners in early nineteenth-century Britain, urbanisation and the rise of the medical pharmaceutical industrial complex saw a decline in the skill by the *fin de siècle*. As Randy Laist argues in his introduction to *Plants and Literature: Essays in Critical Plant Studies*, the devaluation of plant literacy, which lead to the gradual disappearance of advanced green literacy from culture, has been ongoing since the nineteenth century: "when one scans contemporary culture for evidence of plant-based narratives . . . the most dramatic meta-phenomenon is the defoliation of the cultural imagination" (10). If Wordsworth and Whitman "could rely on sharing a botanical vocabulary with readers", incorporating knowledge of local medicinal plants, by the 1890s, plant collectors and "hunters" sought to document, taxonomise, and extract plants from places like the Amazon River Basin as objects of proprietary interest. At the Great Exhibition and in local nurseries, the public saw that the flora of the world was far more complex, and sometimes more deadly, than they had realised. Plants in the Victorian period, however, might be interpellated into the purportedly less dangerous "language of bloom", where flowers became inextricably intertwined with female sexuality. In this discourse based on Linnaean taxonomy, "the bloom of a girl, like the bloom of a plant, is a description of a potential and ephemeral sexual reproduction", the value of which should be treasured for its short season (King, 35). When the cultural problems of plant illiteracy, the potential toxicity of imported plants, and the language of "bloom" meet in *fin de siècle* crime fiction, writers such as the under-studied L.T. Meade locate a potency and vitality in plants that challenges the role of women (and plants) as mere resources or possessions within a masculinist colonial culture.

Indeed, in L.T. Meade's illustrated *Sorceress of the Strand* series (1902–1903), plants enter the *fin de siècle* as vibrant, agentic tools in the hands of a compelling woman criminal, Madame Sara. As a popular phenomenon and contemporary of Arthur Conan Doyle's Sherlock Holmes, the character of Madame Sara taps into anxieties surrounding both the rise in plant illiteracy and the increasing regulation, professionalisation, and even criminalising of medicinal plants in modernity. She contributes, undoubtedly, to an already burgeoning "poison panic" which both feminises and greens crime in late nineteenth-century detection fiction by proliferating the image of a feminine poisoner who can neither be trusted nor stopped. As medicine was increasingly industrialised, herbal knowledge was lost, medical pharmacology was masculinised, and a purely ornamental botanical knowledge was feminised. This shift served to regulate social ties between women and plants and created an incongruous social phenomenon whereby the historical association between plans and medicine was replaced by a late nineteenth-century "poison panic":

DOI: 10.4324/9781003091912-11

a diffuse fear of secret poisoning perpetrated particularly by women (Hempel). At the centre of the fears of the "poison panic" lies the figure of the New Woman, who, in late Victorian popular culture, would often be represented as a criminal and served to feed cultural fears of women poisoners (Miller 145). The New Woman was suspect, entering and occupying masculinist public space and vocations, changing her features, and disguising her age or flaws with cosmetics. Through her connection with cosmetics, and her indeterminate age, Madame Sara is situated firmly within this paradigm of the New Woman.

Madame Sara's criminal work is highly specialised, aided and abetted by her knowledge of plants and her clever navigating of vegetal bodies. Placing toxic plant bodies into human bodies – a trans-corporeal act that shows human vulnerability to plant potency – and escaping detection through her ingenious use of plant camouflage, Madame Sara taps into a contemporaneous yet niche area of women's gardening culture. Namely, a renewed interest in herb gardening, amateur herbalism, and medicinal plants, as indicated by the publication of fashionable texts like Lady Rosalind Northcote's *The Book of Herbs*, published by John Lane in 1903.[1] As Alicia Carroll notes, early-twentieth-century women's culture explored the possibilities for women's professionalisation in plant literacy as well as a resurgence of a historical connection between plants and "magic", both dark and light. This led to women detailing the historical usage of herbs in "spells and sorceries", as well as other plants that "had power in themselves" (Carroll 11). Like such works, Meade's *The Sorceress of the Strand* modernises an alliance between plants and women that threatens male expertise and thus participates in the cultural phenomenon of poison panic which was inspired by a few notable criminal cases and capitalised on by writers such as Meade to drive the tension of the story.

A trained medical professional and capable herbalist, Madame Sara outsmarts the detective Dixon Druce through her deployment of plant literacy, signifying how, after a century of losing literal and metaphorical ground during industrialisation, plants reclaim space in *fin de siècle* culture through the hands of women. Manipulating the classic Victorian metaphor of woman as "bloom", Madame Sara allies with potent plants apparently never taxonomised by Europeans; she occupies a potentially posthuman space while enacting vegetal will and regenerating plant agency (King). As plant and human bodies blur together, Meade's crime fiction explores an anarchic erosion of such ontological boundaries, whilst simultaneously resisting the reduction of women's bodies and plant bodies to mere collectable objects. This act seems to both take advantage of contemporary fears of women becoming dangerous poisoners while subverting the neatly concluded narrative of the female criminal who is caught and faces judgement. Madame Sara, then, participates in popular culture while shifting the narrative to one of female empowerment, no matter the danger in doing so.

Madame Sara is a unique figure in serialised detective fiction, contrasting with Arthur Conan Doyle's Sherlock Holmes, the archetype of masculine intellect and power who most often names and drives the storylines of early detective fiction. In *The Sorceress of the Strand* stories, the female villain Madame Sara is unstoppable. She deploys her seemingly vast stores of knowledge, her intellect, and powers of disguise to outwit the less charismatic detective who pursues her. Her constant success, remaining free and affecting events, affords both women and plants a voice with which to occupy narrative space and take agency for themselves.

Indeed, L.T. Meade's "poison panic" tales reclaim earlier agentic roles for women apothecaries, such as the *venefica*, revitalising them in a period when plant-based poisoning was being increasingly linked to women in popular tales such as Vernon Lee's "The Legend of Madame Krasinska", Una Ashworth Taylor's "Seed of the Sun", and a range of both masculine and feminine poisonings in Arthur Conan Doyle's Sherlock Holmes stores (Hughes and Punter 500). As plant-based or mineral-based poisoning was considered a frightening perversion of women's gardening or food preparation – as well as a secretive form of violence that personified the perceived socialised passivity of women – it was particularly sensationalist, offering a double vision of a vibrant world of female and plant agency suppressed by male power structures, such as pharmaceutical regulation (Hempel). With famous female poisoners like Mary Ann Cotton, a convicted serial poisoner, and Florence Maybrick, a poisoner whose conviction was widely reported, the literary stories of poisoning in Victorian England were spurred on by actual cases of female killers. These few cases inspired widespread fear and, of course, fascination about the very nature of women and plants (Hempel). If, for example, "more than half of wives who killed their husbands" during the nineteenth century used poison, the purportedly benign connection between women, caring, domestic spaces, and gardens seemed suddenly questionable (Robb np). When Madame Sara instrumentalises potent herbs from mysterious gardens, she combines a dangerously unregulated vegetal will with an equally willful female body and mind. Her crimes blur plant and human bodies together, violating the boundaries of women, plants, and even the category of the human, resulting in an anarchic erosion of the ontological boundaries.

Contemporary ecocritics and plant philosophers such as Michael Marder have suggested that plants model an alternative form of agency that may augment the ecological problem of human transcendence and supremacy. Marder encourages ecocritics to imagine the vegetal world inhabited by agentic plants which engage in "plant thinking" and which occupy space as "collective beings" made up of the individual bodies of each plant, taking up commodified human space, and always seeking the collective survival of the species ("Resist like a Plant" 29). Plant-thinking and being, Marder argues, can be disturbing for humans to consider because of its alternative temporality, spatial occupations, and mobility. Marder suggests that plants, often seen as passive and unmoving in their rooted state, are not passive or unmoving at all. Thanks to a constant state of replication, growth, and proliferation, plants exist in an almost anarchic state, their will or plant intention on public display as they grow and replicate simply for the sake of growth and replication. Plants are therefore able to take up space "without appropriating it", pushing back on the British imperialism and capitalism of the nineteenth century (Marder "Resist Like a Plant" 29). "Life is de-centered" for a plant, and so vegetal life is able to grow around obstacles in all directions (Marder "Resist Like a Plant" 29). Because "vegetative growth knows neither an inherent end, nor a limit, nor a sense of measure or moderation", Michael Marder suggests that it may seem inherently "monstrous" ("Plant Soul" 87).

As in *Little Shop of Horrors*, when the fear of vegetal violence due to the growth and seemingly monstrous potential of plants crept into late-nineteenth-century literature, the agency of plants became something fearsome. The suspense of fiction, particularly crime fiction, is driven higher when fear is present. New Women in fiction, similarly, prey on cultural fears of women forsaking traditional roles and the ensuing chaos that such freedom could bring to society. L.T. Meade maximises both types of fear by creating a villain whose intimacy with (and knowledge of) plants imbues her with the power to disrupt the commodified world around her, particularly the marriage market. Madame Sara takes up space in that market, merging human and plant bodies to do so. In the tradition of the New Woman criminal, Madame Sara's beauty is beguiling and her badness appealing, drawing women to her as evidenced by the nickname of one of her victims: "Bee" (Miller). First introduced on board a ship returning from South America, Madame Sara acquires a specialist knowledge of foreign plants which allows her to distil their alkaloids into potent toxins. Her use of plants displays an anarchic energy that overcomes the men who seek to arrest and accuse her. Plants and women are able to find simultaneous agency through the body and actions of Madame Sara and, through her murderous intent, which stands in direct opposition to the portrayal of poisoning as a passive and feminine crime, overcome their cultural fetters and take agency for themselves despite all attempts by the male detective Dixon Druce to know either the villain or her plant compounds.

Madame Sara's Vegetal Violence

L.T. Meade's *femme fatale*, Madame Sara, embodies the concept of the New Woman criminal by seeking power through plant literacy and plant potency. The reader follows her through the tale's detective figure, Druce, a manager of an insurance company that oversees "trade", the financial conditions of British retail agencies, and men's interests in colonial products such as plants, mines and wealthy Brazilian heiresses set to marry British men. Far less charismatic than the villain he pursues, Druce deals straight from the middle class; as a corporate detective, he is tasked with protecting British business domestically and global capitalist/imperial interests. He seeks to protect masculine investments in stories in which Madame Sara seeks power over men, threatening their marriages and dowries, inheritances, and scientific discoveries: in short, their autonomy over various aspects of their lives and financial sources. Madame Sara, then, threatens everything that Dixon Druce and his friend Jack Selby stand for, both professionally as well as personally. When we first meet both men, they are represented as rightful inheritors of green global resources. Ensconced in the heart of green England in "The Meadows, our old family place, is now mine", for example, Dixon's friend Selby has returned from the verdant Amazon (Meade 3). As the scent of the gardens of Madeira wafts over him, he explains he has:

> ...a taste for natural history; that taste took me two years ago to South America. I have had my share of strange adventures and have collected valuable specimens and trophies. I am now on my way home from Para, on the Amazon.
>
> *(3)*

Druce himself has a dream to be "the owner of the best private laboratory in London" (3). Returning home from the Amazon herself, accompanied by two assistants, Madame Sara threatens male expertise and cultural mastery over resources, not to mention the chemical knowledge Druce seeks to obtain.

In the eponymous first story of the series, Madame Sara works to poison two sisters with a mysterious plant alkaloid so that their brother might receive the family inheritance. She is connected to people the brother owes money to, and so she ingratiates herself with the sisters and works to manipulate and kill them. The sisters and Madame Sara are introduced early into the story when Druce runs into an old friend, Jack Selby, who is returning to England with his new wife, Beatrice, and his wife's unmarried older sister, Edith. Selby explains to Druce that the "sisters have an acquaintance on board" the ship, Madam Sara, who is a "a professional beautifier" and who "declares that she can make quite ugly people handsome" (Meade 4, 5). The location of Madame Sara's power in consumer-based beautification blurs lines between public and private space, as she turns the bodies of women into her commodity and trade while still reifying the cultural value of women being tied to their "bloom". As Elizabeth Carolyn Miller notes, Madame Sara exists in a world that is firmly based in consumerism, thus aligning her with the figure of the New Woman. She is also a source of fear and threat to an insurance agent like Dixon Druce, for it seems she has the power to conceal female "flaws", such as signs of age or blemishes on bodies which should rightly be accessed and controlled only by men (Miller 71). Her job of beautifier is shown to rely on unfamiliar or secret plant-based tinctures, sources, and remedies to help her clients. Druce, however, seems incapable of understanding her motives or her work and glosses over all of it at the start of the story, focusing on the powerful figure she cuts. This power imbalance, with Madame Sara holding the upper hand in interactions with Druce, defines their relationship in this story.

For example, when Druce has a chance to interact with Madame Sara, she never lets him gain a solid footing. At first, she questions the meeting itself, asking him "was it really accidental?" (Meade 7). After pushing him for opinions on both her shop and her age, she invites him to visit, but also informs him: "I do hold secrets. I should advise you, Mr. Druce, even in your professional capacity, not to interfere with them" (8). As she tells him this, "the childlike expression faded from her face" and "there seemed to ring a sort of challenge in her tone" (8). The challenge and the loss of that childlike comparison make it clear to Druce, and to the reader, that Madame Sara is not a woman who seeks beauty for the approval of the men in her life. She has power, but it is a power that she does not feel compelled to share with men and so she covers up her own emancipated existence with beauty and a look of "childlike innocence", all of which fade into a confrontational "challenge" when Druce enters her "shop" (8). The surgical implements within speak to her ability to penetrate women's bodies for any number of purposes: beautification, dentistry, or abortion, handing women secret control over their appearance and reproductivity.

When Druce accompanies Edith to Madame Sara's shop, Edith introduces him to the space, saying "we are on the threshold of a magician's cave" (Meade 12). Like a *venefica*, Madame Sara's medical abilities are conflated with magic, a connection similarly established by contemporaneous writers such as Rosalind Northcote. Northcote devotes an entire chapter in *The Book of Herbs* to the history of herbs, exploring and citing why "some herbs were magical" (Northcote 176). By reconnecting herbs to magic, especially in a book written to "lady-gardeners", Northcote revives the historic connection between female healing and the supernatural, a conflation Meade also weaponises in the service of female empowerment. Through her healing, Madame Sara thus poses "a challenge to nineteenth-century medical professionalisation, a trend that had led to the dwindling of midwives and other traditional female medical providers in the Victorian era" (Miller 98–99). She resists the masculine professionalisation and herbal defoliation of the medical field, serving in a position almost like an apothecary. The actual features of her shop are very similar to the features of an apothecary's shop or the storefront of a pre-Victorian medical practitioner. She has "stoppered bottles" of medications, "brushes, sprays, sponges", fine tipped tweezers, and other tools one would expect to find in a shop where someone is dispensing medicines and performing minor surgeries and dental procedures (Meade 13).

Druce, the would-be chemist, enters through Madame Sara's "solid mahogany doors" and perceives a laboratory such as the one he covets himself. He describes her "array of extraordinary looking articles and implements", bottles of "strange medicaments", alongside "delicate needle-pointed instruments of bright steel", seemingly centred around a chair for "administering static electricity" where she kept "dry-cell batteries for the continuous currents and induction coils for Faradic currents" (13). These "strange medicaments" make Druce uneasy, his anxiety compounded by his inability to read or identify them. Rather than simply recognising Madame Sara's tools as those of a practised medical practitioner, Druce treats them like alien implements. This tool of femininity allows her to resist his "criminological gaze" through the body modification and appearance modification practices of women, once again aligning Madame Sara with the "New Woman Criminal" that was emerging in fiction in this period (Miller 27). As a distinctly "foreign" woman practicing on other women alone and without the supervision of men, Madame Sara evokes the codified language of the turn-of-the century abortionist. When Madame Sara threatens to disrupt Selby's inheritance of Bee's fortune, Druce fears that the "situation is a serious one in view of the monetary issues and the value of the lady's life" (13). These seem, however, to be one and the same. Druce reflects: "the more I saw of Mrs. Selby and her sister the more I liked them. They were quiet, simple, and straightforward. I felt sure that they were both as good as gold" (13). As a threat to two women about to share their resources with British men, Madame Sara is a dangerous force indeed, bringing not money to share, but Amazonian knowledge to retain herself in England. Druce mystifies Madame Sara's work, presenting it not as a legitimate medical business but as something foreign and unknowable, cueing the reader to her murderous potential as well as the power she holds within the story.

The atmosphere of Madame Sara's laboratory ties her medical knowledge to her knowledge of plants. As a parfumier, Sara scents her shop with the "scintillating" aromas she sells made from unknown, rich, imported flowers. On entering her laboratory through an imported "mahogany door", one's sensory experience is immediately controlled by her (Meade 13). The text playfully juxtaposes the various forms of plants, as both scents and as solid items, from the door to the "polished oak square table" that sets off Madame's instruments and, in a separate room, a "wooden operating table" (13). These wooden pieces, as monoliths of her work, solidify her presence and power both literally and metaphorically, as Dixon appears suitably horrified by her workspace despite the fact that Sara proclaims herself "innocent". Women clients, however, are drawn in – as "Bee" is – by the dangerous flower that is Madame Sara. The beautiful ageless Madame Sara acts as her own advertisement, offering women the irresistible prospect of lengthening and augmenting their own bloom, to become ageless, like herself. By using floral language to describe the maturation of a young woman's sexuality, her sexual "bloom" and the bloom of flowers became synonymous (King).

In texts of this genre, mentions of flowers or vegetal bodies in bloom are almost always a metaphor for female human bodies on the verge of sexual awakening or availability, at the peak of their potency and fecundity. However, Madame Sara makes women's bloom illegible to men in *The Sorceress of the Strand*. Madame Sara, her actions, her treatment, and her agency all put plants to work to benefit women's chances on the marriage market, which she herself avoids. Madame Sara, defined by her female criminality, embodies social change. Her existence is a subversion which responds to the pejorative social construction of female autonomy as dangerous or criminal. If freedom means a life of crime, then a criminal Madame Sara becomes. This powerful new form of women's agency through literary rebellion emerged alongside new understandings of plant potency. Even her physical body is plant like; in a key illustration (Figure 8.1), Sara seems to meld into an exotic potted palm while threatening Bee's sister Edith. The mystical and uncanny agelessness of Madame Sara seems to mirror a perennial plant which prolongs bloom, regenerates, and re-blooms every year.

The connection between the two is proven in this first story, when Madame Sara is revealed to have access to, and knowledge of, plants that are entirely beyond the scope of the men in power. Against the rule of fair play later developed in Golden Age detective fiction prohibiting the use of unknown poisons, Meade's tale depends upon the widespread rise in an understanding of global plant illiteracy, further characterising Dixon's and friends' ignorance and Sara's expertise as well as characterising the depths of Amazonian or global plant biodiversity yet not understood by Europeans (Knox). When Madame Sara finally completes her control over Beatrice, killing the young woman, she does so by using a plant alkaloid, "a powerful poison, unknown to European toxicologists" (Meade 22). The poison, used to remove Bee and her sister from the line of inheritance, is "very like hyoscine, one of the worst toxic-alkaloids known" (34). From "the old home of the Selby's, nestling amid its oaks and elms", Dixon and Vandeleur – a local medical jurist – are at a loss when it comes to identifying the plant at the root of Madame Sara's criminal intent (34).

The closest that Dixon and Vandeleur can get to understanding the source of the poison that Madame Sara has created is to compare it to what they know, hyoscine, a second metabolite of plants from the nightshade or Solanaceae family (Muranaka et al.). This means that the plants from the Solanaceae family produce a handful of substances that work secondarily to their growth. Some protect the plant from predators and others are used in intra-species plant communications. In this family of plants, though, known as the nightshade family, one of the substances produced secondarily to their growth is an alkaloid known as hyoscine. This alkaloid production is so intrinsic to the plants that the name of hyoscine was inspired by the naming of some of the plants in this family, namely *hyoscyamus niger*, commonly known as henbane. Just as *atropa belladonna*, deadly nightshade and henbane's cousin, is defined by the alkaloid atropine, so too does the presence of hyoscine define *hyoscyamus niger*. This alkaloid, which Vandeleur defines as "one of the worst toxic-alkaloids known", was described in Robert Eglesfeld Griffith's *Medical Botany* as "a valuable sedative and narcotic, when administered in small and repeated doses", fatal in larger doses (485). The alkaloid that Madame Sara uses, however, is not actually hyoscine, stepping outside of what her pursuers know. In naming what they know, hyoscine, Vandeleur and Druce must simultaneously reveal what they do not know, handing all the epistemological power to Sara. This unknown plant alkaloid, even more powerful than hyoscine and beyond their reckoning, hidden in the teeth both Beatrice and Edith had treated by Madame Sara,

Figure 8.1 Page 394 – illustration from *Strand Magazine* vol 24, 1902 – "The sorceress of the Strand" by L.T. Meade and Robert Eustace; illustrator Gordon Browne, RBA.

Source: Available in the public domain. Courtesy of Wikisource.org.

is released in the act of mastication, eroding the plant-based sap, gutta percha, placed over the poisonous substance within the tooth. Vandeleur saves Bee and ascertains how Madame Sara delivered the poison, but she "charms" the judge and eludes conviction, blaming the incident on her Brazilian assistants. As Meade withholds the name of the plant, Madame Sara maintains her "mystery" and may return to strike again. "Mercy on us!" Selby exclaims, "Is this a civilized country when death can walk abroad like this, invisible, not to be avoided? Tell me, Mr. Vandeleur, what I must do" (Meade 34). Vandeleur, however, cannot promise a full identification of the dangerous plant that kills Edith and threatens the more unfortunate Bee who loses only her tooth and not her life.

Engaging the newly piqued cultural understanding of the vast array of toxic plants from the Amazon River basin, Madame Sara exceeds the boundaries of woman's knowledge while her mysterious poison plant also eludes capture by the chemist Druce, who seeks himself to one day exceed his vocation as corporate detective. Early on when Druce tours Madame Sara's workplace, she tells him that "knowledge is power" before letting him see what she confesses to be her "secrets" (Meade 13). What Druce sees is, to him, dark chemistry, illegible through reason or his existing skill set. Later, when Madame Sara is near tears and asking for information about the death of Beatrice, her "brimming tears" suddenly dry "as though by magic" when she discovers she will not be able obtain any more information from Druce (25). Alternately expert and "uncanny", everything about Madame Sara seems shrouded in magic and intrigue, as is the field of plant literacy, which Dixon remains unable to master.

Like her age, Madame Sara's knowledge is not detectable to Druce. Exiting the detective narrative with her plant secrets intact, Sara acts, in a sense, like a plant. She disrupts commodified space with her anarchic presence, resilience, and distinct skill at disrupting commerce to take resources for herself, in this case the marriage market in which Bee is supposed to be only attracted to her husband who alone benefits from her resources. As Sara enters male space, telling Druce, "I am a doctor", she affronts the limits of women as well as the limits of modern pharmaceutical knowledge, suggesting a range of biodiversity not yet understood or commodified (13). Time itself seems unstable, as Meade's *The Sorceress of the Strand* modernises an alliance between plants and women that was once only associated with early modern culture, suggesting increased agency though this partnership rather than through old images of sexual passivity and bloom. Masculinist expertise seems not to have progressed, moreover, to conquer the plant world whose intricacies remain mysterious or unknown. In contrast, Madame Sara's vegetal knowledge stands for new horizons in which women and plant life have unknown capabilities and in which plant literacy and understandings of biodiversity become modern forms of a certain type of criminal agency.

Herbal Growth for the *Sorceress of the Strand*

In "The Talk of the Town", Druce continues to be "haunted" by Madame Sara and decides "hunting her" is good "recreation". This time Druce and Vandeleur find her attached to "a case of poisoning", the agent being "some poisonous alkaloid of the erythroxylon group", or the coca plant, indigenous to South America (Meade et. al.). Vandeleur is able to save Professor Piozzi's life, but Vandeleur and Druce must keep him protected so that he can lecture on his scientific breakthroughs. This poisoning represents how seriously Madame Sara perverts traditional female roles such as motherhood. Of all the items of Piozzi's breakfast to poison, she poisons his milk. It is in this wholesome, feminine beverage whose purpose is to nourish, that Madame Sera places a "poisonous alkaloid of the erythroxylon group" (Meade et. al.). Cocaine and derivatives of its herbal cousins were used in Victorian and *fin de siècle* medicine as a quick and effective analgesic, especially when surgeries were to be performed on mucous membranes (Small). Though the abuse of it was not a foreign concept, cocaine and its herbal cousins were primarily tools of healing which allowed medical professionals, such as Madame Sara, to treat issues with the nose, sinuses, or mouth without the patient having to be etherised. Yet, to Vandeleur this is simply a "poisonous alkaloid". Similar to the alkaloid in "Madame Sara", Vandeleur relies on the scientific or Latin names for the substances, seeming to obfuscate their herbal source in professionalised language that served as a comprehension barrier for the laymen. Their easy use of expert language, however, belies their gap in plant literacy despite Druce's collection of specimens in the Amazon. Madame Sara possesses the plant knowledge "unknown" to the male detectives, exposing their ignorance despite their extensive knowledge of Linnaean taxonomy. Like other women studying medicinal plants at this time – Alicia Amherst, Rosalind Northcote, Agnes Arber – her plant knowledge comes through her own experience and study of them, not from masculine systems of expertise (Carroll NWE). Sara's skill is discounted as "magic" by Druce; however, it is simply plant chemistry to Sara.

While Madame Sara's first plant attack fails, her second dastardly attempt again reveals the failure of male 'experts' to identify plants. As Piozzi stands to give his lecture, he faints. Piozzi is rushed off, after which Vandeleur, with an almost Holmsian flourish, promptly reveals that the man has been poisoned:

> "You see this", said Vandeleur, pointing to the great palm that towered over the table at which Piozzi had stood. "And you see this", he repeated, seizing one of the branches and shaking it.
>
> The long, tapering, green leaves rattled together with an odd metallic sound.
>
> "Look here!" said Vandeleur, and he pointed to the fine tips of one of the leaves. "This plant never grew. It is made – it is an artificial imitation of the most surprising skill and workmanship.

> The pot in which it stands has certainly earth at the top" – he swept away a handful – "but there below is a receptacle which is generating carbon monoxide gas."
>
> He bent and broke one of the branches.
>
> *(Meade et. al.)*

In this scene, Vandeleur not only reveals Madame Sara's expert knowledge of plants, but also his own disregard and violence toward vegetal life. Before Vandeleur even has confirmation that this plant is artificial, he is "seizing one of the branches and shaking it". Vandeleur seems to enact on this plant the very violence that he wishes he could inflict upon Madame Sara, as revealed when "he ben[ds] and [breaks] one of the branches". Although this final snapping of a branch allows him to reveal that it is hollow, he has already discovered that the plant is artificial, so it can equally be interpreted as a sublimation of the violence Vandeleur would like to inflict upon both the plants and the woman which elude him.

In choosing an artificial plant body to camouflage her final attack on Piozzi, Madame Sara foils her hunters again. The pot in which it stands has "earth at the top", but carbon monoxide gas for roots. Even as a technically lifeless thing, this plant is potent and enacting agency as it delivers gas to choke Piozzi. Once again, Madame Sara has used a plant body to invade and penetrate the human body, perversely reversing the physiological function of plants to make oxygen by making the artificial plant a vehicle for carbon monoxide. Suddenly, Piozzi and Vandeleur are victimised in a professional setting by the very objects they seek to know and possess. As Sara crosses gendered lines professionally, she invades their very bodies through the vehicle of her plant knowledge. A small and focused amount of the gas is used, and yet Piozzi is compromised, his masculine prowess effortlessly and openly diminished.

Were this social offence not enough, Madame Sara seems to push back against history, turning Victorian plant hunting, collection, and the rise in botanical knowledge, all signs of masculinist imperialist superiority, around. In fact, "it is no exaggeration to say that the plants that surrounded Victorians were often transported from foreign soil and almost always modified by human actions to form a second, cultivated nature" (Chang 1). This "cultivated nature" defined the Victorian relationship with vegetal bodies as plants became de facto members of the household structure rather than wild growing bodies. This focus on cultivation and domestication brought plants into social focus, but the focus became insular as culture dictated the popular plants to grow. Indeed, Sara takes advantage of an escalating increase in British biodiversity through colonial collection practices alongside the average detective's poor plant literacy and fools them with her fake plant so that only through close study, and too late at that, can the men involved even recognise that the plant is not living. In a culture where plants were imported, cultivated, supported, and became points of pride for the growers, such "alterations are personally and socially significant" (Chang 10). Madame Sara transforms something that seemingly represents passivity, a plant which seems to the empire to do little more than exist and grow, into a tool for killing. Like the imaginary killer alien plants of science fiction stories, Meade's fake plant is given deadly vegetal agency which perverts the plant's purpose in life. The plant, something which produces oxygen naturally, has been fabricated and transformed into something which produces a deadly gas. Like Madame Sara, a woman and a healer who kills, the plant has become a deadly anathema of what it socially and ecologically should be. This deadly act is, of course, carried out at the will of Madame Sara and through the potency of plants which, again, turns the discourse of bloom around, placing women and plants in agentic positions. In the end, Madame Sara lives to plant again, escaping prosecution and heightening anxieties about New Women and plants, two seemingly unknowable and uncontainable agentic beings.

Certainly, the *Sorceress of the Strand* series is a site where such anxieties can coalesce pleasurably, to delight the reader. Sara, a woman who is unnaturally young and beautiful despite being of an advanced age, disrupts the rules of the marriage market in which men seek to ascertain a certain value in a bride. Her bloom should assure the buyer of her youth, purity, and beauty. As men should be the assessors of bloom, women should subject themselves to it without, in a sense, talking back. Despite this elaborate, cultural vegetable code and the knowable world of Linnaean taxonomy, agentic plants and women prove themselves illegible to detection, endlessly fascinating, and potent.

In this situation, Madame Sara's education and practices provide her with "skill" and knowledge, as well as a kind of kinship with plant bodies. With the power to sprout imitation plants, along with her narrative power to elude capture (and return to grow more narrative), Sara herself seems to become plant, establishing a fascinating kinship with their mode of becoming. Because "vegetative growth" possesses this power to perennialise, regenerate, or return, often apart from human intervention or assistance, plants know "neither an inherent end, nor a limit, nor a sense of measure or moderation", and may be perceived, Michael Marder suggests, as inherently "monstrous" (Plant Soul 87). Indeed, Marder suggests that plants, despite their rooted state, are not passive or unmoving at all. Thanks to a constant state of replication, growth, and proliferation, plants exist in an almost anarchic state, their vegetal intentions or will on public display as they grow and replicate for the survival of their species. Plants are, therefore, able to take up space where they like, pushing back on the British Imperialism of the nineteenth century (Marder "Resist Like a Plant" 29). The vegetal anarchy of taking up this space seems to echo the sexual anarchy of the New Woman when she occupies men's public spaces. The interests of plants and women align well enough that they become one another in the form of Madame Sara. "Life is de-centered" for a plant and so vegetal life grows into Madame Sara, occupying her person without appropriating her will or agency (Marder "Resist Like a Plant" 29). The centre of human and plant are shifted so that the lines between Madame Sara and vegetal lives become undefined. In becoming plant, Madame Sara moves towards a new state of being, the perpetual motion of life and growth being represented in the action of the word which resists stagnant binaries in the same way Madame Sara does. She becomes occupied by a vegetal presence so that, like a monstrous plant, she can take agency for New Women and for plants through fear, rather than needing it to be generously bestowed upon her by men. Through this partnership of becoming, both woman and plant transform into more than they are singularly.

Becoming Woman and Vegetal

Herbs and plants appear in unexpected ways in Meade's stories. Once there, their bodies grow and expand to exercise their own anarchic energy in the narrative. Detection and investigation, on the other hand, serve as tools to unearth herbal secrets. However, even the agent of global industrial capitalism, Druce Dixon, and his fellow experts cannot always recognise the plants which Madame Sara handles so expertly. Despite their own collection, lab work, and interest in "natural history", they are outwitted. Madame Sara reverses the directional flow of the classic metaphor of bloom, standing in for vegetal bodies as vegetal bodies stood in for women in Victorian literature. In fact, she seems to step beyond the metaphorical and act as a plant, becoming vegetal and enacting vegetal will. This, then, becomes a significant moment for seeing both plants and women as agentic figures.

Crime fiction, since its emergence in the mid-nineteenth century, repeatedly allies the physical space of the story and the transgressive actions of the villain (Miller 17). The detective serves as a reader, interpreting the surroundings for both material and ephemeral signs which lead to the conclusion of the case (Miller 63). Relying on cutting-edge technology and fluency in social interactions, the detective contains the deviance of the villain and restores order at the end of the tale, creating a sense of closure and resolution. The flexibility of the genre allows it to respond actively to politics, national changes, shifting ideologies, and social perceptions (Miller 17). However, with the contemporary spectre of climate change and environmental crises, crime fiction increasingly faces profound questions, not only regarding the restorative nature of the genre, but also relating to its depictions of law, justice, and criminality.

As early ecocrime fiction, L.T. Meade's *Sorceress of the Strand* seems to take the first steps towards questioning the restorative nature of the genre through Madame Sara's continued evasion of Dixon Druce. Lives are saved, but justice remains unserved as Madame Sara takes on an almost supernatural status. The story of Madame Sara's ecocrime thus shifts the crime fiction narrative, frustrating the drive for resolution and destabilising the status quo in the process. Druce, Vandeleur, and the other men featured in these stories extract and hoard ecological wealth for the good of capitalism and their own advancement. Per the genre of crime fiction, they should be seen as the heroes, and yet, as Madame Sara continues to triumph, their own crimes against ecology and culture begin to shine through. Sara's extant similarities to the men that pursue her reframes British plant collection itself as potentially criminal. Ecologically, Sara points to the wideness of plant biodiversity of the kind that eludes so-called "experts", subverting the masculinist colonial gaze that energises much early detective fiction. She inhabits the liminal ecological spaces of the stories, promoting the power of female and plant bodies whilst simultaneously undermining the systems of wealth and capitalist accumulation that would ultimately lead to deforestation and the current climate crisis. From our current moment of ecological instability, her crimes allow us to reconsider imperialist "plant collection" for what it was and often still is – criminal.

When the legal plundering of the Amazon is linked to criminal intent, the text's parallel description of the Brazilian heiresses "as good as gold" suddenly seems exploitative, and laws relating to a wide variety of collection practices seem deficient. The lines between hero, detective, and villain also become more difficult to draw. In this way, crime fiction like L.T. Meade's *The Sorceress of the Strand* series can be seen as anticipating feminist, postcolonial, and ecological concerns. As British society destroyed global ecologies for the sake of commerce, plant and human lives were extracted and commodified in comparable ways. Through the act of becoming plant, Madame Sara's criminality becomes representative of the active resistance of both women and plants to their treatment by an aggressive patriarchal hegemony. She is, by the law of the land, a criminal. But like an eco-terrorist, her law-breaking points to gaps in legal and scientific discourse and thus calls into question the very typologies of criminal behaviour. In this way, Madame Sara becomes a subversive presence, her body blurring with plant bodies in a feminine-ecological anathema that uses criminal acts to demand, and take by force, the agency that turn of the century society otherwise denied both woman and plant.

Note

1. Transcorporeal refers to something that transcends the boundaries of a body and blurs the lines of bodies and identities (Alaimo).

Bibliography

Alaimo, Stacy. "Trans-Corporeal Feminisms and the Ethical Space of Nature." *Material Feminisms*, edited by Stacy Alaimo and Susan Hekman, Indiana UP, 2008, pp. 237–265.

Amherst, Alicia. *A History of Gardening in England.* U London P, 1986.

Arber, Agnes. *Herbals, Their Origin and Evolution, a Chapter in the History of Botany, 1470–1670.* Cambridge UP, 1912.

"Ate a Poisoned Berry." *New York Times*, 8 Oct. 1900, p. 7.

Carroll, Alicia. "'Leaves and Berries': Agatha Christie and the Herbal Revival." *Green Letters*, vol. 22, no. 1, 2018, pp. 20–30, doi:10.1080/14688417.2018.1438303.

———. *New Woman Ecologies: From Arts and Crafts to the Great War and Beyond.* U Virginia P, 2019.

Chang, Elizabeth Hope. *Novel Cultivations: Plants in British Literature of the Global Nineteenth Century.* U Virginia P, 2019.

Doyle, Arthur Conan. "A Study in Scarlet." *A Study in Scarlett*, 2008, www.gutenberg.org/files/244/244-h/244-h.htm.

Edwards, George. "Case of Poisoning by Belladonna", *London Lancet*, 1 Aug. 1851, pp. 106–107.

French, Roger, and Andrew Wear, editors. *British Medicine in an Age of Reform.* Routledge, 1991.

Gerald, John. *1633 Edition.* Edited by Thomas Johnson, Dover Publications, 1633.

Grieve, Maud. *A Modern Herbal.* Dover Publications, 1971.

Griffith, Robert Eglesfeld. *Medical Botany: Or Descriptions of the More Important Plants Used in Medicine, with Their History, Properties, and Mode of Administration.* Lea and Blanchard, 1847.

Hallissy, Margaret. *Venomous Woman: Fear of the Female in Literature*. Greenwood Press, 1987.
Hempel, Sandra. "James Marsh and the Poison Panic." *The Lancet*, vol. 381, no. 9885, 2013, pp. 2247–2248, doi:10.1016/s0140-6736(13)61472-5.
Houle, Karen L. F. "Animal, Vegetable, Mineral: Ethics as Extension or Becoming? The Case of Becoming-Plant." *Journal for Critical Animal Studies*, vol. IX, no. 1/2, 2011, pp. 89–116.
Hughes, William, and David Punter. *The Encyclopedia of the Gothic*. Blackwell Publishing, 2016.
Karpenko, Laura, and Shalyn Claggett. "Introduction." *Strange Science: Investigating the Limits of Knowledge in the Victorian Age*, edited by Lara Karpenko and Shalyn Claggett, U Michigan P, 2017, pp. 1–18.
Kennedy, Meegan. "Discriminating the 'Minuter Beauties of Nature': Botany as Nature Theology in a Victorian Medical School." *Strange Science: Investigating the Limits of Knowledge in the Victorian Age*, edited by Lara Karpenko and Shalyn Claggett, U Michigan P, 2017, pp. 40–61.
King, Amy. *Bloom: The Botanical Vernacular in the English Novel*. Oxford UP, 2003.
Knox, Dana. "Knox's Commandments – the 10 Rules of Golden Age Detective Fiction, Part II." *The Cozy Mystery List Blog*, 27 Jan. 2015, cozy-mystery.com/blog/knoxs-commandments-the-10-rules-of-golden-age-detective-fiction-part-ii/.
Ledger, Sally. *The New Woman: Fiction and Feminist at the Fin de Siècle*. Manchester UP, 1997.
Lee, Vernon. "The Legend of Madame Krasinska." *Vanitas: Polite Stories*. William Heinemann, 1892.
"Little Girl Takes Poison." *New York Times*, 4 Aug. 1901, p. 2.
Marder, Michael. "Plant-Soul: The Elusive Meanings of Vegetative Life." *Environmental Philosophy*, vol. 8, no. 1, 2011, pp. 83–99, doi:10.5840/envirophil2011815.
———. "Resist Like a Plant! On the Vegetal Life of Political Movements." *Peace Studies Journal*, vol. 5, no. 1, 2012, pp. 24–32.
Meade, L. T. "The Sorceress of the Strand by L. T. Meade – Free EBook." *The Sorceress of the Strand*, manybooks.net/titles/meadeltother06sorceressstrand.html.
Meade, L. T., et al. "The Sorceress of the Strand." *The Sorceress of the Strand*, digital.library.upenn.edu/women/meade/sorceress/sorceress.html#IV.
Miller, Elizabeth Carolyn. *Framed: The New Woman Criminal in British Culture at the Fin de Siècle*. U Michigan P, 2008.
Muranaka, Toshiya et al. "Continuous Production of Scopolamine by a Culture of Duboisia Leichhardtii Hairy Root Clone in a Bioreactor System." *Applied Microbiology and Biotechnology*, vol. 40, no. 2/3, 1993, pp. 219–223, doi:10.1007/bf00170370.
N.A. "L. T. Meade, 1854–1914." *Women in the Literary Marketplace*, 2002, rmc.library.cornell.edu/womenLit/LT_Meade.htm.
———. "Women and Power: Emily Massingberd." *National Trust*, 4 Feb. 2018, www.nationaltrust.org.uk/gunby-estate-hall-and-gardens/features/women-and-power-emily-massingberd.
———. "Eustace, Robert." *SFE: Eustace, Robert*, sf-encyclopedia.com/entry/eustace_robert.
Naeem, Shahid et al. "Introduction: The Ecological and Social Implications of Changing Biodiversity." *Biodiversity, Ecosystem Functioning, and Human Wellbeing: An Ecological and Economic Perspective*, edited by Shahid Naeem et al., Oxford UP, Incorporated, 2009, pp. 3–13.
Nagy, Victoria M. "Narratives in the Courtroom: Female Poisoners in Mid-Nineteenth Century England." *European Journal of Criminology*, vol. 11, no. 2, 2013, pp. 213–227, doi:10.1177/1477370813494841.
Northcote, Rosalind. *The Book of Herbs*. John Lane: The Bodley Head, 1903.
Paterson, Leonie. "In the Footsteps of Sir Arthur Conan Doyle." *Botanics Stories*, Royal Botanic Garden Edinburgh, 4 Apr. 2017, stories.rbge.org.uk/archives/18307?fbclid=IwAR1lIHG0D2CTkoXK5PweoyIAfeMizZKsgraaqjBFAdP356C_ur6vy-S_g2I.
Robb, George. "Circe in Crinoline: Domestic Poisonings in Victorian England." *Journal of Family History*, vol. 22, no. 2, 1997, pp. 176–190, doi:10.1177/036319909702200203.
Sanecki, Kay N. *History of the English Herb Garden*. Ward Lock, 1992.
Small, Douglas. "Masters of Healing: Cocaine and the Ideal of the Victorian Medical Man." *Journal of Victorian Culture*, vol. 21, no. 1, 2016, pp. 3–20, doi:10.1080/13555502.2015.1124798.
Voskuil, Lynn. "Victorian Orchids and the Forms of Ecological Society." *Strange Science: Investigating the Limits of Knowledge in the Victorian Age*, edited by Lara Karpenko and Shalyn Claggett, U Michigan P, 2017, pp. 19–39.
Wilde, Oscar. *The Importance of Being Earnest, The Broadview Anthology of British Literature Concise Edition, Volume B*. Edited by Don LePan, Broadview, 2014, pp. 924–955.

9

"SCORCHED EARTH"

Transgressive Bodies, Historic Criminality, and Colonial Recursions in Louise Erdrich's *The Round House*

Malinda Hackett

Ever since the first work of fiction to be published by a Native American author (John Rollin Ridge's *The Life and Adventures of Joaquin Murieta*), works of Indigenous crime fiction have offered a strikingly divergent interpretation of criminal investigations when compared to those illustrated in the mainstream canon. Novels such as Linda Hogan's *Mean Spirit* (1990), which gives a fictional account of the real-world Osage Indian murders, Sara Sue Hokklotubbe's *Deception On All Accounts* (2003), and Marcia R. Rendon's *Murder on the Red River* (2017) and *Girl Gone Missing* (2019) offer a subversive view of criminality by focusing on crimes affecting tribal communities and reservation lands. By employing Indigenous detectives who find themselves immersed in the criminal investigation, most often as victims of circumstance, these narratives decentre traditional representations of law enforcement, whilst also offering a unique perspective into the experience of tribal communities.

Unlike canonical representations of the crime fiction genre, the Indigenous victims in these stories do not obtain justice in similar fashion to their predominantly white counterparts. This narrative choice accurately reflects the Native American experience, as real-world governing institutions rarely provide just reconciliation for past or present crimes, largely due to historic and systemic prejudices that have both silenced the Indigenous peoples and rendered them insignificant within the realm of law enforcement. As a result, Indigenous crime fiction blurs the lines between definitions of right and wrong while ultimately highlighting Native American viewpoints that humans, and their environment, collectively suffer when a person resorts to psychopathic criminality for individual gain at the expense of a communal good. These beliefs find their impetus in Indigenous spirituality which place high value on the elemental connection between all known things. As such, and as Glen Sean Coulthard articulates,

> Within this system of relations human beings are not the only constituent believed to embody spirit or agency. Ethically, this [means] that humans [hold] certain obligations to the land, animals, plants, and lakes in much the same way that we hold obligations [to] other people.
>
> *(61)*

Louise Erdrich's novel *The Round House* (2012) has been recognised for its important contributions to the Indigenous crime fiction genre for its realistic portrayal of an Ojibwe community and the suffering they endure after two of its members become victims of criminal violence. Erdrich's investigative narrative follows 13-year-old Joe Coutts as he attempts to discover the man responsible for raping

 DOI: 10.4324/9781003091912-12

and torturing his mother, Geraldine Coutts, as well as murdering fellow tribe member, Mayla Wolfskin. Ostensibly, Erdrich's narrative offers a tragic view of the complex legalities involved when attempting to prosecute the violent crimes perpetrated against Indigenous women on borderlands. Geraldine's initial reluctance to reveal the identity of her attacker hinders tribal and local law enforcement's ability to prosecute the crime, while her silence reflects real-world instances where Indigenous women have refused to disclose the crime or the identity of a perpetrator for fear of retaliation or revictimisation. From an ecological perspective, however, Erdrich's *The Round House* offers crucial insight into the intersections between criminality, the environment, and coloniality, especially when considering sexual violence against Indigenous women as symptomatic of a European settler mindset which fostered the gendering of land as a means to justify its civilisation. In *The Lay of the Land: Metaphor as Experience in American Life and Letters*, Annette Kolodny reflects on this colonial phenomenon: "In a sense, to make the new [American] continent Woman was already to civilise it a bit, casting the stamp of human relations on what was otherwise unknown and untamed" (9). When considering the colonial mindset as having stemmed from 'human relations' that subordinated women in the public sphere, Mavis E. Mate offers clarification from the perspective of late medieval English politics, where patriarchally structured systems did not afford married women any "public, legitimated power" (61). This lack of authority extended to widowed and single woman who, even if they owned land, could never become "a major office-holder" (62). Given that men exercised this power over women during this time, that they extended this line of practice to justify their imperial interests in land acquisition seems the logical transition. Keeping these observations in mind, it can be argued that Geraldine's rape signals an earth/human imbalance where the Indigenous female body, and by extension tribal land, exists in a recursive state of exploitation and disposability facilitated by the ongoing imperial interests of settler communities. Writing about ecofeminism, Greta Gaard and Patrick D. Murphy cite Ynestra King's seminal work on the subject: "The building of Western industrial civilisation in opposition to nature interacts dialectically with and reinforces the subjugation of women, because women are believed to be closer to nature" (3). By framing sexual violence against Native American women as a recursive criminality that fractures Indigenous ecosystems, which base themselves on the relationality between the land and all its inhabitants including human and non-human, Erdrich highlights how the colonial imperative of environmental dominion maintains systemic imperialist formations within tribal communities.

Coloniality Criminality as an Ecological Recursive Practice

When considering Erdrich's *The Round House* from the perspectives of both ecofeminism and environmental criminality, it is important to first understand coloniality as a recursive practice that persists on an ecological scale. For this discussion, Ann Stoler's work on imperial duress and present-day coloniality adds fruitful insight into Erdrich's text. In *Duress: Imperial Durabilities in Our Times*, Stoler urges the academic community to shift its view of the postcolonial, arguing that the "scholarly romance with 'traces' risks rendering colonial remnants as pale filigrees, benign overlays with barely detectable presence rather than deep pressure points of generative possibilities or violent and violating absences" (5). For Stoler the term postcolonial invokes a sense of past finality which obfuscates the deeply rooted structures of imperialism that remain embedded in colonised communities, those of which persistently generate new modes of oppression and containment for indigenous peoples especially within the context of ecological resources, land, and place. Stoler describes these new formations by borrowing Michel Foucault's definitions of "retranscriptions" and "recuperations", asserting that these "recuperations" share a "history [that] is marked by the uneven, unsettled, contingent quality of histories that *fold back on themselves* and, in that refolding, reveal new surfaces, and new planes" (26, 30). More specifically, historical recursions "do not occur at the levels at which they once appeared, [nor] on the planes of social relations in which they were once activated" (31). Stoler insists that such colonised spaces remain under "imperial duress", or what she recognises as "the hardened, tenacious qualities of colonial effects [and] their extended protracted temporalities, [along with] their durable, if sometimes intangible constraints and confinements" (7). In other words, for the Indigenous communities who continue to experience the effects of colonisation, as compared to their settler counterparts, the idea of a postcolonial fundamentally cannot exist.

Stoler reasons that part of the power of a postcolonial cultural narrative derives from the fact that these embedded colonial structures have become mostly intangible given their perceived connection to a specific time in history, a misconception the governing structures within settler communities rely on in order to maintain not only their power over Indigenous peoples but also their control over geographical spaces and natural resources. For example, while many view the creation of the reservation as a result of early colonisation and thus situate it concretely in a distant past, settler communities presently depend on this type of forced containment in order to maintain extraction zones that divert natural resources such as timber and water from Indigenous lands. Patrick Wolfe famously defines the structural organisation of colonisation as "settler colonialism" stating that "Settler colonies were (are) premised on the elimination of native societies. The split tensing reflects a determinate feature of settler colonisation. The colonisers come to stay – invasion is a structure not an event" (2). According to Macarena Gómez-Barris, "this viewpoint, similar to the colonial gaze, facilitates the reorganisation of territories, populations, and plant and animal life to extractible data and natural resources for material and immaterial accumulation" (5). This loss of geographic locality and the planetary creation of what Barris refers to as a 'corporate bio-territory' takes into account the severing of Indigenous populations from their ancestral lands which not only affected Native Americans historically but continues to do so generationally. From the perspective of American colonialism, specifically, Robert J.C. Young reiterates this point asserting that in "many colonized countries, settlers created vast farms and estates by driving off those who had traditionally lived on that land, some of whose descendants to this day continue to live in an impoverished landless limbo" (45). The effects of such vast land displacement can be seen throughout Native American populations in the United States and can also be witnessed in Indigenous communities on a global scale.

In *The Round House*, Erdrich contends with recursive coloniality and Indigenous displacement by illustrating tribal land in a perpetual state of ecological disarray. The disharmony between environment and human becomes clear in the opening scene of Erdrich's novel, when Joe explains his attempts to remove small trees from underneath his family's home:

> Small trees had attacked my parent's house at the foundation. They were just seedlings with one or two rigid, healthy leaves. Nevertheless, the stalky shoots had managed to squeeze through knife cracks in the decorative brown shingles covering the cement blocks. They had grown into the unseen wall and it was difficult to pry them loose . . . As the afternoon passed and everything on the reservation grew hushed, it seemed increasingly important to me that each one of these invaders be removed down to the very tip of the root, where all the vital growth was concentrated.
>
> *(1–2)*

Initially, the implications of Joe's reflection appear somewhat obvious; his invocation of the term invader when referencing the trees attacking his family home clearly conflates their destruction with the event of colonisation. Furthermore, the tenacity of the small trees – that, as Joe informs the reader a moment later, lived through "a North Dakota winter" by growing deep and impenetrable roots with very little water, "feeble light", and "a few crumbs of earth" – mirrors the practice of what Wolfe observes as settler colonialism (1). Like the early European colonisers who forced themselves into the Americas and onto Indigenous land while setting up infrastructures to ensure their permanence, the trees have set up a similar style of immovable residence.

Alongside their more overt meaning, however, the image of the invasive trees arguably speaks to several deeper ecological issues that continue to affect Native American populations. Given that Indigenous belief systems and spiritual practices rely on "a sacred relationship between the earth and the people; [where] the people respectfully care for and till the earth, and the earth regenerates the human body" (Adamson 54), it can be said that being confined to reservation land placed Indigenous populations into unnatural states of existence, from an environmental point of view. Taking this a step further, Franz Fanon situates this land displacement within the framework of Indigenous economies, stating that the violence "which governed the ordering of the colonial world" resulted in "the destruction of the indigenous social fabric, and demolished unchecked the systems of reference of the country's economy, lifestyles, and modes of dress" (6–7). It is important to keep in mind that Indigenous economies differ from the more dominant Westernised versions in that they do not centre on the monetary. Rather, "People are wealthy who have many strong relationships with each other and with all persons in the landscape, including nonhuman persons" (Trosper 5). In other words, Indigenous economic infrastructures concern themselves with an overall sustainable and relational symbiosis between the tribal community and the surrounding environment which includes both animate and inanimate entities. This can be referred to as an "economy of affection [which] has a strong subjective component, based on the feeling of community resulting from participation in making and sharing relational goods" (5–6). Linda Child offers a view of these Indigenous economies from a gendered perspective, especially when considering Indigenous mothers where these relational systems "include the natural environment", for example as with the Ojibwe practice of making "wooden frame cradleboards" so women could perform outdoor labour with their new-born babies (16). This integration of sustainable environment, labour, and the needs of tribal women exemplifies the mutual reciprocity fostered by Indigenous economical relational systems. Turning back to the Coutts home, the violence with which the trees attack the foundation points to a troubling reality where the relationality between an Indigenous people, their domestic space, and the land has been severed. Not only does Joe, by his own admission, seek to annihilate the unseemly trees underneath his family home but he also exhibits a noticeable imperial mindset as he attempts to exert dominance over the surrounding plant life.

That Erdrich conflates reservation trees with the act of colonisation speaks to important points concerning the overall environment in the Americas, especially when taking into account how the colonial impact of early Europeans settlers permanently altered its ecological landscape. Graham Huggan and Helen Tiffin contextualise this irreversible destruction within a Western mindset:

> Once invasion and settlement had been accomplished . . . the environmental impacts of western attitudes to human being-in-the-world were facilitated or rein-forced by the deliberate (or accidental) transport of animals, plants and peoples throughout the European empires, [which instigated] widespread ecosystem change under conspicuously unequal power regimes. . . . The genuinely natural ways of indigenous ecosystems were irretrievably undone as 'wild' lands were cleared for farming or opened up to pastoralism.
>
> *(6–7)*

Thus, the invading trees take on additional meaning when considering them as part of a colonial ecological process. Not only do they signal an earth/human imbalance within the context of indigenous environmental belief systems, but, as Erdrich's text suggests, the trees arguably signify an anomalous ecology that has been forced into an even more unnatural existence.

Imagery that highlights these types of ecological recuperations as a recursive consequence of colonialism persists throughout Erdrich's narrative as the text illustrates the tribal environment where the Coutts family resides. When Joe describes how his mother plants flowers in "paper milk cartons", "ice cream buckets", and an "old tractor tire painted white", and how Cappy's father Doe has a deck "filled, as all our decks tend to be, with useful refuse . . . snow tires in black garbage bags, rusted jacks, a bent hibachi grill, banged-up tools, and plastic toys", Erdrich invites the reader to consider the multiple implications of such waste (6, 223). On the one hand, this waste clearly signals Western and capitalistic modes of consumption while also stressing Indigenous sustainability practices that prioritise recycling and repurposing "useful" waste, as the ice cream carton turned flowerpot demonstrates. On the other, by illustrating how Joe and his tribal family and friends live among the refuse, Erdrich also amplifies settler perspectives that devalue Indigenous peoples and the land on which they reside. By highlighting a Native American tribe as coexisting with trash, Erdrich's text points not only to Western viewpoints that normalise littering on tribal lands but also to cultural narratives that view Indigenous peoples as disposable and by association non-human. On a larger ecological scale, in visualising settler practices that perpetuate the disposal of Indigenous communities, and thus rendering them invisible, Erdrich forces the reader to contend with Western perceptions that dismiss environmental destruction and the negative impact it creates for future generations on this earth.

Joni Adamson addresses this crisis when discussing the difference between Western and Indigenous environmental mindsets:

> Not only were other people often regarded as part of nature – and thus treated instrumentally as animals – but also they were forced or co-opted over time into western views of the environment, thereby rendering cultural and environmental restitution difficult if not impossible to achieve.
>
> *(6)*

Adamson addresses a crucial point about the stagnant state of the environmental crisis, and it gains relevant meaning when contextualised within the imagery of 'nature in confinement' that Erdrich locates in *The Round House*. Not only does the vegetation planted in used containers parallel Indigenous peoples who exist in a perpetual and comparable state of containment, but this comingling of refuse and nature can also be viewed as a microcosm of the larger global ecological problem, where the wasteful products of the dominant culture have altered the earth's ecosystem in such a way that restitution, to borrow Adamson's language, may not be possible. These capitalistic practices can be viewed as part of what can now be referred to as a type of ecological criminality, especially given that these corporate behaviours almost always employ environmental racism, which by definition involves the

> poor, often racial minority communities whose residents are subjected to a disproportionate level of many kinds of toxicity, while being denied ecological benefits such as clean water and air, as well as sustainable use of resources for a decent and dignified standard of living.
>
> *(Johansen 1)*

Female Indigeneity and Domesticity

As the Coutts' home suggests, illustrating the severing of an ecological Indigenous relationality plays an important part in conveying recursive colonial criminalities in *The Round House*. Erdrich amplifies these ecological issues by highlighting how violence against Indigenous women also serves as a 'reactivation', to borrow Stoler's term, of colonial practices. In order to recognise how these gender-motivated violent reactivations function in Erdrich's text, a discussion of the nuances of embedded colonialities alongside an understanding of Indigenous women and their role in the tribal community becomes necessary. As Stoler attests, colonialism often maintains power through more nuanced and not so readily identifiable modes of imperialist formation. Linda Gordon expands on the more subversive representations of colonialism in what she defines as "internal colonialism", or "the intimate, social, and cultural aspects of domination" which involve the "economic aspects of the familial and the intimate, such as household labor and reproductive labor", along with "the circulation of resources among kinfolk" (433). Gordon further asserts that these internal colonial structures encompass "vital dimensions of domination and control, those that took place in families, households, and schools as opposed to factories, mines, or fields" (433). This idea of an internal colonialism can be witnessed in the Coutts home when Erdrich's text reflects how the ramifications of Geraldine's rape not only irreparably alter her family's relationships to one another but also how they create a permanent state of foundational instability.

Erdrich's text centres Geraldine within the private space of the Coutts home not as a victim confined to the domestic, but as one who anchors the family through life giving practices. The seriousness with which Geraldine views her place in their domestic space can be seen when Joe relays that the milk in their refrigerator has gone bad while initially visiting his mother in the hospital shortly after the rape:

> Her serene reserve was gone – a nervous horror welled her face. The bruises had come out and her eyes were darkly rimmed like a racoon's. A sick green pulsed around her temples. Her jaw was indigo. Her eyebrows had always been so expressive of irony and love, but now were held tight by anguish. Two vertical lines, black as if drawn by a marker, creased her forehead. Her fingers plucked at the quilt's edge. Sour!
>
> *(Erdrich 23)*

Despite Geraldine's face reflecting the grotesque brutality of the attack, Joe's revelation that the milk has gone sour overshadows the primacy of her physical injuries and her agony gives insight into the family's home space and the significance it holds. According to Joe, his mother "had grown up without refrigeration and was proud of how clean she kept her treasured icebox. She took the freshness of its contents seriously" (23). Therefore, it makes sense that she would display this kind of anger upon hearing that something she takes pride in has been destroyed as a result of her attack. Taking this one step further, however, Geraldine's reaction to the sour milk indicates her understanding of a deeper enmeshment, one that involves the ties between her, the domestic space, and the land. For the Ojibwe in particular, women hold powerful positions within the structural hierarchies of tribal communities, especially because of their ability to extend the private and domestic spheres to the public and economic in order to create cohesive bonds that provide sustainable infrastructures for the tribe. Brenda Child (Red Lake Ojibwe) corelates the significance of Indigenous women in the Ojibwe culture to the natural environment:

> The demanding rituals associated with puberty, which involved considerable time and commitment by the young woman and other female relatives, do not seem to have been primarily concerned with fertility. Instead, the rituals involved in the first menstruation represent a highly meaningful coming-of-age symbolizing a woman's power to give life that is first and foremost associated with the powers of the universe, and therefore linked to the rest of the community. Female rituals acknowledged the sacredness of life and the centrality of women's role in society.
>
> *(7–8)*

Given that the Ojibwe value the connection the female body shares with the natural environment through their reproductive capabilities – while simultaneously refuting Westernised practices that place value on women only if they propagate offspring – it can be inferred that, for the Ojibwe, life giving takes on more nuanced meanings. Life giving can also apply to a woman's ability to utilise their natural environment to strengthen and maintain the bonds of their tribal community through healing, labour, wisdom, and nurture. Thus, for the Ojibwe women, their connection to the land, coupled with their centrality within the domestic sphere, is integral to the overall welfare of not only their immediate family but also the larger tribal community. Unlike settler communities, where capitalist interest severs the worker from the land and the by-products of their labour, such as food and shelter, the Ojibwe value the individual and the work they perform as symbiotic and reciprocal. Further, individual labour contributes to one's understanding of the self and their place in the communal well-being of the tribe as a collective entity. Turning back to Geraldine, while the sour milk might seem trivial, it can be argued that her visceral reaction suggests a metaphysical understanding that the attack has fractured her sense of her indigeneity as it applies to her feminised role in the tribal community (as the bridge between earth and human). To overly simplify, the milk, which derives from the earth, provides sustenance but it has gone sour, which means it can no longer sustain life. Metaphorically speaking, however, the rape has "soured" Geraldine whose physical and emotional trauma renders her unable to fulfil her tribal and communal role.

Erdrich's novel extends its preoccupation with ecological concerns by illustrating how the rape affects not only Geraldine's sense of herself as an Indigenous woman but also how the family suffers collectively because they share a relational way of being. For Joe, the altering of the nightly family dinner which his father unsuccessfully takes over the cooking of after his mother's rape, becomes particularly demonstrative of the unseen consequences of the violent assault on his mother:

> My father beckoned the two of us to sit down. There were potatoes, nearly cooled, way overcooked, disintegrating in an undrained pot. He ceremoniously heaped our shallow bowls. Then we sat looking at the food. We didn't pray. For the first time, I felt the lack of some ritual.
>
> *(Erdrich 35)*

A few moments later, Geraldine physicalises what Joe subjectively understands about their nightly ceremony when she begins eating, "With no ceremony, she picked up her spoon and plunged it into the stew" (35). Note Erdrich's repeated use of the word ceremony along with the conflating of ceremony with the daily ritual of not merely consuming food but the sharing of a meal together. By coalescing female indigeneity with custom and ritual, Erdrich iterates values about Indigenous women and their importance in the ecological make-up of the tribal community which includes the private sphere of its members.

Joe reiterates his understanding of Indigenous ceremony in relation to his family by acknowledging how they have been forever altered while witnessing his mother's subsequent mental breakdown during their nightly routine after she accidentally drops and breaks a casserole dish:

> We were not churchgoers. This was our ritual. Our breaking bread, our communion. And it all began with that trusting moment where my father walked up behind my mother and she smiled at his approach without turning. But now they stood staring at each other helplessly over the broken dish.
>
> *(43)*

The violent shock of the broken dish, which happened because his father took his mother by surprise, arguably echoes back to the violence of the assault on both Myla and Geraldine. Geraldine confirms this when reacting to the shattering glass as if being assaulted; according to Joe, she emitted a "keen, low anguished cry" as she "backed up to the sink, trembling, breathing heavily" (42). On a deeper level, however, the disruption it causes for the Coutts family suggests that a palpable fragility lies underneath the perceived stability of their home space. Essentially, Erdrich's narrative 'retranscribes', to borrow another one of Stoler's terms, the event of the assault on Geraldine within the domestic space of the Coutts home, thus injecting the colonial practice of violence against Indigenous women into an immediate present. Erdrich intensifies this colonial immediacy by accompanying the event of the breaking dish with Geraldine's physical appearance which bears the scars of the assault:

> Deep violet streaks and the yellow of healing contusions still marred her face. The white of her left eye was scarlet and her eyelid drooped slightly, as it would from then on, for the nerve had been tampered with and the damage was irreversible.
>
> *(35)*

Here, Erdrich's text invites the reader to consider not only the insertion of structural coloniality within the private spaces of the Indigenous people, as Gordon would argue, but also the temporal reshaping and inevitable persistence of these dominant colonial paradigms throughout history and into the present day.

Gendered Indigenous Violence and the Environment

In situating the murder of Myla and the attempted murder of Geraldine in the Ojibwe tribe's ancestral round house, Erdrich continues the novel's examination of a recursive ecological criminality by highlighting the interstitial relationship between the female Indigenous body and the environment. Relayed by Joe's grandfather Mooshum, the story at the centre of the round house provides a fundamental understanding of Indigenous feminine ecosystems and their value in Native American spiritual belief and culture. According to Mooshum, the round house signifies "the reservation year", or the year of devastation the Ojibwe experienced directly after colonisation:

> We starved while the cows of settlers lived fat off the fenced grass of our old hunting grounds. In those first years our white father with the big belly ate ten ducks for dinner and didn't even send us the feet. Those were bad years.
>
> *(Erdrich 184)*

In sum, Mooshum tells of how the ancient Ojibwe tribe, crazed with starvation during that first reservation winter, become convinced that one of their members has turned into a windigo. After ostracising Aniikwe from the community, they place the task of killing her, the custom for dealing with a windigo, on her young son, Nanapush. Unable to kill his own mother, Nanapush chooses to free her instead and the two venture off into the snow together where they eventually become separated. Nanapush seeks out the Old Buffalo Woman who allows him to kill her so that he can use her body for shelter and food. Once reunited, Nanapush and Aniikwe feed their tribe with the meat of the Old Buffalo Woman, after which she convinces a disillusioned Nanapush that his altruism will be his tribe's salvation (185–187). She then instructs him to build a structure that will symbolise the reservation year and the Ojibwe people's resilience:

> Your people were brought together by us buffalo once. You knew how to hunt and use us. Your clans gave you laws. You had many rules by which you operated. Rules that respected us and forced you to work together. Now we are gone, but as you have once sheltered in my body, so now you understand. The round house will be my body, the poles my ribs, the fire my heart. It will be the body of your mother and it must be respected the same way. As the mother is intent on her baby's life, so your people should think of their children.
>
> *(214–215)*

This underlaying of Joe's investigation with the mythology of the round house has multiple implications, particularly when considering violence against Indigenous women as a form of ecological criminality. On the surface, Erdrich's subsuming of Indigenous mythology into the present-day criminalised space of the round house functions similarly to the trees attacking the foundation of the Coutts' home and the later event of breaking glass in their kitchen. Once again, Erdrich's narrative retranscribes a specific set of cultural events from the past, in this case white men raping and murdering Indigenous woman, to bring the violence of colonialism to the forefront, all the while refuting the very idea of a postcolonial present. In other words, the criminal violence directed towards both Geraldine and Mayla in the round house does not offer an exact replication of historic colonial violence, in the simplest terms. However, Erdrich's coalescing, or 'refolding', as Stoler attests, of past and present spatiality and temporality speaks to the persistence of an immediate violent coloniality, as opposed to thinking about such violence as only occupying a definitively located past-space.

Turning to the actual mythology of the round house, the appearance of the windigo in its origin story also raises important points about the intersection between gendered violence and the environment. The windigo has deep roots in Indigenous belief systems, especially those concerning moral and social responsibilities to the environment and its inhabitants. According to Adamson: "It is said that people 'go windigo' when a dangerous spirit takes possession of their soul, causing them to become greedy, gluttonous, and have an insatiable desire for human flesh" (Adamson 104), and

> The fear of the windigo is really a fear of excessiveness of any kind. In Chippewa culture, a balance, a sense of proportion must be maintained in all interpersonal relations and activities. Hoarding or any manifestation of greed is discountenanced.
>
> *(A. Irvin Hallowell as qtd. by Adamson 104–105)*

This last note carries significant relevance given its resonance with entitled modes of contemporary excess and consumption which find their impetus in colonialist and imperialist motivations for acquiring land and its natural resources along with the labour of Indigenous bodies, which the colonisers extracted in both physical and sexual forms.

Keeping this in mind, Linden Lark clearly views domination over Native American bodies as a natural ecological state of coexistence as evidenced by his rantings during his attack on Geraldine and Myla. When tormenting both women, Lark states that he is "one sick fuck . . . who hates Indians generally and especially for they were at odds with my folks way back" while expressing his belief that "they have no standing under the law for good reason" (161). The situation that placed the Lark family 'at odds' with the surrounding Native American community involves Bazil and his court ruling against them during a trial that concerned their gas station, which resides on tribal land and where they had been in the practice of overcharging the Indigenous community. By correlating the troubled history of his family's gas station to his present crimes, Lark conflates dominion over the land and its natural resources with white entitlement to the Indigenous female body. Expanding on Kolodny's articulation of the pastoral impulse and colonisation, Anne McClintock offers crucial insight into the colonial practice of gendering land:

> Knowledge of the unknown world was mapped as a metaphysics of gender violence . . . and was validated by the new Enlightenment logic of private property and possessive individualism. In these fantasies, the world is feminised and spatially spread for male exploration, then reassembled and deployed in the interests of massive imperial power.
>
> *(23)*

As such, and according to McClintock, these entitlements include propriety over natural resources for their economic exploitation and included the right to put the Indigenous body in a subordinate and labouring state in order to extract these new colonial commodities. Note too that McClintock frames these imperial impulses within the realm of fantasy. In raping, torturing, and murdering Indigenous women, Lark constructs a fantasy world where he has reversed what he perceives as an unnatural order of things: "Things are the wrong way around . . . [b]ut here in this place I make things the right way around for me" (161). Gretchen T. Legler observes this practice as being essential to settler/Eurocentric forms of domination:

> Constructions of nature as female (as mother/virgin) are essential to the maintenance of this harmful environmental ethic and are essential to the maintenance of hierarchical ways of thinking that justify the oppression of various "others" in patriarchal culture by ranking them "closer to nature" or by declaring their practices "natural" or "unnatural"
>
> *(Legler 228).*

In her speech to Nanapush, The Old Buffalo woman also pays close attention to the regeneration of the Ojibwe people by instructing the tribal community to honour the mother and her future children. The impetus for Lark's violent crime rests on his infatuation with Mayla, whom he learns has birthed a child by another man (her employer). Not only does Lark view himself as entitled to Native American women because of his status as a "white man", but he attempts to further exploit Indigenous female bodies when he blackmails the baby's biological father who, it turns out, took advantage of Mayla in the first place (161, 299, 300). Thus, Erdrich's text forces the reader to view sexual violence against Native American women as an event not confined to the present but as an assault that initially finds its impetus in imperial imperatives from the past. As Mary Paniccia Carden surmises: "As spokesperson for the colonial logic of elimination, Lark propagates and performs the triumphalist model of national history that erases Native peoples from U.S. geographies and cultural narratives" (106). Echoing back to the visible trash on the reservation, Lark takes this one step further when conveying his perception of the gendered Indigenous body as exploitable and disposable: "I don't know what to do with the evidence. Silly me. Maybe I should burn the evidence. You know, they're just evidence" (162). Through the reduction of Geraldine and Mayla to the status of erasable objects, Erdrich points to the subsistence of broader and systemic ecological structures that continue to seek the eradication of Indigenous peoples in the Americas.

Feeling Nature and the Indigenous Detective

Returning to the intersection of criminality and the environment, Erdrich's anthropomorphising of the round house, on the one hand, aligns with the Indigenous spiritual belief of animism. On the other, it lends itself to a specific investigative technique employed by the Indigenous detective that involves attuning with nature as a way to deduce the events surrounding a particular crime. Different from "reading" a scene, this ability definitively sets the Indigenous detective apart from typical codifications of the investigative figure, as they channel what Wolf recognises as an elemental connectedness between all things: "Indigenous cultures around the world have stories of transspecies communications and even kinship, ancestry with earthothers – other species, as well as ecological entities such as plants, waterbodies, rocks and mountains, sky, and wind" (xix). Joe displays this spiritual practice and connection to the elements when he encounters the round house during the course of his investigation:

> The log hexagon was set up on top of a slight rise, and surrounded by rich grass, vivid green, long and thick. I dropped my bike. There was a moment of intense quiet. Then a low moan of air passed through the cracks in the silvery logs of the round house. I started with emotion.

> The grieving cry seemed emitted by the structure itself. The sound filled me and flooded me. Finally, it ceased.
>
> *(Erdrich 59)*

Note how Joe situates the round house in nature; the logs that make up its foundation and the lush grass surrounding it foreground its importance in the geography of the tribal land on which his reservation resides. Joe proceeds to glean details from the night of the attack through an immersive intuition facilitated by his connectedness to the round house, eventually discovering the exact location of his mother and Mayla's assault: "At that moment, a certainty entered. I knew. He had attacked her here. The old ceremonial place had told me – cried out to me in my mother's anguished voice" (60). Then, a few moments later, Joe describes how

> My heart was beating so hard as I followed the action in my understanding that I did not feel the water. I felt his overpowering frustration as he watched the car disappear. I saw him pick up the gas can and nearly throw it after the vanishing taillights. He ran forward, then back. Suddenly, he stopped, remembering his stuff, the car, whatever he did have, his smokes. And the can. He could not be caught with the can. However cold it was that May, the ice out but the water still freezing, he'd have to wade partway in and let water fill the can. And after that, as far out as possible, he had surely slung the water-filled tin and now, if I dived down and passed my hands along the muddy, weedy, silty, snail-rich bottom of the lake, there it would be.
>
> *(61)*

As opposed to the Sherlock Holmes of the mainstream investigative universe, who most famously utilises hegemonic and Eurocentric modes of intellect to "read" clues, Joe relies on his attunement to both the human and non-human environment, which includes Lark himself, to determine what happened to his mother that night. Conversely, the violence of the crime has imprinted itself upon the reservation ecology which, in turn, allows itself to be read by Joe in order that he can deduce the particulars of the crime. Speaking from the perspective of an ecological criminality, the earth becomes the victim of the crime in equal measure to both Geraldine and Mayla and, by extension, the collective tribal community.

Conclusion

When investigating his mother's rape, Joe reflects on his relationship with the women in his tribal community: "I was lucky: I was a boy doted on by women" (Erdrich 25). He then continues to explain the importance of these women by situating them in relation to himself according to their geographical proximity to his family home, thus creating what can be viewed as a feminised metaphysical sense of Indigenous spatial relationality. For example, his father had bought their home and moved it onto a plot of land that used to belong to Geraldine's late uncle. Joe's Uncle Whitey had used the rest of that land to build a gas station where his aunt Sonja also worked. His other aunt Clemence lives up the hill from the Coutts home. By laying the foundation for these geographical and topographical relationalities, Erdrich stresses their importance in Indigenous culture and belief systems. Writing on Lakota philosopher Vine Deloria Jr.'s seminal work on the topic, Glen Sean Coulthard reiterates Deloria's rejection of the more obvious understanding that Indigenous peoples carry strong ties to their homelands. Rather, as Coulthard asserts, land functions as an "ontological framework for understanding *relationships*" (60). In other words, "Place is a way of knowing, of experiencing and relating to the world and others" (61). Contextualising this within the realm of crime and given that "place-based" ways of knowing inherently lend themselves to communal and ecological modes of human bonding, it can be surmised that the criminal violence perpetuated against one tribal member, in this case Geraldine and Mayla, affects the entire tribal community. By calling attention to the notable intersections between gendered violence, indigeneity, and ecology, Erdrich effectively subverts these Eurocentric perspectives of criminality by highlighting how the violence of a white man disrupts an entire ecosystem of human beings rather than a sole individual. By the end of *The Round House*, Joe's tribal community has been irreparably fractured as an immediate result of the rape; Geraldine never fully recovers emotionally or physically, Bazil dies of a heart attack after assaulting a mocking Lark in a corner store, Cappy dies in a car accident, and his aunt Sonja runs off with a large amount of money she initially hides for Joe (under the guise of saving it for his college fund) after he discovers it midway through his investigation.

Erdrich leaves the reader with several crucial moments to reflect upon when considering the notion of a collective ecological suffering that finds its impetus in any mode of violent criminality, no matter how justified. Joe relays a dream he has after he and Cappy murder Lark in an act of revenge:

> We were back at the golf course in the moment I locked eyes with Lark. That terrible contact. Then the gunshot. At that moment, we exchange selves. Lark is in my body, watching. I am in his body, dying. Cappy runs up the hill with Joe and the gun, but he doesn't know that Joe contains the soul of Lark. Dying on the golf course, I know that Lark is going to kill Cappy when they reach the overlook. I try to call out and warn Cappy, but I feel my life bleed out of me into the clipped grass.
>
> *(307)*

Erdrich visualises the coalescing of physical beings to indicate their interconnectedness in relation to their environment, similar to when Joe places himself in Lark's mind at the round house. However, she does so with a more pressing urgency here by envisioning Joe and Lark experiencing interchangeable embodiments. A short time later, Joe becomes violently ill and thus recalls the memory of digging out the small trees from underneath his family home:

> While I was ill, I watched the golden light pass across my walls. I could feel nothing, but my thoughts ran wild. Always I kept going back to the day I dug the trees out of the foundation of our house. How tough those roots had clung. Maybe they had pulled out the blocks that held our house up. And how funny, strange, that a thing can grow so powerful even when planted in the wrong place.
>
> *(293)*

Erdrich's decision to collate Joe's sickness, which he surmises stems directly from his having taken a life, with the plants from earlier in the narrative appears deliberate, particularly given the centrality of recursive coloniality and ecological violence to the narrative. As with the initial descriptions of the invading plants earlier in the novel, Erdrich utilises their presence to signify the initial act of colonialism and, more specifically, the unnatural state of the Europeans on Indigenous land. However, viewing these observations as part of a larger conversation on recursive coloniality, and its interstitial relationship to ecological violence, facilitates a deeper understanding about the current environmental crisis in the Americas and the larger globe. Joe ends his narrative expressing his desire for the time "Before", or what he considers his idyllic existence before Lark's crime shattered his life. However, the question remains if the damage being done to the globe can truly be reversed, especially when, as Erdrich reiterates, these colonial structures remain so deeply entrenched in our collective ways of being.

Bibliography

Adamson, Joni. *American Indian Literature, Environmental Justice, and Ecocriticism: The Middle Place*. U Arizona P, 2001.

Carden, Mary Paniccia. "'The Unkillable Mother': Sovereignty and Survivance in Louise Erdrich's *the Round House*." *Studies in American Indian Literatures*, vol. 30, no. 1, 2018, p. 94, doi:10.5250/studamerindilite.30.1.0094.

Coulthard, Glen Sean. *Red Skin, White Masks: Rejecting the Colonial Politics of Recognition*. U Minnesota P, 2014.

Erdrich, Louise. *The Round House*. Harper Perennial, 2013.

Foucault, Michel. *Society Must Be Defended: Lectures at the Collège de France, 1975–1976*. Picador, 2003.

Frantz, Fanon. *The Wretched of the Earth*. Translation by Richard Philcox, Grove Press, 1963, 2004.

Gaard, Greta, and Patrick D. Murphy. *Ecofeminist Literary Criticism: Theory, Interpretation, Pedagogy*. Urbana, 1998.

Gómez-Barris, Macarena. *The Extractive Zone: Social Ecologies and Decolonial Perspectives*. Duke UP, 2017.

Gordon, Linda. "Internal Colonialism and Gender." *Haunted by Empire: Geographies of Intimacy in North American History*, edited by Ann Stoler, Duke UP, 2006, pp. 427–451.

Huggan, Graham, and Helen Tiffin. *Postcolonial Ecocriticism: Literature, Animals, Environment*. 2nd ed., London, 2015.

Johansen, Bruce E. *Environmental Racism in the United States and Canada: Seeking Justice and Sustainability*. ABC-CLIO, 2020.

Kolodny, Annette. *The Lay of the Land: Metaphor as Experience as History in American Life and Letters*. U North Carolina P, 1975.

Legler, Gretchen T. "Ecofeminist Literary Criticism." *Ecofeminism: Women, Culture, Nature*, edited by Karen J. Warren, Indiana UP, 1997, pp. 227–238.

Lobo, Susan, et al. *Native American Voices*. 3rd ed., Routledge, 2016.

Mate, Mavis E. *Women in Medieval Society*. Cambridge UP, 1999.

McClintock, Anne. *Imperial Leather: Race, Gender, and Sexuality in the Colonial Contest*. Routledge, 1995.

Trosper, Ronald L. *Indigenous Economies: Sustaining People and Their Lands*. U Arizona P, 2022.

Wolfe, Patrick. *Settler Colonialism and the Transformation of Anthropology: The Politics and Poetics of an Ethnographic Event*. Cassell, 1999.

Young, Robert J. C. "Postcolonial Remains." *Postcolonial Studies*, vol. 43, no. 1, 2016, pp. 125–143, doi:10.1002/9781119118589.

10

"ANIMALS TAKING REVENGE"

Imagining Murder as an Ecological Encounter in Olga Tokarczuk's *Drive Your Plow Over the Bones of the Dead*

Andrew Yallop

Literary representations of ecological devastation often feel thematically akin to a pervasive, global crime to which there is no single localised solution (Walton and Walton 3). While some classical detective fiction alludes to environmental harm as a backdrop to individual crime(s), the environment is generally a peripheral concern, if a concern at all. Even in the latter half of the twentieth and early twenty first centuries – where environmental harm looms ever larger as a political and cultural concern in the public imagination – attempts to represent ecological crime directly in literature have faced perennial difficulties. For instance, how can a single detective figure isolate individual victims of a broad Anthropocene violence visited upon the environment? How can an ecological conception of harm be represented and quantified as crime, especially when these harms ripple across species, nations and jurisdictions? And even if a crime can be established, how is guilt to be apportioned? Answering these questions requires a radical shift in our imagination of the environment and humanity's relationship to it. According to Amitav Ghosh, "the climate crisis is also a crisis of culture, and thus of the imagination" and in response, writers must address "the grid of literary forms and conventions that . . . shape the narrative imagination" (9, 7).

In *Drive Your Plow Over the Bones of the Dead*, Olga Tokarczuk works within the detective genre to reimagine environmental harm as both an ecological encounter and an act of vengeful, extra-judicial justice, where non-human creatures are both victims and perpetrators of murder. First published in Poland as *Prowadź swój pług przez kości umarłych* in 2009 and translated into to English by Antonia Lloyd-Jones 2018, *Drive Your Plow Over the Bones of the Dead* has proved rather difficult to classify. Reviewers and critics have referred to the novel variously as an "astonishing amalgam of murder mystery, dark feminist comedy and paean to William Blake" (Perry), a "unique masterpiece, a genre hybrid [and a] socially critical satire" (Kernev-Štrajn 72) and "an eco-critical narrative which hijacks a crime story in order to sell [a] resounding eco-friendly and posthumanist . . . manifesto" (Kowalcze-Pawlik 188). Critics typically situate *Drive Your Plow* in relation to the crime fiction genre at large, either as an experimental yet intelligent game played out within the armature of detective fiction (Wzorek 100), an "unconventional murder mystery with an environmentalist and animal-rights slant" (Franklin), or as a "barbed, shrewd parodic eco-noir" (Toynton). Tokarczuk herself conceives of the novel as a "moral and metaphysical thriller" (qtd. in Kernev-Štrajn 100, 72), formed in the "cake mould" of detective fiction and redolent of postmodern detective texts like Umberto Eco's *The Name of the Rose* (qtd. in Rejter, 30). Through her narrator Janina Duszejko, Tokarczuk imagines a darkly ironic reality, wherein animals take revenge on humans for hunting and killing their kin. Janina (I will

DOI: 10.4324/9781003091912-13

refer to her henceforth by her first name for clarity, despite her professed aversion to it) formulates a theory of murder which bestows agency to both humans and non-humans, challenging the patriarchal status quo of the village hunters. In doing so, she undermines the politics of exploitation which define animals as 'game' to be exploited for pleasure or financial gain. More broadly, her theory pushes back against the pervasive logic of entrenched capitalist and patriarchal power structures by questioning the treatment of non-humans under the law and even reimagining how non-human protections might be designed, policed and enforced. While *Drive Your Plow* captures the idealism of Janina's vision by revealing her as the perpetrator in the final act of the detective plot, Tokarczuk also refuses to endorse the politics of animal rights activists who rely upon in-kind violence to make their point.

Drive Your Plow is framed as a murder mystery set in the remote and wintry Polish village of Kłodzko, where the narrator and protagonist Janina Duszejko, "a kind of eastern European Miss Marple", stumbles across the body of a deceased poacher whom she refers to as Big Foot (Perry). This death is then followed by several others: the police commissioner (the 'Commandant') is found crumpled in a well; the 'President' of the local hunting club is asphyxiated by a swarm of *cucujus maematode* beetles; fox-farm proprietor Innerd is caught in an animal trap; and the village priest, Father Rustle, is burnt alive inside his own church. Initially the police treat the deaths as accidents. Janina, however, develops her own 'Theory', hypothesising that the men were murdered by animals as punishment for their participation in hunting and poaching. By Janina's logic, Big Foot didn't choke on a bone, he was "punished [by the Deer] for killing them in such a cruel way" (Tokarczuk 230); the commandant didn't fall down the well, but was pushed in by Animals, as evidenced by the "hundreds of deer prints at the scene"; and Innerd was deliberately lured to his death by "beautiful, noble, [wild foxes] with wise faces" (Tokarczuk 79, 157). However, the deaths of the president and Father Rustle are less well explained by Janina's theory, and this soon raises suspicions amongst Janina's friends about her potential involvement in the crimes.

Janina's investigation presents the reader of *Drive Your Plow* with a vision of the world in which Animals are "strong", "wise" and sentient beings "with a very strong sense of justice [and] more human than people in every possible way" (Tokarczuk 69, 182–183). The disconnect between the hunters' objectifying and exploitative view of the world and Janina's more democratic conceptualisation of violent crime – where Animals are not 'just things' for people to act upon, but wise creatures who desire justice for harms rendered upon their kin – is at the heart of the novel's political critique. Janina's theory is essentially an extension of her moral worldview, "a primer on the politics of vegetarianism, a dark feminist comedy, an existentialist fable and a paean to William Blake" (Perry). Her affinity for Blake is reflected in her idiosyncratic use of normalisations in the novel, as well as her habit of using capitalisation to create unusual proper nouns, such as the word 'Animals'. One of Janina's most striking idiosyncrasies is her habit of giving both humans and animals names that she "consider[s] suitable and fitting": she refers to Deer as "Young Ladies", an old fox as "Consul" and her two greyhounds as her "Little Girls", while bestowing more impersonal monikers to her male, human neighbours like "Oddball" and "Big Foot", the latter of whom she doesn't "even regard as [a] human Being" (Tokarczuk 13, 91, 31, 6).

Building on the work of Garry Marvin and Susan McHugh, Michael Lundblad defines the field of "human-animal studies" as an exploration of "intertwined interactions, relationships, and becomings that involve human and nonhuman beings" and encourages the reader "to study animals with humans, and humans with animals, never forgetting that we are both animals in general, and humans in particular" (2). In the context of a murder mystery, the notion of death and unlawful killing in *Drive Your Plow* presents a highly charged place from which to consider the intertwined relationships between humans and animals and the ethics of violence, which, in Janina's version of events, muddies the distinctions between human and non-humans. Indeed, as Kari Weil writes in reference to the fiction of Temple Grandin and J.M. Coetzee: "Death is the place where the conceptual and ontological distinctions that language makes possible break down, including the distinctions between human and animal" (91).

As indication of this blurring, Janina will often use human names without qualifying whether the creature is a human or non-human animal. This creates a particularly uncanny effect in the context of her theory regarding Big Foot's death, who she suggests choked on the flesh of one of the "Young Ladies" who live in the forest, and in relation to the death of her "Little Girls", who she suspects have been murdered by unscrupulous hunters. As such, these animal deaths (or murders, as Janina sees it) are the point where the division between human and non-human falls away, revealing the extent to which the boundaries between legal rights – such as the social treatment of animals – are ultimately a human construction. At the conclusion of the novel, however, Janina confesses to the murders of all the men except for Big Foot. After discovering a photo in Big Foot's house in which the bodies of her Little Girls are displayed amongst other dead animals in front of a hunting party, she admits to her close friends Oddball and Dizzy that she needed to punish the men. She recounts how she killed the commandant and Innerd by bludgeoning them to death with a deer's head frozen inside a plastic bag taken from Big Foot's cabin and notes that she got away with it because "[n]obody takes any notice of old women who wander around with their shopping bags" (Tokarczuk 231). But after sprinkling the intoxicated president with insect pheromones stolen from her friend Boros and setting Father Rustle's church ablaze with him inside it, her friends, and the authorities, begin to suspect her involvement. Despite this, Janina believes that the men were killed in a way consistent with her theory, stating that: "I wasn't lying when I kept insisting it was Animals taking revenge on people. That was the truth. I was their Tool" (Tokarczuk 236). The reader is thus left with a moral justification for the murders of the men that Janina considers responsible for the demise of her 'Little Girls'.

Indeed, Janina's murderous spree, which she imagined would "reverse everything Evil", performs the same violence upon the men which they enacted upon their animal victims (Tokarczuk 234). However, in the context of the novel's ostensible framing as detective fiction, Janina's escape, and her lack of punishment at the conclusion of the novel, raises questions about the novel's ethical position. In a letter to the police outlining her 'Theory', Janina delivers a "petition for the Deer and other eventual Animal Culprits to go unpunished, because their alleged deed[s] [were] a reaction to the soulless and cruel conduct of the victims, who were, as I have thoroughly investigated, active hunters" (Tokarczuk 172). Anna Kowalcze-Pawlik argues that Tokarczuk's novel has at least some role to play in "shaping a growing social awareness of the political uses of anger as an 'outlaw emotion', not setting a precedent but creating an imaginary scenario in which women can take justice into their own hands" (189). While this is no doubt true, I wish to make the case that Janina's 'imaginary scenario' of crime and justice also stages murder as an ecological encounter, whereby criminal violence is committed by and enacted upon human and non-humans alike. While Janina is revealed to be responsible for the homicides, the imagined relationship between non-humans and the legal system which Janina presents in in *Drive Your Plow* is precisely the type of imaginative work which Ghosh calls for in response to the climate crisis of the Anthropocene. In this case, Tokarczuk works within and against "the grid of literary forms and conventions", by breaking the 'cake mould' of detective fiction to re-imagine crime as an ecological encounter between human and non-human beings (Ghosh 7).

The 'Cake Mould' of Detective Fiction: Acting Inside and Outside the Law in *Drive Your Plow Over the Bones of the Dead*

In orthodox examples of classical detective and mystery fiction, the reader comes to the text with certain expectations; namely, that there is a crime to be solved, that it will be solved at the conclusion of the novel and that the solution will restore the legal order of the narrative world which was threatened by the aberrant actions of the criminal (Alewyn 65). Thus, the "dynamics of classical detective fiction" typically give rise to a narrative which "move[s] from obscurity to elucidation" and where mystery ultimately "gives way to method, to a demonstration of the essentially simple – though practically complex – way in which analysis can unlock the world" (Kerrigan 59). In *Drive Your Plow*, the representation of crime, the narrator's investigation and the narrative drive toward solution are all inflected by the perspective of the narrator and protagonist Janina. However, unlike most classical detective fiction protagonists, including Agatha Christie's Miss Marple, Janina is uninterested in restoring the status quo as it currently stands, because doing so would affirm a model of law and order under which people who shoot and kill non-human creatures are also acting "within the law" (Tokarczuk 58). In *Drive Your Plow*, the interests of the male hunters are intertwined with a capitalist logic of extraction wherein the exploitation of the natural world is not only condoned but rewarded, and where environmental protections are regarded as a nuisance, even a detriment to economic productivity. When Janina confronts a group of men hunting pheasants, declaring that they "have no right to shoot at living Creatures!", one of the hunters calmly replies that they are acting "within the law" (Tokarczuk 58, 59). When Janina threatens to call the police, the hunter responds: "There's no need [to call] because the Police are already here . . . don't upset yourself [w]e're within the law" (Tokarczuk 59). As she looks around, Janina realises that the pheasant hunters are implicitly endorsed by the presence of "the pot-bellied figure of the Commandant in the distance" (Tokarczuk 58). This interaction illustrates that, for Janina, acting within the law to identify a criminal lawbreaker, and upholding the law by bringing them to justice, is of little help. As a result, Janina feels that she is left with no choice but to operate outside of the law, because legal protections do not currently extend to non-human creatures, and the police force is led by a man who Janina regards as a complicit, possibly corrupt, authority figure.

Instead, in the words of Janina's beloved William Blake, Janina feels she "must create a system or be enslaved by another man's", a sentiment which echoes Tokarczuk's attitude towards crime fiction in *Drive Your Plow* (28). As mentioned, Tokarczuk works within what she refers to as 'the cake mould' of detective fiction, using the avatar of her detective protagonist Janina to imagine what an investigation of crime under an alternative legal system – one where animals exist on an equal footing with humans – might look like (qtd. in Rejter, 30). In doing so, Tokarczuk presents a metaphysical pastiche of the detective form which deconstructs the ontology of human/non-human rights whilst still managing "to install and reinforce as much as undermine and subvert the conventions and presuppositions it appears to challenge" (Hutcheon 1–2). In other words, in order for *Drive Your Plow* to succeed as a detective narrative, Tokarczuk must first successfully invoke the intellectual puzzle of Golden Age detective fiction. This detective frame is comprised of two parallel narratives, "[t]he first, that of the crime [which] is in fact the story of an absence: its most accurate characteristic [being] that it cannot be immediately present in the book [if at all]" and the second narrative, which is one "of detection" (Todorov 46).

In *Drive Your Plow*, Janina reimagines crime as harm and mistreatment of both human and non-human creatures, a curious doctrine which proves entirely incompatible with that of the local police. The disconnect between Janina's view of crime and that of the police also illustrates the extent to which the failure to recognise non-human harm – and the speciesism which underpins relationships between the villagers and the forest creatures – stems from a failure of imagination and empathy for non-human suffering. When Janina reports the mistreatment of Marysia, Big Foot's dog, and outlines her "observations and suspicions" that Big Foot's poaching "pose[d] a threat to many Creatures, human and animal", the police commandant appears unsure whether she "was making fun of him or if he was dealing with a madwoman [t]here were no other possibilities" (Tokarczuk 25). The commandant simply cannot imagine a situation where animal rights are enshrined and enforced by law and responds to Janina's complaint by assuring her that "[w]e had no idea [Big Foot had] been poaching", attempting to placate her by telling her that "[w]e'll see to it [p]lease go home, and don't worry about it. I know [Big Foot] well" (Tokarczuk 25). There is an overwhelming sense that the hunters are all part of a boys' club, the inference being that while hunters occupy positions of power in the village, nothing will change. Even if the police were responsive to Janina's concerns, she is left with little legal recourse, because, in the commandant's words, Big Foot's abuse of Marysia "is not a matter for the Police [a] dog is a dog [w]hat do you expect? Dogs are kept in kennels and on chains" (Tokarczuk 26).

Janina's theory of murder in *Drive Your Plow* is, crucially, an ecological reading of crime. Critical perspectives on ecological crime fiction share an interest in texts which shift the emphasis of crime to "the story of our ecological awareness" (Hollister 9). Whether it is termed "dark ecology", "green noir" or "ecocrime" fiction, this subgenre is built upon representations of the environment as bound up in crime and emphasises the "human depredation of nature" (Rugg 607). While Janina's theory is developed over nearly the entirety of *Drive Your Plow*, the clearest description of 'crime', in her terms, comes towards the novel's end, following her confession and unmasking as the real murderer. In describing the photograph found in Big Foot's dresser, Janina provides the motive for her murders and outlines her vision of ecological harm as "proof of a Crime":

> [I]t was all plain to see in the photograph. The best proof of a Crime that one could possibly imagine. There stood the men in uniforms, in a row, and on the grass in front of them lay the neatly arranged corpses of Animals . . . like little dots, as if those Animal's bodies were a sentence written to me . . . form[ing] a long ellipsis [as if] to say "this will go on and on.". . . In the corner of the picture lay three dead Dogs, neatly laid out, like trophies. One of them was unfamiliar to me. The other two were my Little Girls.
>
> *(Tokarczuk 228–9)*

Janina's vision of "the best proof of a Crime" is the grisly display of animal corpses as hunting trophies, which for her represent victims of murder, including her two "Little Girls". For Janina the "long ellipsis" of sanctioned violence represented by the neatly arranged animal corpses also signifies the cyclical human violence endemic to the Anthropocene, a crime that "will go on and on" unless addressed and stopped (Tokarczuk 228).

The contrast between Janina's reimagination of crime and the myopic view of crime taken by the police commandant is defined by gender biases, homosocial loyalties and the systemic limitations of the law, but also by the commandant's failure to imagine "other possibilities" and motives for murder involving non-humans. When Janina reports a 'murder' after witnessing the "Crime [of a] poor young Boar . . . lying in a pool of brown blood", another victim of illegal poaching, the city guard responds that she has "more compassion for animals than for people" (Tokarczuk 92). Janina's response to the guard provides the reader with a concrete sense of her vision of crime in contrast to the injustices of the existing system:

> "[T]hat's not true. I feel just as sorry for both. But nobody shoots at defenceless people", I told the City Guard that same evening . . . "True. We're a law-abiding country", confirmed the guard. He seemed good-natured and not very bright. "It's Animals that show the truth about a country", I said. "Its attitude towards Animals. If people behave brutally towards Animals, no form of democracy is ever going to help them, in fact nothing will at all".
>
> *(Tokarczuk 92–3)*

Janina's doctrine lays bare the "the truth about [the] country" in which she lives and highlights the inherent disingenuousness of a so-called "law-abiding" society that permits the routine murder of non-human animals. In response, she reimagines a story of detection in order to make ecological violence a detectable crime, because if such killing "become[s] exempt from punishment . . . nobody notices it anymore" and if "nobody notices it, it doesn't exist" (Tokarczuk 96). Following Ghosh's call to action, the first step towards addressing broad scale environmental harm is making that harm visible. In *Drive Your Plow*, Tokarczuk does so by reimagining violence inflicted upon non-humans by the village hunters and poachers as a series of criminal acts which are also murders to be solved.

Janina's Theory of Murder: A Democratic Encounter Between Human and Non-Human Beings

In the context of a society where "[c]rime has come to be regarded as a normal, everyday activity [and] [e]veryone commits it", Janina's investigation of the murders is shaped by her own standard of wrongdoing which lies beyond the limits of the existing law and thus is no longer "a matter for the police" (Tokarczuk 96, 26). Janina's rejection of the "form of democracy" which enables and implicitly condones the murder of animals echoes Timothy Morton's characterisation of ecological thinking, a paradigm that Tokarczuk brings to her reimagination of crime fiction in *Drive Your Plow* (Tokarczuk 96). For Morton, thinking ecologically is not just an abstract exercise which occurs in the mind, but is also

> a practice and a process of becoming fully aware of how human beings are connected with other beings – animal, vegetable, or mineral. Ultimately, this includes thinking about democracy [and what] a truly democratic encounter between truly equal beings [might] look like, what would it be – can we even imagine it?
>
> *(Morton 8)*

The act of "[t]hinking about [what] a truly democratic encounter between truly equal beings might look like" is precisely the sort of epistemological work that Janina undertakes over the course of the novel. By imaginatively reconstructing the murders of the hunters as a series of crimes which treat human and non-human creatures as "truly equal beings", Janina demonstrates that such a vision is not wholly unimaginable at all (Morton 8). Janina's theory that animals are taking active revenge on people implies that animals are endowed with sufficient intelligence to recognise injustices and harms visited upon them by the hunters and poachers of the village (96). In this sense, the crime to be investigated is not simply a question of 'whodunnit', but rather, as Przemysław Czapliński and Bartosz Woźniak write, a question "about the right to kill . . . a disclosure of the law's hidden basis and a counteraction against it" (29, 30).

In *Drive Your Plow*, Janina goes into considerable detail in developing her philosophy of animal vengeance, initially sharing her hypothesis regarding Big Foot's demise with her friend Oddball: "[y] ou know what I think about it don't you? [the] Deer that were standing outside his house when we got there [t]hey murdered him" (79). Although Janina is not sure of the specifics of Big Foot's murder, she theorises that "[m]aybe they just gave him a fright while he was so barbarously eating their sister" and caused him to choke to death (79). Later Janina outlines her interpretation of the commandant's death in a letter to the police, where she argues that "the marks of deer hooves" at the crime scene suggest that "the deceased had been lured out of his car and led into the undergrowth, under which the fatal well was hidden" (171). Thus, in her view, it becomes "highly probable that the Deer he persecuted inflicted summary justice" by "surround[ing] him" and slowly "push[ing] him towards the well" until he fell in (180). In the case of Innerd's murder, Janina describes a scenario that is even "eas[ier] to imagine", one in which Innerd is "enticed into the bushes" by tame foxes and then "falls into [a] trap and is deprived of his life" (171). Janina's use of formal language – "the deceased", "highly probable", "summary justice", "deprived of his life" – evokes the crisp, logical inferences associated with crime and homicide reports, bridging the divide between ecological thinking and the ratiocinative logic of classical detective fiction wherein "[m]ystery gives way to method" (Kerrigan

59). Within Janina's theory, as in Coetzee's *Disgrace*, the killing of animals by humans breaks down and confuses the linguistic distinctions between human and non-human as well as killers and victims.

By inscribing the animals' revenge killings with a cold, calculated logic, where summary justice becomes an eye for an eye, Janina imagines Morton's "democratic encounter" as an equivalent exchange of criminal violence between humans and non-humans. In Janina's description of the murders of Big Foot, the commandant, and Innerd, her use of verbiage also ascribes the animals with a sense of agency that seems almost human; "lur[ing]" the commandant and "inflicting" justice, "enticing" Innerd and giving Big Foot a fatal fright. The degree of agency that Janina ascribes to the animals brings to mind a further implication of Morton's ecological thought: that imagining a truly democratic encounter between equal beings implies a certain degree of intelligence on behalf of our non-human counterparts. Morton elaborates on this argument with a series of checks:

> [D]o nonhumans possess language? Yes. How about imagination? Check. Reason? Copy that. A sense of mind? No doubt. Can they use tools? Indeed. Do they display improved skills and learning over time? Absolutely. Can nonhumans feel compassion? Of course . . . If butterflies have the capacity to make a choice, then surely it's game over for rigid distinctions between humans and nonhumans?
>
> *(70–1)*

Janina's own worldview necessarily acknowledges that animals possess the type of cognitive faculties which would allow them to comprehend exploitative violence as 'wrong'. It is a democratic vision of crime which doubles back on the hunters, who are subsequently positioned as both perpetrators and victims. Via Janina's detective logic, non-human animals are capable of understanding and enforcing 'summary justice', which they do so through spectacular acts of revenge.

Understandably, this is an uncomfortable proposition for most of Janina's peers in the village, as indicated by Oddball's incredulous response to Janina's narrativisation of Big Foot's and the commandant's murders: "[a]re you trying to say it was collusion [t]hat the deer conspired against [them]?" However, Janina's description of Innerd's demise dramatises the hubris born from stubbornly confining one's mind to "rigid distinctions between humans and nonhumans" (Morton 70–1). Janina speculates that when Innerd arrived at his farm on the day of his death, he looked out the window at the forest and saw

> some beautiful, fluffy wild red Foxes [who were not] in the least afraid; they were just sitting there like Dogs, steadily watching him in a challenging way [m]aybe in his small, avaricious heart a hope was born – that here he had chanced upon an easy profit, for such tame, beautiful Foxes could be lured and caught. But how come they're so trusting and tame? he thought.
>
> *(Tokarczuk 157)*

In Janina's reconstruction of Innerd's murder, his failure to imagine the foxes as anything other than creatures to "be lured and caught" and exploited "for an easy profit" is leveraged against him (157). Innerd is distracted by thoughts of turning the fox cub's pelts "into . . . fine fur collar[s]" to sell and sees their "trusting and tame" nature only as a source of personal gain, leaving him vulnerable to the vengeful designs of the "beautiful, noble, Animals with wise faces" (158). In Janina's imagination, Innerd is both an exploitative hunter and a symbol of the voracious forces of neoliberal capitalism which ruthlessly exploit the "trusting and tame" creatures and abundant resources of the natural world in the single-minded pursuit of easy profit.

According to Janina, it is Innerd's blindness to the foxes' intelligence that allows the animals to entrap him and enact revenge in the form of poetic justice. Waiting in wake for Innerd, the animals proceed to play "a strange game with him, vanishing from sight and then reappearing, two, three of them . . . fluffy fox cubs too" as a means of enticing him deeper into the forest (157–158). However, just as Innerd begins to advance towards the cubs, "legs bent, leaning forwards, with a hand stretched out ahead of him . . . his fingers pretend[ing] to be holding a tasty morsel" he becomes caught in a snare (158). Janina's speculative reconstruction of the 'crime' captures Innerd's ironic reversal from hunter to hunted, his death crystallising the fallacy of anthropomorphic visions of nature where the submissive or "sentimental aesthetics" of animals becomes "an obstacle to ecological thought" (Morton 8). Instead, Morton argues that ecological thought must include "negativity and irony, ugliness and horror" because ecological democracy "is well served by irony [which] insists that there are other points of view that we must acknowledge" (Morton 17). Innerd's ironic demise in his own snare, so tight around his leg it was "as if it had grown into the flesh", becomes a grim parable for the self-destructive nature of the capitalist pursuit of profits, which is propelling the world exponentially towards the climate crisis (154, 155).

An untidy *Denouement* and Unresolved Questions of Justice

The hunter who "passed his own death sentence" by stumbling into his own fox trap speaks to the double-edged nature of thinking ecologically, or democratically, about crime in *Drive Your Plow*. More broadly, Janina's reimagination of ecological crime as "a long ellipses" of animal corpses which "will go on and on" lies at the heart of her radical conception of environmental justice (228–9). The ironic nature of Innerd's demise dramatises the hubris stemming from a failure to respect the democracy between beings, in turn echoing the broader self-sabotaging nature of environmental harm associated with the Anthropocene. Janina's theory actively seeks to change both the philosophy towards, and regulation of, environmental harm by effecting a paradigm shift both in the minds of her peers in the narrative world of the novel. In so doing, Tokarczuk also challenges her readers' attitudes towards the environment "by shifting the focus away from an anthropocentric worldview towards a more inclusive, biocentric one", where animals can be both murder victims and murderers (Hübben 2). By reimagining crime as an ecological encounter between humans and non-humans on an equal, even democratic footing, Janina exposes the blindness which underwrites "the Anthropocene, the age of humanity's destruction of the biosphere", which will go on and on if left unaddressed (Rugg 607).

However, Janina's murderous spree, and her subsequent escape from justice, prompts us to question the extent to which her imagination of justice is truly ecological at all. In fact, there is an argument to be made that Janina demonstrates an overly cosy identification with the animals on whose behalf she professes to act, which itself implies the subsistence of a form of species hierarchy. Janina's habit of referring to deer as "Young Ladies" or to wild foxes as "fluffy" and "wise faced" reveals a predilection for a kind of "sentimental aesthetics" which, as I've noted earlier, Morton feels "is an obstacle to the ecological thought" (8). Morton elaborates further on the topic, noting that,

> [w]e can't in good faith cancel the difference between humans and nonhumans. Nor can we preserve it. Doing both at the same time would be inconsistent. We're in a bind [but] [t]he bind is a sign of an emerging democracy of life forms.
>
> *(76)*

At times in *Drive Your Plow*, Janina seems in danger of falling victim to this bind and "cancel[ling] the difference between humans and nonhumans" (Morton 76). When Janina makes an impassioned defence of animal rights in an unjust world "where killing and pain are the norm", Tokarczuk ventriloquises this criticism through the police guard's response (98). The guard accuses Janina of "hav[ing] more compassion for animals than for people" because her feminine "instinct for caring" is foisted upon animals now that she "doesn't have anyone [human] to look after anymore" (93, 98). However, while Janina does have a compassionate, even motherly affection for the deer, the "Young Ladies", as well as her "Little Girls", by treating all animals as both potential victims and as potential killers she makes an interesting counterpoint: that actually she "feel[s] just as sorry for both" humans and animals, however, "nobody shoots at defenceless people" (Tokarczuk 93).

Whether Janina's justification for the murders is an elaborate psychological mechanism to side-step her own feelings of compunction as a murderer, or whether she is a zealous agent of ecological democracy at its most brutal – or both – is not entirely clear. Her 'confession' ascribes all agency for the killings to the 'wise' animals of the forest, specifically

> [t]hose Deer [she and Oddball] saw outside Big Foot's house . . . chose me from among others – maybe because I don't eat meat and they can sense it – to continue to act in their Name. They appeared before me . . . to have me become the punitive hand of justice, in secret. Not just for the Deer, but for other Animals too.
>
> *(Tokarczuk 230)*

Here all intent and culpability rests on the side of the Animals, who are the real criminal architects with Janina merely acting as an intermediary, their 'Tool'. However, Janina's reasoning is reminiscent of the problematic attitude demonstrated by some deep ecologists who consider themselves "empowered to act on behalf of other beings" (Macy 210). Arguably, Janina's claims of "acting in [the] Name" of nature either assumes direct access to the desires and wishes of the Animals or projects her own desires onto them. In either case, she trivialises the real differences between human and non-human beings and her theory, in Val Plumwood's words, ultimately retains "many of the problems of moral extensionism [and] remain[s] in a subtle way human-centred" (197).

Thus, despite *Drive Your Plow's* radial ecocritique, Janina often risks inhabiting the same kind of human-centred imagination of deep ecology to which Plumwood refers, projecting her own vengeful desires upon the animals. Although Janina professes to kill on behalf of the wronged animals and act "in their Name", she sets a trap to kill Innerd because it's what she '*wanted* to happen' (230, 235, my emphasis). Janina confesses that she set off to see Innerd "on purpose [and] had prepared a trap" earlier that day, leading Innerd towards the snare, "assuming that my trap would break his neck, just like a Deers" and admits that's "what I wanted to happen, because he'd fed my Little Girls' bodies to the Foxes [and] [b]ecause he hunted" (235). Janina's second account of the murder, the real story of the crime in the detective fiction framework, pushes animal agency to the background, suggesting that her revenge was really about vengeance for her "Little Girls", rather than an expression of the animals' own intrinsic sense of injustice. Janina's reimagination of "Animals taking revenge on people" stands on its own as an adaptation and reimagination of the detective form as a violent enforcement of ecological democracy. However, Janina's ultimate confession seems to undo the subversive implications of her theory by isolating her as the criminal and letting the hunters off the hook – at least those that Janina hasn't already murdered.

Janina's confession and escape from justice in *Drive Your Plow* also upsets a crucial final element of the classical detective fiction narrative known as the *denouement*, wherein the triumphant detective successfully identifies the criminal. Typically, the announcement of the killer has the cathartic effect of dispersing the "universalized, free-floating guilt" that stems from the implication that we are all potentially murderers "in the unconsciousness of our desire" (Žižek 59; Davis 294). In other words, as Colin Davis puts it, "[t]he detective is a guardian of the law who proves our innocence by isolating the guilt of others" (294). Despite her vivid reconstructions of the murders enacted by wise animals, Janina confesses to the crimes at the conclusion of the novel. However, the question of what fate should befall the other hunters who escaped her brand of vengeful justice is less clear. Janina's confession doesn't exculpate any of the human hunters or poachers, and Janina herself remains at large. It is only the non-humans who remain innocent of wrongdoing. Thus, rather than leaving the reader with the "gift of a tidy world . . . in which all questions have been answered", Janina's confession invokes the subversive and destabilising effects of *film noir*, inducing "attenuated but insistent feelings of guilt (Grossvogel 81; Hollister 1011).

Part of the untidiness of Janina's confession stems from the lingering question relating to forms that meaningful ecological justice should (or could) take. Legally speaking, this relates to the prosecution of ecological crime and what determines the basis of a guilty verdict. Indeed, if Janina was tried, she "would be sentenced by the law which says: it is forbidden to kill a man, because human life is of immeasurable value" (Czapliński and Woźniak 30). However, by reimagining crime to include harm to both human and non-human beings, Janina also lays bare the faulty logic of a legal system in which the hunting and killing of animals is positioned within the law, which implies that "human life is [of] value, because we are not animals" (Czapliński and Woźniak 30). The truly destabilising implication which runs beneath Janina's logic is that the natural world, particularly its non-human inhabitants, are capable of being both victims and criminals in a *legal* sense. Just as Janina's feminine vision of detection unsettles the "boundaries of the status quo" (Makinen 61) by pushing "off-center the whole male/female . . . intellect/emotion, physical strength/weakness dichotomy" (Klein 4), her insistence on treating humans and non-humans as equal 'beings' under the law likewise constitutes "a revision of the entire human order [in a] moral, legal, and metaphysical" sense (Czapliński and Woźniak 31). Janina even provides the reader with a picture of what this form of legal system might look like: according to her, historically, animals have been tried both in "cases of Murder" as well as for crimes "against nature", such as a 1471 Basel suit "against a Hen, which laid strangely coloured eggs" (Tokarczuk 171).

Janina earnestly appeals "to the gentlemen of the Police not to shy away from the idea that the perpetrators of the above-mentioned tragic incidents could be Animals" citing a French case in 1639 where "a court in Dijon sentenced a Horse for killing a man" (Tokarczuk 171). Yet, she is forced to admit that it's been "a long time since we have had cases of crimes committed by these creatures" heard in court (Tokarczuk 171). While her examples may seem absurd, Janina's reimagining of the climate crises as an equivalence of legal rights between humans and non-humans does have a historical precedent, and laws are, after all, subject to change. The lack of resolution in *Drive Your Plow* forces the reader to question the limitations of the law as it currently stands and consider both the implications and possibility of treating humans and non-humans democratically under the legal system. By having both Janina and the hunters and poachers (who weren't murdered) escape justice, the novel echoes the feelings of injustice associated with a real-world system where, individually and collectively, perpetrators of environmental harm remain unpunished by, or entirely outside of, legal systems. Just as Janina imagines an ecological encounter as criminal violence between human and non-human creatures, we might reimagine the violence of the Anthropocene as an endless series of unresolved crimes. This imaginative exercise is no mere abstraction either; the first step to preventing crime is recognising that a crime has occurred in the first place.

Bibliography

Alewyn, Richard. "The Origin of the Detective Novel." *The Poetics of Murder: Detective Fiction and Literary Theory*, edited by Glenn W. Most and William W. Stowe, Harcourt Brace Jovanovich, 1983, pp. 62–78.

Blake, William. *Jersusalem: The Emanaion of the Giant Albion*. Princeton UP, 1991.

Czapliński, Przemysław, and Bartosz Woźniak. "Concatenations: On the Works of Olga Tokarczuk." *The Polish Review*, vol. 66, no. 2, 2021, pp. 8–35, doi:10.5406/polishreview.66.2.0008.

Davis, Colin. "Psychoanalysis, Detection, and Fiction: Julia Kristeva's Detective Novels." *Contemporary French and Francophone Studies*, vol. 6, no. 2, 2002, pp. 294–306, doi:10.1080/718591983.

Franklin, Ruth. "Olga Tokarczuk's Novels against Nationalism." *The New Yorker*, 29 July 2019, www.newyorker.com/magazine/2019/08/05/olga-tokarczuks-novels-against-nationalism.

Ghosh, Amitav. *The Great Derangement: Climate Change and the Unthinkable*. U Chicago P, 2016.
Grossvogel, David. *Mystery and its Fictions: From Oedipus to Agatha Christie*. James Hopkins, 1979.
Hollister, Lucas. "The Green and the Black: Ecological Awareness and the Darkness of Noir." *PMLA/Publications of the Modern Language Association of America*, vol. 134, no. 5, 2019, pp. 1012–1027, doi:10.1632/pmla.2019.134.5.1012.
Hübben, Kelly. "Animals and the Unspoken: Intertwined Lives in Martha Sandwall-Bergström's Kulla-Gulla Series." *Barnboken*, vol. 36, 2013, doi:10.14811/clr.v36i0.149.
Hutcheon, Linda. *Politics of Postmodernism*. Taylor and Francis, 2003.
Kernev-Štrajn, Jelka. "Zoper 'Naravni' Red Sveta (Against the 'Natural' Order of the World)." *Primerjalna književnost*, vol. 39, no. 2, 2016, pp. 71–88.
Kerrigan, John. *Revenge Tragedy: Aeschylus to Armageddon*. Clarendon Press, 1996.
Klein, Kathleen Gregory. *The Woman Detective: Gender and Genre*. 2nd ed., U Illinois P, 1995.
Kowalcze-Pawlik, Anna. "Madwoman at Large: Prophetic Anger and Just Revenge in Olga Tokarczuk's *Drive Your Plow over the Bones of the Dead*." *The Polish Review*, vol. 66, no. 2, 2021, pp. 187–202, doi:10.5406/polishreview.66.2.0187.
Lundblad, Michael. "Introduction: The End of the Animal – Literary and Cultural Animalities." *Animalities: Literary and Cultural Studies Beyond the Human*, edited by Michael Lundblad, Edinburgh UP, 2017, pp. 1–21.
Macy, Joanna. "Awakening to the Ecological Self." *Healing the Wounds: The Promise of Ecofeminism*, edited by Judith Plant, New Society Publishers, 1989, pp. 201–211.
Makinen, Merja. *Agatha Christie: Investigating Femininity*. Palgrave Macmillan, 2006.
Merivale, Patricia, and Susan Elizabeth Sweeney. "The Game's Afoot: On the Trail of the Metaphysical Detective Tale." *Detecting Texts: The Metaphysical Detective Story from Poe to Postmodernism*, edited by Patricia Merivale and Susan Elizabeth Sweeney, U Pennsylvania P, 1999, pp. 1–24.
Morton, Timothy. *The Ecological Thought*. Harvard UP, 2010.
———. *Dark Ecology*. Columbia UP, 2016.
Parry, Sarah. "Drive Your Plow Over the Bones of the Dead by Olga Tokarczuk – The Entire Cosmic Catastrophe." *The Guardian*, Guardian News and Media, 21 Sept. 2018, www.theguardian.com/books/2018/sep/21/drive-your-plow-over-the-bones-of-the-dead-review.
Piotrowska, Agnieszka. "What Does (a Nasty) Woman Want?" *Feminist Encounters: A Journal of Critical Studies in Culture and Politics*, vol. 4, no. 2, 2020, p. 33, doi:10.20897/femenc/8521.
Plumwood, Val. *Environmental Culture: The Ecological Crisis of Reason*. Taylor and Francis, 2001.
Rejter, Artur. "Literature Against the Discourse of Posthumanism. On the Example of Prose by Olga Tokarczuk." *Język Artystyczny*, vol. 1, no. 16, 2017, pp. 27–47.
Rugg, Linda Haverty. "Displacing Crimes against Nature: Scandinavian Ecocrime Fiction." *Scandinavian Studies*, vol. 89, no. 4, 2017, pp. 597–615, doi:10.5406/scanstud.89.4.0597.
Schmid, David. "The Locus of Disruption: Serial Murder and Generic Conventions in Detective Fiction." *The Art of Detective Fiction*, edited by Warren Cherniak, Martin Swales, and Robert Vilain, Palgrave Macmillan, 2000, pp. 75–89.
Todorov, Tzvetan. "The Typology of Detective Fiction." *Modern Criticism and Theory: A Reader*, edited by David Lodge, Longman, 1988, pp. 158–165.
Tokarczuk, Olga. *Drive Your Plow Over the Bones of the Dead*. Translated by Antonia Lloyd-Jones, Text Publishing, 2019.
Toynton, Evelyn. "Written in the Stars." *TLS*, 16 Nov. 2018, www.the-tls.co.uk/articles/tokarczuk-astrology-suffering-animal-rights/.
Walton, Jo Lindsay, and Samantha Walton. "Introduction to Green Letters: Crime Fiction and Ecology." *Green Letters*, vol. 22, no. 1, 2018, pp. 2–6, doi:10.1080/14688417.2018.1484628.
Weil, Kari. "Killing Them Softly: Animal Death, Linguistic Disability, and the Struggle for Ethics." *Configurations*, vol. 14, no. 1, 2008, pp. 87–96, doi:10.1353/con.0.0013.
Wzorek, Anna. "Olga Tokarczuk's Game with the Rules of Criminal Novels: The Case of 'Plough through the Bones of the Dead'." *Respectus Philologicus*, vol. 24, no. 29, 2013, pp. 98–107, doi:10.15388/respectus.2013.24.29.8.
Žižek, Slavoj. *Looking Awry: An Introduction to Jacques Lacan Through Popular Culture*. The MIT Press, 1992.

11

PROTECTING THE RHINOS AND OUR YOUNG DEMOCRACY

Nature and the State in Post-Apartheid South African Crime Fiction

Colette Guldimann

The emergence of crime fiction as the most popular genre in post-apartheid South Africa is testimony to the fact that, as Andrew Pepper and David Schmid point out, the globalisation of crime fiction is something of a fait accompli and "indigenous crime cultures are now emerging from, and speaking to, their own sites of production" (1). South Africa might be taken as an interesting counter-example of Pepper's and Schmid's critique of celebratory descriptions of the globalisation of crime fiction that ignore the role of the state. They stress that the globalisation of crime fiction "should not simply be understood as a one-way process whereby the genre moves to populate the globe" (Pepper and Schmid 3). Instead, they argue that the "global implications" of the crimes on display must be considered, necessitating new modes and "new strategies of representation in order to do justice to a changed and changing world" (Pepper and Schmid 3). Far from the state being seen as irrelevant, crime fiction in post-apartheid South Africa is currently viewed by many critics as the new form of politically engaged fiction, replacing the "literary" novel, which played this role under apartheid (de Kock 34).[1] This is due to crime fiction's rise to prominence in the decade following the first democratic elections in 1994 and, according to critics, its role in facilitating, and responding to, this transition. A popularly held idea in South African crime fiction criticism is a version of postcolonial disillusionment, first propagated by Leon de Kock (2016), who views the predominance of all forms of crime writing in South Africa as a response to the political transition in 1994: the celebration and mythology of the "rainbow nation", which was followed by a deepening sense of disillusionment with the state abandonment of the values of the anti-apartheid struggle on which democracy was founded (a discussion I will return to later in the chapter).

While many critics now recognise South African crime fiction as a form that is capable of responding to political issues, there has been little attention to the question of ecology, despite the fact that several established writers of crime fiction series have produced novels with environmental themes, either woven into the plots or presented as the main problem. One exception to this is Sam Naidu (2014), who examines the work of Deon Meyer, a pioneer of South African crime fiction. When commentators mention environmental themes in South African crime writing, it is Meyer who is most commonly cited, specifically his novel *Blood Safari* (2009).[2] For Naidu, the conventions of the crime thriller – "anthropocentric narratives with the detective occupying centre-stage" – ultimately determine the form of Meyer's novels, with "specific environmental issues" informing the plots "without necessarily becoming focal elements" (60). Despite her reservations about Meyer's texts, Naidu

 DOI: 10.4324/9781003091912-14

concludes that "the inclusion of ecological concerns" consolidates "the status of the post-apartheid crime novel as the political novel of our time" (69). Yet she does not give any indication of how ecological concerns might relate to crime fiction's function as a political novel. This chapter will present the missing piece by locating ecological issues within the debate about South African crime fiction as a form of politically engaged fiction and triangulating the relationship between crime fiction, politics and ecology.

In doing so I am going to draw on Pepper's and Schmid's introduction to *Globalization and the State in Contemporary Crime Fiction*, where they identify three distinct "approaches the contemporary crime novel takes to mapping the relationship between crime, globalization and the state" (10). These involve various degrees of complexity in terms of the relationship between the state and globalisation and the increasing hybridisation of the form required in order to trace these relations. I suggest that Pepper's and Schmid's categories can be usefully adapted to begin to map the trajectory of environmental discourse within post-apartheid crime fiction. I propose to extend Pepper's and Schmid's examination of new typologies to include the various hybridisations of the genre demanded by ecological questions, and I adopt Marta Puxan-Oliva's idea that a serious incorporation of environmental concerns "challenges genre conventions and drives crime fiction towards hybridization" (362). This examination will extend the existing discussion about crime fiction in South Africa to reveal how environmental issues raise questions about the state *as well as* globalisation.

The questions raised about the tension between genre and environment are global. Can the conventions of the genre accommodate ecological concerns in a meaningful way and, if so, what would the South African context have to contribute to an understanding of this? Patrick D. Murphy differentiates between authors whose use of nature seems "a matter of using a convenient topical variation to change the setting or the backdrop for their formulaic plots" and those for whom "nature conservation and environmental justice appear as deep-seated beliefs held by the authors who have found ways to bring their causes into the production of commercially successful novels" (119). Puxan-Oliva agrees, stating that "[n]ot all crime novels dealing with the environment use it for the same ends" (362). She adopts Abhra Paul's suggestion that "in several crime novels environmental issues are marginal to the story's focus, while in others a crime related to the environment is the fictional core problem" (qtd in Puxan-Oliva 362). In what follows, I map the developing trajectory and positioning of environmental crime, in relation to the crime fiction genre, in South Africa.

Solving Individual Crimes While Environmental Crime Remains Unresolved

Pepper and Schmid begin with more traditional models of crime fiction, "those that feature a single protagonist operating within a geographically limited setting" (10). While such novels are less hybrid in their form, they are still able to "generate a complex and multi-layered understanding of the relationship between crime and neo-liberal capitalism" (10). The emphasis in these novels is "less on individual acts of criminality than on the ways in which structural changes to national and global economies produce conditions where crime is all but inevitable" (10). These ideas can be usefully adapted to analyse early post-apartheid crime fiction from the decade of its commencement, the 2000s. The novels presented here are more or less traditional in terms of genre and while presenting individual acts of criminality they seem, to me, to be more interested in structural elements within national and global economies which produce crime. A uniquely South African aspect is that the investigation of a contemporary crime leads back into the apartheid past to reveal the ways in which the apartheid system generated criminality. These novels are typical of a type of crime fiction written in the wake of South Africa's Truth and Reconciliation Commission (TRC), a public exercise in bringing the political crimes of the past to light. In such novels, the detective often becomes an agent of historical correction, exposing crimes that have (and would have) remained hidden and bringing about retrospective justice within, or without, the law. While there has been some critical attention to crime fiction and the TRC, to date nothing has been written about the role played by environmental crime in such novels.

One of the earliest novels to weave environmental crimes into its murder plot is Richard Kunzmann's *Salamander Cotton* (2006). In place of a single protagonist, it is a hybrid police procedural and private investigator (PI) novel, as there are two main characters, a police detective and a PI, working on two interlinked cases. The novel, set in 2004 (though its narratives delve back into the apartheid past), opens with the gory murder of Bernard Klamm, a wealthy retired former asbestos mine owner, and this crime is investigated by Detective Inspector Jacob Tshabalala. Simultaneously, Harry Mason, a former police detective, is hired by Klamm's estranged wife to investigate the disappearance of their daughter some thirty years prior to the murder. As Tshabalala investigates Klamm's past, readers learn of his brutal treatment of workers in the asbestos mines he owned. Klamm mined the very toxic crocidolite, blue asbestos, and the earliest parts of the story date back to the 1950s. In a section dated "October 1958", the narrative is presented from the point of view of eleven-year-old Obed Ditlholelo, whose family work for Klamm, and readers learn of Klamm's cruelty, his role in the death of Obed's father and the subsequent illness and death of various family members as they fall ill with "the cough" (200–202).

While this provides insight into Klamm's racism, Tshabalala discovers that Klamm was responsible for covering up a medical epidemic and environmental disaster caused by the asbestos mining industry. His past actions are explained by the vice health director of the Centre for South African Occupational Health, Dr Elizabeth Reynolds. She relates how, in the 1950s, doctors "demonstrated a link between mesothelioma and asbestos exposure – *any* asbestos exposure" and that according to the research "we had an epidemic on our hands and so needed to move fast" (274). In England, with its health and safety standards, it took workers about eight years to develop asbestosis, while in the Northern Cape, people who didn't even work in a mine were contracting the disease as asbestos dust "coated everything, from growing fruits to porches, to the mosquito nets over people's beds" (277). As an executive of a mining conglomerate, Klamm's intervention ensured that the doctors' report was "rewritten, played down, banned from the media for fear of inciting both mutiny on the mines and panic in the thousands of households into which asbestos had been introduced" (276).

When Tshabalala asks how bad the epidemic *was*, Dr Reynolds corrects the past into present tense: "*Is*. New cases are still on the rise" (277). She claims, "it's the worst environmental disaster this country has ever seen, more so because we have no record of its true extent because so many important documents have been destroyed" (277). Thus, while the plot investigates Klamm's murder, this leads to the exposure of his crimes, which are linked to the structural exploitation and greed of the state apartheid system and the international capital that backed it. The apartheid government wasn't concerned about "black miners and it wanted its position as the world's largest producer of crocidolite . . . protected at all costs" (276). Klamm was able to keep the medical information hidden "until world sanctions against apartheid, and the toxic mineral itself, forced these companies and the government" to finally accept that South Africa's days of international asbestos production were over (276).

Both cases, the murder and the disappearance, are solved by Tshabalala and Mason respectively. The case of the daughter's disappearance also turns into a murder, and the two investigations invoke the structural conditions of apartheid, which made crime inevitable. Klamm's daughter was murdered by Obed in revenge for his father's death and, in fact, Obed had wanted to kill Klamm as well. By introducing Obed's story into the narrative, the resolution of the second case thus questions individual culpability, as Klamm is positioned just as much as a criminal and Obed as a victim. Klamm's murder was arranged by his estranged wife's current partner in order to inherit Klamm's farm, still rich with other mineral deposits. The murder was executed by another former mine worker, José Eduardo Cauto, who was wrongfully accused (by Klamm) and convicted of murdering Klamm's daughter and spent thirty-nine years in prison. This was Klamm's revenge for an interracial affair between Cauto and his daughter during apartheid. The structural conditions that facilitated brutality and the greed of the privileged few thus underpin both crimes.

While the two murder cases are solved, it is clear that the environmental disaster is left unresolved. The "widespread contamination" of the South African environment "due to historical asbestos mining operations that were poorly regulated" is documented as an ongoing concern (Ndlovu, te Water Naudé and Murray 82). This is vividly demonstrated when Mason falls into in an unmarked raw asbestos site while pursuing a suspect and feels that his body is being "contaminated by millions of hairy spiders" crawling "into his mouth, down his throat and into his lungs, their legs pricking him like needles, the way the fibre itself might" (389). The novel thus presents an example of what Puxan-Oliva calls "problematizing the genre through the conventional, fully fleshed and explicit resolution of the murder" (368). The crime is solved "at the individual level", but what remains "unresolved is the environmental crime" (368).

Deon Meyer's *Blood Safari* (2009) introduces his recurring character, Lemmer, a former Ministerial bodyguard, who is hired by Emma Le Roux to act as her "invisible" while she attempts to track down her long-lost brother, Jacobus Le Roux, previously assumed dead. "Cobie", as he is known, has been living under an assumed identity at a vulture rehabilitation centre and has gone missing after shooting a Sangoma (traditional healer) and three local men as revenge for the poisoning of fourteen endangered vultures in order to make traditional medicine. The location of the action in various game parks, as well as the rehabilitation centre, weaves an ecological discourse into the novel (see Naidu 64). *Blood Safari* sets up tensions between conservationists and indigenous groups in the context of democratic South Africa. The first is Cobie's revenge on the Sangoma for killing the endangered birds. The second is the conflict between conservation parks and land reform claims by Indigenous people displaced from what is now the Kruger National Park by colonialism and apartheid. The novel thus introduces questions about ecotourism and ecoambiguity (see Thornber 2012). Far from celebrating national conservation parks, Meyer locates the origin of conservation within the greed of colonialism. This is done through the perspective of black police Inspector, Jack Phatudi, who states that "the Boer" created the Kruger National Park because "the whites, had killed nearly everything and they wanted to save the last few", (115) just as they shot thousands and thousands of elephants because they were "poor and ivory was good money", but that's considered okay "because they were white and it was a hundred years ago" (116). Now the Sibashwa people, who are claiming their original land in the Kruger National Park back, are poor, and it is Phatudi who explains that the problems are socio-economic and suggests a solution: "we need to make jobs for the people, then they will leave the vultures alone" (116).

While the novel raises various political questions regarding conservation, Indigenous land rights and issues of environmental justice, the plot ultimately returns to the apartheid past to explain the mystery of Cobie's disappearance. In 1986, Cobie was an accidental witness to a plane crash, the plane carrying Samora Machel, the president of independent Mozambique. This was orchestrated by aerospace systems developer, Southern Cross Avionics, in order to demonstrate their cutting-edge technology to the apartheid government in exchange for manufacturing contracts. Although Lemmer insists that he is not a detective, the conventional PI narrative takes over the novel. He single-handedly tracks down Cobie and, subsequently, Chair of Southern Cross, Quintus Wernich, the person responsible for killing Machel, Cobie's parents and for multiple attempts to kill Cobie. The structural economic conditions that produce a climate where crime is all but inevitable are revealed in a conversation where Lemmer challenges Wernich about his crimes, including murder. Citing apartheid as a "war", Wernich states that "the fate of nations takes precedence over the individual" and that sometimes "one has to make very difficult decisions, in the interest of the greater good" (363). Lemmer's retort, "Or the greater profit", reveals the criminality built into the state apartheid system (363).

Lemmer cannot act legally against Wernich, and just as the crime was transnational, the resolution comes about in the same fashion. Lemmer tips off Machel's former bodyguard, Raul Armando de Sousa in Mozambique, and Wernich is assassinated. When he visits de Sousa, Lemmer claims that he would like to say he's doing this "because [he] believe[s] in justice", but the truth is that it's "because [he] believe[s] in revenge" (367). Typical of the hard-boiled genre, "justice" is eventually served, but it is only possible outside the boundaries of the law. While there is a form of retribution meted out to the apartheid criminal, the environmental questions are left open-ended. As with Kunzmann, the crime is solved "at the individual level" while the environmental crime "remains unresolved" (Puxan-Oliva 368). Thus, crime fiction's formal conventions highlight "the environmental crime even as it overshadows its non-resolution by drawing on the easy solution of . . . murder" (368).

While the main focus in Kunzmann and Meyer is on criminality brought about as result of apartheid, state crimes are linked to globalisation. In *Salamander Cotton*, the mining company responsible for the medical cover-up is based in London, and they are "less than willing" to assist with the investigation, afraid of the resultant issues of pollution and compensation and Tshabalala "can do little about it" (272–273). Similarly in *Blood Safari*, Wernich admits that Machel's plane was bought down with the assistance of Israeli technology, as they were "working closely with the Israelis on several levels" (362). In these novels, a contemporary crime leads into an investigation of the apartheid state and the environmental questions arise as part of that process. The relatively traditional models of crime fiction used require a resolution which only takes place at the level of individual crime, the murder. The next section shows how a more meaningful incorporation of environmental crimes beings to apply pressure to the genre.

Applying Pressure to the Genre for a More Expansive Spatial Reach

Pepper's and Schmid's second approach features works that, whilst still reliant on traditional features of crime fiction, "use larger geographical settings, such as the border, the state, and the transnational region" (12). A South African writer who, as I will demonstrate, provides crime fiction that applies "pressure to the traditional features of the genre in an attempt to make it more responsive to a rapidly globalizing world", is Jassy Mackenzie. (Pepper and Schmid 12–13). Mackenzie is best known for her Jade de Jong thriller series featuring a feisty female PI and bodyguard who, in the hard-boiled fashion, often takes the law and justice into her own hands. Mackenzie has produced five de Jong novels and yet, despite the fact that the last three contain environmental themes, existing criticism of her work has been limited to questions of gender.[3] Mackenzie's novels discussed here represent a shift away from the country's past towards "larger geographical settings", and this more "expansive spatial reach makes them particularly self-conscious examples of how crime fiction might address the contemporary realities of global capitalism" (Pepper and Schmid 12–13). The conventional PI form becomes increasingly hybrid as the murder mysteries are subsumed by narratives of environmental catastrophe.

The first inclusion of environmental degradation occurs in Mackenzie's third novel, *Worst Case* (South Africa, 2011; published in the United States as *The Fallen*, 2012). Jade goes on holiday to the St Lucia wetlands, an ecotourist destination – proclaimed South Africa's first UNESCO World Heritage Site in 1999 – and breeding ground for the endangered leatherback turtle. When her romantic partner, police Superintendent David Patel, announces a reconciliation with his estranged wife, Jade falls into the arms of fellow holidaymaker, Craig Niewoudt, an environmental ecologist conducting studies of the area. Craig provides information which lays the groundwork for understanding the environmental plot that Jade subsequently stumbles onto. He explains that mining companies want to strip mine the sand dunes and that there is "increasing pressure from developers" to "allow mining to take place in the park" (49). He explains the tensions between capitalism, ecotourism and the local community:

> The argument for declaring the area a national park was that ecotourism could bring as much benefit to the area as mining. But that hasn't ended up happening . . . Meanwhile the local communities . . . are among the poorest in the country and they rely on natural resources to survive.

. . . They're pressuring the authorities to open up areas within the park, because they need more space and materials for subsistence farming.

(49)

As with Meyer, there is conflict between the poverty of the local people and conservationist ecotourism. Yet, here, both are under threat from neoliberal capital.

While investigating the murder of her scuba diving instructor, Jade stumbles, unwittingly, onto a plot by the CEO of Richard's Mining, Patrick Zulu, to sink a tanker filled with extremely toxic used engine oil, bought from Pakistan, in the estuary. Zulu's idea is to "engineer an environmental catastrophe" that would destroy ecotourism activities so that "the industrial and mining sectors would, through necessity, be encouraged to grow" (179, 197). Craig states that it would be "difficult to think of a bigger ecological disaster" (178). Despite requesting assistance from the environmental enforcement officials, the Green Scorpions, Jade, in PI mode, figures out what is being planned and manages to detonate the bomb and blow up the tanker, the *Karachi*, before it leaves Richard's Bay harbour, thus minimising environmental damage to the estuary, which is "extremely vulnerable, because it can retain oil for more than ten years" (177–178). Zulu is killed in the incident.

While this environmental plot occupies a major part of the novel, it is not the main murder mystery. The novel opens with a murder, which remains unexplained until the end of the text, and ends with another impending murder. The environmental degradation is thus sandwiched between these two crimes, which would constitute the actual criminal case in traditional crime fiction. Although Jade is not hired to investigate the mining conspiracy, she solves the case and saves the estuary, while, in a reversal of Kunzmann and Meyer, the resolution of the murder is left open-ended, and it's only in the following novel that readers find out what happened. The ecological case thus subsumes the human murder plot and fills the PI frame. While the specific environmental catastrophe is prevented, readers are reminded of global climate and environmental disaster as Jade watches international news about a forest fire "raging out of control in Australia decimating thousands of acres of indigenous forest" and a story about an oil tanker that sank just off the coast of Sri Lanka "discharging its load of filthy oil onto the beach there" with "footage of oil-soaked sand littered with poisoned fish and dead birds" (75).

In the following de Jong novel, *Pale Horses* (2012), the murder investigation that Jade is hired to conduct turns into an environmental plot, which subsumes the original murder mystery and, in a hybrid twist, concludes with Jade killing her client (who is also revealed to be the murderer). She is hired by Victor Theron to investigate the death of his base-jumping partner, Sonet Meintjies, whose parachute was sabotaged. Meintjies was working for a charity that assists Indigenous people with land reform claims and the development of agriculture. Jade discovers that Meintjies worked on an ancestral land reform claim for the Siyabonga community that resulted in Meintjies' ex-husband losing his farm which "had been in [the] family for generations" (ch. 12). Jade's position on this political issue is made clear as she fights the "impulse to point out that indigenous communities might well have been living there a thousand years before that" (ch. 12). Yet the farm, Doringplaas, has now been abandoned by the Siyabonga: they have all disappeared, the houses "ripped up; their mill was gone; their cattle nowhere to be seen" (ch. 12).

Mackenzie thus also raises the issue of political and agricultural restitution in post-apartheid South Africa, and there is a great deal of interesting information provided in the novel as Meintjies' sister, Zelda, is a journalist writing about land reform. Mackenzie, however, steers the plot away from state politics to address the problem of global capitalism. In a very reductive summary of the plot, Jade's investigation finds that the Siyabonga were provided with a highly experimental variety of genetically modified (GM) maize seed from American conglomerate, Global Seeds. The variety had never been tested, and Global Seeds donated half a ton of seeds "in order to run field tests on it" (ch. 43). It is speculated that the hybrid seed was produced with cancer cells to promote faster growth, while also including "one of the major human tumour protein suppressants" to counteract this (ch. 43). The experiment fails, and the members of the Siyabonga community and their animals develop an aggressive form of cancer and die. Global Seeds are intent on suppressing this story, and they destroy what is left of the Siyabonga lands.

It is Jade's client, Theron – who has murdered Meintjies and is also planning to murder her sister, Zelda – who is complicit in covering up the GM seed story. In an extension of the transnational crime plot, he plans to sell the remaining seeds to parties in the Middle East, who want to contaminate maize crops in the United States as environmental sabotage. Theron dies falling from his balcony in a luxury high-rise building in Johannesburg after a confrontation with Jade. Though not explicitly stated, the suggestion is that he has been pushed by Jade. *Pale Horses* is filled with fascinating information about land reform, the monopoly of GM seeds in South African maize production and the issue of GM farming in general, including toxic dumping in the global South. In this way it shows how the state is both in collusion with, and also affected by, global capital.

There is a shift in these last two novels away from local politics towards environmental degradation as a global issue and both raise questions about South Africa, and the global South, as a dumping ground for toxic waste and experimentation, as elucidated by Rob Nixon in *Slow Violence*. In *Worst Case*, Zulu's cronies are able to obtain toxic engine oil from multiple suppliers in Pakistan where millions of litres were "creating a major pollution problem in the densely populated country" (75). Criminal elements in Pakistan are selling contaminated oil to South Africa, while in *Pale Horses*, Theron is planning to sell the lethal seeds supplied from America to groups in the Middle East. There is thus a circulation of toxic material, driven by capitalism. Mackenzie is less focused on the state's political redress, compensating for the injustices of apartheid, and is more concerned about how this is undermined by the devastating impact of globalised capitalism.

In Mackenzie's texts, the pressure applied to the traditional features of the genre becomes visible, as the plots attempt to become more responsive to a rapidly globalising world. These environmental crimes do not constitute Jade's official cases and are investigated in her personal capacity as a result of her commitment to environmental justice. In both, she has to resort to taking the law into her own hands. These novels begin to show the inadequacy of the solitary investigator and the conventional crime fiction formula when it comes to addressing environmental crimes that are being perpetrated on a global scale. These can only be addressed through much greater hybridity as shown in the final section.

Environmental Concerns Driving Crime Fiction Towards Hybridisation

Peter Hain's *The Rhino Conspiracy* (2020) is one of several recent thrillers set in the context of State capture in South Africa.[4] Hain was born to South African parents and grew up in South Africa but left when he was sixteen and his anti-apartheid activist parents were forced into exile in the United Kingdom. Hain has held various political positions as a Labour MP and Secretary of State in the United Kingdom. *Rhino Conspiracy* was his first work of fiction and has been described, variously, as a "timely", "unusual", "political" and "ecological" thriller. It represents, as I will demonstrate, the move away from traditional crime fiction to more hybrid forms, described by Pepper and Schmid, which enables a "greater geographical scope and a more expansive sense of the ways in which transnational networks or crime and policing operate in today's world" (16). It represents a dramatic shift in terms of crime fiction's representation of crime, the state and globalisation. Furthermore, it presents an example of a novel where "crime related to the environment is the fictional core problem", thus demonstrating how "environmental concerns challenges genre conventions and drives crime fiction towards hybridization" (Puxan-Oliva 362). Its remarkable innovation is that it also explicitly links questions about crime, the state and globalisation to environmental survival.

As opposed to the common focus on the state in state capture, the interrogation of the government in *Rhino Conspiracy* comes about largely as a result of the ecological investigation into rhino poaching, thus making the ecological primary. The characters discover that illicit smuggling of rhino horn is taking place at the highest levels in government, within the president's office, and this becomes symptomatic of the general corruption, immorality and degradation of democracy within the state. Since the rhino horn is sold to Vietnam and China, the novel raises overt questions about global markets that drive crime and acknowledges that the problem cannot be solved at a local level. The struggle to eliminate poaching becomes, simultaneously, a plot against the current corrupt state and the text consciously interweaves human and animal survival. For the future of the rhinos, the corrupt elements in the government must be brought down, and this is linked to the survival of democracy and the Mandela legacy in South Africa, an ideal betrayed by state capture.

The novel's prologue links ecological survival to political freedom. It opens with a character, known only as "the Sniper", lying in wait for rhino poachers in the privately-owned Zama Zama game park, where the rhino action takes place.[5] Readers learn that the last time he played the role of a sniper was to protect Nelson Mandela "on the momentous day when Madiba took the first steps of his long walk to freedom" (Prologue). His mission at the time was "defensive protection for the old gentleman who held the future of the nation in his hands", to spot "assassins and shoot them" (Prologue). Now, protecting the rhinos, he has to become the assassin. Protecting the democracy born in 1994 and the rhinos in the contemporary moment are linked through the narration: "just as his duty then was to protect Madiba", now "he was protecting the legacy of Madiba" (Prologue). The novel draws the political and ecological together through the character of Thandi Matjeke. Thandi visits Zama Zama at the start of the novel and becomes involved in a relationship with game ranger, Isaac Mkhize. Simultaneously, Thandi begins to support a well-known political figure from the anti-apartheid struggle, who is now opposing the current government. He is known only as "the Veteran". As a young black "Born Free" but not a "Born Forget", Thandi is just as disillusioned with the government and its betrayal of the political values that fuelled the transition to democracy (ch. 3). She offers her support to the Veteran and becomes his political apprentice. Involved in both the worlds of conservation and politics, she connects the Veteran and other political figures to those fighting against rhino poachers, and the two plot lines intersect. The Veteran, the Owner of Zama Zama and Mkhize work together to set a trap for the poachers. A tracker is planted in a replica rhino horn, a corrupt game ranger takes the bait, and the horn is passed on to the smugglers. Surveillance of the bugged horn in order to identify the culprits is the main narrative thread. The political and ecological struggles merge when it is discovered that the rhino horn is being smuggled in diplomatic pouches from the president's office to Vietnam and China. The two groups of characters now face a common enemy. The degradation of democracy and the destruction of endangered species are linked to, and through, corruption and greed. As the Veteran declares: "If we can expose that together, we kill two birds with one stone. Protecting the rhinos and protecting our young democracy to preserve Madiba's legacy" (ch. 7).

One of the first moves towards hybridisation is that, despite the "chase" (tracking the horn), there is not a single detective within the novel. The detection is carried out largely by Thandi with the assistance of a range of people with nothing in common except the desire to protect the rhino and the principles of democracy. *Rhino Conspiracy* thus provides a classic example of Murphy's suggestion that critics need to look at the examples of "nature-orientated mystery novels – with or without detectives and perhaps even without murders – in order to understand the degree to which environmental consciousness and nature awareness has permeated popular and commercial fiction" (143). As the detective is usually also the focaliser, this illustrates another adjustment to the crime genre. Discussing Van Gieson's *The Wolf Path*, Murphy claims that she

> Has enabled her readers to choose from a range of characters with whom to identify, without always having to identify with the hero of the novel, while at the same time not having to reject the hero in order to make an identification with another character.
>
> *(137)*

This is uncommon in the popular genre of murder mystery which does not usually facilitate "multiple characters as speaking subjects and internally persuasive environmental discourse structure" (137).

Rather than the oppositional elements of apartheid, or post-apartheid, politics, the principle organising the two opposing sides in the novel is determined by characters' attitude towards the rhinos. The politically motivated Veteran and Thandi are equally committed to the struggle against poaching. The largely apolitical Mkhize is gradually drawn into the political struggle through his developing relationship with Thandi. The narrative also draws characters from the old apartheid system into the new ecological struggle. The Sniper was an assassin for the apartheid South African Defence Force involved in the elimination of "terrorists", who has made the transition to democracy. He is persuaded by the Owner of Zama Zama to put his training to use in a new struggle: "He was a trained killer. Not for satisfaction" but for "the wildlife of his heritage, for humankind, for decent values . . . for what he yearned for his country to be again" (ch. 15). Yet rather than criminalising his actions the text aligns him with the vulnerable rhinos as he realises that he will also be targeted by poachers: next time, "they would come even more heavily armed" and "in greater numbers. Then *he* would be their prey, probably *equally with the rhinos*" (ch. 10, emphasis mine).

While the text positions a former apartheid military sniper on the side of the rhinos, we find a former anti-apartheid activist at the core of the rhino poaching operation: Moses Khoza in the president's office.[6] The character of Moses Khoza presents a composite of the various generations of anti-apartheid activists. He was a youth activist in the 1976 "Soweto" generation and was imprisoned on Robben Island with Mandela and Oliver Tambo. While the '76 generation were alienated from Mandela who had already been in prison for fourteen years, Khoza learns to respect Mandela and becomes a figure of reconciliation between the different generations of struggle activists. He represents honourable anti-apartheid values yet has succumbed to pressure from the president and the state culture of greed. Another former SADF military figure, Piet van der Merwe, is in charge of the poaching operation and supplies the horn to Khoza. Van der Merwe has a background in various racist military forces in Rhodesia and apartheid South Africa, and this past serves to make him irrelevant in a South Africa without rhino poaching: when the trafficking is exposed, he feels as though he's "lost [his] country" and that "maybe he should disappear into one of those Afrikaner white enclaves in South Africa" (ch. 23).

Returning to Murphy's suggestion that we need to "look at the examples of nature-orientated mystery novels – with or without detectives, and perhaps even without murders" (143) – it is significant that there is not a single human murder which drives the plot. The Sniper shoots several people, yet these actions are never positioned as criminal. According to Puxan-Oliva, crime fiction that fully addresses environmental issues inherently pushes the limits of genre conventions, "because it destabilizes the idea that crime is always aligned with murder" (369). She asks how "a crime whose real criminal dimensions transcend our current understandings of violence" might "instigate a reformulation of previous understandings of crime and violence, which are at the core of current definitions of crime fiction" (369). The killing of rhinos for poaching is positioned as criminal and described in the forensic detail that would usually accompany a human corpse. Just as the vulnerability of rhinos, and humans who protect them, is brought together in the novel, the "murder" of rhino and other animal species is presented in human terms. Elise, the Owner's wife, states that "sixty percent of mammals, birds, fish and reptiles" have been wiped out in the last five decades: "A sixty-per-cent cull of humans would be wiping out North America, South America, Africa, China and Oceania" (ch. 23). Referring to the poaching, Mkhize says: "If it was humans we would call it genocide" (ch. 3).

A final question raised by Puxan-Oliva relates to the "resolution" and how "the fact that most environmental crimes cannot be solved in one case, by one individual, or within one legal framework play into the pressure for a final resolution?" (369). An examination of this pressure in *Rhino Conspiracy* leads back to the question of crime fiction as *the* political novel in contemporary South Africa. In *Losing the Plot: Crime, Reality and Fiction in Postapartheid writing* (2016), de Kock attempts to explain why crime writing (both fiction and non-fiction) has come to dominate the South Africa literary landscape. He suggests that the relatively sudden emergence of crime fiction is a response to the social and political upheaval of the transition from apartheid to democracy and that South African crime writing pivots around a "continuingly problematised notion of [that] transition" (2). An initial wave of optimism, "evident in the early phase of upbeat transitional ferment", was followed by a "gradual and deepening sense of 'plot loss' among South African writers and intellectuals" (3). The disillusionment and disorientation experienced in the post-transition years by those "still rooted in a social imaginary that continues to hold dear the founding tenets of the 'new' democracy" sets the scene for post-apartheid literary culture characterised by crime fiction (4). The discrepancy between the ideals on which the new democracy was founded and the outcome in the later post-transition years created the conditions for "wide-ranging investigation into the causes of the perceived inversion, or perversion, of the country's reimagined destiny" (4). This derailing of the tenets of the new democracy has come to be "regarded as criminal", and the crime fiction genre thus comes conveniently to hand since it "typically sets out to pinpoint a culprit", and thus, in implicitly wider terms, the "sources of social and political perversity" (4).

Rhino Conspiracy falls squarely into this category of investigating "plot loss" in the new South Africa. The Veteran describes the transition – the "joy of victory, so long and so determinedly fought for" – that was followed by "the betrayal": "[t]he trashing of their values and their ideals" (ch. 23). This derailing of the tenets of democracy is branded as criminal and the source of the "social and political perversity" of state capture (de Kock 4). Yet *Rhino Conspiracy* adds several new elements to de Kock's model. There is a shift away from a sole focus on the criminality within the state, and its consequences, towards an acknowledgement that local criminal activity is structurally linked to the global. Mkhize points out that rhino horn trafficking is both a local and a transnational problem, which can only be solved at both levels: "Unless we deal with poverty in these communities and stop selfish people in other countries paying so much money for rhino horn, we are fighting a losing battle" (ch. 3). This relationship between local and global legal frameworks is equally at play in the novel's very provisional resolution. The president is eventually arrested, not for his crimes of state capture, but for his role in rhino horn trafficking (ch. 23). Just as the crime is transnational, the president's downfall comes about through a transnational set of actions without a single detective: an expose of the rhino horn trafficking by a British MP, Bob Richards, in the UK parliament leads to "an international outcry", and the president is eventually extradited through transnational legal cooperation, when South Africa's Constitutional Court executes "the demand by the International Criminal Court for the President's extradition" to the Hague (Epilogue). This enacts a shift from a sole focus on state capture, as the political problem is addressed through a transnational legal network working against environmental crime at a global level. The novel demonstrates hybridity in that its resolution is the most provisional of all the texts discussed. Despite the minor victory, it is clear that the future is left undetermined, both politically and environmentally. There is a new president, but the ANC leadership is still split, with corruption and "cronyism remaining deeply embedded" (Epilogue). Despite the lack of a full resolution, the novel gestures towards a new future. In a final section of the Epilogue ("*and even later . . .*"), Thandi is invited by the new president to stand for Parliament. By this time, she and Mkhize are married, and she has become the official spokesperson for the Veteran. This consolidates her role as the character who combines the political and environmental threads in the novel and thus gestures towards new possibilities for the future.

Rhino Conspiracy interweaves human and environmental discourse and survival. In doing this it addresses the divide between deep ecology – "in which the interest of the biosphere overrides the interests of individual species, including the human" (Marland 1512) – and social ecology, "in which the notion of the domination of nature by man stems from very real domination of human by human" (Bookchin 65). It deconstructs the racial and political binary, seen in Meyer and others, between conservation and democracy in South Africa. The novel presents the figure of the black game ranger, Mkhize, who locates conservation within his South Africa heritage:

> This is my land. My ancestors have lived off this soil for centuries . . . I learnt about the bush from my dad: about the tracks that different animals leave and how to trail them. He took me out with him, taught me how fragile nature is . . . he kept explaining how, if we humans don't protect the wildlife and nurture nature, we endanger our own future.
>
> *(ch. 2)*

Due to its status as the new political novel in post-apartheid South Africa, attention to local crime fiction genre innovation has been viewed largely in terms of how the form is representing and negotiating shifts in response to national transition. While the significance of this is not to be understated, I suggest that the genre is being utilised and developed in other, equally important, ways. Investigating this nationally adopted form of writing through the lens of ecology provides a new way of making developments within genre conventions visible. The hybridisation, of both form and content, required by a serious inclusion of ecological crime and environmental justice, opens up new avenues in the discussion of crime writing in South Africa and presents a model of the relationship between the state and globalisation, reconfigured through ecology.

Notes

1. This is part of a larger debate about the emergence of crime fiction in post-apartheid South Africa – after its relative scarcity under apartheid – as well as the initial dismissal of the "genre" form by earlier critics. For a more detailed discussion of this, see Guldimann, 2020.
2. Marta Puxan-Oliva (2020) mentions *Blood Safari*, though she does not elaborate. Another example of attention to Meyer and ecology is Bibi Burger (2018).
3. Sabine Binder identifies Mackenzie taking the genre in new directions by "crafting Jade in ways that allow female readers to identify with and, as women, to feel empowered by her". She further claims that in the South African context "Jade serves to highlight the fact that . . . women require alternatives to the kind of justice that the TRC and the neo- liberal state at large can offer them" (100).
4. According to Neil Arun state capture "describes a form of corruption in which businesses and politicians conspire to influence a country's decision-making process to advance their own interests". In South Africa it is associated with the close affiliation between former President Jacob Zuma (2009–2018) and the Gupta family.
5. Three main characters – the Sniper, the Veteran and the Owner (of Zama Zama) – are not given names. Hain claims that these fictional characters are "based upon real people", and he has "chosen not to give them names to keep a sense of mystery". See the interview with Hain, 2020.
6. A criticism that might be levelled at the novel is that some of the characters seem representative of a specific "type" in South Africa's history.

Bibliography

Arun, Neil. "State Capture: Zuma, the Guptas, and the Sale of South Africa." *BBC News*, BBC, 14 July 2019, www.bbc.co.uk/news/world-africa-48980964.

Binder, Sabine. *Women and Crime in Post-Transitional South African Crime Fiction: A Study of Female Victims, Perpetrators and Detectives*. Brill Rodopi, 2021.

Bookchin, Murray. *The Ecology of Freedom: The Emergence and Dissolution of Hierarchy*. AK Press, 2005.

Burger, Bibi. "Dark Ecology and the Representation of Canids in Deon Meyer's Fever." *Tydskrif Vir Letterkunde*, vol. 55, no. 3, 2018, doi:10.17159/2309-9070/tvl.v.55i3.5501.

De Kock, Leon. *Losing the Plot: Crime, Reality and Fiction in Postapartheid South African Writing*. Wits UP, 2016.

Guldimann, Colette. "A New Beginning for Good People: National Identity and the New South Africa in Deon Meyer's Crime Fiction." *Crime Fiction and National Identities in the Global Age*, edited by Julie Kim, McFarland, 2020, pp. 115–137.

Hain, Peter. "Interview: Peter Hain Talks about the Rhino Conspiracy: Crime Fiction Lover." *Crime Fiction Lover | The Site for Die Hard Crime & Thriller Fans*, 4 Oct. 2020, crimefictionlover.com/2020/10/interview-peter-hain-talks-about-the-rhino-conspiracy/.
———. *The Rhino Conspiracy*. Muswell Press, 2020. Ebook.
Kunzmann, Richard. *Salamander Cotton*. Pan Macmillan, 2006.
Mackenzie, Jassy. *Worst Case*. Umuzi, 2011.
———. *The Fallen*. Soho Press, 2012.
———. *Pale Horses*. Umuzi, 2012. Ebook.
———. *Bad Seeds*. Soho Press, 2017. Ebook.
Marland, Pippa. "Ecocriticism." *Literary Theory: An Anthology*. 3rd ed., edited by Julie Rivkin and Michael Ryan, John Wiley and Sons, 2017.
Meyer, Deon. *Blood Safari*. Translated by K. L. Seegers. Hodder and Stoughton, 2009.
———. *Fever*. Translated by K. L. Seegers. Hodder and Stoughton, 2017. Ebook.
Murphy, Patrick D. *Ecocritical Explorations in Literary and Cultural Studies: Fences, Boundaries, and Fields*. Lexington, 2009.
Naidu, Sam. "Crimes against Nature: Ecocritical Discourse in South African Crime Fiction." *Scrutiny2*, vol. 19, no. 2, 2014, pp. 59–70, doi:10.1080/18125441.2014.950599.
Ndlovu, Ntombizodwa, Jim te Water Naudé, and Jill Murray. "Compensation for Environmental Asbestos-Related Diseases in South Africa: A Neglected Issue." *Global Health Action*, vol. 6, no. 1, 2013, pp. 82–88, doi:10.3402/gha.v6i0.19410.
Nixon, Rob. *Slow Violence and the Environmentalism of the Poor*. Harvard UP, 2013.
Pepper, Andrew, and David Schmid. *Globalization and the State in Contemporary Crime Fiction a World of Crime*. Palgrave Macmillan, 2016.
Puxan-Oliva, Marta. "Crime Fiction and the Environment." *The Routledge Companion to Crime Fiction*, edited by Janice Allan, Jesper Gulddal, Stewart King, and Andrew Pepper, Routledge, 2020, pp. 362–370.
Thornber, Karen. *Ecoambiguity: Environmental Crises and East Asian Literatures*. U Michigan P, 2012.

12

"LOOK AT MOTHER NATURE ON THE RUN"

'The Troubles' in Adrian McKinty's Sean Duffy novels

Bill Phillips

Adrian McKinty is never loath to display his learning, whether it be of music, literature, mythology or late twentieth-century history, and his books are all the better for it. Of particular interest are his six Sean Duffy novels: set in Northern Ireland during the 1980s, they follow the exploits of a Catholic Royal Ulster Constabulary detective inspector as he becomes embroiled in some of the most prominent events of the decade. His investigations lead him not only into the depressed, dangerous and war-torn cities of Belfast and Derry, but into the countryside as well; the farms, the forests and the mountains, all of which prove more deadly than the sectarian strongholds of the cities. The traditional Romantic or – if you prefer – Pastoral representation of an idyllic rural world will often raise its expected head but is almost always undermined by Duffy's world-weary first-person narrative. Meanwhile, Belfast, the city of dreadful night, holds an addictive fascination for Duffy, seducing both the detective and the reader with its constant violence, extreme politics and permissive self-destruction. Such contrasting aspects of rural and urban life were neatly summed up by Raymond Williams back in 1973 in *The Country and the City*:

> On the country has gathered the idea of a natural way of life: of peace, innocence, and simple virtue. On the city has gathered the idea of an achieved centre: of learning, communication, light. Powerful hostile associations have also developed: on the city as a place of noise, worldliness and ambition; on the country as a place of backwardness, ignorance, limitation.
>
> *(1)*

This study of McKinty's 'Troubles' novels will assess these oppositions forearmed with the knowledge that they are heavily skewed towards the negative, the "noise, worldliness and ambition" of the city, and the "backwardness, ignorance [and] limitation" of the country. Of the two, the temptations inherent to the fallen city predominate; neither Duffy, nor the reader, want the excitement to end. The inevitable consequences are death and destruction, and not just in a moral sense; the novels repeatedly make it clear that not only people's lives, but the environment itself – both rural and urban – are blighted. There seems to be an allegory here about the environmental and existential crisis we are facing and our reluctance to halt our headlong flight into the darkness.

DOI: 10.4324/9781003091912-15

The Golden Age

Let us begin with Williams's first categorisation, that of the country and its association with "a natural way of life: of peace, innocence, and simple virtue" (1). This is an ancient concept, associated with the Pastoral, which "provided an alternative to the real world" and "afforded a way of attacking contemporary abuses obliquely, by reference to an idealised vision of what life in the country was like" (Barrell and Bull 6). Wordsworth was sufficiently confident of the truth of this idealised vision of country life to declare that there "the passions of men are incorporated with the beautiful and permanent forms of nature" (245); that in the country, the people are better.

For Wordsworth, and the Romantics in general, such idealisation is also associated with childhood, and this is echoed by Duffy in *The Cold, Cold Ground* as he reminisces about his rural origins:

> I'd been born in 1950 in Cushenden when that part of rural Northern Ireland was like another planet. No phones, no electricity, people still using horses to get around, peat for cooking and heating, and on Sundays some of the crazier Protestants rowing or sailing across the North Channel in little doreys to attend the kirk in Scotland.
>
> (7)

But, of course, like the Golden Age of the Pastoral poets, this idealised place is "like another planet"; it is no longer accessible if, indeed, it ever existed. Duffy's sense of loss occasionally includes – again in Romantic vein – a lament for a lost vision of Ireland: "It was starkly beautiful out here under the austere slopes of Slieve Gullion", he muses in *In the Morning I'll Be Gone:* "This was a hallowed landscape: Cuchulainn's kingdom in the era of *Táin Bó Cúailnge* and in St Patrick's time the *Terra Repromissionis Sanctorum* – the promised land of the Saints" (18). Duffy's appreciation for the countryside is, however, infrequent: "Nice up here in the countryside, surrounded by farms and woods, with the Irish Sea a hazy blue-grey line at the eastern horizon" (217), he grudgingly concedes in *Gun Street Girl* while admitting – at the insistence of his girlfriend Beth – that he was "in a beautiful part of County Antrim overlooking the North Channel and a big chunk of Western Scotland" (*Police at the Station* 71). That three of the last four passages quoted refer to the North Channel or the Irish Sea is no coincidence. As we shall see, Duffy's escape route from Ireland to a better life in Scotland, free of 'The Troubles', lies across the water.

The few times Duffy is prepared to recognise the beauty of nature are invariably associated with water, where an alternative, older, world provides either a path out of the troubles to another land or a prelapsarian refuge from the present. This is occasionally to be found on Lough Neagh, "the largest freshwater lake in the British Isles" where "visiting the villages dotting its shores was often like stepping back into an Ireland of a hundred or several hundred years before" (*In the Morning* 88). The lough is "a vast, still, pale-blue presence with no boats and few birds" where Duffy is briefly able to immerse himself in nature "surrounded by woodland: oak, ash, elm and wild apple trees" (89). But Lough Neagh affords only a temporary respite and, at the end of the sixth and last novel, *Police at the Station and They Don't Look Friendly*, Duffy makes his break across the water and moves to Scotland. His means of getting there could hardly be more romantic; accompanied by his partner, Beth, his daughter Emma and his two faithful colleagues, Crabbie and Matty Lawson, he sails to Scotland in a thirty-two-foot-long ketch to visit his new home:

> The sun sinking behind the Irish coast.
> The yellow dark, the red dark, the deep blue dark . . .
> Stars in swirls. A sickle moon. Silence.

> Between Ireland and Scotland not a ship or plane or another vessel.
> Just the night itself and the flat black sea that makes a noise like singing.
>
> *(338)*

The semi-ruined house in Portpatrick, Scotland, where Duffy's new life is to begin "with its falling gables and ivy-covered windows and overgrown garden and path down to the water . . . is terribly romantic" (334) and, although McKinty leaves open the possibility of a seventh novel in the series by retaining Duffy's links to the RUC and the security forces, this last novel ends with the detective's exile just around the corner. Duffy's anticipated escape from his troubled past into the bosom of his family and bucolic domesticity provides a remarkably conventional conclusion to his picaresque adventures as a Catholic cop in the RUC, a denouement that McKinty only tentatively began to disturb with a completed but so far unpublished seventh instalment.

Yet these idyllic rural moments do not define Duffy. Far more frequently, the beauty of nature, if it is acknowledged, is immediately and decisively undermined with crudity or devastation. *Gun Street Girl* begins with Duffy observing a combined security forces operation to ambush gunrunners at a deserted beach: "The weather is perfect, the moon is up and the tide is on the ebb", he narrates, to which a Special Branch officer, blithely immune to Duffy's Arnoldian musings, replies: "Aye, we have the bastards now!" (2). In *Police at the Station and They Don't Look Friendly* there is "Frost on the Ground. Blue sky above the Antrim hills. Mooing of cows, baaing of sheep, growling diggers as Greater Belfast pushed deeper into the Irish countryside" (136), while in *I Hear the Sirens in the Street* a picturesque, whitewashed cottage is spoiled "By a huge rusting oil tank for the central heating plonked right outside" (47). The fact is, Duffy, unlike his usually more prosaic sergeant, has little time for appreciating landscape and the picturesque:

> "It's some view though, eh?" Crabbie said, turning up the collar on his wool trenchcoat.
>
> Indeed it was. Despite the snow clouds you could see clearly down the chilly lough all the way to Scotland across the even chillier Irish Sea. Belfast lay to the south and east, and beyond the city were the foothills of the Mourne Mountains.
>
> "Forget the view, let's look for some evidence", I said.
>
> *(Rain Dogs 67)*

Duffy prefers to get on with the task at hand, the investigation of a mysterious murder in Carrickfergus Castle, "the centre of Anglo-Norman power for seven hundred years", he reminds us. This is the second time in his career that he is faced with one of the classics of detective fiction – a locked-room mystery – and he is anxious to get back to work.

Crabbie, a dour Presbyterian, though far from having Romantic notions about the countryside is devoted to it, his true vocation being that of farmer, and as the novels progress, so does his commitment to the land. But even this practical approach to rural life is constantly undermined. We learn little of Crabbie's own small-holding other than the fact that, like his family, it steadily increases, largely because Duffy, who provides the first-person narration, delights in his own ignorance of farming matters. "It was well after seven now", he comments at one point, imagining that Crabbie would "be done milking the pigs, or plucking the cows, or shagging the sheep, or whatever it was one did on a farm" (*Rain Dogs* 62). Duffy's occasional, reluctant forays into farmland invariably elicit despair at the hopelessness of Irish farming life, a viewpoint shared at times even by the rustic Crabbie: "there's land in Donegal and land in Donegal", he opines. "You could have a thousand acres and every inch of it sucking bog" (*In the Morning* 89).

In *I Hear the Sirens in the Street*, a murder investigation leads Duffy to pay a number of visits to the widow McAlpine, a woman whose misfortune was not only to lose her husband in a shooting, but to be left with nothing to support her but a broken-down sheep farm. Duffy is characteristically pessimistic: "she would age fast out here in the boglands on a scrabble sheep farm with no husband and no help" (50), he concludes. On a later visit, Duffy makes his way across the fields:

> The pasture was little better than a bog with some tuft grass and sodden heather, and in a few moments my DMs were soaked through. Sheep pellets were everywhere and in a slurry pond there was the carcass of an old ewe, suspended just beneath the surface.
>
> *(165)*

In his ignorance, Duffy advises the widow to grow wheat: "No arable crops will grow here", she replies. "It's a marsh. This whole part of Islandmagee is one enormous swamp" (171). Not surprisingly the farm turns out to be a killing ground in which an American secret agent, the widow, her husband and brother-in-law all die, victims of avarice, rural madness and the long reach of the disastrous John DeLorean car company; a far cry from Wordsworth's assertion that in "rustic life . . . the essential passions of the heart find a better soil" (245). Political and economic mismanagement and a disrespect for the environment, all magnified by 'The Troubles' and Britain's 800-year long colonisation of Ireland, lend a hopelessness to the land and the unproductive exploitation to which it is subjected. Joan Martinez-Alier, in his ground-breaking study *The Environmentalism of the Poor*, argues that "agricultural policy should balance environmental, economic, social and cultural values on different geographical time scales" (149). Ironically, despite the near millennium-long time scale of British colonisation in Ireland, it is short-term economic returns which dominate land use. The problem with modern agriculture, as Martinez-Alier goes on to explain, is its "lower energy efficiency, genetic and soil erosion, and ground and water pollution" (149). The slurry pond with the carcass of the old ewe "suspended just beneath the surface" that Duffy observes on the widow McAlpine's land could not be more eloquent.

But beyond Duffy's lack of interest in picturesque landscapes and farming, there lies a fear of nature as a source of menace. "The trees were filled with wood pigeons and down by the water there were gulls, curlews, oystercatchers", we are told in *In the Morning I'll Be Gone*, only for the scene to be interrupted by "Two wee muckers . . . swinging at one another with wooden swords, yelling obscenities and committing a messy angiocide on the wild flowers" (131). Sometimes, as in this case, the menace is largely symbolic, at other times trivial: "I looked out the window at the starlings for ten seconds. One of the little bastards shat on my morning paper" (120), he laments in *I Hear the Sirens in the Street*. This ongoing war continues throughout the six novels, the milk bottles being the usual target of the starlings' attention. Other birds are more fearsome: "A nasty-looking crow staring at me from the drooping curve of the telegraph wire" (*Gun Street Girl* 221) heralds a lengthy meditation on "Morrigan of the black eye. Morrigan of the sorrows, the great queen, the goddess of battle, fertility and strife" (GSG 319). A figure from Irish mythology, Morrigan often appears as a crow, and in *Gun Street Girl*, she looks down upon the crashed RAF Chinook helicopter "filled with all the top MI5, MI6 and Special Branch agents in Northern Ireland flying into a mountain on the Mull of Kintyre" (318). McKinty's technique in the Duffy novels is to weave historical events into the narrative; in this case the aforementioned crash, which occurred on 2nd June 1994, and caused the deaths of all 29 people on board. In the novel, Duffy's secret service handler and sometime lover was among the dead, and he imagines Morrigan looking "down upon a wounded land and she is content, cawing in satisfaction at the patchwork quilt of Ulster and at the mess on the hillside in the Mull of Kintyre" (*Gun Street Girl* 319).

Hills and woodland are particularly frightening: "I've never liked the woods" (104), he confesses in *The Cold, Cold Ground*, and later in the novel, he expresses his fears in a bout of Gothic prose: "It was damp and dark. Strange white mushrooms were pushing their way through the sodden earth. Giant ferns were growing from the shells of fallen trees. There was a dungy smell, the smell of rotting leaves, autumn, graveyards" (281). Mountains are far from safe either, bursting into flame following bomb blasts (*I Hear the Sirens* 92) or simply catching fire and burning for days (*I Hear the Sirens* 259). Duffy's fears are, however, justified. In *I Hear the Sirens in the Street*, he is attacked and beaten after driving his car deep into the woods; in *Rain Dogs* he is shot at with an AK-47 while pursuing enquiries in the snow-bound forests of Finland and in *Gun Street Girl* he is taken "into the boglands. The high country. The Forest. Miles from anywhere. It was wild up here. You could do anything to anybody" (271). As, indeed, he finds out, as his assailants beat him half to death with a baseball bat. The final straw comes in the sixth and final novel, *Police at the Station and They Don't Look Friendly*, where he is taken, handcuffed, by an IRA active service unit, on a long trek up a wooded mountainside to dig his own grave. Far from being a refuge, a source of innocence and peace, nature has turned deadly.

The City of Dreadful Night

We should not be surprised, then, that Duffy prefers urban life and refuses to move into a new spacious house in the countryside built by his partner's father, preferring instead to remain in his cramped terrace house in Carrickfergus: "I'm happy where I'm living now", he tells the mother of his baby daughter. "I know people. They know me. You know how hard it was to win them over?" (*Police at the Station* 73). Not, of course, that the towns and cities of Northern Ireland are especially attractive and McKinty – rather predictably – invokes James Thomson's 1874 poem:

> City of dreadful night. City of the damned. City of no escape. Rain hisses on the cobbles and the open drains. The city hums and seethes. The black Farset bubbles to the surface of High Street oozing human filth. The rusting giant cranes droop over the empty dry docks like the bones of dead gods. Army helicopters sweep the city with a sick white light.
>
> *(Rain Dogs 327)*

This has been, ever since the industrial revolution (and Romanticism), the conventional representation of the city, as Thomson so melodramatically recreated in his long poem, *The City of Dreadful Night*. Before Thomson, Wordsworth bemoaned the need to lodge "in lonely rooms, and 'mid the din/ Of towns and cities" (114), and before Wordsworth, Blake deplored England's "dark, satanic mills" (481), but it was Engels who condemned the horrors of urban life as savagely as any in *The Condition of the Working Class in England* (1845):

> What is true of London, is true of Manchester, Birmingham, Leeds, is true of all great towns. Everywhere barbarous indifference, hard egotism on one hand, and nameless misery on the other, everywhere social warfare, every man's house in a state of siege, everywhere reciprocal plundering under the protection of the law, and all so shameless, so openly avowed that one shrinks before the consequences of our social state as they manifest themselves here undisguised, and can only wonder that the whole crazy fabric still hangs together.
>
> *(58)*

Engels could be describing 1980s Belfast – a city at war with itself, under siege, the reciprocal plundering tolerated by partisan and corrupt security forces.

In 1980s Belfast, urban blight is not only caused by 'The Troubles'. The post-industrial decay began earlier, the ongoing loss of heavy industry and manufacturing, and the subsequent fall from grace of a once empowered proletariat worsened by Thatcherism's bullying of the economy of the United Kingdom into something new and disturbingly inchoate. In the opening scene of *I Hear the Sirens in the Street*, Duffy and McCrabban investigate a trail of blood in an abandoned factory: "Guano covered the walls. Mildewed garbage lay in heaps. Strange machinery littered a floor which, with its layers of leaves, oil and broken glass was reminiscent of the dark understory of a rainforest" (1). It is not clear when the factory was abandoned, but not, it seems, as a result of sectarian violence, but of economic policy. As Žižek explains in *Violence*, "the fate of whole countries can be decided by the 'solipsistic' speculative dance of capital, which pursues its goal of profitability in blessed indifference to how its movement will affect social reality" (11). Žižek's ironic "blessed indifference" is reflected in how people live. In *The Cold Cold Ground* the detectives: "walked among the drab tenements and crumbling 1960s tower blocks. Everything was achromatic and in ruins less than twenty years after it had gone up. A massive social engineering experiment gone horribly wrong" (*The Cold Cold Ground* 184), while in *In the Morning I'll Be Gone*, a residential block in Derry "stank of vomit, beer, rotting leaves and garbage. The occasional black, shoe-sized stain was not the mould I first suspected but, in fact, dead, decaying Norwegian rats" (63). The human cost of Northern Ireland's ruin is emphasised throughout the novels, with frequent scenes of lost, feral children. In *I Hear the Sirens in the Street* Duffy describes: "Lots of kids running around barefoot, burnt-out cars, shopping trolleys and rubbish everywhere" (213), while in *The Cold Cold Ground*, "Children [are] playing on the rubbish heaps and bombsites, dreaming themselves away from here to anywhere else" (331).

Almost all of these wastelands contain reminders of 'The Troubles'. As the detectives drive into Belfast they see "Burnt-out buses. A wrecked Saracen. A post-office van on fire. Soldiers walking in single file" (*The Cold Cold Ground* 170). Such scenes become apocalyptic: "A fallen world. A lost place. Ruined factories. Burnt-out pubs. Abandoned social clubs. Shops with bomb-proof grilles. Check points. Search gates. Armoured police stations" (*The Cold Cold Ground* 331). This sense of an end time develops into an obsession for Duffy, who sees: "Images from the asymmetric wars of the future" (*I Hear the Sirens* 330); "This is the way it's going to be from now on. Wars in cities. Wars with civilians all around. Make one mistake and you're dead" (*I Hear the Sirens* 331). Drawing on Yeats's "The Second Coming", Duffy foresees the worst: "Couldn't they see the future? Entropy maximising. Neighbour against neighbour. Blood feud. The disintegration of this lost lonely province into warring camps. *The falcon cannot hear the falconer*" (*Police at the Station* 41–2). This is made explicit again, in *Rain Dogs*, where Belfast is described as

> the prototype of a new way of living. In 1801 it was a muddy village, by 1901 it was one of the great cities of the Empire, and now Belfast is the shape of things to come. Everywhere is going to look like this soon enough after the oil is gone and the food goes and the law and order goes. *(327)*

In *Police at the Station* and *They Don't Look Friendly*, the final novel of the series, Duffy predicts that: "All cities will look like this in the far future: ruined and fractured, walled and utilitarian" (293). This doom-laden prophecy underlines Žižek's warning about the "speculative dance of capitalism", which not only blights the city but, as we have seen, the countryside as well. According to Martinez-Alier

> Urbanization increases because of productivity increase in agriculture, coupled with low income elasticity of demand for agricultural produce as a whole. Therefore agriculture expels active population. . . . the ecological critique is that increases in agricultural productivity (which depend today on increasing inputs into agriculture and on the externalization of environmental costs) are not well measured because they do not take into account the decreased energy efficiency of modern agriculture, the genetic erosion that takes place and the effluents produced. So both cities and countryside nowadays tend to push environmental problems to higher spatial scales. *(153)*

The country and the city go hand in hand; capitalism does not spare the countryside any more than it does the towns. Just as urban sprawl encroaches on the green fields, so the imperatives of agricultural productivity blight and destroy. Justin McBrien argues that "Capital does not just rob the soil and the worker . . . it is necrotic, unfolding a slow violence, occupying and producing overlapping historical, biological and geological temporalities" (116), and McKinty's novels provide a sample, a picture, of what is occurring: "[c]apitalism leaves in its wake the disappearance of species, languages and cultures", warns McBrien. "[I]t seeks the planned obsolescence of all life" (116).

Capitalism and the Destruction of the Environment

'The Troubles' may have made things worse in Northern Ireland, but they did not instigate the decay, they are the result of centuries of malevolent colonialism and ill-judged economic speculation. McKinty skilfully illustrates this in *I Hear the Sirens in the Street*, in which the DeLorean car debacle is made central to Duffy's successful investigation into corruption and murder. Unfortunately for Duffy, exposing the sordid hand of governments, international security agencies and social elites leads inevitably to his disgrace and demotion: "You have disobeyed direct orders on several occasions. You have embarrassed the force on foreign soil" (325), he is told by the disciplinary board of old men, their "face grey, their noses blue" (324). Again, Žižek's analysis of violence is to the point: "Therein resides the fundamental systemic violence of capitalism, much more uncanny than any direct pre-capitalist socio-ideological violence: this violence is no longer attributable to concrete individuals and their 'evil' intentions, but is purely 'objective', systemic, anonymous" (11). Duffy has fallen foul of the 'objective', anonymous system, and must pay the price; indeed, it is a price that we are all paying, according to anthropologist David Graeber:

> [W]e are looking at the final effects of the militarization of American capitalism itself. In fact, it could well be said that the last thirty years have seen the construction of a vast bureaucratic apparatus for the creation and maintenance of hopelessness, a giant machine designed, first and foremost, to destroy any sense of possible alternative futures.
>
> *(382)*

Graeber's concern that "any sense of possible alternative futures" is denied us is a reflection of capitalist discourse. "Capitalism", argues Jason W. Moore, "has been able to outrun the rising costs of production by co-producing manifold Cheap Nature strategies, locating, creating, mapping, and quantifying natures external to capitalism but within reach of its power" (114). But, as the planet heats up, the cost of exploiting Cheap Nature has become evident, and Moore concludes "[t]here is nowhere to run" (114). And as the true horror of the consequences of capitalism's obsessive wealth accumulation becomes harder to disguise, dissenting voices must be silenced. Graeber notes that there

> is a veritable obsession on the part of the rulers of the world – in response to the upheavals of the 1960s and 1970s – with ensuring that social movements cannot be seen to grow, flourish, or propose alternatives; that those who challenge existing power arrangements can never, under any circumstances, be perceived to win. To do so requires creating a vast apparatus of armies, prisons, police, various forms of private security firms and military intelligence apparatus, and propaganda engines of every conceivable variety, most of which do not attack alternatives directly so much as to create a pervasive climate of fear, jingoistic conformity, and simple despair that renders any thought of changing the world seem an idle fantasy. Maintaining this apparatus seems even more important to exponents of the "free market", even than maintaining any sort of viable market economy.
>
> *(382)*

In Graeber's timeline, it is not unreasonable to see Northern Ireland as an early manifestation – an experimental laboratory even – of this scenario, and it is no coincidence that behind Duffy's downfall is the shadowy hand of American security forces who have been embarrassed by Duffy's findings. Nor should we be surprised that it is an insubordinate policeman who rocks the boat. The paramilitary organisations, both Republican and Loyalist, are knowingly or unknowingly complicit, providing the perfect excuse for the creation of this "vast apparatus of armies, prisons, police, various forms of private security firms and military intelligence apparatus". Without them, there would be little excuse for such an overwhelming deployment of military and security power. Duffy, meanwhile, in line with much contemporary crime fiction, is a detective with a mission to "challenge existing power arrangements".

It is in *Police at the Station* and *They Don't Look Friendly* that Duffy's sense of approaching apocalypse leads him to reflect on humanity's errors: "We were once creatures of the Savannah", he argues, "whose lives were mapped by the journeys of the great migrating herds across the rift valley. We can't live like this. Stationary, on top of one another" (293). There is a suggestion here that humanity has taken a wrong turn, has taken a direction which is not only disrespectful to the planet but which, more seriously – for Duffy remains steadfastly anthropocentric – is heading for a terrifying future which will end, most likely, in the extermination of the species. His choice of music at one point in *I Hear the Siren's in the Street*, Neil Young's "After the Goldrush" (32), with its line "Look at Mother Nature on the run in the 1970s", seems to confirm the idea that Duffy is sensitive to environmental concerns at a time when few people were aware of there being a problem. Neil Young once confirmed (rather doubtfully) that "'After The Gold Rush' is an environmental song" (Hasted), but Duffy's explanation for his choice of song typically undermines any suspicion that he really cares: "Yeah", he confesses, "I was in *that* kind of mood" (32), and settles down for a melancholy evening alone with his habitual "vodka gimlet in a pint glass" (32) – presumably a homage to Raymond Chandler's 1953 novel *The Long Goodbye* where Marlowe "sat in a corner of the bar at Victor's and drank gimlets" (18). Music, like so many things for Duffy, is about feelings; he reserves his thinking for solving crime.

Rain and Slow Violence

Duffy, however, is not one to despair for long and alternates cosy melancholy with defiance, quoting "the great heretic Martin Luther who said 'If the Apocalypse was coming tomorrow, today I would plant a tree'", which Duffy then paraphrases as "And that's how we win: by sticking up a middle finger to the darkness closing in" (*Police at the Station* 321). It is ironic that such defiance is articulated only at the end of the series. In reference to Chester Himes's Harlem Cycle of novels about police detectives Gravedigger Jones and Coffin Ed Johnson, Andrew Pepper argues that after *Blind Man With a Pistol* (1969), which ends with the blind man of the title indiscriminately shooting the passengers in a train,

> There didn't seem to be anywhere else Himes could go, at least within the genre. The fact that his follow-up novel, *Plan B*, an even bleaker tale in which his two protagonists finally turn on each other, was never finished is perhaps revealing.
>
> *(214)*

Perhaps something similar has happened to Adrian McKinty who, according to *Wikipedia*, wrote a seventh novel, *The Detective Up Late*, in 2019, but which his publisher acknowledges is not yet available amid reports of its poor quality (Goodreads). Duffy's escape to Scotland, together with his insistence on Northern Ireland's imminent final collapse, suggest that there is nothing left within the province worth celebrating or saving.

As we have seen, the natural beauty of Northern Ireland holds little charm for Duffy, and the industrial and social decay, exacerbated by 'The Troubles', has turned Belfast and Derry into urban wastelands. According to Lee Horsley, in hard-boiled detective fiction, "it is the big city that most regularly threatens a protagonist's sense of a discrete self, his powers of understanding, and his physical safety" (71). Yet there is something in this blighted landscape which draws Duffy – and the reader – in. This is not surprising. Raymond Williams, remember, argued that on the city, also, "has gathered the idea of an achieved centre: of learning, communication, light" (1). Elizabeth Grosz expands on this in her article, "Bodies-Cities", in which she defines the city as

> a complex and interactive network which links together, often in an unintegrated and de facto way, a number of disparate social activities, processes, and relations, with a number of imaginary and real, projected or actual architectural, geographic, civic, and public relations. The city brings together economic and informational flows, power networks, forms of displacement, management, and political organization, interpersonal, familial, and extra-familial social relations, and an aesthetic/economic organization of space and place to create a semipermanent but ever-changing built environment or milieu.
>
> *(32)*

This "complex and interactive network" has been seriously disrupted in 1980s Belfast, with its roadblocks, no-go areas, divided communities and cordoned-off commercial centres, but this only adds to the attraction for a crime writer and his fictional detective. The noir genre deliberately plays with the contradictions of urban decay and alienation on the one hand and the possibilities and complexities of city life on the other. Andrew Kincaid argues that the detective "is attracted to the city, to its pleasures and vices, and is able there to indulge his melancholia and homelessness" (46). This is undoubtedly true of Duffy, whose pleasures and vices have little to do with pastoral ideals of innocence and simplicity and much to do with the pleasures of over-consumption, professional and social conflict, the rush of urban life and frequent bouts of self-indulgent melancholy, and it is here that we are most likely to find Duffy "sticking up a middle finger to the darkness closing in" (*Police at the Station* 321). He takes delight, for example in his car, a high-powered BMW driven at speed within and between the towns and cities of Northern Ireland: "I was feeling good", he declares in *I Hear the Sirens in the Street*, "as I drove down the coast road to Islandmagee. I accelerated the Beemer up to seventy and then got it up to a nice 88 mph. I dug out a mix tape and put it in the player" (244). In counterpoint to this, there is nothing Duffy fears more than having his beloved car vandalised or stolen. In Derry, he confesses, "I wasn't enjoying the looks the kids were giving me as I drove my BMW past them. Thirteen-year-old boys with mullets, spiderweb tattoos and denim jackets who would just love to steal and joyride a car like this" (*In the Morning* 63). Not the hard men of the IRA or the UVF, then, but thirteen-year-old boys. Duffy selects "the tallest and meanest" of them and pays him 12 pounds protection money.

As well as his car, Duffy likes a drink, either at home in the form of pint-sized gimlets, or in a pub. In *Rain Dogs* he meets Tony McIlroy, an old friend and former policeman turned security consultant for "the famous fifteen-pub crawl that took you through every bar in Carrickfergus" (269), a challenge that Duffy finds daunting: "I was thinking, what if we got half a pint in every pub instead of a pint? Fifteen pints is going to kill us at our age" (269) he grumbles, to McIlroy's disdain. McKinty cannot resist the dramatic irony, as sundry pubs and several pages later Duffy's fears are realised in a post-pub-crawl shootout in which McIlroy is shot in the heart with the last remaining bullet in Duffy's Glock. Pubs evidently play an important part in a detective's life (and death), both socially and professionally, and alcohol is a staple of detective fiction, especially of the hard-boiled American kind. Andrew Kincaid draws on the specific links between Irish and American noir:

> While Chandler and Hammett solve their cases against a boozy backdrop of whiskey and bourbon, the frenetic worlds of Deborah Parker [author Liz Allen], Jack Taylor [author Ken Bruen],

> and Carl McCadden [author Jim Lusby] are fueled by cocaine. Certainly these Irish characters are not immune to the noir trope of alcohol, but their pace is more reflective of the overworked citizen, always on the run, forever zooming between obligations.
>
> *(51)*

Duffy is not fussy. If he's not in the mood for a vodka gimlet, there is the high-grade Turkish cannabis resin, the cocaine or the heroin he keeps in his garden shed "concealed in a Ziploc bag in a tin of grease" (*Rain Dogs* 282) and which he "liberates from the evidence locker before they'd torched it and a couple of bags of brown tar heroin in a ceremony for the *Carrickfergus Advertiser*" (*I Hear the Sirens* 76).

This combination of hedonism and nihilism complicates Duffy's relationship with the environment. McKinty deliberately and continually disrupts expectations in his Duffy novels either by drawing his reader into a world of Pastoral charm, only to dismiss it, or by invoking Romantic assumptions about the human condition and its search for a peaceful, tranquil life of simplicity and fresh air, only to undermine them. He occasionally hints at an environmental sensibility, as Duffy's speech about the "creatures of the Savannah" demonstrates in *Police at the Station and They Don't Look Friendly*, or even more tangentially, in his selection of the Neil Young song in *I Hear the Sirens in the Street*, but this ultimately becomes something of a tease. Certainly, his novels must be included in the sphere of contemporary crime writing which is urban, witty, clever and socially relevant, but this relevance is precisely that, it is social in the conventional dichotomy of society versus nature so condemned by ecocritics – "no one who is old enough to remember life before antibiotics will march under the new historicist banner 'There is no nature'" (42), thunders Karl Kroeber in *Ecological Literary Criticism* while Lawrence Cope explains in more measured tones the fact that "for various schools – formalist, psychoanalytic, new historicist, deconstructionist, even Marxist – the common assumption has been that what we call 'nature' exists primarily as a term within a cultural discourse, apart from which it has no being or meaning" (2). But McKinty is too slippery for these now ancient battles, intent instead on jarring readers' expectations and preconceptions, and the suspicion grows that the absence of much overt environmental sensibility – a sensibility that the literary and cultural criticisms of Raymond Williams, Jonathan Bate and Carolyn Merchant, among others, have taught us to associate with the Pastoral and Romantic – is somehow deliberate, that the novels are telling us that yes, we are destroying the world – look at all the beautiful places the novels describe in Northern Ireland, the sheep and the hills. Look at all the destruction wrought upon them. And ask yourself, what is it that people really want? And the answer will be: distraction, escapism, consumption or, as in Northern Ireland, something worse, something more akin to Satan's "study of revenge, immortal hate,/And courage never to submit or yield" (*Paradise Lost*, Book 1, 106–8). Short-term gains, pleasures and satisfactions; vendettas, conflict and strong emotions, and never mind the consequences. Humanity seems incapable of concerning itself with the environment, of preventing the loss and destruction of ecosystems, of halting the mass extinction of species, never to be recovered. It is too hard, and we just want to have fun.

In his book *Slow Violence and the Environmentalism of the Poor*, Rob Nixon refines and develops Martinez-Alier's ideas about the economic impact of environmental damage, especially on the world's more vulnerable communities. "Slow violence", he says, is

> a violence that occurs gradually and out of sight, a violence of delayed destruction that is dispersed across time and space, an attritional violence that is typically not viewed as violence at all. Violence is customarily conceived as an event or action that is immediate in time, explosive and spectacular in space, and as erupting into instant sensational visibility.
>
> *(2)*

What Nixon describes as "customarily conceived" violence, "immediate in time and spectacular in space" provides exactly the kind of thrill we expect from a crime novel, and McKinty does not disappoint. As a setting, 1980s Northern Ireland is ideal. Not only are there terrorist bombs and shootings, but flaming riots and corrupt policemen all set within the framework of genuine historic events such as the death of Bobby Sands (*The Cold Cold Ground*), the DeLorean Scandal and The Falklands War (*I Hear The Sirens in the Street*), the Brighton Bombing (*In the Morning I'll be Gone*) or the downing of the RAF Chinook helicopter on the Mull of Kintyre in 1994 with the cream of Britain's secret service aboard (*Gun Street Girl*). But the novels also narrate the continuous drip, drip of slow violence. This is symbolised by the rain. Irish noir, like Tartan noir (William McIlvanney, Ian Rankin and Alan Parks, among many others) and Nordic noir (Maj Sjöwall and Per Wahlöö, Henning Mankell and Jo Nesbo, among others), indulges in the possibilities offered by the rain and snow to reflect the darkness of the human heart and its capacity for extreme violence. McKinty – through Duffy's first-person narration – is incapable of leaving the rain alone, milking its gloomy symbolism for all its worth: "Clouds over the Knockagh monument. A storm over the condemned city of Belfast. Elemental rain: cold and tinged with hail. Lightning, thunder. End of Days stuff . . . the usual" (314), he melodramatically declares in the appropriately named *Rain Dogs*. Later in the same novel he claims that the rain is waging a "long-standing war of attrition against Ireland" (345). But tongue in cheek as ever, in the final novel, Duffy asserts that "the pathetic fallacy has never been my cup of tea and I'm no romantic either" (*Police at the Station* 2). Duffy prefers the lure of fast violence but cannot avoid the slow violence of economic and environmental decay that provides, like the rain, the never-ending backdrop to life in Northern Ireland. 'The Troubles' are merely one episode in the litany of ancient violence inflicted on the Irish. Think of the Great Famine of the mid-nineteenth century, an avoidable disaster caused by oppressive economic and social policies, an enforced reliance on monoculture, and the predictable consequences: blight and mass starvation. The economic collapse of Northern Ireland in the second half of the twentieth century was no doubt worsened by 'The Troubles', but the slow violence was already apparent, as Duffy's forays into the most deprived areas – both rural and urban – of the province reveal. And the violence continues, with business as usual. Naomi Klein explains in *The Shock Doctrine* how capitalism thrives on violence and disaster. In a later book, *This Changes Everything*, she specifically refers to the opportunities that the climate crisis offers security services:

> In a moment of candour, the weapons giant Raytheon explained, "Expanded business opportunities are likely to rise as consumer behaviour and needs change in response to climate change." Those opportunities include not just more demand for the company's privatized disaster response services but also "demand for its military products and services as security concerns arise as results of droughts, floods, and storm events occur as a result of climate change".
> (9)

These "business opportunities" are, effectively, what Duffy spends his time investigating, and for which he receives regular raps across the knuckles. But the greater violence, the slow violence, is much more difficult to combat.

Conclusion

Our small, local world, with its emotions, its concerns with identity, place, rights and justice is more pressing than the unknown and unknowable future, and Duffy's Belfast is paradigmatic. Riven by sectarianism born of centuries of British domination and repression, the beleaguered and oppressed Catholic minority responded to the Bloody Sunday massacre of 1972 when 26 people were shot by soldiers of the British army's Parachute Regiment with understandable anger. This hardened throughout the 1970s and reached something of a climax in May 1981 with the death of the first hunger striker, Bobby Sands (the historical anchor to the first Duffy novel, *The Cold Cold Ground*). In such a context, long-term concerns are overshadowed by the immediate and pressing need for defence, survival and revenge. On the other side, the campaign by Westminster and virtually the whole of the media to convince British citizens of the evils of everything Irish was overwhelmingly successful. As a teenager growing up in Southeast England in the 1970s, I learned from family, school, the radio, television and newspapers that the Irish were terrorists (they seemed not to distinguish between the Irish Republic and Northern Ireland), dedicated to the motiveless murder of ordinary, decent English men and women. Little was told of Unionist violence beyond the general belief that 'that Ian Paisley doesn't help matters'. I remember an Irish family arriving in our village to escape 'The Troubles'. "Are you terrorists?" we asked them on hearing their Irish accents. Years later the same blanket association with terrorism was attached to Muslims, however peaceful and law-abiding they might be. Having lived in Catalonia for the last 38 years I have witnessed the same disdain and refusal to negotiate on the part of the Spanish government and its hate-mongering media. All Basques are terrorists and the sly, pacifist Catalans even worse, is the message from Madrid. How can we concern ourselves with the threat of future environmental destruction when our identity is under assault and our passions aroused? And, more to the point, how much easier it is to put aside the serious business of future survival when the seductive conflagration of sectarianism beckons, deliberately fed by authoritarian governments and an incendiary media. No wonder crime fiction is so popular, reflecting back to us the flames of violence, hatred and division. This is where we want to be, not out in the woods and mountains, worrying about the slow, inexorable extinction of humanity.

Bibliography

Chandler, Raymond. *The Long Goodbye*. 1953. Penguin, 1988.

Coupe, Lawrence. *The Green Studies Reader*. Routledge, 2000.

"The Detective up Late (Detective Sean Duffy, #7)." *Goodreads*, www.goodreads.com/en/book/show/38891853.

Engels, Friedrich. *The Condition of the Working Class in England*. 1845. Granada Publishing Limited, 1982.

Graeber, David. *Debt: The First 5,000 Years*. Melville House, 2014.

Grosz, E. "Bodies – Cities." *Places Through the Body*, edited by H. Nast and S. Pile, Routledge, 1988, pp. 42–51.

Hasted, Nick. "The Story behind the Song: Neil Young – after the Gold Rush." *Louder*, 12 Nov. 2016, www.loudersound.com/features/the-stories-behind-the-songs-neil-young-after-the-gold-rush.

Horsley, Lee. *Twentieth Century Crime Fiction*. Oxford UP, 2005.

Kincaid, Andrew. "'Down These Mean Streets': The City and Critique in Contemporary Irish Noir." *Éire-Ireland*, vol. 45, no. 1, 2010, pp. 39–55, doi:10.1353/eir.2010.0005.

Klein, Naomi. *This Changes Everything*. Allen Lane, 2014.

Kroeber, Karl. *Ecological Literary Criticism*. Columbia UP, 1994.

Martinez-Alier, Joan. *The Environmentalism of the Poor*. Edward Elgar Publishing Limited, 2002.

McBrien, Justin. "Accumulating Extinction Planetary Catastrophism in the Necrocene." *Anthropocene or Capitalocene?*, edited by Jason W. Moore, PM Press, 2016.

McKinty, Adrian. *The Cold Cold Ground*. Serpent's Tail, 2012.

———. *I Hear the Sirens in the Street*. 2013. Serpent's Tail, 2014.

———. *In the Morning I'll Be Gone*. Serpent's Tail, 2014.

———. *Gun Street Girl*. Serpent's Tail, 2015.

———. *Rain Dogs*. 2015. Serpent's Tail, 2016.

———. *Police at the Station and They Don't Look Friendly*. Serpent's Tail, 2017.

Moore, Jason W. "The Rise of Cheap Nature." *Anthropocene or Capitalocene?*, edited by Jason W. Moore, PM Press, 2016.

Nixon, Rob. *Slow Violence and the Environmentalism of the Poor*. Harvard UP, 2013.

Wikipedia. "Adrian McKinty." en.wikipedia.org/wiki/Adrian_McKinty. Accessed 21 May 2021.

Wordsworth, William, and Samuel Taylor Coleridge. *Lyrical Ballads*, edited by R. L. Brett and A. R. Jones, Routledge, 1991.

Žižek, Slavoj. *Violence*. Profile Books, 2007.

13

ENVIRONMENTAL CRIME AND THE DIALECTICS OF SLOW AND DIVINE VIOLENCE IN *POSO WELLS* BY GABRIELA ALEMÁN

Rafael Andúgar

This chapter seeks to outline the relationship between environmental crime, corruption, and slow violence in Latin American crime fiction through the examination of an indicative text: the Ecuadorian crime and satirical novel *Poso Wells* (2007) by Gabriela Alemán. Set at the beginning of the twenty-first century, *Poso Wells* expresses deep political mistrust towards a set of neoliberal economic policies that ultimately resulted in the mass privatisation of Ecuador's public resources. The plot follows journalist Varas as he investigates the strange case of a community of elderly blind people who dwell in isolated tunnels underground the fictitious slum settlement of *Poso Wells*, just on the outskirts of Guayaquil. The blind community have been kidnapping and raping the women of the slum for many years but only come to public prominence after they abduct the political candidate for the presidency, Vinueza, who will eventually form an alliance with the group to help relaunch his electoral campaign. Varas will later discover that Vinueza is involved in a corrupt scheme related to the sale of large ecological reserve lands to Holmes, the evil CEO of a Canadian multinational company specialising in copper mining. At the end of the novel, Holmes' and Vinueza's violent plans are halted when a volcanic eruption unleashes chaos.

In this chapter, I argue that the novel represents the contestation between the two theoretical concepts of violence as outlined by Walter Benjamin: the mythical violence of law and the divine violence of nature. In this case, as a *deus ex machina*, the divine violence will restore environmental justice by repelling the violence of the law. These two concepts intersect with Rob Nixon's ideas of slow violence and offer a way into understanding the dialectics and ethics of Ecuador's corrupt legal system. Furthermore, *Poso Wells* portrays the devastating effects of slow violence both on the natural environment and the lives of the characters, particularly women, as the destruction of the fragile ecology becomes expressive of the broader violence of a heteropatriarchal society. Indeed, one of the most revealing messages of the novel is that one cannot be an ecologist without being a feminist first. Thus, to fully understand the novel, one must consider its relation to ecofeminist praxis. Ecofeminist thought has many divergences and debates but is an essential consideration for analysing *Poso Wells*.[1] Patricia de Souza, a Peruvian ecofeminist, has argued that what unites feminist movements is the necessity to address environmental destruction: "There are many feminisms in the world, liberal, religious, but what unites us all is that we will face the climate and resource problem, and there the division will end. And women must be united".[2] In my analysis of the novel, I combine scholarship on violence by Walter Benjamin and Rob Nixon, arguing that it takes on a greater critical depth when cojoined with the ecofeminist theories of Patricia de Souza, Yayo Herrero, Vandana Shiva, and Maria Mies.

DOI: 10.4324/9781003091912-16

Poso *Wells* and the History Ecuador's Extractive Politics

The publication of *Poso Wells* is closely linked to Ecuador's turbulent political history.[3] Alemán has stated that the novel was principally written to dissuade the public from voting for Alvaro Noboa, a right-wing candidate who was running against Rafael Correa, a leftist, in the 2006 election. The contest is satirised in the novel through the figure of Vinueza, a facsimile of Noboa who positions himself as a messianic candidate decreed by God. This is despite possessing the largest fortune in Ecuador, which he is rumoured to have amassed through corrupt means.[4] The real-life elections followed an event known as "the rebellion of the outlaws" ("la rebelión de los forajidos") which ended the government of Lucio Gutiérrez, also known for his nepotism and corruption. Between 2002 and 2006, there was great social tension and a series of anti-extractivist protests. This is the time when Alemán wrote *Poso Wells*, picking up on environmental anxieties and a broader societal despondency towards political corruption.

The relationship between corruption and environmental degradation resulting from extractivism is a controversial topic. Michael L. Ross asserts that there is a direct correlation between the extraction of natural resources, particularly oil, and corruption in states where these resources form the economic base, also known as petro-states. Ross argues that the revenue from oil enables authoritarian states to maintain power, as it is used for both public spending and private profit and is easily hidden as it is not subject to taxation (Ross 6). However, Thea Riofrancos investigates the social response in Ecuador and emphasises how activists and popular movements resisted leftist governments and economic elites who aimed to exploit natural resources (Riofrancos 31). This resistance is evident in *Poso Wells*.

At the end of the twentieth century, Ecuador underwent a transformation in its approach to using oil for economic development. The country moved from a model that used oil to support developmentalist policies to one that embraced oil extraction from a neoliberal perspective. The difference is that developmentalist policies prioritise long-term development and social welfare, while the neoliberal paradigm prioritises free-market economics and individualism. Guillermo Rodríguez Lara, in the 1970s, promoted the state-owned companies that managed the resources; nationalisations were carried out, and Ecuador joined the Organization of the Petroleum Exporting Countries (OPEC). After this period, in the 1980s and until 2006, there was a change in the extractive model because foreign oil companies and national businessmen joined forces to achieve privatisation and deregulation in order to obtain foreign investment (Riofrancos, 8). Later, during the administration of Rafael Correa, the previous model of national exploitation of resources that had benefited the state returned again, but companies were no longer nationalised. In order to overcome foreign competition, large taxes were imposed, making national competition cheaper and resulting in a huge influx of money that was redistributed socially to alleviate poverty (Riofrancos 10–11). Throughout the twentieth century, there have been different shifts towards neoliberalism that exhibit varying features, such as cutbacks, austerity measures, and the involvement of national extractive companies. In some cases, these companies require foreign investment and sell off large portions of land, leading to the destruction of ecosystems. Conversely, there are also periods where the state manages and benefits from these economic activities more directly.

Poso Wells satirically depicts the final years of Ecuadorian neoliberalism prior to the government of Correa, characterised by frequent corruption and land sales to foreign companies. The lands being sold were territories under Indigenous sovereignty, serving not only as cultural spaces but also as important environmental ones. It became an arena of political struggle for the defence of natural resources, with social movements protesting for a national and democratic control of extractivism, even forming a radical and totally anti-extractivist critique that would later clash with the state interventionism of Correa (Riofrancos 31). This historical context is central to *Poso Wells*, where environmental and heteropatriarchal crime combine in an intersectional way, renewing the detective genre in Latin America.

Colonialism and Slow Violence in *Poso Wells*

Environmental violence, according to Rob Nixon, has a dimension of slow temporality. He bases this on the disparity between human lifespans and the temporal scales of geological processes (Nixon 15). One of the most suggestive ideas of *Slow Violence* is the disjunction that arises between the accelerated times of turbocapitalism – along with the loss of attention that new technologies engender – and the real slow erosion of the environment that the violence of CO_2 emissions in the atmosphere produces (Nixon 8). In other words, one of the greatest challenges of current politics is how to promote social change and to stop the gradual deterioration of the environment that occurs slowly over years and that is difficult to perceive. As Nixon points out, the long-term consequences of industrialisation still have an effect today in countries that are perceived to be in a post-industrial transition:

> The past of slow violence is never past, so too the post is never fully post: industrial particulates and effluents live on in the environmental elements we inhabit and in our very bodies, which epidemiologically and ecologically are never our simple contemporaries.
>
> *(8)*

Nixon's study points to non-fiction writer-activists who are endeavouring to represent this slow violence, whilst also arguing for literature as an ideal vehicle to experiment with, and transmit, these difficult-to-perceive realities:

> In a world permeated by insidious, yet unseen or imperceptible violence, imaginative writing can help make the unapparent appear, making it accessible and tangible by humanizing drawn-out threats inaccessible to the immediate senses. Writing can challenge perceptual habits that downplay the damage slow violence inflicts.
>
> *(15)*

The unrepeatable nature of slow violence calls for a form of artistic expression beyond the traditions and limitations of literary realism. Thus, genres such as the fantastic, weird fiction, science fiction and magical realism are literary springs that allow the reader's perception to be de-automatised, revealing the temporal mismatches between environmental degradation and human activity. For example, *Poso Wells* represents slow violence through the story of a settlement that has emerged on the outskirts of Guayaquil and how this space symbolises an imbalance between the environment and the construction:

> Poso Wells no aparece en ningún mapa. Sería imposible que así fuera. Esa enorme cantidad de lodo ganado al estero era, la última vez que alguien realizó un levantamiento topográfico, parte del río. Y el agua corre, no se parcela. . . . Kilómetros y kilómetros de viviendas de palo, caña y aglomerado construidas sobre aguas servidas y barro podrido. Estacas de mangle hundidas en una tierra fangosa y blanda; un suelo inestable donde aparecen grietas con cada marejada alta o corriente que arrastran los buques de alto tonelaje en su camino al puerto de Guayaquil.
>
> *(Poso Wells, 11)*
>
> [Poso Wells does not appear on any map. How could it? The last time anyone did a topographical survey, that huge mass of mud dredged from the estuary was still part of the river. And water flows. It's not subdivided into lots. . . . Kilometres and kilometres of houses built of sticks and reeds held together by a mix of mud and stones, all resting on a suspension of sewage and mouldy clay. Mangrove posts sunk into soft, unstable soil that cracks open in new places with every tide or current sweeping the high-tonnage ships toward the port of Guayaquil.[5]]

The river and mud are the foundations on which the neighbourhood of Poso Wells is built, and act as symbols of the poverty of the community and of the material struggle to inhabit the natural environment. The slow violence of Poso Wells is allegorised in the form of a community of blind men who have been living for generations beneath the city, terrorising, abducting, and raping women. They come to embody the corruption and brutality of a social system based on heteropatriarchy, colonisation, and extractivism. Specifically, the men are revealed to be direct descendants of Spanish colonisers who found the valley of H.G. Wells' novel *The Country of the Blind*, once a paradisiacal and pastoral place:

> El valle, dijo, tenía todo lo que el corazón de un hombre podía desear: agua dulce, pastos y buen clima, laderas de tierra fértil y rica, con marañas de arbustos que daban excelentes frutos y, colgados de una ladera, grandes bosques de pinos que detenían las avalanchas.
>
> *(Alemán 27)*
>
> [The valley, he said, had in it all that the heart of man could desire – sweet water, pasture, a good climate, slopes of rich brown soil with tangles of a shrub that bore an excellent fruit, and on one side great hanging forests of pine that held the avalanches high.]

Poso Wells narrates how this group of Spaniards were subsequently trapped in the valley because of the eruption of the Cotopaxi volcano. The connection that the novel makes between the pastoral valley of Wells and the irregular and anti-pastoral settlement of Poso Wells is crucial. The novel depicts the Spanish conquistadors' so-called "discovery" of this seemingly "utopian" space prior to its destruction by the volcanic eruption. Here we are faced with a metaphor of the arrival of the Spaniards in America and Hernán Cortés' destruction of Tenochtitlán. Poso Wells too becomes a degraded and dangerous space, as the community of blind people, direct descendants of the Spanish conquistadors, stalk their victims from beneath the earth. The violence of the conquest of America, alacritous at first, and slow later, is a metaphor for how the colonisation of space and population develops. The dialectic of civilisation and barbarism, a rhetoric that reached its peak during the period of Independence of different nations – as in the case of Domingo Faustino Sarmiento's *Facundo* – is a key topic of discussion in the twenty-first century Latin American novel and is explored in *Poso Wells* through its critique of corruption and its questioning of the insidious effects of the law, as well as its focus on the destructive power of nature.

Mythical and Divine Violence

In *Poso Wells*, divine violence is portrayed as being inherent in nature, while the gradual violence inflicted by man-made laws is represented as a form of "slow violence". This metaphorical play between the two creates a narrative of crime and environmental fiction. In order to carry out my close reading of the novel and to reveal the text's political critique, I will employ Walter Benjamin's "Toward the Critique of Violence" (1921). My purpose is to show how *Poso Wells* characterises nature as a divine force that enables a form of environmental justice. The community of blind men, on the other hand, impose a "mythical" violence and their alliance with the presidential candidate represents how law and corruption are violences founded by man.

In his essay "Towards a Critique of Violence", Benjamin explores the essence of violence and its relationship with laws, justice, and morality. He contrasts what he calls "mythical violence" with "divine violence", two concepts that he develops from the interpretation of a Greek myth and an episode of the Hebrew bible. In the context of ancient Greek myths, Benjamin sees in the punishment of Niobe, whose children are killed by Apollo and Artemis in revenge for mocking Leto, an example of what mythical violence might be (Benjamin 55). Benjamin argues that the violence inflicted by gods like Apollo and Artemis in the myth of Niobe serves not only as punishment but also as a way of establishing law.[6] This mythical violence is not just a means of preserving existing laws, but a way of creating new ones. Niobe is punished as she does not accept her fate. The violence in these myths is ambiguous and related to destiny, ultimately reflecting the problematic nature of law-making. Benjamin suggests that the positing of law is an act of immediate violence and that power is closely tied to this process. His contention is that justice is the principle behind divine law, while power is the principle behind mythic law (Benjamin 56). On the other hand, Benjamin contrasts mythical violence with an example of a scene from the Bible, that of Korah's rebellion against Moses. Korah gathers a group against Moses because he doubts that he is God's chosen one. Moses explains this to God but asks him to forgive Korah. God's opening of the earth to swallow Korah's horde, is, for Benjamin, indicative of divine violence, a revolutionary violence that, instead of establishing laws and borders, would destroy them and erase their faults (the opposite of the legend of Niobe) (Benjamin 57).

Benjamin conceives of the relationship between violence and law through the use of the German word *Gewalt*, which has multiple meanings including violence, power, and force. He highlights the close connection between violence, power, legal violence, law enforcement, and the creation of laws and suggests that the origin of law lies in mythical violence. This is exemplified by the myth of Niobe, which shows how a boundary is established through the threat of corrective force, leading to the legitimisation of a power based on violence and the creation of equal rights. In this central fragment, he elucidates on the difference between the divine and the mythical:

> Divine violence designates in all respects an antithesis to mythic violence. If mythic violence is law-positing, divine violence is law-annihilating; if the former establishes boundaries, the latter boundlessly annihilates them [. . .] if the former threatens, the latter strikes; if the former is bloody, the latter is lethal in a bloodless manner.
>
> *(57–58)*

I propose that these ideas from Benjamin's theory of violence can be an archetype for discussing the violence that appears in ecological crime fiction, where the environment becomes an active agent. One of the most suggestive aspects of Benjamin's work is that legal violence is anthropocentric, while divine violence is in another sphere, a plane that would be closer to an agency of a different dimension than that of the human. While crime fiction is a space where Benjamin's mythical violence – the violence of laws and law enforcement – enter into dialectical tension until the resolution, the environment and nature appear as a divine violence that is beyond the anthropocentric terrain and that destroys without considering the laws of man. In addition, there is a temporality in Benjamin's ideas that is also suggestive in relation to the different temporal scales between different agents. Whereas divine violence is unleashed quickly, mythic violence is slow, and its effects endure and persist over a long period.

Heteropatriarchal Violence as Slow Violence in *Poso Wells*

Poso Wells is a novel in which the criminals are men who attack both women and the environment, and the connection between the two becomes a political position that denounces the purposefully obscured heteropatriarchal structures that are fundamental to colonial practices. The blind men who kidnap and rape the women of Guayaquil support the corrupt politician who sells and destroys the land in order to accumulate capital. Men inflict violence with the purpose of making women and nature the producers of life and profit, but what they achieve is the opposite. The blind community is sterile and has no offspring, much in the same way that ecocidal enterprises against the planet threaten the future of generations to come. It is crucial to note that ecofeminism highlights the parallel between the exploitation of women and nature through reproductive labour (Svampa 129). The novel uses metaphor by linking the exploitation of women through sexual violence to the exploitation of land through land grabbing. As Annette Kolodny suggests, "implicit in the metaphor of the land-as-woman was both the regressive pull of maternal containment and the seductive invitation to sexual assertion" (Kolodny 67). This approach is consistent with the trend in literature where the themes of land acquisition and gender violence are intertwined through metaphor, indicating a deeper connection between the two (Murphy 213).

Ecofeminism critiques the way human life cycles and ecological limits are ignored by conventional economics, challenging the fundamental basis of capitalist economic paradigms. It dismantles the theoretical divide between humanity and nature and emphasises the material importance of relationships and the vulnerability of human bodies and life (Herrero 5). According to Maria Mies and Vandana Shiva in *Ecofeminism* (1997), ecofeminism is about interconnectedness and wholeness in theory and practice. It asserts the value of every living thing and views the destruction of the earth and its creatures as rooted in the same patriarchal mentality that denies women their own bodies and autonomy (Mies and Shiva 14).

In *Poso Wells*, gendered violence is linked to political corruption and environmental conflict through two intersecting criminal plots: the abuse perpetrated against women by the underground elders and the ecological harm inflicted by a government endorsed yet exploitative mining scheme. Varas wants to solve the first case, but finds himself faced with a non-existent legal framework to protect the victims:

> La logística de la operación no le preocupaba tanto como las leyes ecuatorianas y la manera arbitraria como se dispensaban, dudaba que existiera una legislación específica que penalice el crimen que se había cometido contra ellas y, si existiera, que alguien diera paso al trámite para hacerla cumplir.
>
> *(159)*
>
> [The logistics of the operation didn't worry him so much as the Ecuadorian legal system and the way it was applied. He doubted there was any specific law against the crime that had been committed against the women, and, even if there was, would anyone enforce it?]

In this fragment, detective Varas makes a reflection that is central to the ecofeminist thinking of the novel and stresses the importance of a wider social discussion about what should be classified as a crime. Varas questions the lack of legal recourse to prosecute the blind community, who, due to living underground, remain invisible to the legal system, meaning that their acts of sexual violence are not codified as criminal. This is one of the ways in which ecofeminist crime fiction denounces existing legal frameworks, highlighting the lack of protection in law for women and the environment itself. Varas's reflection relates to slow and mythical violence, particularly the problems that occur when powerful men agree to laws in service of their own interests.

Poso Wells associates the legal impunity afforded to sex offenders with the extra-legal violence of land acquisition: "Detective fiction has the potential to explore examples of slow violence – subtle yet deeply destructive crimes against marginalised or minority individuals, or natural spaces – crimes that are not defined as illegal by the cultures that these groups inhabit" (Cothran 467). Both slow violence and gender violence share common links to the concept of "structural violence", a term coined by sociologist Johan Galtung. As Rob Nixon suggests, Galtung was concerned with "widening the field of what constitutes violence" by foregrounding "the vast structures that can give rise to acts of personal violence and constitute forms of violence in and of themselves" (Nixon 10). In contrast to Galtung, Nixon sees slow violence as temporally linked to moments in history. In other words, the ways that slow violence manifests are fundamentally linked to historical and temporal structures that then shape the experience of violence (Nixon 11). However, ecofeminist authors help us understand that slow violence is inextricably connected to the extractavist logic of Western heteropatriarchal societies.[7]

In the face of the violence and terror that the mythical community of the blind men inflicts on women, there is an environmental "divine" justice – a volcanic eruption – that punishes the self-serving and corrupt, whilst also challenging the reader's perception of reality. The novel portrays corruption and excessive violence in a manner that is alarmingly similar to real life, thereby blurring the lines between fiction and reality. We can take the case of climate change, which is impossible to predict, and the results are unthinkable and unimaginable. When combined with the detective novel, one of the rationalist genres par excellence, a different space appears within which to reflect on how irrational things, such as environmental destruction and systemic gender violence, are continuously invading reality. Violence against women, feminicide, and sexual violence are rampant, after all. Statistics compiled by official authorities between 2014 and 2017 leave discouraging data: "In Ecuador, 6 out of 10 women have experienced some type of gender violence (in various settings and by various people)", and 1 out of 4 women among those aforementioned has suffered sexual violence (67–68).[8]

In *Poso Wells*, heteropatriarchal violence becomes a mythical violence that establishes law and power. The blind community seeks to create a society and allies itself with Andrés Vinueza, the presidential candidate and satirical figure of corruption. In the latter half of the novel, a group of blind elders appears on television and idolise Vinueza as a messianic figure. Later, they attack a drunk pedestrian who was recently with a sex worker, Sun Yi. She is then kidnapped and locked up with the blind men who intend to rape her. Yi defends herself by releasing a snake that bites the genitals of one of the blind men, who, in screaming, attracts the attention of Vinueza, who opens the door and allows Yi to escape. This scene is deeply symbolic and humoristic as the snake becomes a representative of how divine violence can restore justice. The snake's function is ironic as it forms an alliance with Yi to attack the men who imprison her, becoming perhaps an intertextual comment on how Eve eats the apple to gain wisdom and thus, allegedly, brings the "fall" of man. But there is another layer in the scene, likely linked to the religious syncretism of the Andean region where *Amaru* (snake in Quechua) is a divine symbol representative of the underground world of the dead (*Uku Pacha*), but also of fertility and women: "In most Andean cosmologies, the serpent also embodies a monstrous being from the depths of the earth, invested with supernatural powers; a being that embodies destructive forces of nature that humans are unable to control"[9] (Gil García 14). The snake, in this sense, encapsulates multiple cultural significations and, crucially, offers divine justice to Yi in the face of patriarchal violence.

In *Poso Wells*, gender violence, corruption, and environmental destruction appear as structural and invisible forces that must be named and represented. In Varas' investigation, Bella helps him by identifying women who have disappeared through photographs. It is in this moment that a character, Montenegro, says to Varas: "Si uno se olvida de las cosas, ¿es lo mismo que si no hubieran ocurrido?" (108). ["If you forget things, is it the same as if they never happened?"]. This question highlights the importance of memory in preserving the existence of events and experiences. Systemic violence, due to its slow and relentless nature, is often forgotten due to the speed of modern life. But if memory erodes, so too does our history.

The use of violence by the law constitutes slow violence, as it perpetuates practices that test legality through force, resulting in the establishment of new systems within unlawful territories. In other words, there can be no corruption without law (and vice versa). In *Poso Wells*, the representatives of the law bring that which was illegal into the sphere of legality through violence. The moment of the environmental crime occurs when Vinueza proposes that José María be the manager of the Eagle Copper Corporation, after which they sell 120,000 hectares of copper rich land to a series of businessmen from Canadian international markets (132). The novel satirises corporate and legal language of the kind which formally legitimises and invisibilises criminal actions: "los contratos de la formación de la empresa, aunque dentro del Holding constaremos como accionistas" (129). ["That my name can't appear in the company's founding documents, although we'll be shareholders through the holding company"]. In part 3 of *Poso Wells*, Vinueza also perpetuates an environmental crime by selling 7,000 hectares of the Cotacachi-Cayapas Ecological Reserve for an unlimited exploitation time to Holmes, CEO of Eagle Copper Corporation, for US$18,000 (148). Civil society opposes this sale because of the environmental implications, but Vinueza is only thinking about getting rich:

> La explotación a cielo abierto suponía volver al bosque un desierto. Para remover la tierra necesaria en busca del cobre había que mover sesenta toneladas de suelo al día. Adiós árboles, orquídeas y pajaritos: adiós bosque nublado.
>
> *(147)*
>
> [Open pit mining would turn the forest into a desert. Removing all the soil to unearth the copper would mean shoveling out sixty tons a day. Goodbye trees, orchids, birds. Goodbye cloud forest.]

The vast land grabbing also mirrors reality. Land Matrix compiles that in 2020, large land transactions in Latin America amounted to around 45 million hectares (and half of it is mining, much more polluting than the agricultural industry). Furthermore, 2.2% of farmers own more than 41% of the land (Carrión and Herrera, 2012).[10] Political corruption and land acquisition are intertwined within the legal system, where corrupt states divide up land amongst wealthy buyers. The issue of clientelism and cronyism, where politicians and businesspeople work together for mutual benefit, has become part of this discussion. Noam Chomsky argues that such inequalities of power are fundamental to the ideological framework of capitalism:

> What's capitalism supposed to be? Yeah, it's crony capitalism. That's capitalism, you do things for your friends, your associates, they do things for you, you try to influence the political system, obviously. You can read about this in Adam Smith.[11]

Chomsky shows that corruption is a necessary part of the capitalist system itself. It is normal for there to be abuses of power by politicians and for corporations to try to influence governments because the entire legal and economic system is based on the pursuit of profit. *Poso Wells* uses the linking of the political candidate Vinueza to the CEO Holmes to satirise the corruption, power, and crony capitalism present in Ecuador. By crossing legal boundaries and creating violence to preserve their economic profits, these corrupt men destroy the environment for their own personal gain.

Nature as Divine Violence

In *Poso Wells*, nature and its different elements have a close relationship with divinity for various religious and cultural reasons. For the Indigenous communities of the Andean regions, the terrestrial forces that imbue rivers, seas, mountains, and natural elements are part of *Pachamama*, a goddess that in *Aymara* and *Quechua* translates as mother earth (Tola 194). For the communities of Ecuador, Pachamama becomes a symbol of natural resources that have become the object of various global extractive economies. Notably, Pachamama is protected in Ecuador's constitution, codified in 2008 under the principle of good living (in Quechua *sumaq kawsay*). Good living prioritises social and ecological balance over developmentalist policies that seek economic progress (Tola 198). We see this when Varas wakes up at Montenegro's house and contemplates a sublime landscape:

> El amanecer era glorioso: fucsias y destellos violetas inundaban el firmamento. Era el único momento del día en que Montenegro le hallaba algún sentido a la vida.
>
> *(38)*
>
> [The sunrise was beautiful, with fuchsia and violet sparkles flooding the sky. This was the only moment of the day when Montenegro felt life made any sense.]

The vision of nature brings an existential sense to Montenegro, who understands life as if it were an epiphany. Nature is invested with divine power in *Poso Wells* because it is not an endangered subject that can be attacked without countermeasures and retaliation. In the novel, nature demonstrates its force through the manifestation of natural events such as animal attacks and volcanic eruptions. We see this linkage of nature to the divine further when the blind men explain to José María, the administrative director of the Vinueza Group and Companies, that there are angels in this world. Yet José María understands that the celestial creatures they speak of are simply birds:

> No eran otros que los cientos de aves que abundaban en la zona y que, si se hubieran detenido a verlos, habría notado que eran realmente de otro mundo. Especialmente los tangaras, que más que pájaros parecían apariciones del paraíso.
>
> *(160)*

> [Their celestial beings were in fact the hundreds of birds that flocked to this region. If he had taken the time to study these, he would have realised that they really were from another world. Especially the tanagers, which looked more like apparitions from paradise than earthly birds.]

Upon closer examination of this extract, it becomes apparent that although the birds are not celestial, there is a magical element conveyed through the phrases "were from another world" and "apparitions from paradise". The novel highlights the divine and magical aspects of nature, drawing upon the tradition of Latin American magical realism. In passages like this one, the text acknowledges that marvellous things exist in the natural world. The divine violence of Pachamama is contrasted with the violence of extractivism, corruption, heteropatriarchy, and the Anthropocene.

Towards the end of the novel, nature destroys the extractivist plans of Vinueza and Holmes. After a night of partying, Salém and his henchmen have an accident with a police car while drunk driving, causing both vehicles to crash in the jungle:

> Ambos carros rodaron barranco abajo hasta quedar totalmente cubiertos por el denso follaje del bosque nublado de Intag. Seguramente para nunca volver a ser vistos, salvo por los majestuosos caracaras andinos que comenzaban a menguar en la zona y que inesperadamente recibían como una dádiva del cielo raras delicadezas costeñas imbuidas en alcohol para su deleite carroñero.
>
> *(157)*

> [The crash sent both vehicles tumbling down the cliff-side until they were swallowed up by the dense foliage of the Intag cloud forest. Neither would ever be seen again, other than by the majestic mountain caracaras, birds of prey whose numbers were already diminishing but who this morning received an unexpected gift, as if from heaven, of rare coastal delicacies soaked in alcohol for their scavenging delight.]

The divine violence in *Poso Wells* is ultimately embodied by the eruption of the Imbabura volcano. It should be remembered that *Apus*, the gods of the mountains, are attributed a direct influence on divinity and on the vital cycles of the regions that surround them (Cadena 182). The divine violence of nature in *Poso Wells* has its origins in Andean cosmogonic myths, such as the legend of the god Quilatoa, capable of making all the volcanoes of the world erupt. In the Imbabura region, the cult of Pachamama is widespread, as is the belief in the deity that dwells on Mount Imbabura itself, who, together with his partner Cotocachi, created the world:

> In the mythical conception of indigenous and mestizo Imbabureños, the supernatural power conferred on the elements of nature stands out clearly . . . Mountains, springs, streams, creeks,

> trees, etc., are sacred symbols that have life in themselves, since they are animated by spiritual forces capable of altering the course of human life.
>
> *(Villavicencio 249)*[12]

By making Mount Imbabura and its eruption the focal point of the novel, *Poso Wells* revives the Andean cosmogonic myths and endows the natural world with an animated quality, able to divinely impact human affairs. Quickly the different plots of the novel intertwine, concluding with the blind men running down the slope of the volcano as birds fall from the sky. At the same time, Vinueza receives a call explaining that the people of Junín have taken over the mine installations, and on television it is explained that the elections have been postponed due to the volcanic eruption and its attendant environmental effects (174). The kidnapped women are also found by the policemen Banegas, who publicly denounces Vinueza and the elders (175). The volcano ultimately destroys the blind conspirators and ends the political plot to sell the land for economic exploitation. As a natural force, and a narrative *deus ex machina*, this divine violence thus enacts a form of restorative justice. Benjamin perceived the potential for revolutionary violence in God's divine violence, which atoned for guilt and destroyed Korah's horde. However, the divine violence in *Poso Wells* conveys a more unsettling message that goes beyond the religious sphere. The novel's conclusion speaks to the immensity of environmental issues and highlights how climate change or natural disasters are beyond human capability.

The end of the novel leaves an educational warning. Although divine environmental justice has solved the crimes and stopped Holmes' and the corporation's plans to carry out their extractive enterprise in Ecuador, the novel is not naïve and shows how these enterprises have a transnational character that goes beyond the legality of nation-state systems. Holmes escapes to Santiago de Chile because he expects to profit economically from the lack of legislation around the Andean border between Argentina and Chile: "un país virtual abierto a las compañías mineras multinacionales, una suerte de tierra de nadie donde no existían impuestos ni se pagaban derechos" (177). ["A virtual country open to multinational mining companies, a sort of no-man's-land without taxes or royalties to pay"]. In this way, the novel links local environmental problems with a lack of environmental justice globally, underlining how state and transnational legislation is intertwined in such a way as to allow legal loopholes to be continuously exploited. The novel uncovers Martín Arboleda's idea of the "planetary mine", an integrated and global logistical network that "vastly transcends the territoriality of extraction and wholly blends into the circulatory system of capital, which now transverses the entire geography of the earth" (5). Furthermore, the novel highlights the relationship between space technology and how it is discovering the possibilities of extraction, mapping, and geoengineering of planet earth:

> Pero no había apuro, para eso existía la tecnología satelital y él tenía en su poder un mapa de Ecuador con todos sus depósitos mineros resaltados en una gran gama de colores brillantes y eran muchos. Solo sería cuestión de tiempo, de esperar que llegara el gobierno adecuado.
>
> *(177)*
>
> [But there was no hurry, that was what satellite technology was for. He had at his disposal a map of Ecuador with all the mining deposits highlighted in a full range of brilliant colors – and there were a lot of them. It was just a question of time, of waiting for the right government to come to power.]

Poso Wells' chronicles the environmental corruption that many Latin American countries suffer because of the personal economic benefits provided by transnational extractive industries and how this type of slow violence is codified in legal frameworks. The novel, drawing from a range of literary genres, hybridises crime fiction to denounce systemic heteropatriarchal violence and interlinks land grabbing with gender violence. The fiction presents and valorises the divine violence of natural agents, imbued with an Andean cosmogony of Pachamama, as the only force capable of establishing environmental justice.

Notes

1. According to Yayo Herrero, ecofeminism is a philosophy and movement that explores the intersections and synergies between ecology and feminism (Herrero 1). The term ecofeminism was first used by Francoise D'Eaubounne in *Feminismo o la muerte* (1974), where she believed that there was a deep connection between overpopulation, environmental destruction, and male domination. Ecofeminism posits that the subordination of women to men and the exploitation of nature are two sides of the same coin and respond to a common logic of domination and subjugation of life to the logic of accumulation (Herrero 3).
2. My translation from the interview: de Souza, Patricia. "Estamos construidas mentalmente por hombres". *Pousta*. Entrevista por Marcial Parraquez. 2019–09–21. https://pousta.com/patricia-de-souza-entrevista.
3. María Gabriela Alemán Salvador (Rio de Janeiro, 1968) is an Ecuadorian writer who has lived in different countries throughout her life: she studied in the United Kingdom, got her doctorate in the United States, and worked in Switzerland and in various professions in Paraguay. She belongs to the group of writers of Bogotá 39, an anthology of 39 writers under 39 years of age who show promise, and has published several short stories, novels, and comics.
4. Alemán, Gabriela. Interviewed by Dick Cluster. "Literature Is the Minefield of the Imagination: An Interview with Gabriela Alemán." *Los Angeles Review of Books*. 17 July 2018. Last accessed February 21, 2019.
5. Alemán, Gabriela. *Poso Wells*. City Lights Books: San Francisco, 2018. EPUB. All translations from Spanish were made by Dick Cluster. Original: Alemán, Gabriela. *Poso Wells*. Eskeletra, 2007. Aristas Martínez Ediciones, Madrid: 2012.
6. "Niobe's arrogance conjures up the disaster that befalls her not because it injures law but because it challenges fate – challenges fate to a combat in which fate must triumph, bringing a law to light, if need be, only in its victory . . . Violence thus closes in upon Niobe from the uncertain, ambiguous sphere of fate. This violence is not actually destructive. Although it brings bloody death to Niobe's children, it stops short of taking the mother's life, which it leaves behind as an eternal, mute bearer of guilt and as a stone marking the border [Grenze] between human beings and gods, a life now, through the children's death, more inculpated [verschuldeter] than before" (Benjamin 55).
7. *Poso Wells* appears at a historical turning point in the Ecuadorian nation. Until 2006, gender violence had been a judicial matter that was legally prosecuted only through complaints. However, since the presidency of Rafael Correa, the state has started creating policies to guarantee human rights and work towards a national eradication of this phenomenon (Camacho 17). It was not until 2008 that the rights of equality of all persons without discrimination were recognised in the country's Constitution (Camacho 22).
8. My translation from: Secretaría Nacional de Planificación y Desarrollo. *Agenda nacional de las mujeres y la igualdad de género 2014–2017*. Gobierno Nacional de la República del Ecuador.
9. My translation.
10. For more information, please consult the Land Matrix land transactions and *land grabbing* dossiers and publications. Also: Carrión, D., Herrera, S. *Ecuador rural del Siglo XXI: soberanía alimentaria, inversión pública y política agraria*. Quito, M. B, 2012.
11. Noam Chomsky, "Black Faces in Limousines: A Conversation with Noam Chomsky", Joe Walker Blog (November 14, 2008). https://chomsky.info/20081114/Accessed on September 15, 2022.
12. My translation.

Bibliography

Alemán, Gabriela. *Poso Wells*. Eskeletra, 2007; Aristas Martínez Ediciones, 2012.

———. "Carta de Quito." *Cuadernos Hispanoamericanos*, vol. 713, 2009, pp. 47–53.

Arboleda, Martín. *Planetary Mine: Territories of Extraction under Late Capitalism*. Verso Books, 2020.

Benjamin, Walter. *Toward the Critique of Violence: A Critical Edition*. Edited by Peter D. Fenves and Julia Ng, Stanford UP, 2021.

Brigitte Adriaensen et Valeria Grinberg Pla. *Narrativas del crimen en América Latina. Transformaciones y transculturaciones del policial*. LIT, 2012.

Cadena, Marisol de la. *Earth Beings: Ecologies of Practice across Andean Worlds*. Duke UP, 2015.

Camacho, Gloria. *La violencia de género contra las mujeres en el Ecuador: Análisis de los resultados de la Encuesta Nacional sobre Relaciones Familiares y Violencia de Género contra las Mujeres*. Consejo Nacional para la Igualdad de Género, 2014.
Campisi, Nicolas. "South American Literature and Ecofeminism." *The Routledge Handbook of Ecofeminism and Literature*, edited by Douglas Vakoch, Routledge, 2022, pp. 186–194.
Cothran, Casey A. "Mystery and Detective Fiction and Ecofeminism." *The Routledge Handbook of Ecofeminism and Literature*, edited by Douglas Vakoch, Routledge, 2022, pp. 458–468.
Gil García, Francisco. "La serpiente: dimensiones de una divinidad subterránea en los Andes." *La figura de la serpiente en la tradición oral iberoamericana*, edited by Claudia Carranza Vera et al., Fundación Joaquín Díaz, 2016, pp. 13–16.
Herrero, Yayo. "Apuntes introductorios sobre el Ecofeminismo." *Centro de Documentación Hegoa Boletín de recursos de información*, no. 43, 2015, pp. 1–12.
Kolodny, Annette. *The Lay of the Land: Metaphor as Experience and History in American Life and Letters*. U North Carolina P, 1975.
Lalama Quinteros, Pamela Elizabeth. *La enfermedad como motivo literario en tres cuentos de Gabriela Alemán*. Unpublished Dissertation, 2015.
Mies, Maria, and Vandana Shiva. *Ecofeminism*. Zed Books, 2014.
Murphy, Patrick D. "The Ecofeminist Subsistence Perspective Revisited in an Age of Land Grabs and Its Representations in Contemporary Literature." *Feminismo/s*, no. 22, 2013, pp. 205–224, doi:10.14198/fem.2013.22.11.
Naranjo Villavicencio, Marcelo. *La Cultura Popular en el Ecuador*. Vol. V. CIDAP (Centro Interamericano de Artesanías y Artes Populares).
Nixon, Rob. *Slow Violence and the Environmentalism of the Poor*. Harvard UP, 2013.
Parraguez, Marcial. "Patricia De Souza: 'Estamos Construidas Mentalmente Por Hombres.'" *POUSTA*, 10 Jan. 2019, pousta.com/patricia-de-souza-entrevista.
Rama, Ángel. *Transculturación narrativa en América Latina*. Editora Nómada, 2019.
Riofrancos, Thea. *Resource Radicals: From Petro-Nationalism to Post-Extractivism in Ecuador*. Duke UP, 2020.
Riofrio, John. "When the First World Becomes the Third: The Paradox of Collapsed Borders in Two Novels by Gabriela Alemán." *MELUS: Multi-Ethnic Literature of the U.S.*, vol. 35, no. 1, 2010, pp. 13–34, doi:10.1353/mel.0.0065.
Ross, Michael L. *The Oil Curse: How Petroleum Wealth Shapes the Development of Nations*. Princeton UP, 2012.
Saccomano, Celeste. "El feminicidio en América Latina: ¿vacío legal o déficit del estado de derecho?/Feminicide in Latin America: Legal Vacuum or Deficit in the Rule of Law?" *Revista CIDOB d'Afers Internacionals*, no. 117, 2017, pp. 51–78.
Svampa, Maristella. "Feminismos del Sur y ecofeminismo." *Nueva sociedad*, no. 256, pp. 127–131.
Tola, Miriam. "Pachamama." *An Ecotopian Lexicon*, edited by Brent Bellamy and Matthew Schneider-Mayerson, U Minnesota P, 2019.

PART III

Epistemologies

14

"HOLMES, THAT'S SOME SANTA CLAUS SHIT"

Reading Lydia Millet's *A Children's Bible* as Ecological Crime Fiction

MaKenzie Hope Munson and Kevin Andrew Spicer

Admittedly, Lydia Millet's 2020 National Book Award-nominated novel, *A Children's Bible*, is not ostensibly a work of crime fiction. The novel's genre is just one facet of the book that is incredibly overdetermined: the generic status is unclear, borrowing as it does from many kinds of literature. This phenomenon of overdetermination is not simply a way to describe the work, it is also a methodology Millet uses to introduce her readers to a series of larger questions concerning the climate crisis, the nature of intergenerational guilt (and sympathy), and what forms of thinking are still available to us in the Anthropocene. Given that secondary literature on this book is still nascent, we wish to utilise an exploratory mode, a mode that will attend very closely to the text itself, in the hopes of laying out some territory to be traversed in the future. First, we wish to extract out a very key "crime fiction" element running through the novel, one that we call the "Oedipal detective game". Second, we will trace the larger theological allegory within Millet's novel, which functions as its own kind of "detective game". Third, we read these two games through a key essay by Gilles Deleuze, "The Philosophy of Crime Novels", which lays out two quite divergent kinds of "detective fiction" that we link to the two games. We will conclude with some potential implications for thinking about ways to read the interweaving of these different perspectives as they converge on larger environmental concerns in the novel.

Of Environmental Devastation and an Oedipal Detective Game

A Children's Bible begins by deceptively playing into the incipient aspect of the book's title: "Once we lived in a summer country" (1). Like any classic fairytale, there is a callback to the past, to some distant time that would allow for the contents of the story. To our narrator, this earlier time included countless elements of nature: a lake, the ocean, a marsh, a stream, a river, the beach, sand, etc. Of course, each of these things still exist in one form or another, but there is no doubt that their forms and frequency have been altered and continue to change as the climate crisis worsens. The fairy tale-esque atmosphere seems apropos given the state of the world in the book. The novel chronicles the vacation of twelve teenagers and younger children whose families join together to rent a mansion ("built by robber barons in the nineteenth century", the narrator, Eve, tells us very early [3]) somewhere on the East Coast of the United States. At the time of its building it was "a palatial retreat for the green months", but those green months now seem to be far fewer in number – and certainly the events of the narrative do not occur during such idyllic months as the children slowly watch the world

DOI: 10.4324/9781003091912-18

fall to pieces around them, beginning with an incredibly destructive storm that "[floods] the subway tunnels in New York" and causes "the river in Boston to [overflow] its banks" (96). The world outside is nothing short of a nightmare:

> Downed power lines electrocuted drivers, and cars and garbage cans and pets had been swept away down streets that looked like rushing rivers. We watched video of collapsing houses . . . . Riots, they said. Looting. States of emergency. The president had promised some money.
>
> *(96)*

The recourse to the fairy tale genre is just one of the medicinal remedies that the characters try out, but it is not, for us, the most significant one. Rather, it is the invention of what we would like to call an "Oedipal detective game" that provides one of the most salient coping strategies within the novel. This game is first mentioned on page five of the novel, where the narrator describes the children's "[h]iding [their] parentage as a leisure pursuit, but one [they] took seriously" (5). Indeed, each of the children are tasked with hiding who their parents are, making a competition of who can last the longest in this state of concealment; the parents, unaware of the game in its entirety, play along anyway – this serves as an early sign of their lack of care toward much of anything, but especially their children.

The origin of this Oedipal detective game speaks to a desire to create distance between the children and their parents and crystallises a question that many young people ask themselves amidst climate crisis: who did this to our world? These thoughts naturally lead to disgust and anguish towards older generations, those who had countless chances to avoid these circumstances but chose not to. This generational apathy serves as a basis for the beginning of the book, as the children all navigate ways to both physically and emotionally disconnect from their absent-minded parents. Evident from the very inception of this novel is an attitude of repugnance from the children to their parents, who most notably "liked to drink" (3). The parents are dismissive and selfish in the eyes of the children and while the narrator describes this activity as something that the children do to add fun to their summer, there is also an underlying severity to the game, marked by Eve's incredibly detailed descriptions when each of the children "reveal [their] origins":

> Jen's eleven-year-old brother was a gentle, deaf kid named Shel who wanted to be a veterinarian when he grew up. He suffered a bout of food poisoning just one week in and had to be tended by their parents, so that ID was made. The mother had adult braces and droopy shoulders, the father a greasy ponytail. He picked his nose while talking. He talked and picked, picked and talked.
>
> *(15)*

The children's allegiance to this game may have something to do with an idea proffered by Jeffrey Jerome Cohen in his "Noah's Arkive" essay, which explores connections between the story of Noah and the climate crisis. This text synthesises stories of modern climate disasters with the experience of Noah in the Bible to emphasise the disparities of environmental injustice: "The flood makes evident a lack of affective connection already present, the everyday inability of sympathy to cross boundaries of nation, race, species, class" and, in the case of this novel, generations. While the children struggle to cope with a world on the brink, the parents spend their days drinking away their anxieties. According to Eve and all the other children, the parents are to blame for this – thus necessitating the distance created by playing this game. An interesting point of contention to note is that the parents are explicitly liberals who hardly resemble the typical punching bag for those looking for an outlet to release their climate-based frustrations. Almost every scene captures the parents' left-leaning remarks about whatever situation they face; for example, several fathers bemoan the true cause of the weather disaster they experience near the middle of the book: "They crossed their legs and expressed some thoughts about a workers' paradise. It would have saved us, said one. If anything could, said another. Capitalism had been the nail in the coffin, said a third" (67). It is clear that the parents are not necessarily blind to their circumstances, and the children certainly pull no punches when it comes down to figuring out the true cause of the climate crisis. This makes the nascence of the detective game a bit of a mystery itself, as many young people can only yearn for parents who can understand their anguish at the current state of the world.

Although the origins of the game are somewhat mysterious, our concept of the "Oedipal detective game" can be synthesised most usefully through a careful attentiveness to Timothy Morton's work. His essay, "The Oedipal Logic of Ecological Awareness", provides a great starting point. In terms of environmental and climate concerns, the name of Oedipus is strongly linked by Morton to what he terms "agrilogistics", a term meant to denote the ways in which early agricultural humanity related to the non-human realm by constantly striving to "delete" this region of being. Offering another word for roughly the last 11,000 years, differing slightly from the more common ones of either the Holocene or the Anthropocene, Morton's work not only connects Oedipus and the agricultural society he is "charged with saving", but also to the "noir detective story" in particular ("Oedipal Logic" 16, 15). Furthermore, this intervention puts forward Oedipus as the main exemplar of a form of thinking that marshals a "deletion" of the non-human in service of a picture of humanity as itself autochthonic, as purely *causa sui*, and thus reads Sophocles' titular character as the first of many "climate criminals". Moreover, Morton draws an intimate connection to questions of genealogy and origin: "There is nothing but Oedipus and his parents, Oedipus who thinks he acts autonomously, exemplifying the agrilogistic meme *We came from ourselves*" (*Dark Ecology* 62). This triangulation of Oedipus, questions of climate and criminality, and genealogy give us sufficient warrant to speak of this "Oedipal detective game" as perfectly operative in the novel as a whole. Not only that, but it also makes possible a perspective on the novel that can uncover this game's function: as the children work so hard to conceal their identities and relations, they are only further revealed; the children are themselves caught in this Oedipal contradiction where they act as both the detectives and the criminals. The detective game acts as their coping mechanism and distraction from realising their very real roles in the climate crisis – roles that may even match the ones their parents play.

Although Morton's account of Oedipus gives us some justification for our term, his treatment of the crime fiction/criminality point of the triangle needs tweaking. *Dark Ecology* is quite content to pull out what it claims is the quintessential "noir fiction" element:

> I am a responsible member of this species [humanity] for the Anthropocene. Of course I am formally responsible to the extent that I understand global warming. That's all you need to be responsible for something. You understand that this truck is going to hit that man? You are responsible for that man. Yet in this case formal responsibility is strongly reinforced by causal responsibility. I am the criminal. And I discover this via scientific forensics. Just like in noir fiction: I'm the detective and the criminal! I'm a person. I'm also part of an entity that is now a *geophysical force on a planetary scale*.
>
> *(8–9)*

We cannot help but wonder if the transition between one's "knowledge" and their "responsibility" here is too smooth. As we noted already, the children and their parents are deeply aware of their role in the climate crisis – knowledge here in Millet's novel does not seem to be the major issue, whereas the matter of "responsibility" seems to be a far more pressing concern. Now, to be fair to Morton, *Dark Ecology* grants that things are strange here, strange precisely because there is something deeply "unconscious" about this knowledge of one's culpability, and this factor itself makes the question of responsibility quite knotted, as the case of Sophocles' hero no doubt shows (58). That said, Morton's mention of methods of rationality and science do quite nicely link back up with standard treatments of Oedipus – as he notes: "Oedipus' *hamartia* is his reason, and his hubris is to use his wits to command everything, as if reason could shrink-wrap the universe" (62). The context of all of this within Millet's novel forces one to question whether all (or, frankly, any) of these methods and tools of rationality are functioning within the noir universe. As Jo Lindsay Walton and Samantha Walton have also very insightfully noted, the entangled nature of our ecological being has repercussions for how we might mobilise the detective fiction genre: "The premise of separation and transcendence of the social field, on which the nineteenth century detective was partly based, may also prove incongruous" (3). However, the leap here from Oedipus to the "noir novel" strikes us as too quick – and we would like to turn now to a somewhat little-known text by Gilles Deleuze to help show that, while Oedipus and his use of reason and rationality certainly can work for one particular strain of detective fiction, the noir genre in particular is a somewhat different quarry. We would like to tease out this potential incongruity by turning to Gilles Deleuze.

Deleuze's essay, "The Philosophy of Crime Novels", explores two subsections of the crime genre that Morton slightly conflates: one that he calls the "old detective novel" (81, 83) and another referred to by the name of *La Serie Noire*, a collection of noir novels under the editorship of Marcel Duhamel. As Millet's work seems to prove its malleability over and over, we find aspects of both types within the novel, each playing an important role. Deleuze explicitly mentions the more classical detective novel's precedents, which reach all the way back to Sophocles' *Oedipus Rex* – a claim shared with Morton. It seems quite unobjectionable to say that Millet's novel utilises the Oedipal detective game motif to serve a function that diverges from how this structure typically operates in the "old detective" genre. Deleuze, for his part, locates a quite paradoxical connection between a criminal and detective as a mirror of sorts in *La Serie Noire*, where they "[are] double[s] of [each other], they have the same destiny, the same pain, the same quest for the truth" (82). When the children act as detectives in the novel, they unknowingly admit to their proximity to their parents, both from a relational standpoint and when it comes to the crime of destroying the planet. As we argued earlier, the game is an attempt on the children's part to create some distance between them and the people who raised them – and frequently it serves to further reveal that the two groups might not be so different after all. But the true reality of the game is that it reveals something the parents always knew: they lack connections with their children. While the children ridicule their parents and their vices, they cannot help but fall victim to those same habits by the end of the novel: they play games with the parents to keep their spirits up, and the children always elect for their prizes to be beer and liquor (209). It hardly takes the end of the book for these desires to manifest, though, as the children begin to steal weed and alcohol from their parents from the very beginning of the summer. By denying all relation to their parents and hiding who they came from, the children can feign a lack of similarity between the two groups, only to have these congruences revealed as the story unfolds. This reality makes the crime even more difficult to investigate. This constant displacement of responsibility for the climate crisis onto older generations does nothing to alleviate the guilt of younger people; in truth, no one is immune from contributing to the increasingly precarious situation that is climate change. This makes the distinction between the traditional archetypes of criminal and cop look even more blurry in the text, which is why the game ultimately ends without all the pomp and circumstance that one might have expected.

The *Serie Noire*-logic seems to play a different role in the story, contributing to the overall resolution: that is, there is *no* resolution. The "old detective novel" for Deleuze – who is thinking here of Arthur Conan Doyle's Sherlock Holmes, Émile Gaboriau's Père Tabaret and Monsieur Lecoq, and Gaton Leroux's Rouletabille – utilizes a methodology that Russell Ford describes as "parallel[ing] the metaphysics of platonism" (61); in contrast, the *Serie Noire* genre suggests that there is no Truth to be found at the end of the detective's investigation, nor is the finding of that Truth the main point of the genre. As Deleuze articulates it: "This is because the truth is in no way the ambient element of the investigation: not for a moment does one believe that this compensation of errors aims for the discovery of the truth as its final objective" (83). *A Children's Bible* ends rather mysteriously, with Eve reassuring her younger brother, Jack, that others will be around to experience what happens after they are gone, leaving the reader to assume that Jack will soon pass away. This scene hardly provides any concrete sense of resolution or revelation. For Deleuze, *La Serie Noire* detective novel relies on situations that have no firm answer, no truth to be found or told, no clear lines. This is equally true of *A Children's Bible*. As the children experience more of the harshness of the new and old world, the prospect of finding a clear-cut or even utopian resolution is ultimately negated. While many may go into a novel about the climate crisis hoping for answers, Millet refuses to provide any.

The detective game traverses the first half of the novel, often functioning in the background and revealed to the reader as the discoveries are made chronologically. The children "lose" the game for a variety of reasons; some are directly outed by their parents; others are far too excited to boast their elite status to a group of "yacht kids". Such anticlimactic endings lend great credence to our claim that the logic of the *Serie Noire* is working here: the mystery is not at all "framed in terms of truth" – rather, the "resolution" of the game comes about through what one could call, to paraphrase Deleuze, an enormous cavalcade of errors (82–83). Ultimately, the game ends with Rafe as the winner, but "he barely had the heart to celebrate his victory. The game was done, but its ending hadn't left any of us happy" (101). The children are unhappy about both the circumstances of the game ending and the fact that their tactic to create distance between themselves and the true criminals (in their eyes) has failed. The end of the game marks a transition in the novel. The children now take a more serious approach to handling the storm in response to their parents' seeming apathy. Those who once competed to win a game they did not want to end must now work together to survive. The stakes in this new game are much higher, and that feeling is not lost amongst any of the young people.

Decoding the Biblical Narrative Threads

Although Millet's novel was very well received, critical work has been slow in catching up. Aside from a tiny mention in Matt Bell's essay in Jeffrey Jerome Cohen's and Stephanie Foote's edited collection, *The Cambridge Companion to Environmental Humanities*, scholarly literature devoted to this novel is still nascent (109). Numerous reviews, however, have focused on the theological motifs, allusions, and intertextual connections to the Old and New Testaments – fitting, no doubt, given the book's title. Jonathan Dee's article in the *New York Times* – "An Epic Storm Turns a Summer Holiday Into Potent Allegory" – notes several of these hints and suggestions, many coming from young Jack:

> Jack keeps noticing – and duly pointing out, to the uninterested others – parallels between current events and the stories in "my book". There are more, even, than he catches: There's a birth in a barn, a plague, a Moses, a Cain and an Abel, even a crucifixion. But part of the novel's genius is that these allusions never really lead anywhere – they don't coalesce into some superstructure of metaphor. The baby born in the manger is just a baby. The allusions aren't symbols or clues; they're just faint echoes, like puzzle pieces too few to fit together. They don't mean what we're used to them meaning.

We wish to further flesh-out the theological narrative within Millet's novel; taking Dee's review as a starting point, we argue that the multilayered nature of the theological allegory provides us with not only another perspective from which to understand the Oedipal allegory of the book, but with another set of tools to interpret the climate crisis in the story. In slight contrast to Dee's reading, we argue that it is not quite so clear that "all of these allusions aren't symbols or clues" – indeed, we want to read these theological parallels as very much functioning like clues, at least as far as some of the characters are concerned. That said, Dee does quite correctly note that while Jack's "decoding" of Biblical messages does not "exactly hit the other characters with the force of a revelation", they certainly hit Jack in precisely that way. A suitable attentiveness to how the theological allegory functions for Jack and Shel suggests that the Biblical resonances in the story do greatly impact how readers receive and understand the novel as a whole. Moreover, the combination of the theological layer of the book – which is a layer that seems to be utilised almost exclusively by Jack and Shel and not by any of the older kids – gives us a unique position from which to notice the convergence of concerns about climate change and those that are so typical of the crime fiction genre more generally.

As already noted, the book begins with a tone that instantly reminds one of a fairy tale. Eve makes clear why such a genre might be made to order for the world she inhabits:

> At that time in my personal life, I was coming to grips with the end of the world. . . . Scientists said it was ending now, philosophers said it had always been ending. Historians said there'd been dark ages before. It all came out in the wash, because eventually, if you were patient, enlightenment arrived and then a wide array of Apple devices. That was how we could tell it was serious. Because they were obviously lying. We knew who was responsible, of course: it had been a done deal before we were born.
>
> *(27)*

Eve goes on to say that she

> wasn't sure how to break it to Jack. He was a sensitive little guy, sweet-natured. Brimming with hope and fear. He often had nightmares, and I would comfort him when he woke up from them – dreams of hurt bunnies or friends being mean. He woke up whimpering "Bunny Bunny!" Or "Donny! Sam!".
>
> *(27)*

Eve is understandably wary of breaking the news to Jack, concerned that the reality would end up looking much like the palliative narratives parents so often feed children:

> The end of the world, I didn't think he'd take it so well. But it was a Santa Claus situation. One day he'd find out the truth. And if it didn't come from me, I'd end up looking like a politician.
>
> *(27)*

How exactly is one to save someone all the detective work here; how is one to protect someone from the true state of the world, someone who, as Eve admits, "still couldn't tell time on a clock that had hands" (28)? In strong contrast to the rules of the Oedipal detective game, Eve decides that, in Jack's case, the point is not to learn the "trick" of "hiding" the traces of a crime (48). She must identify the perpetrators of the climate crimes – and she says she must tell him a story that is not like those of *George and Martha* nor those in his *Frog and Toad Treasury*. She must, instead, "tell [him] a new story . . . a real one. A story of the future" (42).

The morning after Eve tells Jack this tale – the story of "polar bears and penguins", as she and Jack both put it – we learn that Jack has been given *A Child's Bible: Stories from the Old and New Testaments* by one of the parents, known to Eve as the "peasant mother". ("I'd been reading a book about medieval society I'd found in the great house library", she confesses, "[t]here were peasants in the book: serfs, I guess. Using the filter of that history, and with reference to her flowing-dress wardrobe, I'd come to see her as the peasantry" [21].) Perhaps finding Eve's "story of the future" insufficient –

or, more simply, maybe because the *Child's Bible* has "stories with pictures in it", perfect for a lover of *George and Martha* and *Frog and Toad* – Jack immediately resonates with this text by beginning his process of decoding, deductively drawing numerous parallels and analogies between him and his world and that of the Bible – all for his youthful companions for whom, as Eve adds, "religious education [had not been a priority]" (43). Starting with something as simple as noticing there's a character that shares Eve's name, these connections, as Dee correctly notes in his review, resonate, multiply, and proliferate over the course of the novel. The allusions seem to speed up in frequency once Jack befriends Shel (younger brother of Jen), who bows out of the Oedipal detective game very early. Shel arguably plays Watson to Jack's Holmes on the level of the theological detective game. That said, Millet's method of overdetermination also has him play Moses to Jack's Aaron, where the former's trademark stammering and noticeable paucity of linguistic and rhetorical prowess (Exodus 4:10) gets translated into Shel's near complete and total lack of speaking. (Indeed, Jack-as-Aaron is further borne out by the way in which he is the main interpreter of Shel's sign language, which the former has taught the latter [56].) In fact, one must wait almost one-hundred-and-fifty pages before hearing Shel speak – and when he does, he opens his mouth in response to Juicy's comment about how the "Bible was written like two thousand years ago . . . Science wasn't even *invented* then". "'You're very ignorant,' said Shel. Out loud" (145).

As one might expect from a work of "climate fiction", the story of Noah and his ark is profoundly utilised here. Jack and Shel draw parallels between stories in the Bible and current events occurring in their world. The most salient of parallels is indeed the ark as they attempt – while the world unravels – to save a fair number of animals from the storm that wreaks havoc on the mansion everyone is summering in over the course of the book. Curiously, Millet's narrative zeroes in on animals that would appear to be quite "incompatible" with one another. A good deal of this storyline deals with Jack and Shel's desire to save a barn owl with a busted wing placed right alongside that animal's key prey, a rabbit (as David tries to tell Jack and Shel, "'You realize rabbits and owls don't really chill together, right?' said David. 'It's not like picture books where woodland creatures put on dresses and square-dance at picnics'" [71]). The story of Noah's ark no doubt possesses a quite special place within ecological thinking – certainly this has been the case ever since Holmes Rolston III described it as being "the first Endangered Species Project" (48). As previously mentioned, Cohen has argued that the key aspect of this story for us may be wholly ethical in nature. Ken Stone, in his *Reading the Hebrew Bible with Animal Studies*, wonders in the final chapter of that text what it might "mean to read the Bible in an age of extinction?" As he goes on to say: "The question might seem improper. On the one hand, species extinctions are a natural part of evolutionary life. Thus, one could ask whether the Bible has ever been read in a time when extinction was not taking place" (164). Reading Cohen and Stone together, it is difficult not to notice the ways in which

> God's judgment in the story falls upon both humans and animals (who are included together in the category of all flesh at several points), so too humans and animals are saved together on the ark that Noah builds at God's command.
>
> *(158)*

Jack and Shel clearly intuit this necessity for care and concern to reach not simply across the boundaries between species, but also across the lines between predator and prey. Not only do they help to save the barn owl with a broken wing and the rabbit, but also an entire hive of bees, carrying it out of the torrential rainstorm and into the basement of the mansion. As Jack responds to Eve's shocked incredulity at their bringing it in out of the rain, "One raindrop can *kill* a bee" (61). Indeed, the world of nature in the novel seems to conspire to make it easier for Shel and Jack to save the bees: "See, they came *out* of the hive! We didn't know they would do that" (62). For Jack and Shel and their profoundly Noah-esque impulse to extend their sympathy to the non-human tenants of the world, it is not quite the case that things do not, as Dee contends, "coalesce into some superstructure of metaphor". Although the narrative may not cohere into an all-cohesive metaphor, the theological allegory thread of the book does result in something like an ethical imperative to care for the non-human elements of the world. Not only do the bees and their hive survive the storm, Jack and Shel manage to transplant them all back outside of the mansion, where all of the bees then return to the hive, safe and sound.

Perhaps the most intriguing theological parallel comes in the form of "an old woman" – we get no more significant details – who is rich and "a hobby farmer"; one of the itinerant groundskeepers for the robber baron mansion, Burl, suggests to the kids that, as the storm worsens, they pack up and leave the mansion to travel to the farm because "it doesn't feel right here. I know this place. We need to get away" (87). In a moment of what could be read as profound sympathy for their parents, the children do attempt to convince them of the rightness of their decision to leave. But the parents are stubborn, saying that they had indeed "damaged" the mansion and property and therefore worry that they

> "won't get back [their] deposit", said a mother. "If the management company hires their *own* contractors, the surcharge will be highway robbery", said a father. "Then there's the breach of the lease agreement. What was the penalty again?" "Seventy thousand, I believe". "At least". "Leaving right now is, frankly, unacceptable".
>
> *(89–90)*

No doubt Millet wants one to wonder here about this line of thinking – is a "breach of the lease agreement" the most pertinent priority and concern when the world is quite literally going to hell? That said, we can again notice Millet's method of overdetermination with regards to a multiplicity of potential Biblical analogues here. Is it best to read this as another version of the "hardened heart" of Pharaoh, who continually watches the God of Moses lay waste to the land of Egypt while stubbornly and obstinately persisting in his refusal to let the Israelites go from slavery? Or are we to simply fashion potential connections with Noah and the Flood itself – setting aside for the moment that the Hebrew text does not explicitly ever say that Noah directly attempts to save all the flesh of the earth from the calamity that he knows is coming?[1] Not surprisingly, the parents do not accompany the children in their Exodus away from the mansion, ultimately finding their way to the hobby farmer's place – an incredibly fortuitous destination as the farm has goats, rich soil for growing vegetables, and a very large silo with all kinds of food and provisions in it, enough to sustain (like the manna in the desert) Burl, the children, and a small cadre of folks who cross their path and who Burl identifies as "trail angels", the term for people who place food and supplies for those who are hiking the Appalachian Trail (110).

Here on the farm the kids and angels begin to reconstitute a world, a society different from the one they had been part of earlier in the summer. Eve says that all the kids "liked the angels. They hadn't brought us into the world – they hadn't brought anyone into it – and in that fact we felt a bond. In that fact we were equals" (147). This degree of equality and trust quickly metamorphoses into the group's setting up a kind of school, where the trail angels – with all kinds of different areas of expertise – start running classes in poetry, history, biology, science. Eve very wistfully describes "[c]hildren who sat there learning from their teachers, full of trust. Secure in the knowledge that an orderly future stretched out ahead of them" (136). Typical of this novel, the somewhat idyllic time at the farm does not last for long. The children had a "custom of keeping watch" from the very tall silo, but the rains return and just at the moment when they give up this habit of watching the borders of the farm – "[s]ocked in the way we were, enclosed by low clouds, . . . no visibility from the silo" (147) – a group of men with guns materialise. The men – "[r]edneck soldiers", Rafe says derogatorily – proceed to ransack the house, "stuffing food in their mouths", spilling "the contents of drawers and cabinets . . . all over the floor. The aftermath of a burglary", as Eve describes it (149). The children and angels are legitimately and understandably concerned about these men finding all of the food locked away in the silo behind a door with a passcode that Burl happens to know. The leader of the group of men, referred to only as "the governor" (with potentially a reference to the character of the same title to that well-known post-apocalyptic TV show, *The Walking Dead*), proceeds to torture Burl and others for the passcode. The governor gives everyone until sunup one day to reveal it – noncompliance will end with Burl's death. As the sun rises, a number of parents manage to show up at the farm – hearkening back to the Oedipal detective game even though it has long been over, the children spend some time determining who leaked their location – Sukey ultimately emerges as the main culprit here (174–175). The parents attempt to handle the situation by recourse to what they seem to know best – and what they know best is what Terry calls an outright "fantasy", their plan: "'We're going to sue the pants off these bastards,' muttered another father. 'When things get back to normal'" (176). The kids' plan to forestall Burl's death sentence also reaches back to the prior world as they offer to bribe the governor with their parent's money and wealth: "'We don't have exact figures yet,' said Terry. 'But we'll find out. There are some ETFs, sir. And money-market funds'" (179). The narrative leaves the reader wondering whether this ploy works – as Eve is aware that all through this conversation the governor was "[v]ery stoned".

The real resolution of this entire situation is pure *deus ex machina* as the children catch a "[b]linding" flash of lights from a helicopter. "Guys in black, like a SWAT team" leap out of the helicopter and proceed to root out the governor and all of his vigilantes. Stepping out of the helicopter after all the soldiers is "[a] woman. In the light I saw her face was calm. She was slight and old" (182–183). This woman happens to be the owner of the farm, and she justifies the fatal response to the governor and his group by saying that she was "afraid they broke the rules" (183). Earlier in the narrative, when the kids first arrive at the farm, Burl and Val together relay to the children that the owner of the farm has a set of rules they must follow: "First rules. Uh. She's the owner. So we gotta do what she says. And also respect her", Val says (102); there is to be no noise on the weekend; tenants are to "respect [their] elders"; "[n]o breaking the law"; finally, "no sex", Val says,[2] and concludes with: "'The other ones are, don't steal her stuff, tell her what's up if she checks in, and don't try to hook up with the neighbors' kids. Or steal the neighbors' stuff'" (102).

Not only are the rules of the theology game clear, but the one overseeing the game appears to have perfect knowledge of the situation. She knows everyone's name – Eve herself observes that the owner calls Jack by his name and wonders if she somehow let that slip without noticing she did so (184). When Eve tries to tell her there are kids trapped in the barn with the angels and that "[t]hey haven't done anything wrong", she replies, "I know" (183). Moreover, this knowledge seems to come with a great deal of authority as well: when she asks Eve to bring her an ashtray at one point, Eve confesses that "[s]he had a way about her. I didn't consider disobeying" (185). But this degree of support and authority do not last long as the owner of the farm does not stay with her tenants – as soon as the vigilante group is disposed of, she vanishes in the night without a word – taking Burl and all of the other "trail angels" with her. Jack narrates that night:

> "We went to sleep", he said. "Beside the heaters. That lady was nice. She made us hot chocolate on a little stove and sat there in her chair smiling. She told us stories in sign language. So Shel could listen too. Then we fell asleep. When we woke up she wasn't there".
>
> *(189)*

On the surface, a farm that is free of the governor and his men would seem to give this part of the narrative an air of a world operating according to the same logic that Deleuze locates in the old detective novel form: the perpetrators have been caught, justice has been meted out. Befitting a world run by the God of covenants and contracts, those who do not abide by those contracts are punished. There is a Truth here that the owner appears to follow – and yet, this restoration certainly does not undo what happened nor do any of the children ever forget what was done to the angels and themselves. Although it might appear that we are far away from the structures of logic we located within *La Serie Noire*, there is no handling "the problem" here. Although the reader feels some degree of satisfaction here, things are ultimately not all that clear for those still hanging onto the older Platonic metaphysics that drives the classical detective genre – and this is especially true for young Jack.

Weaving Together Inconsistent Threads

Having excavated the Oedipal and theological detective games, it falls to us to suggest how one might conceive of them together. One could say, as a start, that the old detective style and *Serie Noire* work together in this novel to highlight the multitudes of nuance that come with any discussion regarding climate change. On the one hand, it can be incredibly easy to launch a fully deductive investigation and point at the found culprit – and in the eyes of many, including some of the fathers, capitalism would be the answer here (a conclusion that the children find to be quite lacking). But as soon as any sort of conclusion is reached, these answers immediately become murky. How can one point to a system as the cause of real, material struggle, then try to think of solutions to those problems? There is no Truth to seek here.

Although there may be no Truth here, it is significant to think through the fact that there are potential "solutions". We find it tantalising to read the heavy use of indefinite articles – which Dee takes as signalling the novel's fairy tale genre quality (we grant this as *prima facie* accurate) – throughout the novel to point to a larger theme related to the paths taken to "solve" this climate crime, coming through this strong overall desire for anonymity that many characters frequently articulate. The parents are always referred to with the help of an indefinite article: "a mother" or "a father". This makes perfect sense when paired with the fact that all the children strive to keep their identities hidden from each other and the reader. However, as the story progresses, the parents only become less and less individual and distinct from one another, even as their relationships to the children are revealed. In fact, by the end of the story, they are all so indistinguishable from one another that their collective actions can be described with just a single plural and definite pronoun:

> They stopped trying to entertain each other with so-called wit. They didn't talk much or laugh, even when they were drinking. And they drank less and less, shocking us. They went to bed early and slept late saying they liked their dreams.
>
> *(219)*

In the end, the parents are left with no personality, no uniqueness, no singularity – every one of the parents is just like every other one; additionally, as the novel progresses, what seems to be easy solutions to what the children are looking for becomes far more complicated and nuanced, just as the parents reveal that perhaps the creation of the game as a coping mechanism was hardly the answer at all. The end of the novel might leave a reader grasping at straws, trying to reach some conclusion that will help them understand what they have read and how to fix it, but Millet leaves this rather open-ended, implying that there cannot be a straightforward answer to all the problems explored here.

Given the Oedipal detective structure and this recourse to indefiniteness, the clearest implication of this is for thinking about culpability. Does the use of the indefinite article to talk about a particular parent manage to diffuse their responsibility for what the children take to be the crime that is the climate crisis? The indefinite article does open things up to a larger multiplicity; there is no doubt that these articles pick some *thing* or some *one* out, but the way in which it picks that thing out can be difficult to grasp. Perhaps all this suggests is that any metaphysical system that relies on the "old detective fiction" of Deleuze's theorisation – and which thus relies on a kind of hidden Platonism – will quickly show the limits of such a model when applied to climate change, precisely because the Platonic cannot quite handle a conception of truth that is not fundamentally unitary and one. If the multiplicity that comes from this indefiniteness also unsettles the very distinction between criminal and non-criminal, then the question of who gets to mete out punishment and how this gets adjudicated becomes more significant. Millet's novel leaves readers with two – again, competing – pictures. In one, we get the sense that the children would love to punish, but Eve admits that it is difficult to pin down the specific crimes – this becomes even more true as the parents lose all their qualities. Moreover, when thinking of the parents, she willingly notes that "[e]ach person, fully grown, was sick or sad, with problems attached to them like broken limbs. Each one had special needs . . . . What people wanted to be, but never could, traveled alongside beside them. Company" (140). The other picture is the owner's, whose perfectly contractual logic is quite sharp, unflinchingly unambiguous: "*Everyone* knows the rules" (186).

It is tempting to read this second picture as the ultimate outcome of the theological game that Jack and Shel play – like the old detective novel's desire for Truth, it is possible to find it through proper decoding of the ancient biblical texts. However, it is curious that the greatest adherent to this mode of detection finds the owner's judiciousness unsatisfying – not to mention the fact that he is quite peeved when she refuses to tell him how the Bible ends. Indeed, Jack's theological decoding game even further reveals his proximity and allegiance to the Holmesian method of thinking – and its underlying Platonism – when he voices his distaste for both of the "solutions" provided to him by the owner and Eve for "how the story will end" (223). He appears to have a distinct attachment to the deductive reasoning he employed to figure out the mystery of the Bible, so much so that he rebuffs answers that present any sort of abstraction. What the novel demonstrates, though, is that this old detective style is not the most effective when it comes to analysing the future of the world amidst climate change. There is no Truth that can be found and there is no *the* Truth that can be formed, leaving us with the melancholy of *La Serie Noire* genre. Jack struggles to see the point of a world without any human beings around: "But we won't be there to see them. We won't be here. It hurts not to know. We won't be here to see!"; however, this sort of open-ended conclusion may have the potential to leave some readers hopeful (223). Indeed, it may be a sort of comfort to know that no matter all the horrible damage humans do, there will still be a planet here after we take ourselves off it. Yes, the story lacks a definite solution that many climate-motivated readers may seek, but Millet posits there cannot be a Truth here. Where every "truth" about climate change is discovered, there seems to be something to muddy the waters around the next bend. Although not the fairy tale ending a reader may have expected from the first sentence of the book, the real view of what to do in a world facing climate change is important.

Such a situation – operating according to two quite different logics – might appear paradoxical given that these two kinds of detective fiction that Deleuze extrapolates seem to be in stark contradiction with one another. However, it may not be all that nonsensical given the novel's focus on the climate crisis. We might note once again that the two different games produce very different ethical engagements: the Oedipal game makes legible profound degrees of disconnection, between child and parent, between human being and world; the theological narrative produces the exact opposite: more connection, more care for both the human and non-human other. In a world where there are no clear-cut answers, we could find this to be expected in a work that engages with the climate crisis – precisely because, as Adam Trexler has argued in *Anthropocene Fictions*, "the narrative difficulties of the Anthropocene threaten to rupture the defining features of genre: literary novels bleed into science fiction: suspense novels have surprising elements of realism; realist depictions of everyday life involuntarily become biting satire" (14). It should also be clear that this situation does reflect back upon the crime fiction genre as well. It is true that Deleuze's theorisation of the genre seems to rely on a setup that supposes *a* single criminal that – although they may be difficult to discover and the methodology used to find them is no longer a movement of Truth but is instead one long cavalcade of errors – is, at the end of the day, still a single individual. But what occurs when one no longer assumes a single criminal but, instead, engages an entire world that is somehow guilty? How should things change if that paradigmatic figure in the classical detective story of Sophocles' had to finger everyone who had any part to play at all in Thebes as accomplices to his own crime?

Jack wants the conclusions drawn from theological parallels to be clear, but the narrative is rather sanguine, suggesting that in a world under the rather desolate-looking and gathering storm clouds of climate change, it is necessary for the Holmesian/Oedipal metaphysics to be opened up to profound heterogeneities and multiplicities – starting first and foremost with the *La Serie Noire*-logic and structure present within the book and our world. What may be needed is multiplicities of logic, heterogeneities of multiple genres, multiple kinds of metaphysics, differing traditions of thinking. Juicy complains that the theological detective game of Shel and Jack is nothing more than "Santa Claus *shit*" – kid's stuff at best, pure make-believe at worst. Despite this, the necessity of being open to the make-believe seems required for hope – a hope that the novel would seem to signal in its very title – a title that already invokes this opening-up through its very own change from the definiteness of "*The* Bible" into a realm of the indefinite: "*A* Children's Bible".

Notes

1. All good readers of the Christian Scriptures know that although the Hebrew text omits any such attempt on Noah's part (although, as Cohen correctly notes in a footnote, it does appear in the Midrash tradition), the New Testament does explicitly draw the conclusion that Noah does attempt to save people given that he is a man of "righteousness". Would a person of righteousness known to "walk with God" really not reach out to all of the others facing such a horrific fate? The key textual warrants for this reasoning are certainly Christ's own description of his own "coming" as akin to that of Noah's time: "For the coming of the Son of Man will be just like the days of Noah. For as in those days before the flood they were eating and drinking, marrying and giving in marriage, until the day that Noah entered the ark, and they did not understand until the flood came and took them all away; so will the coming of the Son of Man be" (Matthew 24:37–39) – further substantiation is readable from 2 Peter 2:5 as well.
2. The responses this "no sex" rule ("commandment" would certainly be better) receives from some of the teens is humorous all on its own: "'*What?*' squealed Jen. 'Puritanical,' said Terry. 'Frigid,' said Juicy. 'Disrespect. You already broke rule two,' said Jen" (102).

Bibliography

Abrams, Jerold J. "From Sherlock Holmes to the Hard-Boiled Detective in Film Noir". *The Philosophy of Film Noir*, edited by Mark T. Conard, UP of Kentucky, 2005, pp. 69–88.

Bell, Matt. "Climate Fictions: Future-Making Technologies". *The Cambridge Companion to Environmental Humanities*, edited by Jerome Jeffrey Cohen and Stephanie Foote, Cambridge UP, 2021, pp. 100–113.

Cohen, Jeffrey. "Noah's Arkive". *Noah's Arkive*, 1 Jan. 1970, www.inthemedievalmiddle.com/2015/03/noahs-arkive.html.

Dee, Jonathan. "An Epic Storm Turns a Summer Holiday into Potent Allegory". *The New York Times*, 8 May 2020, www.nytimes.com/2020/05/08/books/review/a-childrens-bible-lydia-millet.html.

Deleuze, Gilles. "The Philosophy of Crime Novels". *Desert Islands and Other Texts 1953–1974*, edited by David Lapoujade, translated by Michael Taormina, Semiotext(e), 2004, pp. 81–85.

Ford, Russell. "Deleuze's Dick". *Philosophy & Rhetoric*, vol. 38, no. 1, 2005, pp. 41–71, doi:10.2307/40238200.

Millet, Lydia. *A Children's Bible*. W. W. Norton, 2020.

Morton, Timothy. "The Oedipal Logic of Ecological Awareness". *Environmental Humanities*, vol. 1, no. 1, 2012, pp. 7–21, doi:10.1215/22011919-3609949.

———. *Dark Ecology: For a Logic of Future Coexistence*. Columbia UP, 2016.

Rolston III, Holmes. "Creation: God and Endangered Species". *Biodiversity and Landscapes*, edited by Ke Chung Kim and Robert D. Weaver, Cambridge UP, 1994, pp. 47–60.

Stone, Ken. *Reading the Hebrew Bible with Animal Studies*. Stanford UP, 2018.

Trexler, Adam. *Anthropocene Fictions: The Novel in a Time of Climate Change*. U Virginia P, 2015.

Walton, Jo Lindsay, and Samantha Walton. "Introduction to Green Letters: Crime Fiction and Ecology". *Green Letters*, vol. 22, no. 1, 2018, pp. 2–6, doi:10.1080/14688417.2018.1484628.

15

JOHN D. MACDONALD AND THE ADVENT OF ECOCRIME FICTION

Kristopher Mecholsky

One day, it was just there, and it looked like it had been there for some time. Ecological crime fiction (ecocrime fiction) seemed to spring silently out of the blue in such a way that academia did not take much notice until the mid-2010s. Even so, it had only arrived in noticeable numbers some thirty years prior. Why has ecocrime fiction apparently developed so comparatively late in the evolution of crime fiction? How did it commence and what are its features?

The features of ecocrime fiction thus far seem readily discernible. In "The Landscapes of Eco-Noir", Anna Estera Mrozewicz describes a particular but emblematic mode of the subgenre as propelled by the realisation that even as we try to "save" the world, "we are all implicated in . . . [its] ecological destruction" (99). In discussing recent Scandinavian ecocrime fiction, Linda Haverty Rugg argues–by way of Freudian notions of displacement and Deleuze–Guattari's use of the rhizome, a non-hierarchical, self-organised network–for several features that constitute Scandinavian ecocrime fiction specifically. These features also describe the broader genre as it is developing in the American tradition as well. By way of Rugg's argument, then, I posit that ecocrime fiction generally reflects a heightened but guilty environmental consciousness in which collective guilt and its reckoning are displaced (through "dispersal, disavowal, deferral, or reinstatement of the anthropocentric") in a society in which ecocrime is "all-encompassing . . . systemic and rhizome-like (and thus not easily uprooted or destroyed), and that we are its perpetrators as well as its victims" (614). Rugg rightly concludes that "if we accept these claims, it becomes nearly impossible to represent ecocrime, let alone imagine its detection and solution", but the genre struggles to do so anyway (614).

I find many of the aforementioned features prominent in the work of John D. MacDonald from the 1950s on; indeed, MacDonald seems to be the first identifiable author of ecocrime fiction.[1] I want to use this hypothesis to explore the genesis of the subgenre and examine the convergent evolution of ecocrime fiction in two markedly different cultures. While Rugg identifies how Scandinavian ecocrime fiction has naturally emerged from its secular, leftist culture infused with a "right to nature", in this essay, I rely on tools and techniques from evolutionary biology to trace out a buried pre-history of Anglo – American ecocrime narrative through a genealogy of crime in the language and tradition of sin and evil in literature. In doing so, I specify how four literary traditions – literature that explores evil in the world, the Western, naturalism, and crime fiction – intersect in MacDonald's fiction to yield American ecocrime fiction.

There is no obvious lineage for ecocrime fiction as there is for, say, the detective novel. The *notion* of ecocrime fiction itself is apparent enough, though, since ecological concerns arise thematically in

DOI: 10.4324/9781003091912-19

inchoate crime narratives as early as the stories of Cain and Abel and of Oedipus: after all, they hinge on how you treat nature, and how nature treats you. But ecocrime fiction is not simply a catalogue of references to ecological and environmental concerns, which might instead suggest that practically any ecocritical reading of crime fiction would identify the work as ecocrime fiction. While we can and should read crime fiction ecocritically – as the first issue of *Green Letters*'s volume twelve demonstrates substantially – ecocrimes are not the central focus of crime fiction prior to MacDonald's work. Indeed, the development of crime fiction tells us much about our cultural evolution in relation to crime. If we wish to explore the idea of ecocrime fiction, we must explore the idea of ecocrimes.

In legal traditions, crime is defined as a problem of relatively individualised, intentional harm to society, differentiated from other problems endured by society through the notion of 'moral culpability' in the intentional act. Traditionally, crime narratives have not in form focused on crimes against actual nature; it was practically impossible for humanity to even mistreat nature *qua* nature without relation to the supernatural, as with Cain and Abel or Oedipus. The concept of humanity as a collective – or even as individuals – assuming guilt or shame for destroying or polluting the natural world simply did not exist. Polluting nature did not exist. In terms of civilisation broadly, nature itself did not even have a real, measurable identity. A farmer's land was practically carved out of nothing; the city existed in contrast to that which was beyond the frontier; nations sprung forth from vast "wildernesses" as new lands were "discovered".

Writing in his famous 1967 *Science* essay "The Historical Roots of Our Ecologic Crisis", Lynn White argued that humanity's widespread scientific and technological imprint was Western and Christian in character, deriving both from Europe's hostile, expanding, and technologically powerful nations as well as from the striking anthropocentrism of Christianity, wherein "man and nature are two things, and man is master" (1205). As White reasons, "Christianity, in absolute contrast to ancient paganism and Asia's religions (except, perhaps, Zoroastrianism), not only established a dualism of man and nature but also insisted that it is God's will that man exploit nature for his proper ends" (1205). Timothy Morton observes somewhat similarly in *The Ecological Thought* that in opposition to cultures such as Tibet's, "it's the 'West' that fixates on place, thinking that there's this thing called 'place' that is solid and real and independent. . . . [and] fixation on place impedes a truly ecological view" (26). These preoccupations with humanity's relationship to the natural world as rooted in religion are not coincidental. And while religious critique of ecological relationships leaves much room for debate and criticism – for instance, with the problem of fetishising and romanticising non-Western religions, which Morton does later address – White's and Morton's larger points regarding the West's domineering approach to the world have otherwise been stated repeatedly. In his bestseller *Sapiens*, for instance, historian Yuval Noah Harari memorably concludes his sweeping survey of humanity with the reflection that *Homo sapiens* not only conquered all other species of its genus but have emerged the "master of the entire planet and the terror of the ecosystem" (415).

Whatever the causes and persistent conditions of their continuation, humankind's overall technological and imperial developments in the modern era have indicated a general disregard for natural ecological states in favour of relatively indiscriminate land development and waste disposal. As such, while a recognisable and distinct genre of crime fiction arose in the era of imperial–industrial urbanisation, crimes were personal, or at least social and political, but not ecological. In trying to define environmental crime, Mark A. Cohen points out that crimes usually involve intent and that many legal scholars thus distinguish crimes from torts by means of "moral culpability" (1058). Torts are damages to civil society whose causes can be identified (and which may come about from sheer negligence), but crime usually requires intent and culpable responsibility. But stepping back from strict legal discourse, Cohen more cogently observes that crimes are simply activities that society prohibits, but against which it is also willing to enforce (1058). In this regard, then, it is not difficult to see how an idea of ecocrime fiction could not develop before the twentieth century since society broadly had no notion of crimes against the environment – and once it began to have a sense of them, apparently it did not wish to enforce against them right away. Ecocrime fiction was as nonsensical as the idea of ecocrime.

If the idea of humanity committing crimes against the very world it inhabits was a paradox in the modern era, how is an ecocrime conceived in line with our present literary traditions? That is to say, how might ecocrime fiction have even evolved out of our literary traditions? How have its predecessors potentially influenced its further development? I suspect that to determine how we have developed – or can further develop – a literary sense of moral culpability with respect to our collective actions against the environment, we have to look backward to before the modern sense of private culpability and identifiable, individual intent prosecutable by a civil society. Prior to the humanistic and capitalistic turn of the Reformation and the Enlightenment, Western society at least was dominated by a sense that humanity *could* culpably harm not only the world it inhabited but its very self. And it could do so based purely on its own presence on Earth rather than from any particular actions it took. In short, I am suggesting that through MacDonald, ecocrime fiction arises in the twentieth- and twenty-first centuries from a pre-history influenced by the Catholic (and later wholly Calvinistic/Puritanical) view of humanity's original sin and collective, ongoing "evil-doing" against all of God's creation,[2] and that it was finally sculpted by the naturalism of late-nineteenth and early-twentieth-century American fiction.

Genre and the Example of Evolutionary Biology

Genre theory is notoriously amorphous and contentious. In *Kinds of Literature*, Alastair Fowler proposes a hierarchy of terms that ultimately follows guidelines set by another descriptive science: evolutionary biology (a move with heretofore underappreciated and underdeveloped implications). Like species, genres are a foundational unit of analysis (cf. Jackson et al.). But literary critics might employ other foundational units (such as phrases, cultures in a specific period of history, or particular works themselves) just as other biologists might use, say, genes or ecological niches as their foundational units of analysis.

The analogy extends beyond a simple correspondence of analysis. The objects of analytical inquiry in many areas of both biology and literary criticism are complicated complex systems that are irreducible to simple variable testing. While physicists can tightly control the parameters of their research, evolutionary biologists – much like economists, psychologists, literary critics, and other social scientists – must (or should) satisfy themselves in many regards with descriptive inquiry.

Relying on the example of evolutionary biology for new methods of inquiry, then, an inferential method called cladistics naturally suggests itself. Cladistics is used to estimate evolutionary branching sequences of single-ancestral groups (clades) based on shared characteristics. Literary criticism can similarly trace shared characteristics of genres.[3] As Peter Ashlock notes in his survey and critique of cladistics, a cladogram (that is, a proposed branching sequence) is simply a "hypothesis, the best explanation of the distribution of characters, be they morphological, behavioural, or other . . . using all of the facts available. It cannot be proved, although it may be supported by external evidence". So must the literary sciences be content with "best explanations supported by external evidence" (82).

The mechanisms whereby deeply buried "cultural genetics" – cultural information encoded deep within the complex systems that are our brains and families and neighbourhoods and subcultures – are passed along can happen through means other than the kind of reproduction we are accustomed to among living organisms. This may seem obvious, but in fact metaphors of reproduction are rife in discourse around creativity and literature – from the notion of the "Romantic genius" to Harold Bloom's anxiety of influence to standard declarations like "H. G. Wells is 'the father of science fiction'" (Kirby) or "Mary Shelley is 'the mother of science fiction'" (as Megen de Bruin-Molé deconstructs). Indeed, even the idea introduced by Richard Dawkins of cultural memes as the basic unit of cultural reproduction (developed from the analogue that genes are the basic unit of living reproduction) strongly reinforces this view. It is important, then, to be explicitly open to the idea of cross lineage in literary development – of hybridisation and horizontal memetic transfer.

The lineages that might comprise contemporary ecocrime fiction could best be identified by examining the most recent common predecessor of all ecocrime fiction: John D. MacDonald. As I mentioned at the outset, the characteristics of ecocrime fiction in MacDonald's fiction reflect its Scandinavian expression, as described by Linda H. Rugg: namely, evidence of a heightened but guilty environmental consciousness, in and through which collective guilt over omnipresent, systemic ecocrime is displaced. Ecocrime fiction depicts a society in which ecocrime is ubiquitous and self-directed: "we are its perpetrators as well as its victims" (Rugg 614). MacDonald's work does not feature ecocrimes in the same manner as traditional, personal crimes. The latter necessarily motivate thrillers, but the former we can expect (and indeed find) presented in a less individualistic, more communal way.

Genre Influence in MacDonald's Ecocrime Fiction: Marlowe, Hawkeye, and Oedipus

In his crime fiction, MacDonald focuses more often on widespread political and business collusion and corruption and, with his most widely known creation (Travis McGee), puts forth an ironically self-conscious Don Quixote, a "shabby knight errant" (Copper 50), to protect vulnerable individuals from that corruption. Modern, tarnished "chivalric" heroes contesting widespread social corruption, as opposed to just highly individual wrong-doers, were a long-standing development in MacDonald's fiction, as Lewis D. Moore (8), David Geherin (19–35, passim), and Edgar W. Hirshberg (42) all point to in their studies of his fiction. MacDonald ultimately fashioned McGee as a self-exiled fatalist wandering at the fringes of a civilisation he loves to hate, lifting its veil from the outside while he alleviates its inevitably doomed victims, regarding his society somewhat anthropologically as a collective of human animals. Yet these wanderers are not ascetic, truly chivalrous heroes; they are lusty participants in humanity, rather relishing the allures of the natural world. These Quixotes see, and embrace, their Aldonzas/Dulcineas for all they truly are. MacDonald refuses to separate humans from nature, in terms of both the causes and the consequences of their actions.

His depiction of crime is thus reminiscent of Rugg's description of Scandinavian writers like Maj Sjöwall and Per Wahlöö and Peter Høeg, who deviate from the usual concerns of crime fiction and tend to consider guilt in a far less individualistic way, rejecting "classical crime fiction's idea of individual guilty parties in order to reassign guilt to a failure of society at large" (601). While MacDonald's wandering knight-errant certainly echoes Chandler's famous exhortation for a "man of honour" – "the best man in his world and a good enough man for any world" (18) – MacDonald's approach is far more fatalistic and comprehensive. Rather than reflect a chivalric example for the rectification of the world, MacDonald's hero is often just simply getting by, often cynically, amid the unavoidable corruption of a doomed society. The antagonists he meets and (sometimes) bests may have specific psychological motivations. Not unusually, though, MacDonald dispenses with a motivation and justification, as in the case of *Dead Low Tide*'s Roy Kenney or *The Executioners*'s Max Cady; however, they are always implicated in hostilities far bigger, and far more permanent, than their individual roles. The MacDonald hero may survive individually and may even protect his family and friends momentarily, but the victory is known to be temporary. In terms of heroism and malignancy, MacDonald's fiction eschews the traditional dichotomy of humanity and nature – as described earlier through White, Morton, and Harari – and instead embraces a Mortonesque enmeshment of humanity and nature that reveals ecocrime as systemic and organic, characterised by what Rugg calls an "ambient poetics" that helps to "shift perception of environmental depredation" (605). Indeed, these worldviews correspond with the legalistic difficulties of defining ecocrimes themselves, as indicated earlier.

Regardless of the ways MacDonald's ecocrime fiction deviates from other subgenres, crime fiction itself emerges most obviously as the crucial contribution to MacDonald's overall development. McGee owes much to Marlowe. Since crime fiction is unsurprisingly the most salient progenitor of ecocrime fiction, its analysis in conjunction with MacDonald's work will suggest the other necessary and most recent taxa (or literary trends) that produced it. Given MacDonald's anthropological approach to humanity in nature, in which society is revealed to be no more than a flimsy façade for a Hobbesian wilderness, MacDonald's crime fiction seems to derive from two traditions related through anxiety over urbanisation and industrialisation: the Western genre and naturalism.

Detective stories mirror the Western, as Frederic Svoboda argues, in their depiction of a "new gunfighter who risks himself as the messiah of order but who, even if martyred, can never hope to establish a permanent order" (558). The detective, he notes, roams a moral frontier, which has long existed in America, "from the nightmare forest of 'Young Goodman Brown' down through the 1920s 'Poisonville' of Hammett's *Red Harvest* . . . [to] MacDonald's Florida" (559). As he wanders, Svoboda suggests, "the detective hero must acknowledge with the cynicism of a disappointed romantic that established civilisation is no more than a veneer, that an unwinnable frontier still divides the civilised and savage elements of modern life" (558).

In truth, MacDonald views this moral frontier as a literal frontier of humanity and nature, too, both internally and externally. Jack E. Davis substantially demonstrates this in "Sharp Prose for Green" by detailing the similarities between MacDonald's "Gaia understanding of the world and. . . . acknowledgment that 'man is a part of nature'" in his fiction and in his real-life environmental activism in Florida (508). MacDonald's equation of moral and physical frontiers relied on his observation that

> No sophistry . . . can hide the fact that by thinking of mankind as able to control the environment, we are coming ever closer to the point of no return – wherein the environment will no longer be able to sustain and support mankind as a lifeform.
>
> *(qtd. in Davis 508)*

Thus, the ethically dubious land deals of characters like Elmo Bliss in *Flash of Green* (1962) or Tucker Loomis in *Barrier Island* (1986) may drive the conflicts of those novels, but MacDonald makes perfectly clear that the moral quandary is no abstract problem: individual families are buffeted by complex social forces that antagonise and instigate natural forces which will further threaten, even doom, them. Civilisation is at best an elaborate mechanism within which the human animal competes and ekes out an existence, but it is doing so shamefacedly and at its own detriment.

American literary naturalism thoroughly explored this anxiety over late-nineteenth-century urbanisation and industrialisation, evolving as a reaction to Romanticism and thus as an inverse of the Western. In fact, it replaced the Cooperian Western, as Richard Lehan demonstrates, mimicking the disappearance of the wilderness frontier and the simultaneous emergence of the urban/industrial frontier (236). Whereas in the Western, characters retain "a certain control over the environment" as they combat its evils, in naturalism "the environment [and its evils control] . . . the characters" (229). There is nothing to be done about them. Thus, in *The Virginian*, the titular hero successfully punishes those in discord with the natural world – cattle thieves, horse whippers, and murderers – and establishes order in Wyoming, while in *Sister Carrie*, the eponymous hero is inevitably corrupted by the world as she seeks status in it, her lovers and family collaterally damaged as she eventually finds herself alone and unhappy even with material success.

For a disenchanted, industrialising New World, naturalism offered a secularised, modern dissection of the evils of the world. As John G. Cawelti (68) and Lehan (230) maintain, the Western manifested the Calvinist belief that understood the world as inherently imbued with evil and sin. In a similar way, but from a wearied Enlightenment perspective worn to a kind of fatalism, naturalism depicted the universe as hostile (or at best wilfully indifferent) to humanity in a way that corresponds with the Calvinist tradition – and with a Greek worldview, too.

In *Out of Eden*, Paul W. Kahn suggests that "what terrorized the Greeks was the world, not the inner life" (21). Oedipus, he observes, is polluted but not evil (21). His monstrous reality is a "rent in the universe" that leaves him terrified at the difference between appearance and reality; his horror is "metaphysical, not ethical" (25). He is shocked, as we are, to discover that his virtues are mere appearances, "that at the centre of his existence is a violation of the order of the universe" (26) through no fault of his own. Kahn ultimately argues that all understandings of Oedipus's fate ultimately "converge on the idea that man's knowledge is not adequate to the field upon which he acts" (28). This is MacDonald's point, too.

The Judeo–Christian tradition, though, Kahn moves on to argue, emphasised a personal, inherent, and inherited idea of evil rooted in a notion of identity that is a "product of . . . free will" (34), as opposed to a result of the inherently unknowability of the universe, as Oedipus illustrates. Whereas Oedipus's wandering is a shown-forth lesson to humanity of the "terrifying truth of their own condition before the gods" (28), Cain's wandering shame is a warning to others to act right before one god. For Oedipus, that early detective figure, right action turns out to be impossible, but for Cain and his parents, all of the knowledge they need has been given to them. Right action is possible and a matter of choice. Kahn reasons that sin was the Judeo–Christian internalisation of the pollution of the world: the evils of the world depend on free will and its ordered exercise in relation to a God who governs a reasonable universe. But in the liberal age of post-Enlightenment, "sin" has finally been turned into "crime", which is reason-as-will disordered and directed against the rational project of civilisation and society, instead of God or the gods. The encountered "evils" of the world have been thus conceptualised, then: first as pollution (a revelation of humanity's ignorance of the ways of the universe), then as sin (a disordered free will out of alignment with God's morality), and finally as criminality (twisted reason in defiance of the rationality of civilisation).

In modern crime fiction, sometimes the causes of this criminality are due to social forces, but often they are psychological. MacDonald, however, adopts a naturalistic and Grecian approach by identifying society and civilisation, and even psychology, as mere appearance. Humankind is a humble animal of nature, blind and relatively unknowing of itself and the power of nature, and to some degree without much genuine free will. The will of humans is akin to the will of animals: limited and circumscribed by far greater forces. In short, we are Oedipus. Within the actions MacDonald's heroes and villains take, choices may be made that reveal psychological causes and suffering, but in the end, MacDonald clearly demonstrates that society's odds are stacked against the individual, and nature's odds are stacked against humanity. Particular triumphs or defeats within those contexts come down to chance and will-power in a few moments, but the final outcome is dim.

The Ambient Poetics and Pagan–Calvinist Roots of MacDonald's Ecocrime Fiction

So far, we have explored the characteristics of ecocrime fiction and identified their uniqueness and lineal directions to determine the necessary "taxa" of the genre. MacDonald's ecocrime fiction arises by rearranging the elements and emphases of an adolescent crime fiction, and in doing so it draws on a tradition that explores sin and evil, recently expressed in the Romanticism of the Western and subsequently secularised and inverted by naturalism. The trajectory of MacDonald's overall fiction trended for commercial reasons toward more traditional aspects of the detective story, including the detective's inevitable triumph over individual antagonists, but from at least 1953 on (with the publication of *Dead Low Tide*) to his last, posthumous novel in 1986 (*Barrier Island*), MacDonald experimented with how to depict ecocrime in fiction, offering his most substantial development in 1962 (through *A Flash of Green*). MacDonald's McGee novels usually stand as his most substantial contribution to crime/detective fiction, and while they do not generally reflect best his ecocrime fiction, they still reveal his development of it from naturalism, the Western, and literary traditions of evil through crime fiction and gesture toward his later use of ecocrime fiction as better expressed in *Flash of Green* and *Barrier Island.*

As I have shown elsewhere, MacDonald's McGee books can be classified in several ways, but most of the novels present a criminal threat to the freedom of innocent citizens. In the novels set in Florida, the threat and background is always tied to land development abuse (Mecholsky; see also Moore 115). Summarised this way, the McGee tales' relation to the Western is apparent. These Western-indebted plots unfold in novels like *Bright Orange for the Shroud* (1965), *The Girl in the Plain Brown Wrapper* (1968), *A Tan and Sandy Silence* (1971), and *The Dreadful Lemon Sky* (1975). In depicting these developments through intricate business deals and political calculations, MacDonald also involves himself in a tradition honed by Arthur Conan Doyle, James M. Cain, and Elmore Leonard that celebrates supreme competence and mastery of technique – and which echoes Cooper's depictions of Natty Bumppo's prowess. In recent decades, this mastery has expressed itself most often through a technocratic celebration in "competence porn", as one television writer put it, wherein "competent people banter and plan" (Rogers). Additionally, the villains in many McGee stories – most notably Junior Allen in *Deep Blue Good-by* and Boo Waxwell in *Orange* – derive completely and expressly from the literary tradition that explores evil in the world, as MacDonald proclaimed in the 1980 inaugural issue of *Clues*:

> There exists in the world a kind of evil which defies the Freudian explanations of the psychologists, and the environmental explanations of the sociologists. It is an evil existing for the sake of itself, for the sake of the satisfactions of its own exercise. . . . For me it is less satisfying to say that this is the action of a sad, limited, tormented, unbalanced child than it is to see that this is a primordial blackness reaching up again through a dark and vulnerable soul, showing us all the horror that has always been with mankind, frustrating all rational analyses.
>
> *(69)*

MacDonald here emphasises an express tradition that he participates in – a defiantly anti-modern, Grecian approach to exploring evil in the world, which particularly eschews both the rationality celebrated by crime fiction as well as the "environmental explanations" characteristic of naturalism. Rather than specifically, or only, drawing on a Calvinist ethos that encourages a Romantic–Western response, MacDonald's work synthesises a Greek acquiescence to an unknowable universe with the secular fatalism of naturalism to convey the fact that there is "just a meanness in this world", as Bruce Springsteen sings in "Nebraska" (3:45–52).

In addition to this secular resurrection of a pagan concept of evil, conveyed through plot points revolving around his deft handling of realistic business machinations, MacDonald relies on the ambient poetics that Morton extols to weave in meditations on how these deals have affected and will continue to affect the natural world around those involved. For instance, in *Bright Orange for the Shroud* (1965), when the target of a legal con that seeks to extort hundreds of thousands on the promise of profiting off a land investment syndicate, McGee tracks the con artists to a small Florida town, leading him to reflect on it:

> Marco Village saddened me. The bulldozers and draglines had gotten to it since my last visit. The ratty picturesque old dock was gone, as was the ancient general store and a lot of old weatherbeaten two-story houses which had looked as though they had been moved down from Indiana farmland. They had endured a half century of hurricanes, but little marks on a developer's plat had erased them so completely there was not even a trace of the old foundations.
>
> *(Orange 60)*

Marco's history is not traditionally vital to a thriller, but such observations are crucial to McGee's worldview and his motivation across the series. They may not seem related to specific actions McGee or others take, but they have in fact motivated the antagonists' initial greed as well as McGee's response. These reflections are a foundation for what drives McGee to help and combat those he does. Thus, in *The Dreadful Lemon Sky* (1975), despite the central drug smuggling conflict, McGee sees the city at the heart of the action, Bayside, for what it is: a piece of nature that has been engineered to be a playground for the greedy. As nature itself is steamrolled over in the name of progress, so are the innocent and guilty individuals caught up in the schemes.

MacDonald's approach in these novels obliquely reflects the naturalism that influences his writing, replacing the Romanticism of the Western's influence on detective fiction, and melds it by degrees with a Calvinist-influenced, Grecian-adjacent depiction of natural, unknowable evil in the world and a dysfunctional, myopic relationship between humankind and the natural world. Nevertheless, more often in the McGee series itself, traditional elements of crime fiction, detective fiction, and the thriller tend to dominate the overall action, so particular crimes tend to serve as a portal for contemplating a particular community's wrongdoing toward the natural world or humankind's broader crimes against it, expressed in an ambient poetics that emerges mutedly. MacDonald's most thoroughly ecological novel, *A Flash of Green*, does not rely on the flashy and engaging violent offences nor the snappy dialogue and description of so much crime fiction, including his own. Instead, it is dominated by an ambient poetics of ecocrime itself. As a result, the 1962 novel is rarely even considered crime fiction. But it is just these features that solidify it as an exemplar of ecocrime fiction.

In the novel, a smattering of important citizens in Palm City, Florida clash over a revived bay development project. Ostensibly, two sides fight over whether to develop the area around Grassy Bay for resorts and businesses. On one side are concerned environmentalists, members of the Executive Committee of Save Our Bays (S.O.B.'s [*sic*]), who had previously foiled an attempt by outsiders to privatise and fill the bay, while on the other side are a group of locals joined together as the Palmland Development Company who seek to develop the bay anyway. In reality, two significant players stand behind the scenes and generate the real conflict and outcomes: Jimmy Wing, a local journalist, and Elmo Bliss, a powerful local politician and construction magnate. Jimmy is a self-described loner and observer who maintains that he wants to stay out of the fight. As a county commissioner, Elmo is publicly against the bay's development but has in fact engineered the whole deal, even bringing Jimmy into his inner circle to dig up dirt on the S.O.B.'s in order to blackmail and discredit them. Over the course of the novel, the tactics that Bliss's groups take to ensure passage of the bay's development increase in severity, beginning with social pressure and vague threats of possible public embarrassment, ultimately leading to armed blackmail and battery.

But *Flash* takes great pains to be a paragon of ecocrime fiction. First, there are very few actual crimes committed in the novel. As Rugg suggests, ecocrimes require "a much larger scope than the murder of individual human beings" and "demand . . . a different kind of narrative canvas, in which the interdependence of people, things, and nature can take shape" (611). Thus, the few crimes depicted are much more tempered and even abstracted than they are in most crime fiction. As the orchestrator of the bay development, Elmo Bliss is quite content to simply dishearten the S.O.B.'s since he's certain (and proved to be right) that the city council will permit the bay to be filled. Instead, he initially wants to rely on the personal information Jimmy uncovers to merely hint of improprieties so that opposition to the fill does not ultimately disparage his chances at higher public office. Later, he unleashes forces he may or may not have control over when he involves religious fanatics to persecute the sexual mores of some of the S.O.B.'s and their family, inadvertently resulting in the brutal whipping of Jackie Halley, one of the members. But for most of the novel, the S.O.B.'s are harassed in a self-organised way by the community based on biased, but not false, information in the newspaper or through rumours in town. The committee is slowly but naturally worn down by the community who is generally for the project, and it is only toward the end of the novel when a few identifiable crimes take place, and they seem to be perpetrated by tangential figures acting independently. But the central ecocrime of the novel is obvious. Midway through the book, a biologist states flatly to Wing, "filling the bay would be a criminal act. It will take away forever something which cannot be replaced or restored" (169). Not only is the act never actually a legal crime, by the novel's end, Jimmy is harassed by a security guard as he looks at bay's filling: "'Anything you want, mister? . . . . This is private property.' 'I know. And that's the trouble, isn't it?'" (335). What was once public land is now a predicate for trespassing and legally justifiable prosecution.

Indeed, there is no outright, outsized "villain" in the novel, like Boo Waxwell or Junior Allen. Elmo Bliss and his lackeys are slick antagonists to a degree, but in fact they are not too different from the novel's hero, Jimmy Wing. More often than not, Wing directly contributes to the obstacles the S.O.B.'s face because Bliss convinces him that if Wing did not do it, someone else would, and far more destructively. In fact, far from wanting the bay to be filled, Elmo actually agrees with the S.O.B.'s, "in more ways than you imagine", he tells Jimmy.

> If this whole coast could be just as it was when I was a boy, I'd be happy. . . . It's all a sad thing the way it changes, all the wild things and wild places going, one by one, but you and me, we can't change it or keep it from happening. All we can do is get in there and get our piece of it. *(94–95)*

Where Elmo differs from the S.O.B.'s – indeed, where Jimmy too differs from them – is whether any effort should be made to try to slow wilderness destruction in the wake of humanity's greed.

While *Flash of Green* deviates from general crime fiction in its lack of an identifiable, truly threatening villain, it instead clearly conveys that while there is something that could be called "evil" in the world – not to be confused with "sin" – the actions of the individuals around it are not part of any stable moral universe. Instead, actors on all sides decide their own moral stance, which may or may not even be justified, and the outcome of events does not inevitably lead toward good rewarded and evil punished. As Jimmy begins to come to terms with the truth of his involvement with Elmo Bliss, that he is actually aiding in blackmail, he recognises the banality of evil within its undeniable existence, in stark contrast to the romance of fiction:

> It [blackmail] had overtones of melodrama which gave it a twist of comedy. It needed wax on the tips of the mustache. . . . Then they would set a trap for him, and the slug would smash him against the wall. He would fall, twitch and die, as the music came up and the good guy embraced the lovely girl. But he had learned that most people who do questionable things are as unremarkable as the people who don't. Most people who are thrown into a cell for good reason are vastly astonished to find themselves there, because they look and feel like anyone else. *(115)*

There's no Snidely Whiplash here, and no white-hatted Gary Cooper to order things once and for all. After this reflection, Jimmy visits the accounting office of Chet Rand to follow up on a blackmail lead. Jimmy encourages Chet to give dirt on one of the most influential members of the S.O.B.'s, Dial Sinnat, a personal enemy of Chet's. Chet ferrets out Jimmy's intention and suggests some ways to blackmail Sinnat, noting that there are many ways to really ruin a person that are perfectly legal: "Bastardliness is relative, friend" (118). Moments like these throughout the book steadily condition Jimmy to reconfirm his worldview, particularly as he meditates on his animal reality:

> He stared at his own hand resting on the steering wheel, a long hand . . . grasping the wheel with indifferent simian competence. . . . The hand is the animal reality . . . for blows and tools

> and caresses. Morality is an unreal conjecture . . . the conflict of rationalizations. . . . If they were peeled away quickly they did no harm. But the longer they remained the more difficult they were to dislodge. . . . And so truth, forever out of focus in any case, was prey to these further distortions assembled over the years. They made a comfortable muffling, a padded toughened wall, as opposed to a Calvinist rawness.
>
> *(119)*

Jimmy Wing is no Chandlerian "man of honour", let alone a white-hat, but he wholly commits himself to the truths he sees and, for MacDonald, his commitment itself is honourable. He refuses to be comforted by false rationalisations of morality, even though he struggles with how to live in a way he still finds ethical in relation to the few he cares about. Morality is a "comfortable muffling", but MacDonald's ecocrime protagonist lives in a Calvinist rawness and tries to find some integrity therein. What he notices all around, from those filling the bay and those fighting to save it, is a heightened but guilty environmental consciousness and evidence everywhere that ecocrime is omnipresent and "we are its perpetrators as well as its victims" (Rugg 614). In Wing's case, it's literally true.

The Ecological Gothic

Beyond its depiction of interdependent crimes and villain-less antagonisms, at all turns, *Flash of Green* reflects Morton's call for an ambient poetics of the world it divulges. Through the narrator, Wing, and various other characters, MacDonald repeatedly situates the world of the novel in the fullness of nature and time. Portraying a car accident that Wing covers for his newspaper, for instance, the free indirect discourse of MacDonald's narration indicates the way human society will parse out blame and quantify the suffering of the accident, but feels compelled to ask, "who was responsible for a road too narrow for the traffic, or for shoulders scoured down by summer rains?" (99). No one, of course, and everyone. Humankind is caught up in nature, whether or not it recognises so in its civilisation games. And when Wing interviews a biologist associated with the S.O.B.'s, he is told in no uncertain terms that not only is filling the bay a crime against nature, but humankind's very existence is a crime:

> We are in a very short time of natural history when we have a plague of men. . . . With science he has suppressed too many natural enemies. He is too numerous. He is poisoning the air and waters of the earth. He is breeding beyond reason. He is devouring the earth and the other creatures thereon. But it will come to an end, of course. . . . Geometric growth is insupportable.
>
> *(170)*

The journey of the ecocrime detective – gunfighter is markedly different from those of his predecessors. It relies much on the influence of naturalism and a Calvinist expression of paganism: he struggles with the fact that there is no particular and individualised crime to fight, that humankind's relationship to the natural world is doomed, that there is no morality. All of these initially inspire him to stay out of unpleasant events; at times, he may even encourage them along and get his cut in the meantime, since morality is but a rationalization, and the end is obvious. For the ecocrime fighter, meaning is derived from the fight itself and from the value attributed to personal relationships.

As the foregoing accounts indicate, characteristics of MacDonald's fiction reveal indebtedness to a few eighteenth- and nineteenth-century literary trends. Crime fiction itself developed inextricably in reaction and relation to the Enlightenment, through the Gothic and through the Western particularly. In twentieth-century strands of it, naturalism influenced its evolution. While MacDonald arguably contributed the first works of ecocrime fiction – followed soon thereafter by Ross Macdonald, as Nathan Ashman demonstrates – the degree to which the particular trends his fiction participated in continued in ecocrime fiction that followed is highly debatable. Nevertheless, as Ashman observes of Macdonald's writing, and as Rugg does of Scandinavians', manifestations of the return of the repressed figure prominently in practically all ecocrime fiction. I think it is the Gothic, then, which is at the very heart of the subgenre, as it is at the heart of crime fiction generally, and it is this shared trait that manages to keep them in relation to each other, different as they are. As Jerrold E. Hogle argues in the introduction to the *Cambridge Companion to Gothic Fiction*, Gothic fiction helps us address and disguise desires and anxieties related to the modern era's realisation, as Leslie Fiedler suggested, that the Enlightenment's promises could just as easily end in horror (Hogle 4; Fiedler 136). In fact, ecocrime fiction is so difficult to define, in some ways, because it is not nearly as suited to depicting ecocrimes as the gothic mode. As Timothy Morton aptly puts it: "We have guilt because we have shame. We have shame because we have horror" (*Dark Ecology* 117).

Notes

1. I want to be very clear that I am quite open to disproving this hypothesis. Given the features I describe here, I am genuinely eager to see if ecocrime fiction pre-dates MacDonald's works in any language tradition.
2. This is a wide-ranging, long-lasting debate, but I am referring to the Augustinian notion derived from Paul's epistles, and which Ronald Paulson describes as "ineradicable human sin" (6) – corrupted goodness (Godness, in fact) that distinguishes humanity and which Paul and Augustine passed down to Martin Luther and John Calvin (Paulson 25; Schneewind 33) and the Puritans (Paulson 15): total depravity.
3. I am leaving out my more thorough defense and explanation of this method, literary cladistics, for a different publication. In the meantime, inquiries regarding it are welcome.

Bibliography

Ashlock, Peter D. "The Uses of Cladistics". *Annual Review of Ecology and Systematics*, vol. 5, no. 1, 1974, pp. 81–99, doi:10.1146/annurev.es.05.110174.000501.

Ashman, Nathan. "Hard-Boiled Ecologies: Ross Macdonald's Environmental Crime Fiction". *Green Letters*, vol. 22, no. 1, 2018, pp. 43–54, doi:10.1080/14688417.2018.1431139.

Cawelti, John G. "Prolegomena to the Western". *Western American Literature*, vol. 4, no. 4, 1970, pp. 259–271, doi:10.1353/wal.1970.0008.

Chandler, Raymond. "The Simple Art of Murder: An Essay". 1950. *The Simple Art of Murder*, Vintage, 1988, pp. 1–18.

De Bruin-Molé, Megen. "Hail, Mary, the Mother of Science Fiction". *Science Fiction Film & Television*, vol. 11, no. 2, 2018, pp. 233–255, doi:10.3828/sfftv.2018.17.

Fiedler, Leslie. *Love and Death in the American Novel*. Dalkey Archive Press, 1960.

Fowler, Alastair. *Kinds of Literature: An Introduction to the Theory of Genres and Modes*. Oxford UP, 1982.

———. "The Formation of Genres in the Renaissance and After". *New Literary History*, vol. 34, no. 2, 2003, pp. 185–200, doi:10.1353/nlh.2003.0017.

Harari, Yuval Noah. *Sapiens: A Brief History of Humankind*. London: Vintage Books, 2015.

Hogle, Jerrold E. "Introduction: The Gothic in Western Culture". *The Cambridge Companion to Gothic Fiction*, edited by Jerrold E. Hogle, Cambridge UP, 2002, pp. 1–20.

Jackson, Nathan D., et al. "Species Delimitation with Gene Flow". *Systematic Biology*, vol. 66, no. 5, 2016, pp. 799–812, doi:10.1093/sysbio/syw117.

Kirby, Alice. "H.G. Wells: The Father of Science Fiction". *Oxford Open Learning*, 11 Nov. 2015, www.ool.co.uk/blog/h-g-wells-the-father-of-science-fiction/.

Lehan, Richard. "Literary Naturalism and its Transformations: The Western, American Neo-Realism, Noir, and Postmodern Reformation". *Studies in American Naturalism*, vol. 7, no. 2, 2012, pp. 228–245, doi:10.1353/san.2012.0016.

MacDonald, John. *Dead Low Tide*. Gold Medal Books, 1953.

———. *The Executioners*. Gold Medal Books, 1957.

———. *A Flash of Green*. Fawcett, 1962.

———. *Bright Orange for the Shroud*. Gold Medal Books, 1965.

———. *A Deadly Shade of Gold*. Gold Medal Books, 1965. Reprint, Random House, 2013.

———. *The Girl in the Plain Brown Wrapper*. Gold Medal Books, 1968.
———. *The Dreadful Lemon Sky*. Gold Medal Books, 1974.
———. *The Empty Copper Sea*. J. B. Lippincott, 1978.
Mecholsky, Kristopher. "Ecological Anarchy and Philosophical Anarchism in the Florida Crime Fiction of John D. MacDonald". *Detecting the South and Fiction, Film, and Television*, edited by Deborah Barker and Theresa Starkey, LSU Press, 2019, pp. 186–204.
Morton, Timothy. *The Ecological Thought*. Cambridge UP, 2010.
———. *Dark Ecology: For a Logic of Future Coexistence*. Cambridge UP, 2018.
Mrozewicz, Anna Estera. "The Landscapes of Eco-Noir *Reimagining Norwegian Eco-Exceptionalism in Occupied*". *Nordicom Review*, vol. 41, no. s1, 2020, pp. 85–105, doi:10.2478/nor-2020–0018.
Paulson, Ronald. *Sin and Evil: Moral Values in Literature*. Yale UP, 2007.
Rogers, John. "LEVERAGE #204 'The Fairy Godparents Job' Post-game". *Kung Fu Monkey*, 11 Aug. 2009, kfmonkey.blogspot.com/2009/08/leverage-204-fairy-godparents-job-post.html.
Rugg, Linda H. "Displacing Crimes Against Nature: Scandinavian Ecocrime Fiction". *Scandinavian Studies*, vol. 89, no. 4, 2017, pp. 597–615.
Schneewind, Jerome B. *The Invention of Autonomy: A History of Modern Moral Philosophy*. Cambridge UP, 2012.
Springsteen, Bruce. "Nebraska". *Nebraska*. Columbia Records, 1982.
Svoboda, F. "The Snub-nosed Mystique: Observations on the American Detective Hero". *Modern Fiction Studies*, vol. 29, no. 3, 1983, pp. 557–568. JSTOR, jstor.com/stable/26281379.
White, Lynn. "The Historical Roots of Our Ecologic Crisis". *Science*, vol. 155, no. 3767, 1967, pp. 1203–1207, doi:10.1126/science.155.3767.1203.

16
CHOKING TO DEATH
True Crime and the Great Smog

Anita Lam

While true crime is not a single, monolithic genre (Biressi), it can be broadly conceived as past newsworthy stories of "real events, shaped by the teller and imbued with his or her values and beliefs about such events" (Murley 6). Typically, true crime narratives focus on murder and prioritise retellings that invoke sensations of horror, pain, fear and frustration in readers (Punnett). While documentary treatments of human-on-human criminal violence have been popular since the mid-sixteenth century (Wiltenburg), true crime assumed its modern form with the publication of Truman Capote's nonfiction novel *In Cold Blood* in 1966 (Schmid). Despite some generic adaptations to enter new markets, true crime texts form a coherent body of writing because of writers' continued adherence to a specific set of conventions, narrative techniques and themes. More precisely, true crime authors have mined the idea that violent crime is "an act that can fundamentally reshape a community and create or lay bare unspoken fears between members of that community" (Browder "True Crime" 121). When murder is represented, human actors are typically foregrounded as both victims and criminals, with greater narrative emphasis being placed on psychological explanations for why people kill. Because booksellers note that "no title on Jack the Ripper ever gathers much dust" (qtd. in Weyr), true crime stories have predominantly focused on a specific kind of killer: the serial killer. As such, the genre has been credited with creating and bringing to life the psychopath as an important pop cultural icon (Murley) and figure of monstrous horror (Ingrebretsen). As hidden threats that can cause spectacles of gory violence, true crime's psychopaths leave behind them a series of graphically destroyed bodies.

These acts of devastation are made all the more terrifying because they occur in environments characterised by unremarkable weather reports. "[T]rue crime", after all, "has its own weather", and much of that weather is described as unexceptional (Seltzer 1). As a result, ordinary weather reports contribute to the construction of a conventional, "normal scene" in the world of true crime. While normality is rendered through the inclusion of everyday scenes in which nothing is out of the ordinary, these scenes are "more exactly, *abnormally normal*" (Seltzer 42, my emphasis added). They are scenes shaped by childhood fantasies of innocence and clichés of the everyday. To establish these too-normal scenes in true crime, Mark Seltzer observes that "there is a lot of weather" because "weather indexes what is left of nature – what is left to mark the limit or 'outside' of the social order as an order and a system" (60). For him, weather serves as a sign of "the natural", distinct from the social order; it appears as a naturalised and atmospheric backdrop against which abnormal, human offenders wreak havoc. When we assume that weather is a mere scene-setting device or background sign of normality, we ignore, however, the important ways in which it, too, can enact harm or violence.

 DOI: 10.4324/9781003091912-20

What some have historically mistaken for "normal" weather has now been understood as a *normalised abnormality* and reread as evidence of human-made environmental degradation. As a clear example of an abnormal weather phenomenon that has long been normalised, we can consider London fog. Because London increasingly became defined by its "big smoke" in the nineteenth century, gaining international recognition for a romanticised vision of its atmospheric pollution, the normalisation of London fog obscured the way fog could kill and concealed how air could become an invisible, non-human killer (Fuller 204). As the first industrial nation to become predominantly urban, Britain – with its regular "pea-soupers" – was notably the place in which the modern idea of pollution was invented (Thorsheim 2). In 1905, the words "fog" and "smoke" were combined to form a neologism – "smog" (des Voeux) – naming, arguably for the first time, a climatic phenomenon manufactured by both humans and industrial modernity. Far from natural, the smog revealed instead the breakdown of nature (Taylor). Although appeals were made to regulate smog by multiple smoke abatement committees since the late-nineteenth-century, coal fires continued to burn in London, culminating in the Great Smog of 1952. From December 5 to December 9, 1952, the Great Smog smothered the city, covering around a thousand square miles. Due to an unexpected temperature inversion, the city's coal smoke remained trapped at ground level and could not be dispersed. As a result, this "perfect storm of freak weather patterns and environmental ignorance" came together to create a "silent disaster" (Corrigan), which has since been described as the "Great Killer Fog". Causing death rates in London to increase by 2.6 times, the smog was officially estimated to have killed nearly four thousand people in 1953 (Logan). More recently, the estimated death toll has been revised to about "12,000 excess deaths from December 1952 through February 1953 because of acute and persisting effects of the 1952 London smog" (Bell and Davis 389).

The Great Smog is remarkably transformed into a true crime event in Kate Winkler Dawson's *Death in the Air: The True Story of a Serial Killer, the Great London Smog, and the Strangling of a City*. Through a close textual analysis of *Death in the Air*, this chapter investigates how the Great Smog has been constructed as not only a product of human handiwork, but also a silent, mass strangler. Juxtaposed against this environmental killer, Dawson documents the parallel story of infamous serial killer, John Reginald Christie. Responsible for the deaths of at least eight female victims, Christie – like the smog – asphyxiated many of his victims, by choking them to death. In line with true crime's preoccupation with serial killers, Dawson hooks readers by promising to trace Christie's heinous actions against the mass murdering effects of a toxic smog that swirled through post-war London. Unlike the typical, homicidal violence represented in true crime – that is, visible violence in the form of criminal events or actions that instantaneously and explosively erupt – the slow violence of the Great Smog is not spectacular; it occurs gradually and mostly out of sight; it does not always produce immediate effects but rather shapes an aftermath characterised by staggered and delayed destruction (for more on slow violence, see Nixon). The relative invisibility of slow violence presents a representational challenge when media images and narratives, including those in the genre of true crime, favour the immediate and the spectacular. How, then, can slow-moving disasters that have been long in the making, ones "that are anonymous and star nobody" (Nixon 3), be represented in dramatic enough terms to warrant public attention and political intervention?

This chapter addresses this question by examining how the slow violence of a historical killer fog can be represented through the conventions of true crime. In what follows, I explore how Dawson deploys these conventions to provide both continuity and structure in her articulation of the Great Smog as a mass murderer. In so doing, I consider how *Death in the Air* offers a symmetrical analysis of two killers when it uses the same conceptual repertoire to make sense of their killings. A symmetrical analysis can help dissolve the great divide between humans and non-humans, by investigating and describing non-humans in the same terms as their human counterparts (Latour). By treating the Great Smog in the same analytic way as a notorious London serial killer, Dawson pursues answers to questions that have only been typically asked of human killers: Who was victimised? How did the killing occur? Who was responsible for the harm? When situated in true crime's genre-specific conditions of possibility, the romance of London fog dissipates, as the fog is transformed into a horrifying threat that seems to disproportionately affect female victims. When reframed as a quiet accomplice to Christie's murders, smog comes into focus as a non-human agent with its own toxic propensities. Although the London fog is represented as a symbiotic extension of criminal figures, such as Christie or Robert Louis Stevenson's Mr. Hyde, it emerges and endures because of the actions of "Dr. Jekylls" (i.e., those who hide their responsibility behind an air of respectability). By examining the representation of the Great Smog as a mass killer, this chapter considers the ways in which crime, agency and responsibility intersect in the formation and representation of deadly air pollution.

Who Is Victimised? Gender and the Death of Romance

While true crime stories can read like, or even look like, crime fiction, their appeal depends upon their ability to "bend fiction toward fact" (Seltzer 16). By playing with the line between fact and fiction, true crime's factual claims are themselves bound up in the presentation of what Laura Browder calls a "dystopian romance" ("Dystopian Romance" 928). In contrast to fictions that can romanticise and gloss over predatory tendencies and behaviours, true crime's "truths" have the power to strip away these fictions in order to reveal a terrible reality, one in which romance can mask actual harm and toxicity. Consequently, true crime stories demonstrate how the seductive dreams offered by killers are, in reality, nightmares. Shaped by the gender dynamics of true crime – both in terms of readership and representations of spokesvictims (i.e., victims who serve as spokespersons because they have lived through a horrific experience) – *Death in the Air* transforms the romance of London fog into a horror story. To establish generic conventions, this section first considers how true crime typically genders human killers and their victims before analysing how these gender dynamics inform the representation of a killer smog.

Despite the prevalence of male killers and female victims in true crime texts, publishers assume that the majority of readers are middle-class women (Weyr). For female readers, true crime texts can offer them a means for coping with their own past experiences of patriarchal violence and present feelings of fear (Browder "Dystopian Romance"). Like romance novels, another genre with a female-dominated readership, true crime stories reassure readers that happy endings are possible, even in cultures that leave women vulnerable to victimisation by the very men who are supposed to protect them. While happy endings typically take the form of marriage in romance novels, they materialise in true crime when evildoers are caught and punished. According to Browder, these two genres are in conversation with each other because many true crime narratives explore what happens to women who take romance novels too seriously. Indeed, a subgenre of true crime follows the tragic trajectories of supposedly gullible women who are seduced by a psychopath's romantic promises.

It is this subgenre that partially grounds the "meet-cutes" in *Death in the Air*, whereby women meet John Reginald Christie and are subsequently invited into his home in Notting Hill. In contemporary popular culture, the idea of "Notting Hill as a setting for romance" circulates internationally largely because of the British-American romantic comedy film *Notting Hill*. Named as one of the twenty-five best romantic comedies of all time by *Vanity Fair* magazine, the 1999 film "unfolds like a modern-day fairy tale" with eternal appeal (Bradley). After all, the heroine is simply "just a girl, standing in front of a boy, asking him to love her". While girls were not standing in line asking John Reginald Christie ("Reg", according to Dawson) to love them, he nonetheless lured them to his home, so that Notting Hill becomes a site for romance-gone-wrong in *Death in the Air*. In 1920, Reg married Ethel Simpson after a short courtship; at the time of their marriage, "[s]he was twenty-two and quite attractive; he was twenty-one and still painfully shy" (Dawson 147). Years later, the "quiet, meek" Reg lived at Rillington Place with "his skittish, clingy wife" in a "filthy flat" that made him feel "suffocated" (Dawson 5–6). Despite being married, Reg met twenty-one-year-old Ruth Fuerst in 1943. He bragged that he was a special constable with the War Reserve Police. She thought "he seemed charming" and admired the way he looked in "his sharp blue uniform and official cap"; Ruth engaged in an affair with Reg because she "sensed the responsible war constable might be willing to help her, unlike the other men in her life" (Dawson 21). Instead of offering the sought-after protection, Reg strangled Ruth to death during sex, hiding her body under the floorboards in his parlour before moving it to his private garden. After this first kill, Reg charmed, raped and strangled at least six other women, including his wife. He poisoned some of his victims with coal gas and entombed some of their bodies in his coal cupboard.

Since *Death in the Air* extends its narrative about the death of romance to the Great Smog, it is important to examine how the true crime text treats the romance of London fog. At first glance, the London fog "might feel romantic to a couple clutching each other in the mist, like a scene from *Casablanca*" (Dawson 77), but in reality, it can be considered a hellish sign of air pollution. Notably, the romance of London fog has been heavily shaped in the imagination by literary writers, poets and painters, so that the fog has come to evoke both heaven and hell. For the Romantics, clouds could be compared to thoughts as both became a means of expressing the abundance of the human imagination (Harris). At the same time, the Hell most familiar to the Romantics was formed by the experience of smoggy London. Describing London in his 1661 anti-smoke treatise *Fumifugium*, John Evelyn emphasised the "Hellish and dismall Cloud of SEA-COAL" (5) that shrouded the city in an "Evill . . . so epidemicall" ("Dedicatory"). Evelyn's depiction drifted into John Milton's images of Hell in *Paradise Lost* and was also conjured when Percy Bysshe Shelley wrote that "Hell is a city much like London/A populous and smoky city" (*Peter Bell the Third*). While coal burning caused dense smogs in London that were recognised as horrible smoke nuisances, the deadly fog was harnessed by Victorian authors as a literary device, woven into plots by Charles Dickens and setting the scene for Sherlock Holmes' adventures. When Claude Monet visited a wintry London at the turn of the twentieth century, he arose in the morning hoping to see fog. Passionate about how the fog made London beautiful, the French Impressionist declared, "[w]hat I love more than anything in London is the fog" (qtd. in Smith). "It is the fog", according to Monet, "that gives [London] its magnificent breadth"; it is the "mysterious cloak" within which otherwise regular buildings become grandiose (qtd. in Dawson 77). Living in the imagination, "pea-soupers" shaped fictional worlds; in reality, these "London particulars" were so accepted as a part of city life that Londoners missed them when they disappeared (Mirzoeff 224). Created by coal fires burning in the hearths of many homes, the resulting "[s]wirls of fog were romantic and beguiling to Londoners, whose affinity for an open fire was virtually a requirement for being British" (Dawson 42).

Yet the romance of fog, like the romantic elements in true crime, can easily turn deadly for women – especially for women who are caught in it. In the nineteenth century, the fog was deemed dangerous to middle- and upper-class women who desired to walk unaccompanied around London (Corton *London Fog*). In the gloomy deneness of London's smog, women would be unable to discern the presence of strangers, and as a result, were potentially vulnerable to victimisation by predatory men – none more dangerous and ingrained in the imagination of true crime than Jack the Ripper. Indeed, no popular visualisation of Jack the Ripper today would be complete without the presence of fog (e.g., Smith *Jack the Ripper*), despite the fact that the real Ripper is likely to have committed his crimes under clear skies (Corton "Beyond the Pall"). In true crime accounts of Ripper's murders, the fog foils law enforcement's pursuit of the murderer (e.g., Begg and Fido) and is "blended seamlessly" with the serial killer into a part of British folklore (Cook 8). Because Jack the Ripper "just seemingly disappeared into the London fog, never to be heard from again" (Haynes), he coexists with the fog to the point that his is also a story about the London fog.[1] When we juxtapose true crime's serial killers with fears about the London fog's effects, we notice a few striking similarities. Both killers are typically gendered masculine, and both seem to disproportionately victimise women.[2] As a result, they both threaten the domesticity of middle-class families by transgressing boundaries set up to demarcate private spaces. To bring alive the gendered threat and experience of victimisation for readers, *Death in the Air* highlights the plights of female spokesvictims in the following three ways.

First, a generic female spokesvictim is rendered in photographic form on Dawson's book cover. Beneath a title ("DEATH IN THE AIR") written not only in capital letters, but also in the colour of dried blood, a woman looks directly at the reader. The bottom of her face is obscured by a scarf, worn to mitigate the effects of inhaling dirty air. While Big Ben and the Houses of Parliament appear as hazy outlines in the background, the woman is clearly visualised in the centre of the grayscale image. The rough texture of the woman's tweed coat is contrasted with the smooth polish of her pearl necklaces. Tangled at her neckline, the strands of pearl imbue the spokesvictim with the glow of class distinction; they also draw readers' attention to her neck. As a result, the book cover visually foreshadows how women were threatened by strangulation and breathtaking violence in 1952 by both human and non-human killers. As an image, it brings to mind other cinematic images of horror that have lingered over the tragic ends of female victims. Even if killers in horror films have been variously figured over time as humans, birds, slime and fog, the victim has remained "eternally and prototypically the damsel" (Clover 205). The image of the distressed female victim endures, not only because it visualises Edgar Allan Poe's claim that the death of a beautiful woman is "unquestionably, the most poetical topic in the world" ("The Philosophy of Composition"), but also because audiences fear more for women in peril than for men placed in similarly terrifying situations (Clover 206).

Secondly, smog's spokesvictims are gendered feminine because they are most intimately connected with the hearth. The hearth has long been represented as a domestic, feminine and private space. Because "'Hearth' and 'Home' are inseparable concepts" (Wormleighton 52) that have been ideologically entangled with meanings of social and familial stability, they are ideals that have contributed to the romantic image of families gathering together around the warmth of a hearth fire. However, hearth fires came with much smoke, and that smoke was generated by the burning of coal, especially in post-war Britain. By 1952, there was at least one coal fireplace in each home because coal was the only major source of domestic heating in London at that time. Most families, however, could only afford to buy cheap "nutty slack", the sea coal that wasted most of its energy making suffocating smoke rather than heat. When burnt, the soft, bituminous coal released high levels of sulphur dioxide, carbon dioxide and soot into the air, all of which can worsen respiratory illnesses, destroy lungs and claim lives over time.

Notably, the toxic effects of burning coal have been known and visualised since the late nineteenth century. On April 3, 1880, *Punch* magazine introduced King Fog, a hooded, demonic figure cloaked in black with broad enough batwings to obscure the sun. In the cartoon, the hearth is labelled "the cradle of fog", revealing how King Fog was born from the Victorian ideal of domesticity. In November 1880, *Punch* elaborated on its representation of the Fog Demon, by depicting its relationship to King Coal. In so doing, John Tenniel's illustration attached a face and body to the sovereign force behind the foggy Grim Reaper. Playing on the nursery rhyme "Old King Cole", the jolly "Old King Coal" sat upon a hearth-like throne, enjoying a bowl of carbonic acid and smoking his pipe. Doing harm in the clouds above England, King Coal and the Fog Demon were strikingly given form as masculine agents, linking them to the male evildoers that typically populate true crime texts. In post-war Britain, King Coal continued to rule because coal was one of the country's few thriving industries, serving as a key export and source of employment. As such, the economics and politics of Big Coal muffled, for a time, politicians' environmental concerns about air pollution. Consequently, coal continued to circulate in domestic and public spaces in ways that tangled distinctly human and masculine agency with the action of markets, political alliances and fog itself. Moving coal into domestic spaces, the circuit of coal distribution was made possible by coalmen. In contrast to the cleanliness of the housewives that received them, coalmen were described in *Death in the Air* as constantly "filthy and sweaty"; they had blackened faces and nails "chipped and filled with soot"; they secreted "a chemical smell – not smoke, but unburned fuel" when they trudged through kitchens while housewives cringed at their presence (Dawson 127). Depicted as dark and dirty figures in Dawson's book, coalmen triggered disgust from the women that they encountered, appearing as horrific minions of King Coal.

Lastly, the horror of coal-burning smoke in *Death in the Air* is tied to the visceral, bodily reactions of female spokesvictims. As a body genre, horror is devoted to the arousal of bodily sensations (Clover), such as disgust, terror and fear. Titled "The Shiver", Dawson's prologue introduces readers to Rosemary Sargent, a teenager playing hide-and-seek in "the corpse of her neighbor's house" while the mist thickened into a fog (12). From the outset, readers can anticipate feeling a shiver as they will come to fear for Rosemary, a young damsel (soon to be) in distress. Young, female children are typically represented as "ideal victims" in media narratives, according to criminologists and critical victimologists. When portrayed as ideal victims, girls are considered wholly innocent and blameless by society, and as such, are deserving of both sympathy and assistance (Carrabine et al., Christie).

With Rosemary's sympathetic perspective dispersed across chapters in *Death in the Air*, readers piece together the horror of the Great Smog's dislimning effects through her plight. On the first day of the Great Smog, for example, the foggy "stuff seeped into [Rosemary's] skin, like a thin layer of lotion that refused to be scrubbed away with soap flakes"; the heavy cloak of fog "scorched her eyes and her nose, and then squeezed her throat" causing her to struggle for breath (Dawson 29). On the third day of the Great Smog, "the fog had oozed through the windowsills upstairs and into thirteen-year-old Rosemary Sargent's bedroom in southeast London" before killing her father (Dawson 73). Here, Dawson's description of the fog brings to mind how Bram Stoker imagined the piercing chill of vampiric presence in *Dracula*. As Mina Harker lay still in her bed, she, like Rosemary, endured the "heavy and dank, and cold" fog, "which had evidently grown thicker and poured into the room" despite the closed windows (Stoker 215). The climate of Dracula's London was not unlike the climate of smog that characterised post-war Britain. In both instances, smog is represented as a polluting, unnatural presence. Without a clearly defined body, the smoke's formlessness becomes a troubling source of horror because its insidiously slow form of violence cannot be limited or contained by architectural or bodily barriers. The fog can envelope but not itself be enveloped. It does not respect any distinct boundaries between the public and the private, especially since it can penetrate bodies and creep into safe spaces.

How Did the Crime Occur? Smog as a Criminal Accomplice

> [L]ook out this window, Watson. See how the figures loom up, are dimly seen, and then blend once more into the cloud-bank. The thief or the murderer could roam London on such a day as the tiger does the jungle, unseen until he pounces, and then evident only to his victim.
>
> *(Conan Doyle qtd. in Dawson 49)*

In "The Adventure of the Bruce-Partington Plans" (1908), Sherlock Holmes entreats Dr. Watson to consider how the fog can conceal criminal activities from both casual and sustained observation. Used as an epigraph for chapter three in *Death in the Air*, the passage presents the fog's presence as a challenge, even to the great Sherlock Holmes, "the most perfect reasoning and observing machine that the [Western] world has seen" (Conan Doyle 161). In this section, I examine how this epigraph helps stage for readers a conjectural return to the scene of the crime. Highlighting the role of accomplices, I explore epigraphs as paratextual accomplices while also paying attention to how the fog can become a criminal's accomplice.

To begin, epigraphs, such as the one that opened this section, are paratexts. According to Gérard Genette, the paratext is a threshold or "an 'undefined zone' between the inside and the outside, a zone without any hard and fast boundary on either the inward side (turned toward the text) or the outward side (turned toward the world's discourse about the text)" (2). In this sense, the paratext shares with the fog a set of elastic edges that can blur hard and clear boundaries. In contrast to the asphyxiating effects of smog, the paratext "provides an airlock that helps the reader pass without too much respiratory difficulty from one world to the other, a sometimes delicate operation, especially when the second world is a fictional one" (Genette 408). As an orienting device, the paratext aims to smoothly transport readers into the author's textual world without causing a hitch in their breathing. In *Death in the Air*, an opening quotation by Conan Doyle helps readers transition from the foggy world of Sherlock Holmes'

London to the smoggy confines of post-war Britain, generically moving from classical detective fiction to true crime. Because Sherlock Holmes emphasises how the "cloud-bank" can foil detective work, the epigraph also serves as Dawson's accomplice when it comes to shaping readers' interpretation of the Great Smog as a criminal accomplice. As a dark pall, the Great Smog reduced visibility and removed the possibility of police surveillance. Indeed, the "blackout of pollution" enabled criminals to burgle homes, snatch purses, attack people on the streets, rob a post office and smash shop windows while police officers could barely stumble back to their patrol cars (Dawson 41).

Despite the escalated number of calls for assistance during the Great Smog, the Metropolitan police realised that they "faced an adversary that was defending criminals. The fog was everywhere now" (Dawson 53). Echoing Charles Dickens' description of "[f]og everywhere" in *Bleak House* (qtd. in the epigraph of Dawson's prologue 1), Dawson allows readers to observe, with their own eyes, the effects of an all-pervasive fog by featuring a series of black and white photographs in *Death in the Air*. As another set of paratextual accomplices, these photographs illustrate how the smog created a cover of darkness. These images not only heighten the sense of truthful reality that true crime readers seek, but they also provide evidence of how "a choking smog took ahold of London, turning day into night" (photograph caption provided by Dawson). Because these images appear alongside crime scene images associated with Christie's murders, readers are directed to examine photographs of the Great Smog as though they, too, are crime scene images. When reviewed in this light, the photographs highlight the *noir* aesthetic of crime scene photography, a style of visualisation that has linked darkness to both crime and anthropogenic fog since the nineteenth century (Lam).

In staging a speculative return to the scene of the crime, whether through photography or fictive recounting, true crime imitates the classic detective novel when it positions readers as observers in the reconstruction of the murderer's acts and motives. In scenes that narratively focus on John Reginald Christie's criminal performances, readers observe how fog enters the picture as an accomplice to murder. For instance, they behold as Christie fictively ruminates about the fog when it blankets his private garden on the first day of the Great Smog:

> If [he] were to glance down from his kitchen window that night, as he often did, his small plot would be almost indistinguishable, smothered by the fog. The swirls born from the coal were bewitching. *And the fog was once his accomplice, such a reliable conspirator. He drew inspiration from its noxious fumes – it had helped him murder her*, the second woman, after all. The back garden's broken pots, dead plants, and general rubbish usually visible out his back window vanished in the fog – the fence, which could barely stand on its own, faded like a lost memory.
>
> *(Dawson 43–44, my emphasis added)*

While fog in this passage might be construed as merely an atmospheric element for creating an air of mystery, I would like to seriously consider the smog as a criminal accomplice. When smog is transformed into a partner in crime – that is, when it becomes more than a concealing cloud-bank or a source of imaginative inspiration – it assumes a degree of material agency that is rarely attributed to non-human matter. As Jane Bennett has cogently argued, we have a habit of ascribing agency and vibrancy to living human beings while denying these same qualities to matter. By characterising matter as passive, dull or inert stuff, we are prevented from detecting a fuller range of vital, non-human powers that circulate within and around our human bodies. To explore the animacy of things, such as smog, is to consider their capacity to "act as quasi agents or forces with trajectories, propensities, or tendencies of their own" (Bennett viii). Not only can non-humans impede or support the designs of humans, but they can also collaborate, interact and cooperate with humans to *act*.[3] In short, they, too, can *do* things by producing effects, making a difference and altering a course of events. When fog is described by a serial killer as "a reliable conspirator", it is worth examining how the human killer conspires with toxic, non-human matter to engage in murder. In general, conspiring denotes the capacity "to combine in action or aim" and "to unite in producing" an effect (*Oxford English Dictionary*). While the fog does not agree to Christie's reprehensible plot in the way a human conspirator can, it does, by its very presence, cooperate with him to effect the results that he desires. When the fog is united with Christie, they kill. Here, it is useful to revive the Latin roots (*conspīrāre*) of the verb "to conspire", in order to resuscitate its meaning "to breathe together". It is by breathing together that the fog helps Christie murder his second victim, Muriel Amelia Eady.

In October 1944, Christie killed Muriel with his own invention in what he considered to be "[a] really clever murder" (qtd. in Dawson 45). His method of murder took advantage of London's affinity for coal-burning stoves, where coal gas pipes – like motor exhaust pipes – could churn out lethal levels of carbon monoxide. Lured by the promise of having her cough cured by Friar's Balsam, a compound used to treat breathing issues, Muriel inhaled through Christie's homemade breathing mask. The refashioned gas mask was attached to two rubber tubes, one led to steam water infused with Friar's Balsam and the other to a coal gas pipe. When Muriel breathed through the mask, she was knocked unconscious by the deadly carbon monoxide pumping out of the gas pipe. As a result, she was unable to fight back when Christie strangled her to death. Christie required a weakened victim because his own "body wasn't healthy – he had loads of aches all over and he was recovering from bronchitis" (Dawson 47). Before Christie even met Muriel, both criminal and victim had unknowingly been breathing together the noxious fumes of smog, an invisible poison gas circulating in London during World War II and heightened in 1952 (Fuller). Both their bodies were physically weakened over time by their long-term exposure to dirty air. Smog was already secretly doing damage inside both of their bodies, intensifying their experiences of bronchitis, coughing and other respiratory issues.

Who Is Responsible? Criminal Cover-Ups and Holding Jekyll Accountable

> A great chocolate-coloured pall lowered over heaven . . . there would be a glow of rich, lurid brown . . . and here, for a moment, the fog would be quite broken up, and a haggard shaft of daylight would glance in between the swirling wreaths.
>
> *(Stevenson qtd. in Dawson 8)*

Drawing on Robert Louis Stevenson's observations about the "swirling wreaths" of London fog, the epigraph for *Death in the Air*'s first chapter begins to establish the symbiotic links between representations of smog and images of criminality. Following the representational tradition set down by Stevenson's Gothic novella *The Strange Case of Dr. Jekyll and Mr. Hyde*, Dawson transforms John Reginald Christie into a Hyde-like figure, by emphasising a shared shape-shifting ability. When literary critics focus on Hyde's description as "dwarfish", "ape-like" and "troglodytic" (Stevenson 10, 15, 10), Hyde is read as a fictional representation of Cesare Lombroso's atavistic, "criminal man" (e.g., Arata, Mighall, Reid). Hyde's criminality becomes linked to animalistic urges and beastly violence. Media depictions of Christie similarly focused on animal imagery to make sense of his criminality. Journalists were fascinated by his transformation from an "insignificant little mouse of a man" to the "Beast of Rillington Place" (Dawson 268). "[L]ike Jekyll and Hyde", Christie hid his murderous actions behind a mild-mannered demeanour as well as within a larger story about the London fog (Dawson 268). Here, it is instructive to consider Hyde's strangely amorphous form – what we might even call a "fog-like" shape. Like urban air pollution, "Hyde is described as 'shifting, insubstantial mists,' and like Hyde, the fog is 'embattled,' warring with winds that 'continually charg[e] and rou[t]' the foul mixture as it 'lower[s] over heaven,' clasping London in its choking embrace" (Manning 193). Created out of a concoction of impure materials, Hyde is born as pollution (Taylor), and in this way, can be placed in symbiosis with the London fog. The origins of his existence are strikingly similar to those that created the London smog, which is also a grimy product of human intervention.

While Stevenson treats Hyde as a metaphorical embodiment of pollution, Dawson argues that Christie and the Great Smog are separate killers that coexist in time and space and can interact to greater devastating effect. They are two killers that share a similar source of toxicity (e.g., burning coal) and *modus operandi* (i.e., suffocation). While Christie embodies the smog, in the sense that his body bore the effects of long-term exposure to air pollution, the smog does not owe its existence to Christie. To identify those responsible for the smog, Dawson returns to Stevenson's tale for a lesson on revealing cover-ups. For her, Stevenson "used the fog itself to show the mutation of identity – the nightmarish landscape protected the murderous Mr. Hyde as he committed his crimes. When the fog drifted away the civilised Dr. Jekyll reappeared" (42). In *Death in the Air*, Dawson reveals what has been concealed, by making Jekyll appear in clear focus in the aftermath of the Great Smog. Through her "braided narrative", she demonstrates how one serial killer was constructed as "more terrifying to Londoners than a deadly fog that strangled thousands – a naiveté that only benefited . . . the politicians who tried to cover up the choking smog that was largely of their own creation" (7). Because media attention, both in the aftermath of the Great Smog and even today, has been disproportionately placed on (re)covering Christie's crimes, a lack of sustained coverage on the political forces behind the Great Smog have enabled those responsible to avoid intense public scrutiny. As the paratextual praise from other authors reiterates, Dawson's "narrative poses a powerful moral question: who's the worse killer – a madman who strangles seven women and a baby, or government officials whose staggering indifference allows thousands to die in the great London smog of 1952?" (Glenn Frankel, Pulitzer-prize winning journalist, qtd. in Dawson).

To address this moral question, readers are encouraged from the outset to seek Jekyll as much as Hyde when they examine an illustrated map of post-war London. As another paratextual element, the map locates Christie in reference to his "Murder House" at 10 Rillington Place. The deadly fog, however, escapes a single, localisable site because London is the scene of the Great Smog's "crimes". Ecological damage, when conceived as crime, modifies the notion of crime itself as an act attached to discrete location points. Instead of pinpointing all the places of death and victimisation caused by lethal pollution, Dawson's map identifies instead London's centres of political calculation, such as the Palace of Westminster and the Whitehall Government Offices, where parliamentary decision-making not only failed to prevent the Great Smog, but also served to minimise the smog's effects. Like *Strange Case, Death in the Air* examines the criminals that hide behind a public image of respectability. More specifically, it documents the following three cover-ups perpetrated in the aftermath of the Great Smog and in the name of King Coal.

One, the Conservative government, under Winston Churchill, attempted to deny responsibility for marketing nutty slack in a way that contributed to the Great Smog. Ten days before the start of the Great Smog, the government had begun to promote nutty slack as a cost-saving option. As small nuts of brown coal mixed with dust, nutty slack was of lower quality than the government's rationed coal. However, it was affordable to most Londoners during the desperate post-war years. In the government's advertisement campaign, nutty slack was peddled to domestic consumers as ration-free and unrestricted fuel. More than one million tons would be made available, and families could purchase as much as they wanted for thirty percent less than the cost of rationed coal. As "the cheapest solid fuel freely available" (Minister of Fuel and Power, Geoffrey Lloyd, qtd. in Dawson 98), the dirty fossil fuel burnt relentlessly during those five cold days in December 1952, contributing to the poisonous smoke that produced an abnormally high death rate. On February 24, 1953, however, the Ministry of Fuel and Power denied responsibility for facilitating the Great Smog, by falsely claiming that the distribution of nutty slack "had not begun at the time of the fog" (qtd. in Dawson 217). Three months later, nutty slack, the government argued, was not the only cause of the fog; rather, the fog was "an enduring and central tenet of London life itself . . . and it had been for hundreds of years" (Dawson 238). Because the fog was rationalised as a "normal" and constant presence in the city, Conservative politicians concluded that no one was to blame for the fog.

Two, the government also sought to be held blameless for the Great Smog by implicitly engaging in victim-blaming. It attempted to deny that the smog caused great harm to minimise its political responsibility for the health of Londoners. By characterising "fog victims" as physically vulnerable (e.g., elderly or sick "due to cold and damp, quite independently of the fog") and likely to die anyway, victim-blaming was partially embedded in the assumptions made by ministers, such as Minister of Housing Harold Macmillan (qtd. in Dawson 186), about who was most affected by the fog.

Three, the Conservative government changed the parameters for establishing the number of fog victims to further downplay the severity of the Great Smog's effects on Londoners. While the Great Smog could not be conceived as a victimless incident, it could be reconceived as having harmed fewer victims. To that end, W.P.D. Logan, the government's chief medical statistician, recalculated the number of "fog deaths" in early 1953, by shortening the window of time for calculating these deaths to two weeks instead of nearly four weeks. With the narrower timeframe, Logan was able to announce a "new", reduced death toll for the Great Smog, which counted two thousand fewer fog deaths than the Ministry of Health's original estimate. In so doing, "the government was essentially covering up thousands of dead victims" (Dawson 188). In contrast to the government's calculations, Dr. E.T. Wilkins, the officer-in-charge of pollution investigations at the Fuel Research Station, revealed a second spike in deaths in the fog's aftermath between January and mid-March 1953. By factoring in the long tail of air pollution's effects, Wilkins estimated that the death rate was fifty percent higher than the government's estimate, making the death toll closer to twelve thousand. When the Ministry of Health released its report in 1954, the government claimed that the additional deaths in 1953, such as those discussed by Wilkins, were caused by a flu epidemic rather than by smog. This false claim prevailed for decades.

According to Dawson, these cover-ups amount to malfeasance and an eschewing of responsibility for the health and lives of Londoners. Despite growing evidence that the fog was deadly in 1952, lawmakers "refused to blame pollution – and coal – for turning the clouds toxic", demonstrating how "the fog was [also] smothering [politicians] – air pollution was too difficult to reduce and too expensive to eradicate" in the early 1950s (Dawson 158, 200). In 1953, politicians and lawmakers seemed more alarmed by Christie's confessions than by the smog's casualties. The fog's mass murder was overshadowed by the actions of a Notting Hill serial killer. As a result, toxic air remained an invisible killer, and those responsible remained mostly hidden behind an air of respectability.

Conclusion

In the year that John Reginald Christie was tried and punished, *The Interim Report of the Committee on Air Pollution* was released. The report blamed industries, domestic consumers as well as local and national governments for creating and maintaining London smog. It also gave rise to the world's first piece of legislation on air pollution in 1956 – the *Clean Air Act*. While the *Clean Air Act* placed regulatory controls on home fires and made coal merchants and households responsible for the use of clean coal, its enactment did not completely absolve a government from refusing to act promptly in response to the documented harms of air pollution. In concealing the slow violence of smog, politicians covered up the true toll of air pollution and failed to account for thousands of fog victims. Although the smog caused fatal breathing for thousands, both media and political attention were placed instead on the Rillington Strangler, demonstrating society's long-standing appetite for crime stories. To work with this existing (human and media) interest in spectacular forms of criminal violence, this chapter examined how the slow violence of London's polluting fog could be translated into the sensation-driven conventions of true crime. As such, it aimed to change the ways that we have been culturally conditioned to peer *through* the fog (Lam). Encouraged by detectives, such as Sherlock Holmes, to see through the fog, we do not typically look *at* the fog. Yet in recent years, the fog has served as an important clue for investigating the "invisible crime" of climate change (Brisman).

In *Death in the Air*, Dawson transforms the Great Smog into a highly visible, non-human, mass killer that could be fictively narrated as a masculinised serial murderer. The London fog actively shaped a dystopian romance, one in which Jekylls were arguably more dangerous than any Hyde. To explore the foggy edges of this true crime text, we examined paratextual elements, revealing the ways in which true crime borrows from and is in constant conversation with other kinds of horror texts and visual forms, such as cartography and photography. Like the fog's ability to muddle clear boundaries, true crime blurs the boundaries between different genres, such as romance and classical detective fiction, and between fact and fiction. Because of its enduring popularity and generic adaptability, true crime can offer a vocabulary of crime, agency, and responsibility for articulating environmental degradation in ways that resonate with wider audiences. Given its preoccupations as a genre, true crime can resituate environmental destruction on moral grounds, making even "normal" weather appear both strange and suspicious. By laying bare political and economic agendas, it has the potential to reshape and redirect fear towards the slow violence of environmental catastrophe.

Notes

1. When Alfred Hitchcock adapted Mary Belloc Lowndes' fictional novel *The Lodger*, the most successful account of the Ripper murders before World War I, he visualised serial murder as "a story about London fog" (the 1927 film's subtitle), framing the city's fog as a dangerous cloak for deviance and moral degeneracy (Corton *London Fog*).
2. There does not appear to be data that the Great Smog disproportionately affected women; rather, the toxic air harmed both men and women, especially if they were elderly (Logan). Consequently, the emphasis on female (spokes)victims can be construed as a specific representational choice in *Death in the Air*
3. In Bruno Latour's terms, matter, whether human or nonhuman, can be described as an *actant*. Because actants rarely act alone, they are often brought together as an assemblage with other bodies and forces to become a powerful source of action.

Bibliography

Arata, Stephen. "The Sedulous Ape: Atavism, Professionalism, and Stevenson's *Jekyll and Hyde*". *Criticism*, vol. 37, no. 2, 1995, pp. 233–259.

Begg, Paul, and Martin Fido. *The Complete Jack the Ripper A-Z*. Kings Road Publishing, 2015.

Bell, Michelle L., and Devra Lee Davis. "Reassessment of the Lethal London Fog of 1952: Novel Indicators of Acute and Chronic Consequences of Acute Exposure to Air Pollution". *Environmental Health Perspectives*, vol. 109, no. 3, 2001, p. 389, doi:10.2307/3434786.

Bennett, Jane. *Vibrant Matter: A Political Ecology of Things*. Duke UP, 2010.

Biressi, Anita. *Crime, Fear and the Law in True Crime Stories*. Palgrave, 2001.

Bradley, Laura. "The 42 Best Romantic Comedies of All Time". *Vanity Fair*, 13 Feb. 2023, www.vanityfair.com/hollywood/2018/08/best-romantic-comedies-list.

Brisman, Avi. "Representing the 'Invisible Crime' of Climate Change in an Age of Post-Truth". *Theoretical Criminology*, vol. 22, no. 3, 2018, pp. 468–491, doi:10.1177/1362480618787168.

Browder, Laura. "Dystopian Romance: True Crime and the Female Reader". *The Journal of Popular Culture*, vol. 39, no. 6, 2006, pp. 928–953, doi:10.1111/j.1540-5931.2006.00328.x.

———. "True Crime". *The Cambridge Companion to American Crime Fiction*, edited by Catherine Ross Nickerson, Cambridge UP, 2010, pp. 121–134.

Carrabine, Eamonn, Paul Iganski, Maggy Lee, Ken Plummer, and Nigel South. *Criminology: A Sociological Introduction*. Routledge, 2004.

Christie, Nils. "The Ideal Victim". *From Crime Policy to Victim Policy*, edited by E. A. Fattah, Palgrave Macmillan, 1986, pp. 17–30.

Clover, Carol T. "Her Body, Himself: Gender in the Slasher Film". *Representations*, no. 20, 1987, pp. 187–228.

Conan Doyle, Arthur. "A Scandal in Bohemia". *The Penguin Complete Sherlock Holmes*, Penguin Books, 2009, pp. 161–175.

Cook, Andrew. *Jack the Ripper*. Amberley Publishing, 2009.

Corrigan, Maureen. "'Death in the Air' Revisits 5 Days When London was Choked by Poisonous Smog". *NPR*, 17 Oct. 2017, www.npr.org/2017/10/17/558316674/death-in-the-air-revisits-5-days-when-london-was-choked-by-poisonous-smog.

Corton, Christine. "Beyond the Pall . . . How London Fog Seeped into Fiction". *The Guardian*, Guardian News and Media, 31 Oct. 2015, www.theguardian.com/books/2015/oct/31/beyond-pall-how-london-fog-seeped-into-fiction.

———. *London Fog: The Biography*. The Belknap Press of Harvard UP, 2015.

Dawson, Kate Winkler. *Death in the Air: The True Story of a Serial Killer, the Great London Smog, and the Strangling of a City*. Hachette, 2017.

Des Voeux, Henry Antoine. "Fog and Smoke". *Coal Smoke Abatement Society*. Public Health Congress, July 1905.

Evelyn, John. *Fumifugium*. The Rota at the University of Exeter, 1976, https://archive.org/details/fumifugium00eveluoft.

Fuller, Gary. *The Invisible Killer: The Rising Global Threat of Air Pollution – And How We Can Fight Back*. Melville House, 2018.

Genette, Gérard. *Paratexts: Thresholds of Interpretation*. Translated by Jane E. Lewin, Cambridge UP, 1997.

Harris, Alexandra. *Weatherland: Writers & Artists Under English Skies*. Thames & Hudson, 2015.

Haynes, Suyin. "Jack the Ripper's Victims and the True-Crime Trap". *Time*, 9 Apr. 2019, time.com/5566076/jack-the-ripper-true-crime-victims/.

Ingebretsen, Edward J. "The Monster in the Home: True Crime and the Traffic in Body Parts". *The Journal of American Culture*, vol. 21, no. 1, 1998, pp. 27–34, doi:10.1111/j.1542-734x.1998.2101_27.x.

Lam, Anita. "Establishing Shots: Detecting Anthropogenic Fog in Modern Crime Scene Photography". *Criminal Anthroposcenes*. Palgrave Macmillan, 2020/2021, pp. 51–105.

Latour, Bruno. *Reassembling the Social*. Oxford UP, 2005.

Logan, W. P. D. "Mortality in the London Fog Incident, 1952". *The Lancet*, vol. 261, no. 6755, 1953, pp. 336–338, doi:10.1016/s0140-6736(53)91012-5.

Lombroso, Cesare. *Criminal Man*. Duke UP, 2006.

Manning, Pascale McCullough. "The Hyde We Live in: Stevenson, Evolution, and the Anthropogenic Fog". *Victorian Literature and Culture*, vol. 46, no. 1, 2018, pp. 181–199, doi:10.1017/s1060150317000389.

Mighall, Robert. *A Geography of Victorian Gothic Fiction: Mapping History's Nightmares*. Oxford UP, 1999.

Mirzoeff, Nicholas. "Visualizing the Anthropocene". *Public Culture*, vol. 26, no. 2, 2014, pp. 213–232, doi:10.1215/08992363-2392039.

Murley, Jean. *The Rise of True Crime*. Praeger, 2008.

Nixon, Rob. *Slow Violence and the Environmentalism of the Poor*. Harvard UP, 2011.

Poe, Edgar Allan. "The Philosophy of Composition". *The Raven and the Philosophy of Composition*. Paul Elder and Company, 2017, www.gutenberg.org/files/55749/55749-h/55749-h.htm.

Punnett, Ian Case. *Toward a Theory of True Crime Narratives*. Routledge, 2018.

Quinn, Anthony. "London Fog Review – A City in the Thick of It". *The Guardian*, Guardian News and Media, 13 Dec. 2015, www.theguardian.com/books/2015/dec/13/london-fog-the-biography-christine-corton-review.

Reid, Julia. *Robert Louis Stevenson, Science, and the Fin de Siècle*. Palgrave, 2006.

Schmid, David. "True Crime". *A Companion to Crime Fiction*, edited by Charles J. Rzepka and Lee Horsley, Blackwell Publishing, 2010, pp. 198–209.

Seltzer, Mark. *True Crime: Observations on Violence and Modernity*. Routledge, 2007.

Smith, Clare. *Jack the Ripper in Film and Culture: Top Hat, Gladstone Bag and Fog*. Palgrave Macmillan, 2016.

Smith, P. D. "London Fog by Christine Corton – The History of the PEA-Souper". *The Guardian*, Guardian News and Media, 27 Nov. 2015, www.theguardian.com/books/2015/nov/27/london-smog-christine-corton-review.

Stevenson, Robert Louis. *The Strange Case of Dr. Jekyll and Mr. Hyde*. Dover Publications, 1991.

Stoker, Bram. *Dracula*. Wordsworth, 2000.

Taylor, Jesse Oak. *The Sky of Our Manufacture*. U Virginia P, 2016.

Thorsheim, Peter. *Inventing Pollution: Coal, Smoke, and Culture in Britain Since 1800*. Ohio UP, 2006.

Weyr, Tom. "Marketing America's Psychos". *Publisher's Weekly*, vol. 240, no. 15, 12 Apr. 1993, pp. 38–41.

Wilkins, E. T. "Air Pollution and the London Fog of December, 1952". *Journal of the Royal Sanitary Institute*, vol. 74, no. 1, 1954, pp. 1–21, doi:10.1177/146642405407400101.

Wiltenburg, Joy. "True Crime: The Origins of Modern Sensationalism". *The American Historical Review*, vol. 109, no. 5, 2004, pp. 1377–1404, doi:10.1086/530930.

Wormleighton, Alison. *Victoria: Decorating with a Personal Touch*. Hearst Books, 2004.

17

"EVERY CRIME HAS ITS PECULIAR ODOR"

Detection, Deodorisation and Intoxication

Hsuan Hsu

In Spider Robinson's Hugo-Prize winning speculative narrative, "By Any Other Name" (1976), an environmentalist who has become disillusioned with industrial modernity develops a virus that radically enhances humans' olfactory sensitivity, leaving humanity with "a sense of smell approximately a hundred times more efficient than that of any wolf" (Robinson, 29).[1] This "hyperosmic plague" (ibid.) brings about the end of industrial civilisation: a fifth of the world's population is killed (or self-immolates) as a result of sensory overstimulation concentrated in cities and industrial sites; others survive by emigrating to rural settlements and developing advanced nose plugs. The novel's hyperosmic humans experience olfaction as both capacity and debility: on the one hand, they can use it to track the scent trails of human and non-human creatures, including hitherto unperceived atmospheric entities called "muskies" (31); on the other hand, the sensory overstimulation can devastate their minds, and synthetic cleaning chemicals become deadly weapons.

Robinson's novella draws together an eclectic range of influences: his own experiment with living in the woods as a young man, his job as a night watchman guarding New York City's pungent sewers in 1971, the synergies between 1960s counterculture and environmentalism, and the 1970 Clean Air Act spearheaded by Maine Senator Edmund Muskie.[2] Hyperosmia connects all these threads, precipitating a massive worldwide decline in anthropogenic emissions that have suddenly become unbearable. Enhanced olfaction also attunes the novella's characters to the differential distribution of noxious smells in dense urban communities like Harlem, where the African American narrator's mother and brother die almost instantly. Robinson's work exemplifies the critical potential of hyperosmic narratives: in addition to zeroing in on smells that usually float near – if not below – the thresholds of sensation, attention, and cultural value, a heightened sense of smell amplifies patterns of premature death already present in differentiated atmospheric geographies. Hyperosmic narratives experiment with a radical redistribution of the sensible, simultaneously inverting the hierarchy of the senses and drawing attention to the visceral, trans-corporeal environmental exchanges inherent in olfactory perception. In the novella, this reorganisation of the human sensorium has immediate, transformative effects on nearly every aspect of political, economic, social, and environmental activity.

"By Any Other Name" also exemplifies the ambiguity of olfactory perception, which blends cultural associations with biochemical materiality. Robinson's hyperosmic humans are not directly harmed by airborne toxins: they self-immolate when their olfactory bulbs are overwhelmed by urban and industrial smells. If this difference implies a vital distinction between industrial odours and putatively natural ones, it also suggests that the difference may have to do more with cultural perceptions

 DOI: 10.4324/9781003091912-21

of smell than with the chemicals that comprise those smells. After all, why should the scent of bleach or asphalt be more overwhelming to a hyperosmic than the scent of the rose invoked by the novella's title? In the course of imagining olfaction as a site of trans-corporeal vulnerability, Robinson paradoxically privileges smell's semiotic dimensions. In his hyperosmic novella, industrial odours are not unpleasant because they're harmful: they're harmful because they're unpleasant. As one of the hyperosmic plague's inventors puts it, "all the undesirable by-products of twentieth century living . . . quite literally *stink*" (25, emphasis in original). Paradoxically, Robinson's engagement with olfaction functions both to dramatise the geographies of atmospheric violence and to translate them into the moralising, dematerialised language of pleasant and unpleasant sensations, fragrance and "stink".

To the extent that they draw attention to intersections between olfaction and risk, hyperosmic narratives dramatise the material ambiguity of olfactory perception. Because our olfactory impressions are simultaneously semiotic and biochemical, perceptions of olfactory risk may be based on cultural conditioning, embodied experience, or a combination of both. To what extent can cultural and moral discourses about smell – frequently oriented by imperatives of deodorisation – be disarticulated from chemical toxicity? This chapter explores how detective fiction – a genre that frequently features hyperosmic characters and plots – engages with the cultural and chemical components of olfactory risk perception. Positioned as agents of deodorisation, hyperosmic detectives endeavour to stigmatise and eradicate odours perceived to be culturally and morally deviant; yet their very dependence on smell renders detectives vulnerable to olfactory intoxication and the uneven dynamics of differential deodorisation. I will move from the figure of the deodorising detective established in nineteenth-century narratives to later works that juxtapose the sniffing out of clues and culprits with what I call *environmental detection*, or the detection of unevenly distributed material presences.

Detective Fiction and Deodorisation

In an incisive reflection on the sniffer dog as an instrument of state surveillance, the political theorist Mark Neocleous asks:

> Why, with a critical intellectual culture saturated with analyses of biopolitics, biosecurity, biosurveillance and biometrics, has so little been said about the smell of power? Why is the state's "nosiness" still understood almost solely through the ocular and the aural?
>
> *(167)*[3]

Like biosurveillance more broadly, the genre of detective fiction is commonly associated with visual surveillance and ratiocination. In Edgar Allan Poe's formative detective tales "Murders in the Rue Morgue" (1841) and "The Purloined Letter" (1844), C. Auguste Dupin intersperses his detective work with improvised lectures on ratiocination and optics, explaining his methods through analogies with the judicious employment of sidelong glances by astronomers, and the perspectival shift required to discern a word stretched across the surface of a map in an "excessively obvious" fashion (Poe, "Murders" 545; "Purloined" 990). But "Rue Morgue" juxtaposes vision and ratiocination with another influential motif: the figuration of the detective as a sniffer dog. When Dupin boasts that "The scent had never for an instant been lost", he initiates a mode of olfactory detection that at once positions the detective as an agent of deodorisation and undercuts the notion of the detective as a disembodied mind or "private eye" (Poe, "Murders" 553). This motif of the deodorising detective extends across nearly two centuries of detective fiction, from the olfactory methods staged by Arthur Conan Doyle and Rudolph Fisher (both discussed later) to the olfactory hypersensitivity of contemporary characters like Rudolfo Anaya's Sonny Baca, Thomas Pynchon's Conkling Speedwell, and Artyom Litvenenko's "Sniffer". Although it has received little critical attention, the trope of the "nosy" detective is so well established within the genre as to have inspired devices such as the Conkling Speedwell's olfactory laser or "Naser" and punishments such as the slitting of Jake's nostrils in Roman Polanski's *Chinatown* (1974).[4]

The sense of smell – one of the most pervasive metaphors and methods for the detective's virtuosity – is also a site where bodily, mental, and affective integrity gives way to chemical intoxication. Thus, the deodorising detective is a deeply ambivalent figure that renders one of the most prominent cultural icons of rational deduction dependent upon olfaction – a sense that is inherently trans-corporeal, immersive, and (for post-Enlightenment Westerners) notoriously difficult to describe. While nineteenth-century detective fiction acknowledges the presence of intoxicating atmospheres (for example, the "curling eddies of smoke that oppressed the atmosphere of [Dupin's]", the "opalescent London reek" described outside the windows of Sherlock Holmes's apartment in "The Adventure of the Abbey Grange", or the tobacco and cocaine that animate the thinking of both these detectives), its plots underscore the detective's role as an agent of deodorisation who sniffs out and expunges deviant odours (Poe, "Purloined" 974; Doyle, 711). The legal scholar Sarah Marusek's discussion of the connections between law and olfaction aptly characterises smell's normalising function in detective plots: "Through smell, law normalises bodies, place, and expectations through [*sic*] the exclusion of the deviant, the noncompliant, and the disempowered" (Marusek, 41).

Sherlock Holmes stories build on Poe's metaphor of detection as "scenting". In Holmes's debut novel, *A Study in Scarlet* (1887), Watson describes the detective at work as "a pure-blooded, well-trained foxhound, as it dashes backward and forward through the covert, whining in its eagerness, until it comes across the lost scent" (Doyle, 22). Elsewhere in the story, Doyle literalises this metaphor when Holmes deduces the use of poison from the "slightly sour smell" of a dead man's lips (61). *The Sign of the Four* (1890) introduces Toby the dog, whom Holmes employs to track the faint scent of creosote tar from the scene of the crime. Holmes refers to Toby as "a queer mongrel, with a most amazing power of scent. I would rather have Toby's help than that of the whole detective force of London" (83). (In "The Adventure of the Creeping Man" Holmes reports that "I have serious thoughts of writing a small monograph upon the uses of dogs in the work of the detective" – a monograph that would presumably dwell on the dog's olfactory capacities [65]). Holmes's olfactory virtuosity in *The Hound of the Baskervilles* (1902) suggests a connection between the detective and the titular hound:

> I held it within a few inches of my eyes, and was conscious of a faint smell of the scent known as white Jessamine. There are seventy-five perfumes, which it is very necessary that a criminal expert should be able to distinguish from each other, and cases have more than once within my own experience depended upon their prompt recognition.
>
> *(550)*

Olfaction plays a crucial role in a later story, "The Adventure of the Retired Colourman" (1926), where the killer attempts to cover up the smell of gas used to murder his wife and her lover by re-painting his house (1095–1106). In Doyle's work, scent is both a metaphor and a sensory method of detective work – a powerful tool for policing the boundaries of class, religion, race, and nation.

In his influential study of literary engagements with the London Fog, Jesse Oak Taylor reads Holmes as a sort of amateur meteorologist perfectly attuned to variations in London's thick, anthropogenic atmosphere. Taylor explains that the "opalescent" London fog that suffuses the Sherlock Holmes stories is not an obstruction to vision but the very medium of detection in the Anthropocene's irrevocably transformed atmospheres:

> Holmes sits "amidst the teeming millions, with his filaments stretching out and running through them" ("The Cardboard Box", 1113). Those filaments exist in a more 'prosaic and material'

> form in the patterns and curling wreaths of the fog outside the windows, infiltrating all of London, its alleys and byways, the crevices of its windows and the bodies of its inhabitants. Sherlock Holmes gains access to this pervasive climate of interconnection by manufacturing a malodorous and disordered atmosphere of his own.
>
> *(162)*

Holmes deploys his hyperosmic sensitivity as a hypersensitive atmospheric instrument for identifying and expunging deviant odours. In the course of doing so, he both navigates London's dramatically differentiated atmospheres and obscures them by targeting deviant bodies rather than disparities. In "The Adventure of the Three Gables" (1926), Holmes claps his hand to his pocket upon seeing a "grim and menacing" black prize-fighter named Steve Dixie, who asks "Lookin' for your gun, Masser Holmes?" Holmes's reply: "No, for my scent bottle, Steve" (*Doyle*, 1063). Instead of physical violence, Holmes protects himself by using his personal air freshener to ward off Dixie's supposedly intolerable odour. In "The Adventure of the Devil's Foot" (1910), Doyle stages a material confrontation between Holmes and a toxic West African root whose powder produces both "a thick musky odour, subtle and nauseous" and a fatal sense of terror in anyone who inhales it. As Watson and Holmes learn first-hand, the root's odour is a direct threat to self-possession:

> At the very first whiff of it my brain and my imagination were beyond all control. A thick, black cloud swirled before my eyes, and my mind told me that in this cloud, unseen as yet, but about to spring out upon my appalled senses, lurked all that was vaguely horrible, all that was monstrous and inconceivably wicked in the universe.
>
> *(793)*

After solving the murder through this experiment in "immersive toxicology" Holmes sends the killer – along with his knowledge of the root – back to central Africa (Taylor 153).

Classic detective fiction frequently deploys olfaction as a tool for detecting transgressive, racialised bodies and for controlling their atmospheric incursions on the spaces of white respectability.[5] Anticipating Holmes's encounters with black atmospheres, the Bornean Ourangutan figuratively sniffed out by Poe's Dupin indexes a common stereotype associated with black and brown bodies by practitioners of scientific racism. The racial implications of the deodorising detective echo the role of bloodhounds used to discipline fugitive slaves. Given Poe's Southern origins and the racialised Ourangutan at the heart of his formative detective story, we might consider whether the genre was influenced as much by the slaveholder's bloodhounds as it was by James Fenimore Cooper's narratives of wilderness tracking (as Walter Benjamin famously argued) (Benjamin, 439). Mark Twain gestures toward this line of influence in "A Double-Barreled Detective Story" (1902) – a parody in which Sherlock Holmes's rational method is eclipsed by the tracking skills of a detective born with a preternatural sense of smell. In Twain's story, the detective's hyperosmic condition is referred to as "the gift of the bloodhound" because it was supposedly acquired when bloodhounds were set upon his mother shortly before his birth (298). Smell's narrative function as a means of detecting, avoiding, or removing blackness reflects its powerful role in historical arguments for segregation: as the sensory historian Mark Smith has documented, advocates of segregation claimed not only that they could make racial distinctions based on the smell of black bodies, but also that desegregation would be repugnant because "Negroes have a smell extremely disagreeable to white people" (qtd in Smith, 83). The racial underpinnings of olfactory detection persist in the use of drug sniffing dogs by contemporary police: in 2011, the *Chicago Tribune* reported a vast discrepancy between the success rate of drug sniffing dogs used at traffic stops for "Hispanic" and other drivers (Hinkel and Mahr). When influenced by the conscious or unconscious biases of their handlers, sniffer dogs may be just another mechanism for racial profiling.

In nineteenth- and early twentieth-century narratives of the deodorising detective, Holmes's and Watson's intimate encounter with the Devil's Root is the exception to the rule. Dupin and Holmes are seldom threatened by their (recreational) intoxicants, and, despite his regular and frequently deliberate exposures to London's panoply of smells, Holmes never experiences their cumulative effects. As Taylor notes, London's "abnatural" atmosphere *enables* Holmes's detection; with the exception of the Devil's Root, the atmosphere does not debilitate him. The nineteenth-century detective narrative's tendency to reduce criminality to the actions of individual perpetrators assumes a teleological, procedural model of deodorisation: the detective traces the scent to its deviant source in order to contain or banish it. For many in the nineteenth century, however, unpleasant odours were recognised as environmental health threats – miasmas resulting not from individual crimes, but from structural (and, often, infrastructural) inequities. Although miasma theory gave way to the germ theory of disease transmission in the later part of the century, the perception of smells as potentially debilitating substances persisted in popular health discourses. Whereas early detective fiction frequently frames the detective as an agent of olfactory surveillance and control, twentieth-century narratives of hyperosmic detection channel the genre's olfactory obsessions into critical accounts of differential deodorisation in which the work of detection is confronted with the distribution of smells as a problem of environmental justice.

Environmental Detection

If hyperosmic sensitivity enables the work of detection, it also makes trans-corporeal material agency a central concern of detective fiction – particularly as twentieth-century writers shifted the genre's focus to increasingly stratified and polluted settings. As detectives become intoxicated, they register a model of environmental entanglement that questions the viability of approaches to deodorisation that prioritise the removal of deviant bodies. Conflating detection with exposure, smell unsettles the ideas of ratiocination, bodily immunity, and interpretive control inherent in classic detective fiction. Instead of interpretively reconstructing past events from the traces they leave behind, smell introduces plots of environmental detection in which the threat, or crime, is materially present in the atmosphere – and thus already present in the detective's body. As the art historian Caroline Jones writes, olfaction calls forth new modes of thinking: "Tracing the path of smells requires thinking by sniffing, tracking the logic of stench in trajectories of the self" (13).

The concept of environmental detection indicates a tendency found (frequently in suppressed or marginalised moments) throughout detective fiction that anticipates the contemporary subgenre of eco-detective fiction represented by novels such as Percival Everett's *Watershed* (1996), Paolo Bacigalupi's *The Water Knife* (2015), and Donna Leon's *Earthly Remains* (2017) – as well as investigative documentaries that bring olfactory detection (alongside other senses) to bear on risk perception, such as Judith Helfland's and Daniel Gold's *Blue Vinyl* (2002), Josh Fox's *Gasland* (2010), and Jon Whelan's *Stink!* (2015). Because of its trans-corporeal, chemical qualities, olfaction's connection with the environment opens onto an "interactionist" ontology wherein the detective's body is co-constituted by the environments s/he is investigating. The object of the hyperosmic's environmental detection is not simply a transgression that has already occurred, but harmful toxins circulating between atmospheres and bodies: not an absent event traced by clues left behind, but the atmospherically dispersed agents of slow violence. The fantasy of environmental immunity that underpins the detective conceived as a "private eye" gives way to a model of the detective as "public nose" – an investigator whose sensory and cognitive capacities cannot be extricated from olfactory exchanges with differentially deodorized public atmospheres.

"John Archer's Nose" (1935), the last story published by the Harlem Renaissance author Rudolph Fisher, stages the tension between two interpretations of smell: on the one hand, as a sign of individual morality and behaviours; on the other hand, as a material constituent of risk-laden atmospheres. Fisher's text features a multilayered olfactory plot: if olfaction enables the story's detective figures to solve a murder, noxious air remains as a dispersed material agent of slow violence that exceeds the detective plot's denouement. In the story, detective Perry Dart is assisted by the acute olfactory sense of his friend Dr. Archer, who eventually connects a peculiar smell in the bedroom of a murdered boy with the "evil-smelling packet" of medicinal roots he saw around a dead baby's neck earlier that day (Fisher, 187). Dr. Archer traces both deaths to the supposed ineffectiveness of root medicine: he believes that the baby died (of suffocation due to untreated status lymphaticus – "literally choking to death in a fit") because its parents relied on folk "superstition" rather than modern medicine and X-ray treatments, and he deduces that the baby's grieving father murdered the son of his root medicine provider as an act of retribution (186).

As Dart and Dr. Archer discuss the smells encountered at the crime scene, their banter presents a remarkable meta-commentary on the complex connections between odours, language, literary genres, and black urban geographies:

> "M-m. Peculiar – very. Curious thing, odors. Discernible in higher dilution than any other material stimulus. Ridiculous that we don't make greater use of them".
>
> "*I never noticed* any particular restriction of 'em in Harlem".
>
> . . . "Odors, *should* be restricted", [Dr. Archer] pursued. "They should be captured, classified, and numbered like the lines of the spectrum. We let them run wild – "
>
> "Check".
>
> "And sacrifice a wealth of information. In a language of a quarter of a million words, we haven't a single specific direct denotation of a smell".
>
> "Oh, no?"
>
> "No. Whatever you're thinking of, if [*sic*] it is an indirect and non-specific denotation, linking the odor in mind to anything else. We are content with 'fragrant' and 'foul' or general terms of that character, or at best 'alcoholic' or 'moldy,' which are obviously indirect. We haven't even such general direct terms as apply to colors – red, green, and blue. We name what we see but don't name what we smell".
>
> "Which is just as well".
>
> "On the contrary. If we could designate each smell by number – "
>
> "We'd know right off who killed Sonny".
>
> "Perhaps. I daresay every crime has its peculiar odor".
>
> "Old stuff. They used bloodhounds in *Uncle Tom's Cabin*".
>
> "We could use one here".
>
> *(193–4)*

While Dr. Archer's desire for a comprehensive taxonomy of odours in the service of surveillance takes centre stage in this passage, Dart's incongruous responses shift our attention to the racial implications of olfactory surveillance. Dart's comment about the lack of odour "restriction" in Harlem invokes the neighbourhood's disproportionate exposure to problems with sanitation, ventilation, crowding, and industrial pollution. Rather than reducing Harlem's exposure to odours, Dr. Archer's wish to "restrict" odours would only involve classifying them with words or numbers – a dream of total olfactory rationalisation and control. This would support the detection of individual crimes according to their "peculiar odor" – a process that Dr. Archer concedes is comparable to the use of bloodhounds to track fugitive slaves. The doctor's nose traces crime to individual rather than social causes – to the "peculiar odor" rather than the intractable background atmospheres to which most Harlem residents have become desensitised by prolonged low-level exposures.

Ironically, the "peculiar odor" that exposes the killer in this case is the smell of African American folk medicine. Aligning what he perceives to be backward racial beliefs with smell, Dr. Archer's analysis directly contrasts the deodorising influence of modern medical technologies with the "evil-smelling" roots. Dr. Archer's eagerness to use "X-ray treatments" to melt away the baby's inflamed thymus reflects Fisher's own career as a successful radiologist; ironically, however, the diagnosis and X-ray treatment of an "inflamed thymus" were controversial in the 1930s and eventually proven to be "as mythical as the therapeutic effects of fried-hair charms" (Crafton, 73).[6] Meanwhile, the rootwork that is so vehemently rejected by Dr. Archer invokes the long tradition of conjure and Hoodoo in African American culture. In the 1920s and 1930s, mail-order curio companies made these practices more readily available to growing black urban communities. In her study of the African American spiritual products industry, Carolyn Morrow Long describes a range of scented products – from perfumes and powder sachets to the spicy scents of High John the Conqueror Root and Van-Van Oil (an oil scented with vervain or lemongrass) – that circulated through the mail-order catalogues and spiritual supply stores of the 1930s (99–126). Although the magical properties attributed to these products are frequently dismissed as superstition (as in Crafton's comment about "fried-hair charms", earlier), hoodoo and rootwork promised recent black migrants to the city the possibility of taking an active role in air conditioning. The powers attributed to these materials – which included enhancing charisma, warding off enemies, cleansing household spaces, and inspiring love – may not have been entirely unfounded: smell's capacities to evoke collective and individual memories, along with its influence on affects and behaviours, may exert considerable (though not easily measurable) effects on physical and mental health. Among the antebellum antecedents of these mail-order conjure materials were practical interventions such as "powders designed to aid runaways by throwing tracking dogs off their scent" (Anderson, 84).[7]

If Fisher's hyperosmic detective stigmatises rootwork as a harmful and retrograde superstition, the story also offers a critical counter-narrative that complicates the figure of the deodorising detective. Although Fisher never directly acknowledges it, the true agent of violence in "John Archer's Nose" may be the health effects of atmospheric stratification. It turns out that the murdered boy had a terminal case of tuberculosis – a contagious respiratory disease that disproportionately affected African Americans (the tuberculosis rate in Harlem was five times greater than the rest of Manhattan's [Lewis, 144]), and that is strongly correlated with poor conditions of housing and ventilation. It is thus doubly significant that the killer in this "locked room" mystery entered the boy's room through the building's air shaft, a common ventilation feature in Harlem that Fisher elsewhere depicted in graphic detail reminiscent of naturalist description: "An airshaft: cabbage and chitterlings cooking[;] waste noises, waste odors of a score of families, seeking issue through a common channel; pollution from bottom to top – a sewer of sounds and smells" (Fisher, "City", 38). In order to single out an individual perpetrator, Dr. Archer has to navigate and suppress Harlem's broader panoply of odours: in other stories, Fisher describes Harlem's atmosphere as "vile – hot, full of breath and choking perfume"; "Waste clutters over it, odors fume up from it, sewer-mouths gape like wounds in its back" ("High", 111; "Fire", 142). If "Lenox Avenue is for the most part the boulevard of the unperfumed 'rats'", it is because the atmosphere has already been corrupted by urban planning, immiseration, and negligent landlords. Whereas Sherlock Holmes could deodorise his encounter with a black man by resorting to his scent bottle, Fisher's black detective figures are confronted with the systemic problem posed by infrastructures that permeate Harlem's air with risk. As Bruce Robbins writes:

> Infrastructure smells . . . because attention is not paid, because it is neglected. And it is neglected because it belongs to the public domain, all other tokens of belonging effaced, owned in effect by no one. The smell of infrastructure is the smell of the public.
>
> *(26)*

Fisher's story ironically dramatises the tension between Harlem's panoply of olfactory burdens and the respiratory health of the hyperosmic detective: as Dr. Archer puts it, "I'm going to locate that odor if it asphyxiates me" ("John Archer's Nose", 213).

Fisher's attention to infrastructural racism resonates with the "air of fatality" that Sean McCann discerns in hard-boiled crime fiction – a genre that reinvigorated the detective story in the 1930s–1960s (3). Whereas classic detective fiction purged deviant odours in order to restore a transparent social order, hard-boiled stories – influenced by literary naturalism's portrayals of humanity amid a swirl of environmental forces (Cassuto 296) – remained cynical about the prospect of "clear[ing] the atmosphere".[8] To be sure, hard-boiled authors sometimes indulge in moments of olfactory racialisation, as when Dashiell Hammett notes the "unmistakable . . . smell of unwashed Chinese" or when Raymond Chandler writes, "He had a sort of dry musty smell, like a fairly clean Chinaman" (Hammett, 172; Chandler qtd in Wolfe, 16).[9] But they are distinguished by their interest in depicting an atmosphere of generalised corruption: as Chandler puts it, "It is not a fragrant world, but it is the world you live in" ("Simple Art", 59). In the amoral world of hard-boiled crime fiction, smell does not just provide the detective with clues – it manifests as a trans-corporeal index for social corruption and atmospheric stratification. In *The Big Sleep* (1939), for example, Chandler juxtaposes the cloying scent of General Sternwood's orchid hothouse with the pungent smell of the oil sump holes that had made the Sternwoods rich. In addition to indicating divisions of space and class, odours have potential consequences for health and cognition: in an auto garage, "The smell of the pyroxilin paint was as sickening as ether"; down by the Sternwoods' oil wells, "The smell of that sump would poison a herd of goats" (185). Rather than enabling the detective to sniff out criminals, smell threatens Marlowe with physical and mental debilitation. Far from serving as a clue, the ether-like smell of paint, which "drugged the close air of the garage", dulls Marlowe's attention enough for him to be caught off guard and captured (187). Chandler's detective is not a dispassionate cartographer of the city's ambient smells, but a porous subject co-constituted by the atmospheres he traverses.

In his hard-boiled Harlem detective series, Chester Himes dramatises the contradictions between the policing of individual crimes and the subtler workings of structural violence. *The Heat's On* (1961) stages a provisional climate of hyperosmia, as a heat wave intensifies Harlem's "atmospheric pressures"[10] and ambient odors:

> Heat was coming out of the pavement, bubbling from the asphalt; and the atmospheric pressure was pushing it back to earth like the lid on a pan.
>
> . . . An effluvium of hot stinks arose from the frying pan and hung in the hot motionless air, no higher than the rooftops – the smell of sizzling barbecue, fried hair, exhaust fumes, rotting garbage, cheap perfumes, unwashed bodies, decayed buildings, dog-rat-and-cat offal, whiskey and vomit, and all the old dried-up odors of poverty.
>
> *(30)*

Himes frames insomnia, gambling, knife-fights, and "evil" itself as a result of the neighbourhood's lack of air conditioning: "It was too hot to sleep. Everyone was too evil to love" (ibid.). Sustained by a convergence of structural inequities, these everyday "odors of poverty" become indistinguishable from "smell of [inadequate] infrastructure" (Robbins 25).

Himes would revisit this passage in his unfinished, final Harlem novel, *Plan B* (1993), expanding its olfactory prose into two pages of baroque excess. Early in that novel, a long paragraph catalogues Harlem's indoor stinks, culminating in

> yearly accumulations of thousands of unlisted odors embedded in the crumbling walls, the rotting linoleum, the decayed wall paper, the sweaty garments, the incredible perfumes, the rancid face creams and cooking fats, the toe jam, the bad breath from rotting or dirty teeth, the pustules of pus.
>
> *(51)*

The following paragraph explains that, despite people's beliefs in the "fresh air" of the outdoors, Harlem's outside air was no better:

> Outside there were all the impurities generated by their worn-out automobiles, their brimming garbage cans, the dog shit and cat shit, the putrefying carcasses of rats and cats and dogs and sometimes of meat too rotten even for the residents to eat.
>
> *(51–2)*

Himes's catalogues of stenches insistently connect atmospheric "impurities" with the economic and political issues embodied by poorly maintained homes, rotten food, inadequate dental care, substandard automobiles, and failures of municipal waste removal. Even cancer – a possible long-term consequence of all these stenches – becomes another source of stench: "It stank from . . . body tissue rotten from cancer" (51).

In *The Heat's On*, the Harlem detectives Grave Digger Jones and Coffin Ed have a conflicted relationship with the detective's deodorising profession. Early on, Himes alludes to an earlier novel in the series in which Coffin Ed shot a kid who threw perfume at him, mistaking it for acid.[11] On two occasions, the detectives unsuccessfully try employing a dog to "sniff around" (*HO* 99, 147). A later scene literally deconstructs the figure of the sniffer dog. Believing that a key has been hidden in the dog, the drug dealer Sister Heavenly chloroforms it and systematically dissects it, releasing new odours into an already putrid hotel room: "The hot poisonous air inside of the room, stinking of blood, chloroform and dog-gut, was enough to suffer the average person. But Sister Heavenly stood it" (*HO* 171). Like the sniffer dog, Himes's detectives fail to locate the missing shipment of heroin that drives the novel plot: they identify and stop (often by killing) individual culprits, but eventually discover that the five kilograms of heroin were unwittingly thrown into an incinerator. As the novel concludes, this incinerated heroin disperses into Harlem's atmosphere, blending with both the catalogue of stenches and the intoxicating atmospheres of opium, incense, and marijuana described earlier in the novel. Not only do the detectives (who frequently rely on extra-legal violence) become morally associated with the criminals in Harlem's generalised atmosphere of corruption, but their interventions actually increase the atmosphere's toxicity by contributing the suffocating smells of cordite (from gunshots), the pheromonal "smell of terror . . . like a sickening miasma" evoked by Coffin Ed's menacing presence, and the fumes of burning heroin (*HO* 174).

Blood Shot (1988), Sara Paretsky's fifth hard-boiled novel featuring the female detective V.I. Warshawski, leverages the genre's dual modes of olfactory representation – smell as both clue and airborne risk factor – to stage the economic and political machinery of environmental slow violence. Although the novel's central characters are white working-class women, its account of environmental violence is set in South Chicago – an area whose population is 93% black. As the environmental sociologist David Pellow notes, South Chicago has been described as "one of the greatest ecological disasters in the history of North America", where residents "breath in an estimated 126,000 pounds of toxic pollutants emitted into the air each day and are surrounded by the most landfills per square mile in the United States" (Pellow 68; see Moore). The environmental justice activist Hazel Johnson "has often charged that environmental racism in [South Chicago] is 'another form of genocide'" (Qtd. Pellow 68).[12] Although Paretsky disingenuously downplays these racial disparities in order to underscore the class dynamics of environmental injustice, her novel offers a vivid sensory account of everyday atmospheric violence in South Chicago. *Blood Shot* begins with detective Warshawski reencountering the long-forgotten smell of the neighbourhood in which she was raised:

> I had forgotten the smell. Even with the South Works on strike and Wisconsin Steel padlocked and rusting away, a pungent mix of chemicals streamed in through the engine vents. I turned

> off the car heater, but the stench – you couldn't call it air – slid through minute cracks in the Chevy's windows, burning my eyes and sinuses.
>
> *(Paretsky 1)*

Whereas writers frequently deploy smell to evoke powerful place-based memories, the smell described here draws attention only to itself. The detective is no sooner introduced than her perceptual faculties are debilitated by an indecipherable and inescapable "pungent mix of chemicals".

Some of these airborne chemicals are at the centre of the novel's plot, as Warshawski investigates a series of cover-ups and a murder aimed to suppress the occupational and environmental health effects of Xerxine, an industrial cleaning solvent manufactured in the South Side. As a retired company doctor explains:

> The way they used to make it, it left these toxic residues in the air. . . . if you breathe the vapors while they're manufacturing it, it doesn't do you a whole lot of good. Affects the liver and kidneys and central nervous system and all those good things. . . . You know, they didn't run the plants to kill the employees, but they weren't very careful about controlling how much of the chlorinated vapors got into the air.
>
> *(132)*

After being left for dead in a pungent, chemical-filled marsh, navigating a "thick mist carrying the river's miasmas" (318), and taking down a gangster and a corrupt politician in a chemical plant, Warshawski worries about the effects of her many exposures to airborne toxins over the course of the narrative:

> I shut my eyes, but I couldn't keep out the clamor, or the murky Xerxine smell. What would my creatine level be after tonight? I pictured my kidneys filled with lesions – blood-red with black holes in them, oozing Xerxine.
>
> *(333)*

The novel's title, *Blood Shot*, turns out to refer not to bloodshot eyes or to gunshots but to the slow violence that toxic chemicals introduce into the bloodstreams of workers, local communities, and the detective herself.

The most recent incarnation of the deodorising detective is the eponymous protagonist of the *The Sniffer* (2013–) – a popular Ukrainian television series directed by Artyom Litvinenko and internationally distributed by Amazon Prime and Netflix. Although its provenance is located at some distance from the U.S. detective fiction (and the unavoidable influence of Doyle) that I have discussed so far, the show's many references to Sherlock Holmes and the conventions of Hollywood police procedurals position it as one of the latest, international instalments in the hyperosmic detective tradition. Working with the Special Bureau of Investigations, the show's reclusive police consultant, Käro, relies on his hyperosmic talents to reconstruct crime scenes. Blending olfactory data with his extensive knowledge of chemistry and related fields, he is able to deduce the age, gender, recent contacts, smoking habits, food preferences, weapons, and countless other characteristics of those present at the crime scene. When he cannot solve cases *in situ*, the Sniffer returns to his home – a hermetically sealed apartment where he conducts olfactory experiments in a fully equipped laboratory. At the centre of the show are the Sniffer's virtuosic olfactory capacities, which are often dramatised through the formal innovation of staging his analysis of crime scenes with CGI animated gaseous bodies and props. These aspects of *The Sniffer* underscore how hyperosmia empowers its protagonist to perceive invisible material traces and to leverage these traces in the service of surveilling and containing a motley collection of terrorists, kidnappers, murderers, thieves, art forgers, sex criminals, and traffickers.

The Sniffer's hyperbolic rendering of the olfactory detective stages the immense architectural and social efforts necessary to prop up the hyperosmic detective as a figure of pure rationality. For, as the series slowly reveals, Käro suffers from acute environmental hypersensitivity: his surly personality is not (or not only) the product of machismo but the result of the difficulty and strain with which he tolerates the barrage of city smells outside his home. In a rare moment of self-reflection, he offers this description of his condition: "Imagine a person who lives without a skin, like a snail without its shell" (Litvinenko). When the show is not concerned with crime-solving, *The Sniffer* reflects on how its protagonist compensates for his condition of radical olfactory exposure. His hypersensitivity – reflected in his irritability, solitude, and inability to eat impure foods or to tolerate the presence of others without commenting on their odours – compromises his relationships with his ex-wife, son, and new love interest. His home – a pristine apartment that appears to be the only inhabited unit in a high-rise building – is accessible only by a private elevator in which visitors must undergo an ultraviolet decontamination protocol. In order to employ hyperosmia without succumbing to its intoxicating effects, the Sniffer must occupy the difficult position of a misanthrope sealed from the world in a deodorized bubble – ironically, a bubble with a laboratory chock full of synthetic chemicals that somehow do not affect him. As in most serial detective fiction, the Sniffer's body bears no cumulative traces of its past exposures or chemical body burden: despite being "a snail without a shell" when it comes to atmospheric exposure, his body seems repositioned as a blank slate at the beginning of each new episode. The serial form's tendency toward bodily renewal at the beginning of each new episode resolves the tension between hyperosmia's everyday debilitations and its status as an extraordinary crime-solving ability. This enables *The Sniffer* to reframe masculinity itself – not as an invulnerable body distanced from its surroundings, but as a body that can both leverage and manage its environmental entanglements. While *The Sniffer* takes the hyperosmic detective's environmental detection to its logical endpoint by dramatising Käro's sometimes debilitating environmental sensitivity, it simultaneously glosses over the chronic symptoms and pervasive exposures that distinguish illness narratives by people with Multiple Chemical Sensitivity (MCS).[13]

Simultaneously an organ of olfactory knowledge and olfactory invasion, the detective's nose catalyses diverse responses to modernity's atmospheric risks: nineteenth-century efforts at deodorisation and olfactory control, the systemically poisoned atmospheres of hard-boiled crime fiction and Fisher's and Himes's black detective narratives, and Paretsky's and Litvinenko's chemically sensitive detectives figures who employ their noses to sniff out systemic environmental risks rather than criminalised individuals. The hyperosmic detective indicates how trans-corporeal modes of embodied knowledge and ecological relation have always haunted the genre of detective fiction. In addition to situating detective fiction as an "environmental" genre, the literary history of the hyperosmic detective illuminates the tension between individualised criminality and environmental violence. In many twentieth-century and contemporary narratives, the activity of sniffing out criminals draws detectives into new, intimate understandings of environmental toxicity. Through olfaction, intoxication – frequently stigmatised as a sign of irrationality and criminality – becomes an invaluable yet potentially debilitating epistemological tool for detectives navigating modernity's stratified atmospheres.

Notes

1. Robinson subsequently expanded this novella into the full-length novel, *Telempath* (1976).
2. Robinson's "muskies" – gaseous entities that feed on anthropogenic air pollution – reference Senator Muskie's comparatively robust approach to regulating air pollution, which focused on severely polluted working-class communities.

3. For a broad discussion of the ongoing recruitment of nonhuman animals as agents of olfactory surveillance, see Marks.
4. For a fascinating refiguration of the hyperosmic detective in a fantasy world influenced by African cosmologies (rather than by deodorization initiatives), see James.
5. On U.S. detective fiction's central concern with racial heterogeneity, see Huh and Robinson.
6. Fisher died at the age of thirty-seven from abdominal cancer, probably as a result of his experimental research with X-ray technology.
7. Anderson cites a WPA interview from 1937 in which John Barker reported that a powder derived from horned toads "was to be applied to the bottoms of shoes in order to throw dogs off the trail of escaped slaves" (qtd in Anderson 84).
8. The editors of the *Annotated Big Sleep* note a significant atmospheric disjunction between the novel's dust jacket synopsis and its actual content: "The dust jacket synopsis of the first hardcover edition assured readers that the detective 'clears the atmosphere and leaves the reader content that justice, though of an unexpected sort, will be done.' Our hero seems to think otherwise" (Hill, Jackson, and Rizzuto, 455n17).
9. For a discussion of olfactory discourses targeting the Asian diaspora, see Hsu, Chapter 4.
10. Mawani, n.p.
11. See Himes, *The Real Cool Killers*.
12. Qtd Pellow 68.
13. For a discussion of detective narrative motifs in illness narratives by people with chemical sensitivity, see the longer chapter from which this essay is excerpted: Hsu, 27–55.

Bibliography

Alaimo, Stacy. *Bodily Natures: Science, Environment, and the Material Self.* Indiana UP, 2010.

Anderson, Jeffrey. *Conjure in African American Society*. Louisiana State UP, 2005.

Barad, Karen. *Meeting the Universe Halfway: Quantum Physics and the Entanglement of Matter and Meaning*. Duke UP, 2007.

Benjamin, Walter. *The Arcades Project*. Translated by Howard Eiland and Kevin McLaughlin, Harvard UP, 1999.

Cassuto, Leonard. "The American Novel of Mystery, Crime, and Detection". *A Companion to the American Novel*, edited by Alfred Bendixen, Blackwell, 2012, pp. 291–308.

Chandler, Raymond. "The Simple Art of Murder". *Atlantic Monthly*, no. 174, Dec. 1944, pp. 31–46.

———. *The Big Sleep*. Vintage, 1992.

Crafton, Robert. *The African American Experience in Crime Fiction: A Critical Study*. McFarland, 2015.

dir. Litvinenko, Artyom. *The Sniffer*, Netflix, 2017.

Doyle, Arthur Conan. *The Original Illustrated "STRAND" Sherlock Holmes: The Complete Facsimile Edition*. Wordsworth Editions, 1998, p. 711.

Fisher, Rudolph. "Blades of Steel". *City of Refuge: The Collected Stories of Rudolph Fisher*, edited by John McCluskey, Jr., U Missouri P, 2008, pp. 159–171.

———. "City of Refuge". *City of Refuge: The Collected Stories of Rudolph Fisher*, pp. 35–47.

———. "High Yaller". *City of Refuge: The Collected Stories of Rudolph Fisher*, pp. 111–127.

———. "Fire By Night". *City of Refuge: The Collected Stories of Rudolph Fisher*, pp. 142–158.

———. "John Archer's Nose". *City of Refuge: The Collected Stories of Rudolph Fisher*, pp. 185–219.

Hammett, Dashiell. "Dead Yellow Women" (1927). *The Dashiell Hammett Story Omnibus*, edited by Lillian Hellman, Cassell, 1966, pp. 150–195.

Hill, Owen, Pamela Jackson, and Anthony Rizzuto. *The Annotated Big Sleep*, edited by Owen Hil, Pamela Jackson, and Anthony Rizzuto, Vintage, 2018.

Himes, Chester. *The Heat's On*. The Chatham Bookseller, 1966.

———. *The Real Cool Killers*. Vintage, 1988.

———. *Plan B*. U Mississippi P, 1993.

Hinkel, Dan, and Joe Mahr. "Tribune Analysis: Drug-Sniffing Dogs in Traffic Stops Often Wrong". *Chicago Tribune*, 6 Jan. 2011, www.chicagotribune.com/news/ct-xpm-2011-01-06-ct-met-canine-officers-20110105-story.html.

Hsu, Hsuan L. *The Smell of Risk: Environmental Disparities and Olfactory Aesthetics*. New York UP, 2020.

Huh, Jinny Huh. *The Arresting Eye: Race and the Anxiety of Detection*. U Virginia P, 2015.

James, Marlon. *Black Leopard, Red Wolf.* Riverhead, 2019.

Jones, Caroline A. "The Mediated Sensorium". *Sensorium: Embodied Experience, Technology, and Contemporary Art*, edited by Caroline Jones, MIT Press, 2006, pp. 5–49.

Lewis, David Levering. "Reading the Harlem Renaissance". *The Harlem Renaissance*, edited by Harold Bloom, Chelsea House, 2004, pp. 123–148.
Long, Carolyn Morrow. *Spiritual Merchants: Religion, Magic, and Commerce*. U Tennessee P, 2001.
Marks, Amber. *Headspace: On the Trail of Sniffer Dogs, Wasp Wardens and Other Dumb Friends in the Surveillance Industry*. Virgin Books, 2008.
Marusek, Sarah. "Reasonable or Offensive? Smell, Jurisdiction, and Public Right". *Non Liquet: Law and the Senses Series, Smell*, Apr. 2015, pp. 32–41.
Mawani, Renisa. "Atmospheric Pressures: On Race and Affect". Unpublished paper cited with author's permission (nd).
McCann, Sean. *Gumshoe America: Hard-Boiled Crime Fiction and the Rise and Fall of New Deal Liberalism*. Duke UP, 2000.
Moore, Natalie. *South Side: A Portrait of Chicago and American Segregation*. Picador, 2016.
Neocleous, Mark. "The Smell of Power: A Contribution to the Critique of the Sniffer Dog". *Radical Philosophy*, vol. 167, no. 9, 2011, pp. 9–14.
Paretsky, Sara. *Blood Shot: A V. I. Warshawski Novel*. Random House, 1988.
Pellow, David Naguib. *Garbage Wars: The Struggle for EJ in Chicago*. MIT Press, 2002.
Poe, Edgar Allan. "Murders in the Rue Morgue". *Collected Works of Edgar Allan Poe*, edited by Thomas Ollive Mabbott, Harvard UP, 1969–78, pp. 521–574.
———. "The Purloined Letter". *Collected Works of Edgar Allan Poe*, edited by Thomas Ollive Mabbott, Harvard UP, 1969–78, pp. 974–997.
Pynchon, Thomas. *Bleeding Edge*. Penguin, 2013.
Robbins, Bruce. "The Smell of Infrastructure: Notes Toward an Archive". *Boundary 2*, vol. 34, no. 1, 2007, pp. 25–33, doi:10.1215/01903659-2006-025.
Robinson, Michelle. *Dreams for Dead Bodies: Blackness, Labor, and the Corpus of American Detective Fiction*. U Michigan P, 2016.
Robinson, Spider. "By Any Other Name". *Analog*, vol. 96, no. 11, 1976, pp. 14–67.
Smith, Mark. *How Race is Made: Slavery, Segregation, and the Senses*. U North Carolina P, 2006.
Taylor, Jesse Oak. *The Sky of Our Manufacture: The London Fog in British Fiction from Dickens to Woolf*. U Virginia P, 2016.
Tuana, Nancy. "Viscous Porosity: Witnessing Katrina". In *Material Feminisms*, edited by Stacy Alaimo and Susan Hekman, Indiana UP, 2008, pp. 188–213.
Twain, Mark. "A Double-Barreled Detective Story". *The Writings of Mark Twain. Vol. 23. The Man that Corrupted Hadleyburg and Other Essays and Stories*, Harper and Brothers, 1906 (1902), pp. 293–367.
Wolfe, Peter. *Something More than Night: The Case of Raymond Chandler*. Bowling Green Popular Press, 1985.

18

IN PAOLO BACIGALUPI'S ENVIRONMENTAL SCIENCE FICTION, IMMORAL AND CRIMINAL ARE NOT SYNONYMOUS

Patrick D. Murphy

Since 2008, contemporary author Paolo Bacigalupi has won numerous awards for his environmentally oriented science fiction, such as the Hugo, Nebula, World Fantasy, and Michael L. Prinz prizes, as well as being a National Book Award finalist (Windupstories.com). All of his work, whether short fiction, young-adult stories, or the two novels treated in this chapter, address environmental issues, including some likely to occur in the relatively near future. In both of his novels, *The Windup Girl* (2009) and *The Water Knife* (2015), Bacigalupi relies on plots typical of crime fiction to advance the narrative. As Richard Crownshaw, for example, notes, "*The Water Knife* is a hybrid of science and detective fiction" (891). Bacigalupi also deploys common near future climate fiction (cli-fi) settings and crises as platforms for the investigation of, and commentary about, the environmental issues confronting their readers in the here and now.

It is appropriate to think of science fiction as a mode of representation in which the diegesis relies on a world setting that is not that of the present-day consensual reality, regardless of the specific characteristics of plot. Crownshaw sees science fiction as being particularly suited to addressing climate change because of its "scalar capaciousness", which is needed for grasping the scope of the Anthropocene's environmental impacts. (900). In contrast, crime fiction can be characterised as a type of literature driven by its plot structure. Unlike crime fiction, which can happen anywhere at any time, setting is crucial to the conceptual framework of science fiction (SF), so much so that many highly successful SF novels take their plot structure from other types of literature, such as horror, western, and crime stories. Often, they even retell stories from those genres with little more than the window dressing of future settings, other planets, or alien contact. Likewise, while crime fiction almost invariably relies on a plot that depicts a near perfect criminal plan undone by unanticipated flaws in the planning or execution, science fiction adds a setting that, when falsely perceived as passive, introduces an environment of uncontrollable variables.

Unlike some ecologically-oriented science fiction, such as Kim Stanley Robinson's *New York 2140* (2017), Bacigalupi's thematic commentaries are presented diegetically through the perspectives and observations of multiple character focalisers within the story world, rather than through a non-participant narrative commentator external to that world who appears as an authoritative voice within the text. Likewise, unlike some crime fictions, such as those written by Arthur Conan Doyle and Edgar Allan Poe, there is no all-knowing detective who resolves the events and motives at the end.

DOI: 10.4324/9781003091912-22

Thus, Bacigalupi relies on internally persuasive and conflicting environmental points of view to spur readers to consider the moral and ethical dilemmas that the novels present. He allows readers to determine for themselves the degree to which these are cautionary tales about current human behaviour and political and economic directions and trends. Therein lies the greatest strength of his writing beyond his deft handling of dialogue and the elaborate detail of his settings.

The Criminal and the Immoral: *The Windup Girl* and *The Water Knife*

In *The Windup Girl* (2009), set roughly two hundred years from now, rising sea levels due to climate change, the end of oil as fuel, and food diseases resulting from corporate control and genetic modification combine in a crisis over access to heirloom seed stock. Bacigalupi situates this novel in Bangkok, a city already suffering from sea level rise and frequent flooding and draws on the history of Thailand as a nation-state that has never been colonised by Europeans (in World War II Thailand allied itself with Japan so that technically Japanese troop deployments did not colonise or occupy it either [The Globalist]). The crime plot thus unfolds around the efforts of foreigners to gain control of the country's seedbank, crop production, and market for imported foods, as well as the Thai government officials' actions to abet or thwart those efforts. Like the climatic dimensions of its setting, *The Windup Girl* addresses an environmental justice conflict already well underway in the twenty-first century: the control of seed stock (see Shiva and Kinchy). The title of the novel, though, comes from the term used to describe one of the main characters, a Japanese designed-female android whose existence raises additional ethical questions about biotechnology. *The Water Knife* is set closer to the present day at a point in time when more fortunate Millenials are dying off from old age rather than starving or dying from heat exhaustion. The novel's setting – a megadrought ridden American Southwest – is deeply rooted in the reality of the region's long battle with water shortage (Kesslen; CBS News; Reisner; Woodhouse). Added to this history are the already documented impacts of climate change on the region, such as reduced annual precipitation and higher annual temperatures (Baum; deBuys). He also relies on, and frequently has characters reference, Marc Reisner's *Cadillac Desert* (1986), which warned the nation that the Colorado River Basin would not be able to provide adequate water supplies to support the population growth of such cities as Denver, Salt Lake City, Los Angeles, San Diego, and Phoenix.

It is in this last city that much of the action takes place. As a result, the crime plot centres around efforts among the competing state interests of California, Nevada, and Arizona to secure as much of this reduced water supply as possible by both legal and extra-legal means. Further, the federal government is presented as relatively weak, and the economy of Texas has collapsed producing a flight of refugees westward, reminiscent of the Dust Bowl Okies in John Steinbeck's *The Grapes of Wrath* (1939). Bryan Yazell suggests that these refugees form an essential part of the novel because "the public imagination in the Global North around an issue like climate migration already lags behind the reality" (161). The title of *The Water Knife* also refers to one of the main characters, a mercenary working for the Southern Nevada Water Authority, and thereby placing his moral ambiguity centre stage.

Before discussing the novels in terms of the interstices of crime and ecology in each, it is useful to consider how Bacigalupi addresses the distinctions between a crime and an immoral action, between what is legal and what is ethical or moral. American heroes in a range of novels and films often commit illegal actions in the course of realising justice, gaining revenge, or apprehending deviant people. Their criminal actions are generally not considered immoral by readers and viewers along the lines of an ends-justify-the-means sort of moral code.

Many criminal actions are perceived as immoral only when the ends they are meant to achieve are considered reprehensible, and not in terms of the actions themselves. Often heroes engage in criminal activity reluctantly or only on a case-by-case basis. Immoral characters, in contrast, often commit criminal actions consistently or on a massive scale, such as genocide or the destruction of an entire village of people. Their goal may sound noble or even make environmental sense, such as reducing the human population to the planet's carrying capacity, but the means by which they seek to achieve the goal they envision, is perceived as immoral. Bacigalupi complicates any simplistic binary of good/bad or criminal/lawful in the course of both novels. He does so, in part, because in the realm of environmental destruction, much of what occurs is perfectly legal and sometimes even sanctioned or encouraged by governments and various agencies. Former President Trump's rolling back of environmental regulations made certain reprehensible types of pollution perfectly legal, for instance (Popovich, Albeck-Ripka, and Pierre-Louis). And in Brazil, President Jair Bolsonaro has facilitated the destruction of the rainforest. As Flávia Milhorance has noted: "Indigenous leaders in Brazil and human rights groups are urging the court [International Criminal Court – pdm] to investigate the Brazilian president over his dismantling of environmental policies and violations of indigenous rights, which they say amount to ecocide" (Milhorance).

The federal government has wrestled for well over a century with the problem of trying to determine when irresponsible corporate behaviour becomes illegal ("Corporate Criminal Responsibility"), and ethicists have tried to delineate when irresponsibility at the individual, corporate, and government levels becomes immoral and when "just following orders" fails to provide an excuse for individual actions. The Nuremberg Trials famously had to address that issue after World War II.

Agribusiness and the End of Oil in *The Windup Girl*

As mentioned before, Bacigalupi relies on an internally persuasive rather than authoritative narrative structure and focuses each of the novels through the perspectives of multiple characters on both sides of the plot's conflict (who are also tangentially implicated in it). Through this technique, he foregrounds a complex entanglement of ethics, morality, and criminality whilst juxtaposing crimes ostensibly committed for laudable goals with evil actions justified by positive results. It is particularly striking that in both novels, Bacigalupi initiates the narrative with a chapter focalised through the point of view and internal thoughts of a character on the "bad" side of the crimes committed, even though, or perhaps especially so, readers will initially identify empathetically with these characters.

Thus, *The Windup Girl* opens with the reader inside the head of Anderson Lake, an American corporate operator, who is amazed by a new fruit suddenly appearing in Bangkok's open air market stalls. His amazement in part arises not only from the novelty of the fruit but also its apparent immunity to the global pestilence that has become conjoined with the marketing of American genetically modified and seed patented foods: "the agricultural plagues of the calorie companies" (2). Quickly the reader's attitude toward Lake, after initially sharing the joy of his flavourful fruit discovery, changes when he thinks to himself: "A unique gene that resists a calorie plague or utilizes nitrogen more efficiently sends profits sky-rocketing" (3). The discovery is about capital gain, and shortly in the chapter the reader learns Lake's true purpose for being in Thailand: gaining access to the country's seedbank for the benefit of his U.S.-based transnational corporation, one of the "calorie companies", as they are called throughout the novel. His day job managing a "kink springs factory" to generate energy from toxic algae is revealed as just a cover story for his real mission as a corporate spy (5).

His factory produces a non-functional product and, in the process, uses up scarce energy resources in this carbon depleted world where the computers are supplied juice by treadles like nineteenth-century sewing machines. In this first chapter, Lake is informed of contaminated algae tanks that stop production, a problem that remains relatively submerged in reader awareness until much later in the plot. Like good crime fiction, in the first chapter Bacigalupi is providing breadcrumbs for the reader to anticipate plot developments. The fraudulent nature of Lake's corporate factory front, then, is a second crime introduced in the first chapter. The first is genetic theft for control of food markets, and this one is the toxic wasting of limited energy resources. The first can certainly be considered malevolent when, as one of the other characters observes: "people starve just the same" (6). But what about the phony factory? While certainly irresponsible corporate behaviour it is not actually a crime and so is there anything immoral about it?

The end of the first chapter sets up another conflict in the novel when it becomes clear that there are government forces at work to limit the calorie companies' importation of equipment and potentially contaminating foodstuffs. But before introducing readers to the point of view of those actors, Bacigalupi provides the perspective of Lake's factory assistant, Hock Seng, in Chapter 2, and that of Emiko, the "windup girl", in Chapter 3. Hock Seng's chapter mainly fleshes out the conflict with the Environment Ministry and provides some initial insight to his status as a climate refugee, which Bacigalupi will elaborate throughout the novel. While various people today refer to anthropogenic climate change as an existential crisis, it is a complex problem to define its production as malicious when its agents are so dispersed.

Likewise, the creation of climate refugees too often appears as collateral damage. Only when a particular group attacks another group and drives them out, as happens in the novel to the Chinese, like Hock Seng, who are killed or forced to flee Malaysia, is there an apparent criminal group to bear responsibility. Yet, are they ultimately the responsible party or are they too victims of a global machinery that incessantly places profits above people and turns one group against another as ancillary parties diverting attention from those who have or have had the power to generate a different set of outcomes? Climate refugees are the post facto result in the future of anti-environmental actions being taken today, actions that alter the climate to the extent that one region drowns in excessive rainfall while another region, as Bacigalupi shows in *The Water Knife*, has its crops wither away without the rain that used to fall there.

As readers are introduced to Emiko, Bacigalupi presents her thoughts about herself and her situation, setting up an empathetic proximity as strong as, if not stronger, than one might feel for other characters in the novel. Like Hock Seng, her present-day life consists largely of suffering. While he was once rich in Malaysia, she was once a well-treated servant of her Japanese owner. Both are cast out, discarded if you will, by economic forces beyond their control. The question raised for readers is whether Emiko, as an android, is an abomination – as the religious Grahamites in the novel argue – or if she is the logical result of biotechnological genetic experimentation and should have standing as a "new person"? As with the brute force evolution of crops and animals that are genetically modified for relatively narrow results, such as multi-annual yields or increased milk production, Emiko's design has significant functional flaws that appear to be unintended consequences: her pores are tiny to make her skin more attractive, but as a result she cannot adequately ventilate and cool down her body when exerting herself, for instance (40).

Certainly, her treatment as the object of a sadistic bondage display at a bar is dehumanising, but not a crime since she is a possession rather than a person under the law. But what about her very creation? Should it be a crime to create sentient beings who are denied basic "human" rights? Is it a crime to design inadequately functional components that hinder her ability to move in the same way that the humans after whom she is modelled move? Would animal rights activists apply the same criteria of moral standing to her that they apply when protesting dairy farmers using growth hormones to increase milk production or the penning of calves to produce veal? Stephanie Peebles Tavera, in her focus on the cyborg posthuman aspects of *The Windup Girl*, makes the case that "all matter matters, and nonhuman subjects such as mollusks and bacteria are no less important than humans" in a posthuman ecological system (34).

Most crime novels contain some type of a detective, although not always a police officer or military person. They might be a private investigator, a journalist, or just a concerned citizen who discovers clues to some conspiracy. While Lake is initially introduced as just such a character, he is clearly an anti-hero. Bacigalupi introduces readers to his antitheses, two Environment Ministry officers in Chapter 4, Jaidee and Kanya. They are at a landing field for cargo dirigibles destroying contraband. The senior officer, Jaidee, provides the focal point for these chapters until he is killed by the forces of the Trade Ministry, and then Kanya takes his place. Jaidee likens the continuous crime of smuggling GMO products that will contaminate natural Thai crops to the effects of climate change: "The Thai kingdom is being swallowed. . . . They are being swallowed by the ocean" (47).

While Jaidee and Kanya try to stem this tide of foreign goods, Lake is looking for the man he believes to be the creator of the new fruit that he discovered in Chapter 1, Gibbons. A scientist of dubious ethics, he is a genetic engineer who once worked for one of the calorie companies but now is a protected prisoner of the Thai government. Lake believes that "Gibbons is out there, hiding. Crafting his next triumph. And wherever he hides, a seedbank will be close" (65).

Much of the middle of the novel is taken up with the conflict between the Trade and Environment Ministries. That conflict is then exacerbated by the machinations of the calorie company representatives, who facilitate the Trade Ministries to the point of militarily supporting a coup. Alongside this main plot development run the efforts of Hock Seng and Emiko to survive, defend themselves, and find a way out of their current circumstances. Hock Seng doesn't realise that Lake is not really interested in making functional kink springs, and so he sees his way out to involve the crime of stealing their designs and selling them for enough money to purchase a cargo ship and leave Thailand. Emiko is told by the bar owner who controls her that there are villages of New People in the mountains and that if she ever earns enough to buy her freedom she could flee there. With Anderson Lake becoming smitten with her and moving her to his apartment, she plots her escape.

Meanwhile, as the conflict between the ministries explodes into full blown warfare after the demotion and murder of Jaide, the toxic algae baths at Lake's factory – a mixture of plant material and pig DNA fed fish meal – spawn a deadly and highly contagious virus that first infects some of the factory workers and then spreads through the city, including, ironically enough, killing Anderson Lake. As Heather Sullivan pointedly remarks about this deadly toxin: "Bacigalupi brings the aggressive and unconquerable plant life to the forefront of the novel's challenges" (162). Before Lake dies, though, as part of his efforts to secure his company's alliance with the Thai government, he betrays Emiko, who he claims to be protecting, by turning her over to the sadistic Somdet Chaopraya. In the end he treats her as another means to an end, an item of exchange rather than a sentient being with personhood. But this betrayal becomes as ironic as Hock Seng restarting the toxic algae baths to please Lake because Emiko realises her untapped potential for speed and violence and kills the Queen's protector rather than suffer abuse at his hands. Murder is no crime when it is an act of self-defence, but to view it as self-defence requires readers to grant Emiko legal standing as a "person", not merely a possession. Of course, no one imagines her freedom of choice any more than they realise her ultranatural physical abilities, and the Environment Ministry is blamed instead, resulting in the coup that defeats the forces of the Environment Ministry and places the head of the Trade Ministry in charge of the country.

As Kanya takes over as focaliser in the wake of Jaidee's death, readers learn that she has been accepting bribes from the head of the Trade Ministry to provide insider information about the Environment Ministry. Thus, it is no surprise to readers that she is put in charge of that ministry's police forces, nor that she is the commanding officer assigned to escort the AgriGen calorie company representative to the seedbank that the Environment Ministry has so zealously guarded up to this point in time. All of these actions direct readers toward thinking that she too may be an anti-hero. What may surprise readers, then, is not that the Environment Ministry loses its battle against the forces of global capitalism and transnational biotech companies, but that, at the moment of utmost abjection (the entry of the foreigners into the seedbank), Kanya shoots the AgriGen representative in the head and has her officers kill the entire entourage. Listening to the imaginary voice of Jaidee in her head, Kanya redeems herself.

In the Epilogue, told from Emiko's point of view, the pumps keeping the rising ocean at bay are blown up and the city floods. At the same time, monks enter the seedbank and carry away the seeds to the four corners of the kingdom, decentralising its genetic wealth to protect it from expropriation. Does this represent the destruction of the kingdom or long-term, relatively autonomous, survival for its citizens? Regardless of how strongly readers may cheer this act of resistance to the calorie companies, the Epilogue focuses on Emiko and her meeting up with the genetic engineer Gibbons, who promises that he can use his biotechnological knowledge to engineer children for her. In a gene hacked world, are "natural" *homo sapiens* as doomed as unmodified wheat and corn? Are Gibbons' actions immoral or the logical outcome of human scientific experimentation run amok? If matter self-organises into increasing complexity, then wouldn't synthetic matter with consciousness seek reproductive capability, even when engineered otherwise? One recalls the lessons of *Jurassic Park*: "life finds a way". But even so, like other non-human organisms, Emiko must live in an anthropogenically damaged world.

Climate Change and a Southwestern Megadrought in *The Water Knife*

As with *The Windup Girl, The Water Knife* begins with a chapter told from the point of view of one of its villains, Angel Velasquez, who, like Lake, is an operative working for a corporation. Unlike the treatment of Lake, however, Bacigalupi does allow readers to hear from Angel's boss: Catherine Case of the Southern Nevada Water Authority (SNWA). She is the strategist who directs his illegal, as well as technically extra-legal, mercenary actions. In this opening chapter the environmental stakes of the novel become immediately clear, as Angel is ordered to lead a commando unit to destroy a water processing plant that will render Carver City, Arizona, uninhabitable. Through legal chicanery, the SNWA has a twenty-four-hour window in which to destroy the city's water plant. That is Angel's task, and he carries it out meticulously. Yet he also shows a glimmer of humanity when he saves a supervisor's life by preventing him from running back into the plant that Angel's team is about to blow up.

Although Bacigalupi never uses Case as a focaliser for his chapters, he nevertheless makes excellent use of her to present moral quandaries to his readers. Early in the novel, she has extended conversations with Angel in the chapters in which he is the focaliser. Thus, she is able to present a *realpolitik* perspective on the water rights conflicts and map out a lifeboat ethics view justifying the destructive actions she directs against Arizona for the benefit of Nevada. It is worth noting that a controversial environmentalist, Garrett Hardin, is credited with the development of "lifeboat ethics" (Hardin, "Lifeboat"; Hardin, "Living"). Simply put, the earth as lifeboat cannot hold the entire human population anymore due to its resource consumption beyond planetary carrying capacity. Only two options exist to achieve environmental sustainability: reduce everyone's consumption to a balanced mean or reduce the population below an excess number of consumers. In *The Water Knife*, the latter approach is being taken by one corporate-state entity choking off water supplies to another. This de facto war of attrition generates survival anxiety at every level of society. The poorest characters worry continuously about securing daily water rations. Angel worries whether or not there will be a place for him in one of the arcologies. Even his boss, Case, worries that a misstep on her part could get her tossed from the corporate lifeboat on which she relies.

The practice of lifeboat ethics is on full display in the novel through the depiction of arcologies: in-ground habitats built for those rich enough to purchase or rent. A result of Chinese engineering and multinational capital, they depend for their construction and maintenance on the labour of those not allowed to live inside them. As Maria Isabel Pérez Ramos points out: "These buildings are not envisioned as contributing to diminishing urban environmental degradation; they merely are the shelters of the wealthiest and most powerful in a world gone astray" (54). They thereby not only intensify inequality for the living but also guarantee the dying off of a portion of those left outside who cannot afford the price of water at the pump, which rises with demand until the poorest are priced out in a type of "eco-apartheid" (Ramos 60). Those who cannot pay no longer get a seat in the lifeboat. The foot dragging of the world's wealthiest nations in reducing anthropogenic pollution, primarily CO_2 and methane, while poor nations bear the brunt of rising sea levels and climate change intensified "natural" disasters, demonstrates the lifeboat approach in action today. Likewise, pseudo-solutions, such as recycling or the transition to electrified forms of personal transportation, demonstrate the efforts of those who imagine they are already in the lifeboat and intend to maintain their standard and style of living in the face of the outsized environmental footprint of their daily excessive consumption. The spaceship that will save a select few as the earth is destroyed, or as its habitable environments wither away, is the epitome of lifeboat ethics, as envisioned in numerous novels and movies, such as the recent *Don't Look Up*.

While both novels open, then, with an antagonist's point of view, *The Water Knife*, introduces readers much more quickly to the protagonist. Chapter two introduces Lucy Monroe, a digital journalist in Phoenix documenting the slow death of the city and its environs due to the regional megadrought. As is currently the case, Arizona faces a certain slow death because it has the most junior water rights relative to California, Nevada, and Colorado. Already in May of 2021 news articles are pointing out that the precipitous decline of water levels in Lake Mead and Lake Powell will lead to water rationing in 2022, and Arizona cities will be the first to have their taps tightened ("Lake Mead"; "Lake Powell"; Robbins; Torres; "Water Supply").

Lucy is feeling rather despondent, sensing that her journalism has decayed in quality to an unacceptable level: "Her critics said she was just another collapse pornographer, and on her bad days she agreed" (26). But suddenly, with the gruesome death of her friend Jamie, a water rights lawyer always seeking a big score, Lucy gains a renewed sense of purpose: investigate and solve Jamie's murder. That his murder is related to the control of water seems apparent to readers because Lucy recalls their last conversation, which was focused entirely on water rights, and how even in 1850 John Wesley Powell knew the flow of the Colorado River "wouldn't be enough to cover everything" (30). And Jamie claims he has access to "senior rights you could take right up to the Supreme Court" (33). These water rights will become the golden fleece for which numerous competing interests will fight and die, with Lucy in the middle of the firefight all the way to the end of the novel. Angel will be by her side: an unreliable character who will have to navigate the moral dilemma of determining what constitutes good and evil when it comes to controlling water and do so without an ounce of altruism.

Chapter Three introduces readers to Maria, a Texas refugee, much like Hock Seng, although not as well connected and with fewer resources for escaping her plight, who tries various hustles just to survive. Bacigalupi's characterisation of her also includes some aspects of victimisation similar to the abuse experienced by Emiko. Her focalisation enables readers to get a sense of the desperate existence of millions of people barely surviving at the ragged edges of a collapsing civilisation, the result of environmental irresponsibility well underway in the present day of the novel's readers.

But she also plays a crucial role in the plot. Having accidentally crossed paths with a hydrologist named Ratan as part of a sex party organised by her friend, Maria comes into possession of his copy of *The Cadillac Desert*, which has stashed within its pages the water rights Jamie was discussing with Lucy. She doesn't know they are hidden there and so is unable until the novel's end to use them to her advantage even though she carries the book around with her, not quite knowing why she feels the need to keep it. With Angel sent by Case to ferret out the location of these water rights, and Lucy seeking to uncover the reason for Jamie's murder, their paths quickly cross. They meet up at the city morgue, each trying to find out about Jamie and both recognise the other as someone more than he or she might claim to be and someone who knows something each needs to find out.

As Angel seeks out information on Lucy's journalistic interests, he thinks about Phoenix, so that readers can understand why Bacigalupi has chosen this city for his setting and how the author views the future of the United States reeling from the effects of anthropogenic climate change:

> The only difference between Phoenix and a dozen dying cities in Texas and Alabama and every coastal city around the world was that Phoenix had taken hits not just from climate change and dust storms and fires and droughts but also from a competing city.
>
> *(140)*

These lines are prescient in light of the 2021 drying up of Lake Mead. No doubt Nevada and California will take action to avoid sharing any of the water to which they have senior rights, as indicated by a 1968 court case over seniority ("Water Supply").

When Angel comes calling on Lucy to find out what she knows, he spies a copy of *Cadillac Desert* on her bookshelf and remarks that his boss makes all her employees read it: "She likes us to see this mess isn't an accident. We were headed straight to Hell, and didn't do anything about it" (160). In other words, in Catherine Case's mind, and Angel's, the real crime was committed long ago, back even before the reader's present day, when different decisions could have been made about seeking forms of sustainable development and population control for the permanently *arid* southwestern United States. As a result, whatever they do to secure their lifeboat-style survival is justifiable in the face of climate change driven resource scarcity.

The plot thickens as competing interests step up their violent and criminal efforts to gain possession of the water rights Maria does not even realise she is carrying around. Lucy is kidnapped by Julio, another Nevada water knife who has decided to freelance. It later turns out that he and his partner are the ones who killed Jamie and Ratan but have come up empty handed and hope they can torture useful information out of Lucy. Information she does not have. Eventually, Angel frees her. At this point, Lucy reveals more than she had previously told him: the water rights are from the Pima Tribe and refer to water from the Colorado River, dating back further than any other rights (232–33). They could save the city, but, of course, in the process would choke off water to Vegas and LA Thus, recovering them does not address the water crisis itself but only changes who gets to ride in the lifeboat.

Although Angel and Lucy keep helping each other, the fundamental conflict that intensifies two-thirds of the way through the novel is the tension between her desire to help Phoenix and Angel's contractual obligations to Nevada. Through Lucy's focalisation, readers are given the impression that Angel has a moral centre that could trump his mercenary function. As he states at one point: "Nobody survives on their own" (329). But that slogan doesn't translate into everyone surviving.

Because of Julio's actions, Catherine Case suspects Angel of double crossing her, so when he calls in his location to request extraction, he knows enough not to be there and watches from nearby as a missile annihilates the safe house. Bloody and beaten, and requiring Lucy's help to walk, it appears that he no longer has an obligation to Case or Nevada, and thus could turn and assist Lucy in her altruistic effort to save Phoenix. But, if reading *The Windup Girl* should have taught readers of Bacigalupi's fiction anything, it is that this author does not go for simple resolutions.

Once Angel and Lucy realise that the water rights are paper documents, the kind that might be stored between the pages of *The Cadillac Desert*, they set off in pursuit of Maria, who, it turns out, is on her way to cross the Colorado River illegally into Nevada. Lucy and Angel catch up with her before she has attempted the crossing. While chapters 43 through 46 of *The Water Knife* alternate between Lucy's and Angel's points of view, the final chapter is told from Maria's perspective. These chapters open with Angel and Lucy trading recriminations about Maria, the one person who has no idea of the value of the papers tucked in a book in her knapsack. Bacigalupi has Lucy realise about the water war determining the upward or downward fates of millions of people, that "*They have no idea what they are doing. These are the people who are supposed to be pulling all the strings, and they're making it up as they go along*" (344; italics in original). This crucial statement reinforces Bacigalupi's recognition of the distributed nature of the crimes and immoral actions at the root of our anthropogenic climate crisis. As Bacigalupi tells Adam-Troy Castro in an interview quoted by Scott Selisker, "Our crises are all distributed ones, where everyone shares a little blame but can't credit that our small actions add up to a devastating whole" (fn3, 515). It explains why he refuses to provide a reductionist plot structure whereby there is some single evil individual or organisation whose conspiracy need only be thwarted to save the day. It is in effect an anti-cathartic statement that sets readers up for the conclusion of the novel. Bacigalupi's writing thereby works at cross purposes to the typical detective novel, wherein some immediate problem or crisis is solved, and a single perpetrator or organisation is destroyed or brought to justice. For example, in Clive Cussler's *Trojan Odyssey*, Dirk Pitt captures the villain and foils a Chinese plot to disrupt the climate of North America and Europe through slowing the Atlantic Multi-decadal Oscillation Current (AMOC). Near the novel's end, the narrator reassures readers that the current will continue to flow as it always has. Cussler thus discourages readers from being concerned about climate change during the decade when climatologists were raising concerns that AMOC was being altered, concerns recently raised anew (Broecker; Temple).

As they search for Maria, Angel remarks that Phoenix cannot be saved despite Lucy's desires and that his only chance for survival rests with turning the water rights over to Case, securing a return to her good graces and a seat once again in the lifeboat (345). Lucy, meanwhile, is running on a deep sense of guilt: "I think the world is big, and we broke it. . . . we saw what was coming and didn't do anything about it" (348). That "we" clearly includes the readers of the novel.

In the penultimate chapter, with Angel as the focaliser, Lucy attempts to take the papers at gunpoint to save Phoenix, despite knowing that it will cost Angel his life. Rather than shoot her, Angel lets her go, wishing she would save him and not Phoenix, "but he knew she wouldn't" (367). Does Angel display his moral compass out of a sense of compassion for Phoenix or, more likely, out of affection for Lucy? Readers do not get to find out. On the last page of his chapter, "a gunshot echoe[s] flat across the river" (368).

The final chapter has to be given to Maria as the novel's focaliser because she shoots Lucy. She wants Angel's deal to get her to Las Vegas and into a condo in a sustainable arcology. Maria seeks only immediate survival. Cynically, she remarks about Lucy, "She thinks the world is supposed to be one way, but it's not" (371). And then Maria makes the claim that all of them are in the same boat, by which she means the three of them need to get to Vegas to survive. But readers can easily glean a larger message that Maria's *realpolitik* perspective is precisely the kind of short term, immediate results mentality exacerbating anthropogenic climate change in the present day of readers. Fighting over a seat in the lifeboat does not change the fact that the boat is sinking regardless of what seat a person occupies.

Morality and Legality When Science Fiction Is Also Crime Fiction

Immoral and criminal anti-environmental actions of the future depicted in both novels are the result of predominantly legal environmentally destructive and irresponsibly short-sighted actions well underway today. As Crownshaw puts it when discussing *The Water Knife* as "future-oriented historical fiction" in an essay on "speculative memory": "This is what *happened* when we didn't do what we needed to do earlier about mitigating *and* adapting to climate change for society as a whole, but only for the rich and powerful" (889). And as Scott Selisker,.writing about *The Windup Girl* notes, "It is the world organization of our time in general . . . that is 'responsible' for the future crisis" (502). For the novels to depict more positive resolutions, Bacigalupi clearly suggests, especially in *The Water Knife* with the use of Reisner's book and in *The Windup Girl* with the Epilogue focused on Emiko, very different behaviour toward genetic experimentation, agricultural technology, urban planning, resource consumption, and anthropogenic causes of climate change would have to have already begun in earnest on the part of the societies and economies of the people reading his novels. What we are doing today environmentally in regard specifically to food production and freshwater consumption ought to be understood as criminal even when legal. While everyone contributing to the intensification of these avoidable problems are guilty to varying degrees of complicity and wilful ignorance, responsibility is difficult to pinpoint because of the global distributed network of our participation.

Yet, readers need to understand that there are villains, such as Lake, Case, and Gibbons, and there are people running governments and corporations perpetrating evil actions that need to be redefined and legislated as criminal to bring them to justice and end the destructive practices they promote and enable, even when they reside just outside the scope of the narrative. Bacigalupi's two science fiction novels are also crime fiction, cautionary tales, and contemporary environmental morality plays all in one because the three have become inseparable in daily life. As Bacigalupi remarked in 2013: "It does sort of feel like the decisions we make (or fail to make) right now are going to define everything for our descendants. I suspect they won't thank us" (Millikin 53).

That may be so, but does any kind of fiction, whether a detective or a science fiction novel or some combination of the two, have any impact on those decisions or lack of them? Another award-winning science fiction writer, Kim Stanley Robinson, would like to believe so. According to Joshua Rothman, who interviewed him for his essay in *The New Yorker*, "Can Science Fiction Wake Us Up to Our Climate Reality?" Robinson argues that literature can contribute to an alteration in the cultural structure of feeling that will then lead to people making different choices and taking different actions than they would without that emotional alteration. Certainly, detective novels that wrap everything up in a tidy little package at the end, no matter the difficulties the hero may have faced in the course of the unfolding of its plot, will reinforce cultural inertia rather than challenge it. Likewise, science fiction novels that are basically thinly developed plots on which technological spectacles are hung will also encourage inertia by promoting the belief that the scientific breakthrough just around the corner will provide a global cornucopia and nothing needs to change right now. These are not the stories that Bacigalup writes. Instead, he encourages both an unease in readers that things will not work out spontaneously if anthropogenic alterations of the climate are not addressed and warns that the current course of excessive consumption will not rescue anyone. The novels have the potential to influence readers as part of a larger cultural movement to realise what Barbara Kingsolver's protagonist in *Flight Behavior* comes to recognise: "she saw the pointlessness of clinging to that life raft, that hooray-we-are-saved conviction of having already come through the stupid parts. . . . There is no life raft; you're just freaking swimming all the time" (394).

Bibliography

Bacigalupi, Paolo. *The Windup Girl*. Nightshade Books, 2010.

———. *The Water Knife*. Alfred A. Knopf, 2015.

Baum, Rudy. "Climate Change and Drought in the American Southwest". *Climate Institute*, Oct. 2015, climate.org/effects-of-21st-century-climate-change-on-drought-disk-in-the-american-southwest/.

Broecker, Wally. *The Great Ocean Conveyor: Discovering the Trigger for Abrupt Climate Change*. Princeton UP, 2010.

CBS News. "'Mega-Drought' Depletes System That Provides Water to 40 Million". *WTOP News*, 7 June 2021, wtop.com/weather-news/2021/06/mega-drought-depletes-system-that-provides-water-to-40-million/.

"Corporate Criminal Responsibility". *Encyclopedia.com*, 18 Mar. 2023, www.encyclopedia.com/law/legal-and-political-magazines/corporate-criminal-responsibility.

Crownshaw, Richard. "Speculative Memory, the Planetary and Genre Fiction". *Planetary Memory in Contemporary American Fiction*, 2019, pp. 35–58, doi:10.4324/9781351026185-3.

Cussler, Clive. *The Trojan Odyssey*. G. P. Putnam's Sons, 2003.

deBuys, William. *A Great Aridness: Climate Change and the Future of the American Southwest*. Oxford UP, 2001.

Don't Look Up. Dir. Adam McKay, Paramount/Netflix, 2021.

Gilding, Paul. *The Great Disruption: Why the Climate Crisis Will Bring on the End of Shopping and the Birth of a New World*. Bloomsbury Press, 2011.

The Globalist. "Thailand: Southeast Asia's Buffer Country". *Thailand: Southeast Asia's Buffer Country How Thailand Avoided European Colonization*, The Globalist, 16 Aug. 2022, www.theglobalist.com/thailand-southeast-asias-buffer-country/.

Hardin, Garrett. "Lifeboat Ethics: the Case Against Helping the Poor". *Psychology Today*, Sept. 1974, rptd. by The Garrett Hardin Society, www.garretthardinsociety.org/articles/art_lifeboat_ethics_case_against_helping_poor.html.

———. "Living on a Lifeboat". *BioScience*, Oct. 1974, rptd. in *The Social Contract*, Fall 2001, pp. 36–47, www.garretthardinsociety.org/articles_pdf/living_on_a_lifeboat.pdf. Accessed 5 May 2021.

Jurassic Park. Dir. Steven Spielberg, Universal Studios, 1993.

Kesslen, Ben. "Drought Is Here to Stay in the Western U.S. How Will States Adapt?" *NBCNews.com*, NBCUniversal News Group, 11 June 2021, www.nbcnews.com/news/weather/drought-here-stay-western-u-s-how-will-states-adapt-n1270248.

Kinchy, Abby. *Seeds, Science, and Struggle: The Global Politics of Transgenic Crops*. MIT Press, 2012.
Kingsolver, Barbara. *Flight Behavior*. Harper Collins, 2012.
"Lake Mead Water Level". *LakesOnline.com*, http://mead.uslakes.info/level.asp. Accessed 14 June 2021.
"Lake Powell Water Level". *LakesOnline.com*, http://powell.uslakes.info/Level/. Accessed 14 June 2021.
Milhorance, Flávia. "Jair Bolsonaro Could Face Charges in the Hague Over Amazon Rainforest". *The Guardian*, Guardian News and Media, 23 Jan. 2021, www.theguardian.com/world/2021/jan/23/jair-bolsonaro-could-face-charges-in-the-hague-over-amazon-rainforest.
Millikin, Patrick. "Running Out of Water in the West: *PW* Talks with Paolo Bacigalupi". *Publishers Weekly*, vol. 262, no. 13, 30 Mar. 2015, p. 53.
Murphy, Patrick D. *Persuasive Aesthetic Ecocritical Praxis: Climate Change, Subsistence, and Questionable Futures*. Lexington Books, 2015.
———. "Hydrocarbon Enslavement and Fantasies of Freedom". *Transportation and the Culture of Climate Change: Accelerating Ride to Global Crisis*, edited by Tatian Prorokova-Conrad, West Virginia UP, 2020, pp. 103–122.
Palumbi, Stephen. *The Evolution Explosion: How Humans Cause Rapid Evolutionary Change*. W.W. Norton, 2001.
"Paolo Bacigalupi, Multi-Award-Winning Author". *Windupstories.com*, 29 Dec. 2020, windupstories.com/author-info/paolo-bacigalupi/.
Poe, Edgar Allan. *The Purloined Letter*, poestories.com, https://poestories.com/read/purloined. Accessed 3 June 2021.
Popovich, Nadja, et al. "The Trump Administration Rolled Back More than 100 Environmental Rules. Here's the Full List". *The New York Times*, 16 Oct. 2020, www.nytimes.com/interactive/2020/climate/trump-environment-rollbacks-list.html?mtrref=undefined&gwh=05BD9F26DDF727B195B44E2169FA8BD7&gwt=regi&assetType=REGIWALL.
Ramos, Maria Isabel Pérez. "The Water Apocalypse: Utopian Desert Venice Cities and Arcologies in Southwestern Dystopian Fiction". *Ecozon@*, vol. 7, no. 2, 2016, pp. 44–64.
Resiner, Marc. *Cadillac Desert: The American West and Its Disappearing Water*. 1986 rev. ed. Penguin Books, 1993.
Robbins, Jim. "In Era of Drought, Phoenix Prepares for a Future Without Colorado River Watee". *Yale Environment 360*, 7 Feb. 2020, https://e360.yale.edu/features/how-phoenix-is-preparing-for-a-future-without-colorado-river-water,
Robinson, Kim Stanley. *2312*. Orbit, 2012.
———. *New York 2140*. Orbit, 2017.
Rothman, Joshua. "Can Science Fiction Wake us up to Our Climate Reality?" *The New Yorker*, 24 Jan. 2022, www.newyorker.com/magazine/2022/01/31/can-science-fiction-wake-us-up-to-our-climate-reality-kim-stanley-robinson.
Selisker, Scott. "'Stutter-Stop Flash-Bulb Strange': Gmos and the Aesthetics of Scale in Paolo Bacigalupi's *the Windup Girl*'". *Science Fiction Studies*, vol. 42, no. 3, 2015, p. 500, doi:10.5621/sciefictstud.42.3.0500.
Shiva, Vandana. *Stolen Harvest: The Hijacking of the Global Food Supply*. U Kentucky P, 2016.
Steinbeck, John. *The Grapes of Wrath*. Penguin, 2006.
Sullivan, Heather I. "Petro-Texts, Plants, and People in the Anthropocene: The Dark Green". *Green Letters*, vol. 23, no. 2, 2019, pp. 152–167, doi:10.1080/14688417.2019.1650663.
Tavera, Stephanie Peebles. "Utopia, Inc.: A Manifesto for the Cyborg Corporation". *Science Fiction Studies*, vol. 44, no. 1, 2017, p. 21, doi:10.5621/sciefictstud.44.1.0021.
Temple, James. "The Atlantic's Vital Currents Could Collapse. Scientists Are Racing to Understand the Dangers". *MIT Technology Review*, 29 July 2022, www.technologyreview.com/2021/12/14/1041321/climate-change-ocean-atlantic-circulation/#:~:text=It's%20very%20likely%20that%20the,in%20more%20than%201%2C000%20years.
Torres, Jorge. "Arizona's Water Supply Could Be Impacted by Continued Drought Conditions". *ABC15 Arizona in Phoenix (KNXV)*, ABC15 Arizona in Phoenix (KNXV), 26 Apr. 2021, www.abc15.com/weather/impact-earth/arizonas-water-supply-could-be-impacted-by-continued-drought-conditions.
"Water Supply and Water Rights". *Central Arizona Project*, www.cap-az.com/departments/planning/colorado-river-programs/water-supply-and-water-rights.
Woodhouse, Connie A., et al. "A 1,200-Year Perspective of 21st Century Drought in Southwestern North America". *Proceedings of the National Academy of Sciences*, vol. 107, no. 50, 2010, pp. 21283–21288, doi:10.1073/pnas.0911197107.
Yazell, Bryan. "A Sociology of Failure: Migration and Narrative Method in US Climate Fiction". *Configurations*, vol. 28, no. 2, 2020, pp. 155–180, doi:10.1353/con.2020.0009.

19

FROM CRIME SCENE TO ANTHROPOCENE IN 2010S ARGENTINIAN NARRATIVE

David Conlon

In a 2014 essay entitled "Floating Bodies", Adrian Lahoud proposes "that structural violence is criminal", from which it follows that "the genesis of structure is a criminal event, the space of genesis its crime scene" (496). However, while crime narratives have sometimes found inspiration in ecological crime scenes, the genre is still only in the early stages of reckoning with the temporal, causal, and epistemological problems posed by the criminal codification of structural phenomena such as climate change, the destruction of ecosystems, ocean acidification, and toxic drift. These 'events' are conduits for what Rob Nixon calls 'slow violence', "a violence that is neither spectacular nor instantaneous, but rather incremental and accretive, its calamitous repercussions playing out across a range of temporal scales" (2). Given what Nixon identifies as the "quandary of how to convert into dramatic form" threats that unfold in slow motion across decades (209–10), it is perhaps unsurprising that ecological detective stories have historically tended not to focus on slow violence, and instead usually fall into the category of what Patrick D. Murphy calls "nature-oriented literature", which, while directing "reader attention toward the natural world and human interaction with other aspects of nature within that world" (90), ultimately subsumes nature within the mandates of a more delimited, anthropocentric project. As Samantha Walton avers, the genre's traditional predicate of "corpse and crime scene" means that critiques of societal injustices "still tend to be connected to instances of individual moral failing" (124).

With this chapter, I examine three recent Argentinian ecological crime narratives that precisely reverse this process of subsumption. Instead, they allow the world-altering magnitude of ecological crime to operate as a vortex on the internal logic of the plot, reconfiguring the epistemological coordinates of detection and undermining some of the core precepts of the genre. As a result, these narratives expose the problems inherent in the task of delineating the ecological crime scene in the aftermath of the realisation that existing paradigms of culpability, causality, and temporality have become obsolete. In different ways, this will lead each of these texts to stage a transition from crime scene to radically more expansive '-cenes': chief among them the Anthropocene, or the era defined by the detectable impact of human industrial activity on the Earth's systems at a macro level (Glaser et al. 6), but also the Capitalocene, which underlines the degree to which capitalism has become integrated into the "web of life" (Moore 10), and (perhaps particularly relevant for processes of detection and unveiling) the Agnotocene, Christophe Bonneuil and Jean-Baptiste Fressoz's coinage to denote the "production of zones of ignorance" whereby discrete sites of disciplinary knowledge remain unintegrated, facilitating (for example) the "illusion" of an economy that is dis-embedded from natural constraints (198; 211).

DOI: 10.4324/9781003091912-23

The texts I examine were all first published throughout the middle years of the 2010s, as material evidence of ecological collapse became increasingly difficult to ignore, and I argue that they each provide original modes of narrativising ecological catastrophe through the prism of crime and detection. First, I look at Ricardo Piglia's *El camino de Ida* [*The Way Out*] (2013), which centres on an investigation into a series of bomb attacks (loosely based on the crimes of Ted Kaczynski). As I will show, although the novel is plot-driven in some regards, the story gradually shades into a meditation on ecological paranoia which ultimately serves not only to problematise Kaczynski's culpability, but also to evoke the inherently diffuse nature of environmental guilt in the Anthropocene. The second text I examine, Samanta Schweblin's *Distancia de rescate* [*Fever Dream*] (2014), is also loosely inspired by true events, namely the devastating health impacts of unsafe pesticides deployed by Argentina's soy industry. The novella aligns with Todorov's structuralist description of the detective story as consisting of two stories: the crime and the investigation (139). However, the 'investigation' (which takes the form of an extended dialogue between two victims) focuses almost exclusively on incidental factors, repeatedly making reference to 'the important thing' while simultaneously seeming to studiously skirt around the question of culpability. My suggestion is that we understand these protagonists as operating under the assumption of a rogue causality, which in turn can be interpreted as a symptom of the epistemological morass occasioned by the attempt to investigate what Rachel Carson calls the "death-by-indirection" that is the consequence of pollution (32). Finally, I look at the short story "Bajo el agua Negra" ["Under the Black Water"] from Mariana Enríquez's collection *Las cosas que perdimos en el fuego* [*Things We Lost in the Fire*] (2016). Here, a tale of hard-boiled investigation into police brutality towards residents of a Buenos Aires shanty town morphs into a Lovecraftian account of ecological and societal apocalypse, emanating from the polluted river that abuts the neighbourhood.

While each of these narratives centres on specific events and locations, I contend that each also entails an epistemological probing of existing paradigms of rationcination, troubling and reconfiguring the conventions of the detective genre in the process. As I suggest, the denouement of the latter story posits a radical reframing of environmental narratives that will emanate in the Global South – not by the state or the establishment, but from a space of precarity and on terms mandated by the disenfranchised who stand to lose most from ecological breakdown. More generally, this concern with reframing, reclassifying, and reconfiguring existing paradigms is the common thread that runs through these narratives, not only in terms of how they address environmental crime, but in terms of how they reflexively comment on the writing of crime fiction itself and on the deeper problem of how to frame ecological catastrophe.

The Way Out: Diffuse Culpability

Ricardo Piglia's *The Way Out* is set in the 1990s and is narrated from the perspective of a writer and academic named Emilo Renzi (Piglia's recurring alter-ego) who has been invited to teach literature for a semester at an exclusive New Jersey university. At the core of the story is Ida Brown, a high-flying literary scholar who initially mentors Renzi as he adjusts to life on the campus (loosely modelled on Princeton, where Piglia himself taught for many years). They soon embark on a romantic affair, though the infatuated Renzi is dismayed to learn that Ida prefers to keep the relationship a secret from their colleagues. Ida is then killed in a mysterious car accident, which also inexplicably leaves her with a badly burnt hand. The FBI conducts an investigation into her background, as they believe that her death may be in some way connected to a series of homemade bomb attacks targeted at academics. Renzi himself falls under suspicion, and he hires a private detective to conduct a counter-investigation into the FBI's operations.

The Way Out is a fictionalised retelling of the true story of Ted Kaczynski, aka the Unabomber. Kaczynski had been a high-achieving Harvard undergraduate before dropping out and going to live in the Montana wilderness in the late 1970s. There, in response to what he came to view as the destruction of the natural world by modern industrial society, he initiated a decades-long campaign of mail bombing targeted mostly at airline employees and academics working in the sciences. As in the real-life Kaczynski case, in Piglia's retelling the bomber is eventually apprehended by the FBI when the *New York Times* agrees to publish an article he has written in exchange for the cessation of the bomb attacks. The article itself amounts to an anarchist manifesto calling for a return to pre-capitalist and pre-industrial modes of existence; the author is identified as Thomas Munk when a family member recognises his style of writing. Yet Munk's involvement in Ida's death is never definitively proven. Still mourning her loss, Renzi arranges to meet with Munk in prison, and while he does not endorse Munk's actions, he does appear to become curious (if not sympathetic) towards his underlying radical eco-politics.

The Way Out fulfils many of the criteria associated with the crime genre: it features murder victims, an investigation, and the revelation of a lone killer. However, the novel's continuation for several chapters beyond the unmasking of the culprit, its insistence that Renzi and hence the reader reflect on the validity of Munk's claims, and its withholding of a clear resolution concerning Ida's death, all contrive to trouble the work's generic coordinates. In the process, the novel realises a proposition contained at the end of Piglia's preceding novel, *Blanco nocturno* [*Target in the Night*] (2010), in which Renzi suggests that "[s]omeone should invent a new detective genre, paranoid fiction it could be called. Everyone is a suspect, everyone feels pursued" (252).[1] That Munk's criminality is founded on his (albeit violent) incrimination of the industrialised world at large inevitably means that the faltering moral case against Munk does the work of apportioning guilt to everyone else.

The coordinates of this vague civilisational crime scene are at once everywhere and nowhere throughout the novel. The elite university campus, the mall, Renzi's temporary home in an affluent suburban neighbourhood, the temperature-controlled hotel rooms where Renzi and Ida have their clandestine meetings, and the "unceasing caravan of cars" that congest the motorways between these locations, all serve as metonyms, both for the credit-fuelled post-war "disciplinary hedonism" of the consumerist "Phagocene" (Bonneuil and Fressoz 157) and also for what the same authors call the "Thermocene", which underlines the role played by politics and institutions in the promotion of carbon dioxide emissions, and therefore in ushering in the impending climate cataclysm. As Bonneuil and Fressoz argue, these designations help to "denaturalize" histories of energy, production, and consumption and therefore to articulate the thinkability of the kind of crime scene that Lahoud proposes (107).

Once this crime scene has been posited by Munk, readerly attention is drawn to seemingly uneventful though conspicuously carbon-fuelled sequences such as when, for example, Renzi takes a flight from Newark to San Francisco and then leaves a hotel TV on all night. His suggestion that "tragic events can cause any detail to become significant" (177) speaks to his own increasing paranoia, but this is redoubled in the interpretive gaze of the reader, for whom Renzi himself becomes a cipher for civilisational culpability, while the milieu he occupies – according to the terms proposed by Lahoud – becomes reconfigured in a criminological register. The gradual unveiling of this crime scene is abetted by disquieting encounters with animals (notably, a wild bear that has wandered from the mountain down to the suburbs, and a shark that the faculty dean keeps in his basement aquarium), constituents of a non-human 'outside' about whom, as Renzi suggests, "[w]e no longer know how to talk . . . unless they're domesticated" (37). In the context of Renzi's gathering culpablity, these uncanny encounters operate as accusatory, and Renzi himself occasionally feels 'seen' by animal others: during his first clandestine meeting with Ida, he experiences the hallucinatory sensation of embodying a monkey pressed against the ventilation window, observing himself from above. He later rescues and adopts a cat, which first appears to become domesticated, but then reverts to the wild, escaping and attacking Renzi when approached. Later again, Renzi encounters a graduate student named Nancy Cullers who is writing a thesis on Alfred Hitchcock's *The Birds*, whose message she interprets as analogous to Munk's: "an example of eco-terrorism . . . nature can rise up at any moment and the world will become an inferno" (199).

The threatening non-human outside also materialises in the extreme weather events that occur throughout the novel: a storm (39), a snowstorm (59), an allusion to "the great dust storm of 1851" (115), and later "one of those brief and violent summer storms" (225). While Ida's death is never directly pinned on Munk, it is linked to a storm by way of a traffic alert that had caused her to deviate from her usual route, in turn precipitating the accident. According to Stephen Kerns, murder "compared with other acts . . . is exceptionally vivid and important and in most cases sharply focused in time and space"; "it is strongly intentional, highly motivated, full of meaning", and "an act that remains relatively consistent over time" (2). If Ida's undecidable accident/murder stands in as another cipher and a foreshadowing of the "inferno" predicted by Nancy Cullers – Renzi notes that the word *ida* in Spanish signifies "the way out, the journey of no return" (48) – then it is also a harbinger for a category of mass death that has been seeped of definition, motivation, and meaning. Under the auspices of what Bruno Latour labels the "new climatic regime", causality and therefore culpability become newly contested topics. As causality is made more diffuse and ambiguous, the power of the individual clue is diminished, but the potentiality of every item to be a clue is enhanced. For Latour, civilisational responsibility means not only that "[n]o one in isolation is responsible", but also that "everything that was symbolic is now to be taken literally" ("Waiting for Gaia" 4;30). Whereas Renzi and Ida arriving at their clandestine meeting in separate cars and adjusting the thermostat in their hotel room to control the temperature might previously have served as a *metaphor* signifying dominion over nature, the same carbon-reliant actions are now figured as *metonymically* contiguous with the Thermocene at large. Lahoud, for his part, hypothesises a violence without event, enjoining his readers to "conceive of a world become weaponized . . . in which the very facts of everyday life, the basic habits and simple routines, by the sheer cumulative weight of their repetition, begin to consume, overwhelm, and finally destroy their host". In this context, we are dealing with "a violence with no weapon and no witness, passing invisibly through the human sensorium – leaving no memory, no testimony, and no record" (496).

At the penitentiary to visit Munk, Renzi finds himself framed by floor and wall markings, metallic wire mesh, and blinding spotlights, all of which serve to suggest the idea of Renzi as the bearer (or one of the bearers) of this undecidable culpability. Clearly, under existing paradigms, there can be no organisational or institutional reckoning with the uncanny guilt incurred. It seems important to note, however, that *The Way Out* does not outflank or render obsolete the crime genre as such, but instead mobilises some of its latent tendencies. In blurring the boundary between pursuer and pursued, Piglia's text might even be said to hearken back to Edgar Allan Poe's "The Man of the Crowd" (1840), a foundational text for paranoid detective narratives without clearly defined crime scenes. Insofar as the novel renders late twentieth century suburbia as an Anthropocene mega-text that is literally permeated with clues, it also amplifies the genre's mania for signification, succinctly identified by Chesterton's 1902 "A Defence of Detective Stories", wherein "every fantastic skyline of chimney-pots seems wildly and derisively signalling the meaning of the mystery" and where "there is no stone in the street and no brick in the wall that is not actually a deliberate symbol" (75).

Fever Dream: The Important Thing Already Happened

Samanta Schweblin's novella, *Distancia de rescate*, was published to critical acclaim in 2014, with Megan McDowell's translation (as *Fever Dream*) appearing in 2017. The story takes place in Argentina's rural interior and against the backdrop of the environmental and health damage wrought by country's poorly regulated soy industry. It opens *in medias res*:

> *They're like worms*. What kind of worms? *Like worms, all over*. It's the boy whose talking, murmuring into my ear. I am the one asking questions. Worms in the body? *Yes, in the body*.

> Earthworms? *No, another type of worms.* It's dark and I can't see. The sheets are rough, they bunch up under my body. I can't move, but I'm talking. *It's the worms. You have to be patient and wait. And while we wait, we have to find the exact moment when the worms come into being.* Why? *Because it's important, it's very important for us all.*
>
> (1–2)

It becomes apparent that the words in italicised type are spoken by a young boy named David, that his interlocutor (the narrator) is a woman called Amanda, and that they are in a ward in a rural hospital. Amanda has recently become gravely unwell; David has evidently been unwell for a much longer period and is now trying to prompt Amanda to reconstruct the events leading up to her illness. Through a combination of dialogue and Amanda's interior recollections, this is slowly pieced together: a short time beforehand, Amanda had come to the countryside for a stay in a holiday home along with her daughter, Nina. Amanda befriends a local woman named Carla, David's mother. Over the course of their conversations, Carla reveals that David was poisoned while playing near a stream a few years previously and has been unwell ever since. Amanda starts to intuit that her daughter is not safe and decides that they must return to Buenos Aires. However, before they can leave, Amanda and Nina are both poisoned, apparently by the same toxic substance that had previously poisoned David. Amanda is now spiralling towards death, while her daughter, it appears, has made a recovery and is brought back to the city by her husband.

Schweblin describes genre as "above all a series of limits (in the best sense) amounting to a set of rules and possibilities that are imposed on a particular text, but which in turn allow for one-off original innovations" (Dixon & Schweblin). Although not marketed as a detective story, the novella adopts the structure of the whodunit as outlined by Tzvetan Todorov: it consists of two superimposed temporal series, in which the story of the crime (temporally anterior to the narrative) is relayed through the story of its reconstruction (139), with David and Amanda fulfilling the detective function, and the poisoning with pesticides of the story's protagonists serving as the scene of the crime. However, the epistemological limitations of this project, as well as the resultant emphases that are brought to bear on the case by David and Amanda (who never once allude to the possibility that a criminal event might have occurred), serve to trouble and distort this generic framework. John Scaggs underlines "the central significance of the chain of causation in mystery and detective fiction", in which the solution depends on "the narrative importance of cause and effect" (34). As Dennis Porter argues, the genre inaugurated by Edgar Allan Poe posits "a highly visible question mark" at the outset (86), which, as Scaggs explains, draws the detective and reader "to retrace the causative steps from effects back to causes" (34). The problem, as O'Rawe shows, is that *Fever Dream* is structured around an absent or unknowable cause. While Schweblin's novella appears to confront the causal and temporal specifics of how and when the events leading up to Amanda's sickness and eventual death came about, the 'investigation' serves only to raise the *notional* possibility of marking out an ecological crime scene, while almost simultaneously foreclosing that possibility as being well beyond the realm of epistemological organisation.

The germ of this outcome is already seeded in the opening exchange cited earlier, in which David identifies the 'exact moment' when Amanda began to feel that her skin was beginning to crawl as the 'important' thing. He later says to Amanda: "[w]e're looking for worms, something very much like worms, and the exact moment when they touch your body for the first time" (43). This fetishisation of the spatial and temporal coordinates where weapon and victim coincide can be read as a vestige of the ratiocinative detective story but, as David later appears to concede, his mission is complicated in the context of poisoning caused by pesticides. Amanda's reconstruction of events gets "closer and closer to the exact moment" (55); we finally appear to be 'in' the moment of the important thing when Amanda recounts seeing a plastic drum which has been unloaded from a truck and left in a shed doorway. As

Amanda recalls, Nina sits in the grass and her clothes begin to feel damp, which Amanda assumes to be caused by dew. This seemingly innocuous occurrence begins to recede from view, and David, now increasingly resigned, asserts that "the important thing *already* happened" and that "[w]hat follows are only consequences" (107; emphasis mine). The way the 'event' of contamination itself is circled around and eventually approached supplies only the illusion of its overall legibility. Ultimately, the potential crime scene is denied with statements like "[t]he poison was always there" (139); indeed, as David elsewhere explains, many of the local children "were born already poisoned, from something their mothers breathed in the air, or ate or touched" (124).

The novel's original title in Spanish translates as 'rescue distance', alluding to interrelated questions of ratiocinative calculability and effective agency: rescue distance, as Amanda explains, is her way of conceptualising the relative danger that her daughter is in at any given moment and is calculated as a function of the proximity of the danger and the length of time it would take Amanda to intervene. Hence, Amanda visualises a 'thread' or cord that connects her to her daughter, and which at various points in the story becomes slackened, taut, extended, or in danger of unravelling altogether. Two significant moments serve to pressurise and then implode the viability of these protocols of calculation and intervention. First, David reveals that Amanda's husband has recently been to the area to retrieve their daughter Nina. He visits the home where Carla and David live (along with David's father Omar), where he finds that many of the items inside the house are connected with reams of sisal rope, including photographs of people and animals, but also seemingly random household objects. Carla has already told Amanda of David's increasingly strange behaviour after having been poisoned, and so a plausible inference is that David himself has articulated these rhizomatic connections between items in his home, which, when placed in contrast with the more clearly legible and actionable 'threads' that Amanda visualises as connecting her to her daughter, is suggestive of a far more demanding and complex attempt to cognitively map the ecological crime scene. While David's effort eschews an ethics of proximity in its attempt to more comprehensively map the rogue causalities at play, it can also be taken to constitute a mirror image of David's troubled mind and, more broadly, as a hermeneutic token of illegibility itself.

The sheer improbability of a spatially and temporally contained ecological crime scene is a topic that is comprehensively discussed in Rachel Carson's 1962 classic *Silent Spring*. Laurence Buell credits Carson with having initiated "toxic discourse", one of whose features concerns "the shock of awakened perception" around the ubiquitous threat posed by chemical pesticides (35). Where culpability is concerned, however, this newly acquired perceptiveness does not always have a clear object. On the one hand, the toxic discourse to which Buell refers sets up "David versus Goliath" scenarios (40), whereby corporate interests are pitted against vulnerable and infirm individuals; on the other, the culpability of these corporate behemoths can only be addressed via the language of "allegation or insinuation rather than of proof" (48). Toxic illness, Carson suggests, manifests in effects that are seemingly "unrelated to the cause, appearing in a part of the body remote from where the original injury was sustained" (189). The identification of the agent behind what Carson describes as this "death-by-indirection" depends on a patient "piecing together of many seemingly distinct and unrelated facts developed through a vast amount of research in widely separated fields" (32, 189). As Joshua Schuster explains, Carson's detective work performs the negative manoeuvre of uncovering "ways of understanding the gaps, displacements, and delays between cause and effect as much as the empirically verifiable linkage between things" (141); for Peter van Wyck, Carson's work "marks a moment when the very equipment of causality changes, when causality itself becomes ecological" (ix). In the context of crime fiction theory, then, Porter's formulation that detectives must deal with "effects without apparent causes, events in a jumbled chronological order, [and] significant clues hidden among the insignificant" incurs extra weight, since at least some parts of this complicated picture are destined to remain unresolved, and perhaps to become more (rather than less) jumbled. If the detective's remit is "to re-establish sequence and causality", it follows that the ecological crime scene, which Schweblin's novel hints at without ever fully delineating, must therefore abandon or at least temper its post-Enlightenment causal principles in favour of something considerably more complex and less robust (Porter 30).

From the microscopic worms in the opening lines onwards, threads and ropes serve as a presiding metaphor for the fundamental connectedness of things, but the extent to which this conceptual truism is detectable or mappable is in question. The organisational and epistemological frame keeps scaling up, and arguably progresses beyond the reach of available narrative protocols in the novel's final lines. As Nina's father drives back to the capital and to safety, we are told that "[h]e doesn't see the soy fields, the streams that crisscross the dry plots of land, the miles of open fields empty of livestock, the tenements and the factories as he reaches the city". Nor does he notice that the traffic has stalled on the asphalt, "smoking and effervescent". Ultimately, "he doesn't see the important thing: the rope finally slack, like a lit fuse, somewhere; the motionless scourge about to erupt" (150–51). These final lines invoke a moment of extra-diegetic recognition of the rural and urban fabric as horizontally contiguous, thereby explicating the already-present implication that the events witnessed have taken place in a hinterland of the global Capitalocene. The corresponding implication that the proximate and visible can only be understood when indexed to the remote and *in*visible serves to legitimise David's paranoid reading of seemingly unrelated events and objects as interconnected across a multi-dimensional crime scene. The continuing presence of the 'rope' metaphor viewed in symmetrical juxtaposition with the 'worms' in the opening lines of the novel serves as an additional, vertical axis running from micro to macro, and from the microbial to the infrastructural. By mobilising the many layered dimensions of toxic discourse in this way, and by underlining Nina's father's failure to read the urban text in which he finds himself, the novel ends by lending credence to Latour's claim that "[t]here is no single institution able to cover, oversee, dominate, manage, handle, or simply trace ecological issues of large shape and scope" ("Waiting for Gaia" 21).

'Under the Black Water': From Anthropocene to Polemocene

Like *Fever Dream*, Mariana Enríquez's short story "Under the Black Water" (from the 2016 collection *Things We Lost in the Fire*) participates in the toxic discourse that concerns the long-term effects of pollution, with cause separated and even alienated from effect. Whereas both *The Way Out* and *Fever Dream* reckon with questions of culpability, temporality, and causality that hinder the possibility of an ecological crime scene, thereby frustrating genre expectations, Enríquez's story takes a different approach. The story commences in a recognisably hard-boiled urban noir mode but midway through switches to Enríquez's preferred (hybrid) genre, an amalgam of Argentina's fantastic tradition with Stephen King-inflected horror, culminating in the appearance of a Lovecraftian monster. The story's radical shift in register is such that it inevitably recruits the monster as a figure of rupture itself, i.e., the failure of an epistemological system to apprehend a reality that has overwhelmed collective capacities for ratiocinative detection. The story marks a progression of sorts insofar as it seems to herald the outright implosion of the detective genre. Rather than signifying the obsolescence of the genre when read in this ecological register, however, I would suggest that Enríquez means to communicate something about the epistemological organisation of reality in a more fundamental sense.

In a 2018 article entitled "Climate Change is Not a Problem", Campbell, McHugh, and Ennis justify the contentious proposition contained in their title, not by arguing that climate change does not pose a challenge, but rather that it poses an escalating, increasingly complex series of problems that will rapidly elude any existing organisational paradigms. For Campbell et al, the 'Anthropocene' is primarily of interest not on account of its usefulness as descriptor of reality, but rather in the context of its status as the latest in a radically escalating series of attempts to 'frame' climate change. As the authors explain, climate change was originally described in terms of a "tragedy of the commons" narrative, with subsequent framings foregrounding "risk", "war" (on climate change), "crisis", and "catastrophe"; the most recent frame, i.e. the Anthropocene, "massively expands both the temporal logic and the pervasive spatiality of climate change, situating the category in deep time and on a planetary scale". However, the authors go on to identify "the most important meta-observation that can be drawn" from their analysis, namely that "the more each field discovers about climate change the more it seems to grow in scale" (733). This description of rapidly spiralling and successively obsolete epistemological frameworks vaguely recalls perhaps the most famous Argentinian detective story, Jorge Luis Borges's "Death and the Compass" (1942), about a detective who attempts to resolve a murder mystery through a sequence of widening geometrical patterns, only to find, at the end, not only that he has misread the clues at the first murder scene, but that he has fatally misrecognised the paradigmatic coordinates within which the culprit is operating. This, I would suggest, is a possible precursor for Enríquez's story, which likewise features an ill-fated excursion to the south of Buenos Aires by a detective; in this case, toxicity is fatally misread as a 'problem', but eventually subsumes and reconfigures the story world in its own image.

The story is focalised through Marina Pinat, a public prosecutor who is investigating the deaths of two fifteen-year-olds named Emanuel López and Yamil Corvalán. The victims are both from Villa Moreno, a slum on the banks of the Riachuelo river which enters the sea at the southern edge of central Buenos Aires. One of the bodies (that of Corvalán) has washed up on the banks of the river; the trainers of the other (Emanuel López) have also been recovered. Pinat suspects a police officer named Cuesta of pushing the two boys into the river and, in the opening section, she interrogates Cuesta based on a recording she has obtained in which he seems to clearly incriminate himself. However, Cuesta is unmoved by her arguments that he should confess to lessen his sentence, as he is not convinced that he will suffer serious consequences. His implicit argument hinges on the poverty and therefore the relative unimportance of this part of the city and of the people who live there, evidently an "unimagined community" in the terminology developed by Nixon, who describes these as communities whose literal and figurative occlusion from public view "becomes indispensable to maintaining a highly selective discourse of national development" (150).

Approximately five million people live in the Matanza-Riachuelo basin, which a 2013 feature in *Time* magazine identified as one of the ten most polluted places on the planet. In the neighbouring slums, a quarter of all children have lead in their bloodstreams, and incidences of respiratory and gastrointestinal illness are reportedly even higher still (Blitzer). In a 2020 essay entitled "Riachuelo", Enríquez (who grew up in nearby Lanús) discusses how the river became part of her "intimate landscape", which she evokes through a series of colours and odours pertaining to rotting meat, burnt garbage ("the madeleine of poverty"), and fuel (26). But along with disgust, Enríquez recalls that "[t]he Riachuelo scared me. Its protuberances of animal fat, machine grease, and bottles formed fantastic creatures always just about to show themselves" (26). In Enríquez's telling, the river becomes the focal point of a grim mythos nourished, in part, by the long history of slaughterhouses and meat-processing plants that were located along its banks, with reports of animal blood and viscera reaching the water dating from as early as 1820 (the name of the more inland portion of the river, 'Matanza', means 'Slaughter'). The river's mythos is further enhanced by the yellow fever epidemic of 1871, which was blamed both on the stagnant water of the river itself and the over-crowding of people along its banks. This was when the area gained a reputation for its 'toxicity' among the wealthy, and Enríquez strongly implies that this toxicity is broadly applied not just to the fetid conditions of the water but also in the form of an enduring stigma to the surrounding community. Enríquez's essay highlights government failures of planning and oversight in respect of industrialisation and later de-industrialisation as fundamental to the area's problems, which are further impacted by climate change. The river is habitually subject to periods of flooding, turning the street into a "black river", but Enríquez recounts that these have gotten worse in recent years due to heavier and longer periods of rainfall (28). The discourse around this is obfuscatory, with the people who suffer the effects of the bi-monthly flooding "told that 'it's been raining a lot', as if it were a misfortune and not a definitive change for which no one is prepared" (30). In sum, Enríquez's essay sets up the Matanza-Riachuelo basin as the confluence of multiple '-cenes', including the Anthropocene, the Thermocene, the Capitalocene, and, insofar as the region's problems stem from a complex of institutional blindspots, the Agnotocene.

"Under the Black Water" is contiguous with the toxic discourse contained in "Riachuelo". The body of Yamil Corvalán has appeared one kilometre away from the bridge where he is believed to have been pushed in and where the Riachuelo "is calm and dead, with its oil and plastic scraps and heavy chemicals, the city's great garbage can" (158). A little later, we are told that Marina has previously been involved in cases taken by families directly impacted by chromium and other toxic materials discharged by a tannery, causing illness, death, and birth defects. After Cuesta leaves Pinat's office, she is visited by a heavily pregnant teenager from the slum, who alleges that the presumed dead Emanuel López is still alive. She claims: "He came back from the water. He was in the water the whole time. . . . He's holed up in one of the houses back behind the tracks. He lives there with his friends", and now he wants to meet Pinat (162–63). Unconvinced, Pinat assumes that the police accused of the murders have recruited the girl to report the story and thereby muddy the waters of the investigation. The following day she goes into the slum to investigate further. Having been abandoned by her taxi driver, who refuses to bring her all the way, Pinat is surprised by the deathly silence and by the absence of popular music and shrines to saints in the normally lively area. Heading in the direction of the church where she hopes to encounter her only contact in the slum, Father Francisco, she is followed by a child.

While the teenage girl's bizarre claims might already seem to signal the incursion of the fantastic, Pinat is still able operate under the illusion of epistemological orthodoxy; her subjection of the girl's claims to a rational explanation means that it is still possible to figure events within the rubric of the detective genre. However, at this point the story completes its pivot into horror. Pinat finds that the church has been covered with graffiti depicting seemingly meaningless sequences of letters, repeated several times. The same graffiti covers the walls inside the church, which has been stripped of its Christian icons and which (in place of the altar) contains the head of a cow impaled on a wooden pole in a metal flowerpot. While these objects and symbols hint at an order of signification that is beyond her comprehension, even now Pinat tries to organise and establish dominion over the scene's illegibility: she is relieved to hear the sound of drums, assuming that a *murga* group is preparing for the upcoming carnival period, and she reassures herself that the cow's head must have been left there as a threat to Francisco from a local drug gang. Meanwhile, however, Francisco has appeared in a state of delirium. He claims that the missing Emanuel has woken something which had lain dormant beneath the water, covered up with layers of pollution.

Pinat decides that they must leave, but Francisco grabs her gun and shoot himself, dying instantly. Out in the street, a religious procession is passing; the residents from the slum are marching, with part of the crowd carrying a bed with an idol laid on top. Pinat realises that the human-sized idol with discoloured flesh appears to be alive. She flees, "trying to ignore the fact that the black water seemed agitated, because it couldn't be . . . she covered her ears with her bloody hands to block out the noise of the drums" (175). Even now, Pinat fulfils the detective function of organising information rationally, which in this case means resorting to "trying to ignore"; in these final lines, she serves as a cipher for the Agnotocene, or more specifically as one of the "institutional zones of ignorance" described by Bonneuil and Fressoz. The coming of the monster can be interpreted as a physical manifestation of the social and ecological ills that had been hitherto officially ignored. Or to use the term employed by Rob Nixon, it is an embodiment of the institutionally non-codified slow violence that has been inflicted upon and accumulated by the community over the course of several decades.

Nixon's coinage has the objective of articulating the connection between cumulative acts and their ultimately destructive payload; slow violence describes the suffering and illness, borne primarily by the global poor, wrought by climate change, deforestation, oil spills, and the environmental aftermath of war, and which takes place gradually and often invisibly. As Nixon explains, whereas the traditional conception of violence involves "a highly visible act that is newsworthy because it is event focused, time focused, and body bound", slow violence repels narrativisation due to its dilated temporality and limited visibility (3). The challenge, says Nixon, is in "how to devise arresting stories, images, and symbols adequate to the pervasive but elusive violence of delayed effects" (3). In their different ways, *The Way Out*, *Fever Dream*, and "Under the Black Water" all engage with the challenge posed by Nixon.

In other respects, the severe genre trouble that is evinced by this final story can be viewed as symptomatic of the paradigm-busting, escalatory organisational chaos that is identified by Campbell and co-authors in respect of the 'problem' of climate change and of other proliferating ecological cataclysms. The institutionally and professionally delimited task of marking out crime scenes and of apprehending criminality must reckon with the epistemological and definitional challenges posed by the evolving meaning of crime in the Anthropocene. The conclusion of "Under the Black Water" presages the arrival of a different kind of '-cene' in which the authority that is represented by Pinat and her colleagues is sidelined, and whose coordinates will not be drawn up, much less underwritten, by existing institutional actors. Given the story's depiction of the re-emergence of a drowned adolescent as a kind of chthonic creature spawned, perhaps, by the toxic soup of heavy metals and animal viscera that fester in the depths of the Riachuelo, it is tempting to think of this as what Donna Haraway calls the arrival of the "Chthulucene", defined by Haraway's call for new 'tentacular' kinships and alliances between human and non-human, entangling "myriad temporalities and spatialities and myriad intra-active entities-in-assemblages – including the more-than-human, other-than-human, inhuman, and human-as-humus" (101). Read in a more directly political register, the incomprehensibility of the final image in the slum might be taken to herald a radically new stage in what Bonneuil and Fressoz call the "Polemocene", or an environmentalism of the poor that takes place outside of official organisational channels (253). Though this concept is domesticated by the authors through inscription within a longer history of anti-Establishment resistance and activism, I would contend that the current unthinkability in respect of how this '-cene' will evolve means that it is better registered as the site of a social movement-to-come from an as-yet unimagined community. Relatedly, I would suggest that in "Under the Black Water", in concordance with Campbell et al, "what is being revealed is not a problem that can be framed, but something entirely different, a World that generates, but is not commensurate with, problems, and which no current organisational form can address" (735).

In the meantime, however, it might be the case that the detective genre, concerned as it can be with dichotomies of blindness and insight, is at least somewhat suited to the task, if not of suturing the vast gaps inherent in what Latour calls "the total *disconnect* between the range, nature, and scale of the phenomena and the set of emotions, habits of thoughts, and feelings that would be necessary to handle those crises", then at least of negatively articulating those gaps even though, in doing so, it must proceed by way of warping the precepts that condition the genre. Even though there is, as Latour adds, "no ground control station anywhere to which we could send the help message" ("Waiting for Gaia" 22), the existing rhetoric of policing still provides a vocabulary with which to "imagine a crime without a criminal" and "[a] violence without coordinates on which transgressions might be plotted" (Lahoud 496). The exposure of the limits or failures of justice in respect of slow violence can contribute towards a wider, long-term project of critically reframing conceptualisations of criminality (King: 1247).

Indeed, some of the subversive moves that I have discussed can be perceived to have emerged organically out of detective fiction and particularly out of noir forms; for some, such as Deborah

B. Rose and Timothy Morton, the noir protagonist is particularly suited to "investigating a supposedly external situation . . . only to discover that she or he is implicated in it" (Morton 16–17) and can therefore "uniquely articulate our condition in this dark (Anthropocene) era" (Rose 6). In her classic analysis of the noir thriller, Lee Horsley identifies fear, anxiety, paranoia, ambivalence, and "failures of agency" as typical of the genre (11), and, as Lucas Hollister rightly argues, noir's ways of mapping the world are "fundamentally, attempts to change the scale of politics or political interventions" and are therefore "not ill-equipped to address the scalar political problems of the Anthropocene" (1021).

Arguably, some of these synergies have already been activated in Argentina's crime fiction tradition due to successive military dictatorships between 1955 and 1983, resulting in what Celeste Fraser Delgado identifies as moments of "irresolvable crisis in the definition of crime", an "unsettling of the Argentine scene of crime", "dispers[als of] the notion of criminality", and "suspen[sions of] the certainty of truth" (56). Ricardo Piglia's most famous novel, *Respiración artificial* [*Artificial Respiration*] (1980), was written and published while the notoriously violent and repressive 1976–83 military government was in power, and serves as a case in point. Nominally concerned with an investigation into the whereabouts of the narrator's estranged uncle, the novel serves as an oblique commentary on the crimes of the regime and marks an attempt to reproduce in the reader the paranoid protocols of reading that were fostered by the covert and arbitrary nature of the military's clandestine operations. These operations relied on the tacit approval and therefore widespread complicity of sectors of the population, and *Artificial Respiration* is just one example of innovation in the crime fiction that was generated as a result. Viewed in the context of this tradition, it is perhaps not surprising that Argentinian crime authors in the past decade have been particularly adept at reworking the parameters of the genre as they rise to the pressing and ongoing challenges of deciphering, exposing, mapping out, and articulating ecological crime scenes.

Note

1. Translation by Sergio Waisman. All quotes in English from *El camino de Ida*, *Distancia de rescate*, and "Bajo el agua negra" are taken from published translations by Robert Croll (*El camino de Ida*) and Megan McDowell.

Bibliography

Blitzer, Jonathan. "Life Along a Poisoned River". *The New Yorker*, 25 Oct. 2016, www.newyorker.com/culture/photo-booth/life-along-a-poisoned-river.

Bonneauil, Christophe, and Jean-Baptiste Fressoz. *The Shock of the Anthropocene*. Translated by David Fernbach, Verso, 2016.

Buell, Lawrence. *Writing for an Endangered World: Literature, Culture, and Environment in the U.S. and Beyond*. Harvard UP, 2001.

Campbell, Norah, Gerard McHugh, and Paul Dylan-Ennis. "Climate Change is Not a Problem: Speculative Realism at the End of Organization". *Organization Studies*, vol. 40, no. 5, 2018, pp. 725–744, doi:10.1177/0170840618765553.

Carson, Rachel. *Silent Spring*. Houghton Mifflin, 1962.

Chesterton, G. K. *The Defendant*, edited by Dale Ahlquist, Dover Publications, 2012.

Dixon, Arthur and Samanta Schweblin. "Samanta Schweblin: 'La Literatura Es La Posibilidad De Probar Los Caminos De Nuestras Guerras Más Profundas y Volver a La Vida Real Con Información Vital': Una Conversación Con Arthur Dixon". *LALT*, 12 Nov. 2022, latinamericanliteraturetoday.org/es/2019/02/samanta-schweblin-literature-possibility-testing-paths-our-deepest-inner-wars-and/.

Enríquez, Mariana. *Things We Lost in the Fire*. Translated by Megan McDowell, Portobello, 2017.

———. "Riachuelo". *Tales of Two Planets: Stories of Climate Change and Inequality in a Divided World*. Translated by Megan McDowell, edited by John Freeman, Penguin, 2020, pp. 23–32.

Fraser Delgado, Celeste. "Private Eyes in Argentina: The Novel and the Police State". *Latin American Literary Review*, vol. 22, no. 44, 1994, pp. 49–73.

Glaser, Marion, et al. "New Approaches to the Analysis of Human-Nature Relations". *Human-Nature Interactions in the Anthropocene: Potentials of Social-Ecological Systems Analysis*, edited by Marion Glaser, Beate M. W. Ratte, Gesche Krause, and Martin Welp, Routledge, 2012, pp. 3–12.

Hollister, Lucas. "The Green and the Black: Ecological Awareness and the Darkness of Noir". *PMLA/ Publications of the Modern Language Association of America*, vol. 134, no. 5, 2019, pp. 1012–1027, doi:10.1632/pmla.2019.134.5.1012.

Horsley, Lee. *The Noir Thriller*. Palgrave Macmillan, 2009.

Kern, Stephen. *A Cultural History of Causality: Science, Murder Novels, and Systems of Thought*. Princeton UP, 2006.

King, Stewart. "*Crimate* Fiction and the Environmental Imagination of Place". *The Journal of Popular Culture*, vol. 54, no. 6, 2021, pp. 1235–1253, doi:10.1111/jpcu.13083.

Lahoud, Adrian. "Floating Bodies". *Forensis: The Architecture of Public Truth*, edited by Eyal Weizman and Anselm Franke, Sternberg, 2014, pp. 495–518.

Latour, Bruno. "Waiting for Gaia. Composing the Common World Through Arts and Politics". *What is Cosmopolitical Design? Design, Nature and the Built Environment*, edited by Albena Yaneva and Alejandro Zaera-Polo, Routledge, 2015, pp. 21–32.

———. *Down to Earth. Politics in the New Climatic Regime*. Translated by Catherine Porter, Polity Press, 2018.

Moore, Jason W. *Capitalism in the Web of Life: Ecology and the Accumulation of Capital*. Verso, 2015.

Morton, Timothy. *The Ecological Thought*. Cambridge UP, 2010.

Murphy, P. D. *Ecocritical Explorations in Literary and Cultural Studies: Fences, Boundaries and Field*. Lexington Books, 2009.

Nixon, Rob. *Slow Violence and the Environmentalism of the Poor*. Harvard UP, 2011.

O'Rawe, Ricki. "La política espeluznante en *Distancia de rescate* (2014) de Samanta Schweblin". *Bulletin of Contemporary Hispanic Studies*, vol. 5, no. 2, 2023.

Piglia, Ricardo. *Target in the Night*. Translated by Sergio Waisman, Deep Vellum, 2013.

———. *The Way Out*. Translated by Robert Croll, Kindle ed., Restless Books, 2018.

Porter, Dennis. *The Pursuit of Crime: Art and Ideology in Detective Fiction*. Yale UP, 1981.

Rose, Deborah B. "Anthropocene Noir". *People and the Planet 2013 Conference Proceedings, edited by Paul James, Chris Hudson, Sam Carroll-Bell, and Alyssa Taing, Global Cities Research Institute*, 2013, search.informit.org/doi/epdf/10.3316/ielapa.860610859642687. Accessed 1 Nov. 2022.

Scaggs, John. *Crime Fiction*. Routledge, 2005.

Schuster, Joshua. *The Ecology of Modernism: American Environments and Avant-Garde Poetics*. U Alabama P, 2015.

Schweblin, Samanta. *Fever Dream*. Translated by Megan McDowell, Oneworld, 2017.

Todorov, Tzvetan. "The Typology of Detective Fiction". *Modern Criticism and Theory: A Reader*, edited by Nigel Wood and David Lodge, 3rd ed., Routledge, 2000, pp. 137–144.

Van Wyck, Peter C. *Signs of Danger: Waste, Trauma, and Nuclear Threat*. U Minnesota P, 2005.

Walsh, Bryan. "Urban Wastelands: The World's 10 Most Polluted Places". *Time*, 4 Nov. 2013, science.time.com/2013/11/04/urban-wastelands-the-worlds-10-most-polluted-places/slide/matanza-riachuelo-argentina/.

Walton, Samantha. "Studies in Green Studies in Green: Teaching Ecological Crime Fiction". *Teaching Crime Fiction*, edited by Charlotte Beyer, Palgrave Macmillan, 2018, pp. 115–130.

20

ECOLOGEMES IN CONTEMPORARY AUSTRALIAN CRIME FICTION

The Case of Outback Noir

Katrin Althans

> I love a sunburnt country,
> A land of sweeping plains,
> Of ragged mountain ranges,
> Of droughts and flooding rains.
> I love her far horizons,
> I love her jewel-sea,
> Her beauty and her terror –
> The wide brown land for me!
>
> *(Mackellar 9–16)*[1]

Dorothea Mackellar's poem "My Country" is emblematic of how different the ecology and natural environment of Australia is in contrast to that of England. We find a number of ecological characteristics which, taken together, define a particularly Australian landscape identity alien to Europeans. It is, writes Tony Hughes-d'Aeth, "the imagery of national identification" (302). A similar national identification by means of place has become part and parcel of crime fiction marketing, which has given rise to national subgenres – Nordic noir from Scandinavian countries including Iceland and Finland, Tartan noir from Scotland, Mediterranean noir from countries such as Italy, Spain, and Portugal as well as other national varieties. This includes a more recent addition to the canon of regional crime fiction, that which has variously been termed "Outback noir", "Aussie noir", "bush noir", or "sunburnt noir" (Hammer, 'The New Class of Australian Crime Writers'), i.e., contemporary crime fiction from Australia. Those designations not only directly reference Mackellar's "sunburnt country", but also refer to settings commonly associated with Australia, the bush, and the outback. Descriptions of place have always been stock elements of crime fiction, from the locked-room mystery to the cities of the hard-boiled tradition. It is indeed "place", as Stewart King writes, which "gives the crime meaning" (211). Place also takes centre stage in environmental crime fiction, even though here it is "the global nature of environmental crime" which is of interest (Puxan-Oliva 362). The emphasis here, however, is on man-made culture and society or at least on the environment in relation to humans, i.e., on the "physical, cultural, political, economic, environmental, social and . . . legal circumstances of the place" (King 211). Such an understanding of place as closely tied to humans ignores the many ecological characteristics and the non-human environment found in regional crime fiction varieties, yet it is them which put the sunburnt in the noir. This subgenre

DOI: 10.4324/9781003091912-24

of Australian crime fiction is replete with genuinely Australian ecological entities, entities I will suggest calling ecologemes.

To explore this idea of ecologemes in contemporary Australian crime fiction for this chapter, I will borrow from translation studies and draw on histories of Australian crime fiction from the vantage point of ecocritical and geocritical approaches, strands I will connect through a unique reading of ecological entities as necessary elements of genre analysis in an econarratological fashion. However, my reading goes beyond classifying those ecological entities as parts of the setting; instead, I will show the extent to which, in Outback noir, they have become a new narratological category defining the genre. As such, I suggest, they are not merely markers of a genuinely Australian national (crime fiction) identity, but also indicators of the specific ecological reality of climate change facing contemporary Australia. I will analyse these issues by focusing on texts by two contemporary Australian crime writers: *The Dry* (*TD*) and *The Lost Man* (*TLM*) by Jane Harper and *Scrublands* (*SL*) and *Opal Country* (*OC*) by Chris Hammer.[2]

(Australian) Crime Fiction and National Identity

Although primarily used for marketing, and thus selling, purposes, the categorisation of crime fiction in terms of national varieties also serves the need to assert a national identity – both at home and abroad (Anderson 222). This national identity is established through any kind of cultural specificity, be it the "habits, prejudices, foods, and idiosyncrasies of the locals" (Kim 3), judicial systems and political structures (Anderson 223), or a nation's social fabric and history (Erdmann 12). This, however, is nothing new, as the countryside of England together with the tranquillity of village life found in Golden Age mysteries or the intimidating cityscapes of US-American metropolises and the corruption of its institutions in hard-boiled detective fiction, not to mention Sherlock Holmes's upper-class London, all attest. Setting in the sense of place (and the characteristics and anxieties of its inhabitants) has always been constitutive of crime fiction and has influenced the style of writing. Crime fiction texts are, write Alistair Rolls, Marie-Laure Vuaille-Barcan, and John West-Sooby, "pointedly anchored in their social and geographical realities" and therefore "contain extremely precise descriptions of place" (140).

This is particularly evident in the case of Nordic noir, arguably the most internationally successful of national crime literatures. It is also the one regional variety which has attracted the most scholarly attention so far and often serves as a blueprint for discussions of national identity in crime fiction (cf. Hansen and Waade; Hayward and Hall; Hedberg; Hill and Turnbull; Nestingen and Arvas; Nilsson; Stougaard-Nielsen). In the global literary market, Nordic noir "is associated with a region (Scandinavia), with a mood (gloomy and bleak), with a look (dark and grim), and with strong characters and a compelling narrative" (Hill and Turnbull). Its noir aesthetics are combined with a criticism of the social realities of the welfare state. However, that which readers have come to associate with Nordic noir (and to expect when making their buying decisions), Louise Nilsson argues, is but the representation of "a cosmopolitan imaginary of the *north*" (542). Still, as a brand's unique selling point, the imagined national identities transported in crime fiction are vital for both national and international audiences. They both affirm "a sense of national or regional belonging" (Anderson 222) and serve "as a cultural mediator . . . or as a tourist guide for the armchair traveller" (Pezzotti 94). At the same time, those fictional representations of this imaginary perpetuate national stereotypes and clichés as much as they help familiarise the reader with foreign places and cultures. Additionally, crime fiction finds itself torn between "the allure of a more exclusive national identity" and the effects of globalisation, as Julie H. Kim writes in her introduction to *Crime Fiction and National Identities in the Global Age* (2). This fraught relationship between the local and the global has also been addressed by Eva Erdmann, who argues that the national is superseded by the intercultural even though individual works are recognisably regional (24–25). Barbara Pezzotti, on the other hand, considers national crime fiction in terms of transnationality, arguing that although the importance of national literatures as markers may be waning, the national is still an important frame of reference, as "transnationalism crosses borders and acknowledges them, fuses separate places and recognizes their separation" (95). In other words, one needs the national in order to understand the transnational. Here, however, we have come full circle and returned to Nilsson's "cosmopolitan imaginary". Although intrinsically tied to a certain locale and its characteristics, crime fiction is transnational in its journey across borders as a commodity of the international book market, for the audience of which it creates this "cosmopolitan imaginary" (Nilsson 541). "Thus", Nilsson reasons, "crime fiction should be understood as a discursive field – a network consisting of elements and nodal points that are connected to and build on each other" (541).

Nilsson's idea of a network is that of a dynamic, ever-evolving, and ever-changing system, much like the human brain and its synapses, with new nodes and the links connecting them continually emerging (and disappearing). The very history of crime fiction and its publication in Australia readily lends itself to a reading in terms of a network, and such an understanding of crime fiction very well captures the situation of contemporary Australian crime fiction. Even a cursory glance at the tables of contents of Stephen Knight's 1997 and 2017 editions of his seminal work on Australian crime fiction shows a number of such genre-defining nodes within an international network – be they characters, settings, or themes. Knight both contextualises Australian versions of well-known generic markers in their international relations and considers elements unique to the genealogy of Australian crime fiction, thus inserting its nodes in a three-dimensional network of fictional and historical connections. In terms of international relations, he structures his overview in terms of well-known stock elements of global crime fiction, like mysteries or the private-eye model (Knight, *Australian Crime Fiction* 50, 94). On a national scale, this includes Australia's imperial legacies such as the legal stipulations of the "Traditional Market Agreement", which excluded US-American publishers from moving into the Australian book market, as well as socio-historical events such as the Australian gold rushes (68). As such, Australian crime fiction is a mirror of Australian history and society and has thus "represented and realized the changing contexts and social foci of Australian life" and "embodied contemporary ideas of right and wrong, justice and liberty, including positive responses to criminals themselves and, finally, the long-overlooked indigenous people" (Knight, *Australian Crime Fiction* 12–13).

The one socio-historical nodal point which makes the Australian continent particularly suited for crime fiction is its convict history. First considered an "embarrassment" (Knight, *Australian Crime Fiction* 2), Australia's convict origins soon became "a badge of honour" and thus a common icon of Australian identity both at home and overseas (Reed 24). This gave rise to a certain streak of anti-authoritarianism in Australian crime fiction which, for example, shows in a general reluctance to embrace the police detective as main character, especially in fiction by male writers (Knight, *Australian Crime Fiction* 4–5). Stephen Knight considers these kinds of stories zero-policing stories, i.e., stories in which no detective features or brings about justice ('Crimes Domestic' 24) and thus the resolution is brought about without elements of detection (*Continent of Mystery* 124). In his exhaustive overview of Australian crime fiction from its beginnings to 2017, Knight shows that Australian authors are in fact late adopters of the police detective: It was not until the 1980s that the police detective became a recurring figure across Australian crime fiction; before that, it was almost exclusively women writers of psychothrillers who would use police detectives as heroes in their fiction (Knight, *Australian Crime Fiction* 93, 134, 162). Instead, it is the larrikin[3] and an idealised (and idolised) version of the bushranger who loom large as characters in Australian crime fiction (Shaw 35–36). Inspired by real-life criminal figures such as the notorious bushranger Ned Kelly, who himself "remains for many a national hero" (Knight, *Australian Crime Fiction* 3) and after whom Australia's most prestigious awards for crime writing, the Ned Kelly Awards, are named, such outlaw characters can be found throughout Australian crime fiction (Shaw 37).

National Identity, Culturemes, and Ecologemes

A good starting point from which to analyse those nodes which are anchored in national identities (as fleeting and imagined as those are) is to look at translation studies and the concept of culturemes. Introduced by Fernando Poyatos in 1976, the term cultureme is based on linguistic vocabulary and refers to elements which carry cultural information (Lungu-Badea 18, 69). For Vermeer and Witte, such elements would be any cultural specificities which are endowed with a culturally specific meaning in a given culture and, in turn, are not found in other cultures (137). Culturemes are, writes Michel Ballard, constitutive of cultures and civilizations (149). They therefore depend on the cultural context within which they exist and the meaning ascribed to them within this cultural context. As the translation of culturally specific elements which have no direct equivalent in the target language poses a number of challenges for translators, the theory of culturemes has gained some traction in translation studies, especially in Germany and Romania. This issue is particularly relevant when it comes to different translation strategies, in this case the binary opposites of domestication vs. foreignisation. Whereas, in the words of Lawrence Venuti, domestication means "locating the same in a cultural other" and thus minimising any content foreign to the reader, foreignisation means "locating the alien in a cultural other" and thus making the reader aware of the foreignness of the text (266). Translation strategies dictate the handling of culturemes in translation (and vice versa) and thus also shape the image of cultural identities of foreign places that readers create in their minds. Culturemes, however, are not specific to translations from one language into another, but also occur from one cultural readership to another, even within the same language. They are, e.g., one aspect the readers' tourist gaze is directed at when reading national and regional varieties of crime fiction for their cultural idiosyncrasies. Very obvious examples of such cultural specificities would be certain regional dishes, as discussed by Jean Andersson in French and Icelandic crime fiction (228). Other examples include types of buildings and their surrounding spaces, as shown by Sarah Reed (76–77), or simply different forms of greetings.

This connects back to the broad spectrum Stephen Knight has identified as "contexts and social foci of Australian life" in Australian crime fiction, i.e., aspects ranging "from convictism to modern urban professional and gender-complex life, not overlooking less serious *national* formations, like horse racing [the Melbourne Cup] . . . and Pacific holidays" (*Australian Crime Fiction* 12, emphasis added). Culturemes, however, only refer to characteristics of the human cultural sphere. What is left unconsidered is another, to my mind, crucial aspect of what makes crime fiction genuinely local – that of the non-human environment. Bending the term cultureme to my needs, I suggest calling such characteristics ecologemes. By ecologemes, I mean units which carry ecological information of non-human place and environment as well as units of place-, space-, and time-specific ecological realities which are, despite being sometimes universal concepts, region-specific. In the case of Outback noir, those ecologemes most prominently refer to place in the sense of landscape, the weather, and climate. In Outback noir, they have become a new element defining the genre.

Let's continue where we left off with the idealised and idolised bushranger: The natural habitat of the bushranger is, as the name suggests, the bush, a landscape which itself is deeply ingrained in the national consciousness of Australia (Schaffer 52). It is both revered and feared, immortalised in the writings of Henry Lawson and A.B. "Banjo" Paterson as well as Barbara Baynton. Its narrative representations, together with those of the outback, have become "a dominant trope in Australian cultural narratives", write Jane Stadler, Peta Mitchell, and Stephen Carleton (34). Stadler, Mitchell, and Carleton also emphasise the schizophrenic relationship white Australia has in regard to its frontiers, with representations of the bush and outback oscillating from "a dark vision of the Outback", as in the case of Kenneth Cooke's *Wake in Fright*, to a "romantic portrayal", as in Baz Luhrmann's *Australia* (37). Emma Doolan succinctly summarises that "the Australian national character has been constructed on and through an essentialised Australian landscape characterised as either a productive paradise or a harsh and hostile enemy to be endured, if it cannot be overcome" (41). This harks back to traditions of the pastoral and anti-pastoral, which Terry Gifford considers concepts rather than distinct genres (17). Talking of the pastoral as concept rather than narrowly defined genre allows me to also refer to generic elements of the pastoral (and of its various anti-, post-, and counter-variants) and their subversions when analysing distinctly environmental and ecological entities – ecologemes – in Australian crime fiction, as it often is the pastoral which regionalises crime fiction. Place in crime fiction is, it seems, characterised by its relation to the countryside: What comes to mind here is the nostalgic pastoral of the Golden Age, especially Miss Marple's rural St. Mary Mead, but also the yearning for a pastoral idyll (and the ensuing representation of the anti-pastoral) in the American hard-boiled tradition (Ashman 45). The pastoral in crime fiction is closely linked to representations of landscape but adds to these a sense of the idyllic, either as sought, lost, or illusory.

The landscape of Australia is rich in differences, including the coastal regions and urban centres in the East, the tropical Queensland Hinterland, the bush, the outback, and many more. They can be defined by means of physical, human, or social geography; by means of agrarian or vegetal conditions; or by means of the national imaginary. That landscape features prominently in the national imaginary has already been established (cf. Schaffer 22). It has, however, also become a problematic category due to the contested nature of land ownership both prior and even more so after the High Court's *Mabo* decision (Hughes-d'Aeth 292–93): Although in *Mabo v Queensland [No. 2]* it was held that Native Title had not been extinguished by colonial dispossession, two different titles to land in Australia coexist – which makes land a notoriously contested entity. In Australian crime fiction, however, there hardly seems to be a trace of this. Although cognisant of Indigenous contributions to Australian crime fiction, as well as the representation of Indigenous characters in the broader history of the genre, Stephen Knight's work is principally concerned with the question of land from the perspective of white Australians, even though subheadings like "Taking the Land" and "The Lost Land" in his book *Continent of History*: A *Thematic History of Australian Crime Fiction* would seem to suggest otherwise. As Knight himself writes in the second and more chronologically than thematically structured revised edition of his account of Australian crime fiction, "a real representation of indigenous-related issues in crime fiction" has only occurred, and "somewhat slowly", too, in the past few years, i.e., in the twenty-first century (*Australian Crime Fiction* 11). He does, however, discuss the history, practices, and ramifications of Aboriginal dispossession when he elaborates on the avenging character of the land itself in another text. Here, he contextualises the different ways of representing the Australian landscape and reads them as "cultural colonisation" ('Crimes Domestic' 20).

This lack of representation of both Indigenous characters and issues is also mirrored in the texts I am discussing in this chapter. All four novels are written by white Australian authors and stand in the tradition of a Western literary genre, and one which in its Australian tradition is struggling with incorporating Indigenous presences – as are the novels I am analysing. They do not feature Indigenous protagonists, indeed hardly any Blak Australians at all: The current Indigenous inhabitants of the fictional outback town of Balamara are mentioned in passing in *The Lost Man* (*TLM* 125), and in *Opal Country*, the only black character is Stanley Honeywell, who is of "Islander, Chinese, Indian, Aboriginal" descent (*OC* 78) and described as "a gentle giant" (*OC* 55) – an ex-con (and murder suspect) who suffered brain damage in a prison brawl (*OC* 164). Other references are purely historical and suggest both colonial violence and the annihilation of Aboriginal people, as when Aboriginal remains are mentioned in *Scrublands* (*SL* 188) or the eponymous nineteenth-century stockman from the stockman's grave in *The Lost Man* is said to have "raped an Aboriginal girl and got himself killed for it" (*TLM* 291). As one reviewer puts it for *The Lost Man*, "it's all so painfully white" (Forrester). Those novels, however, are hardly alone in their omission of Blak characters, something which is, argues Birri-Gubba author Nicole Watson, due to Whiteness and white cultural values being considered the neutral norm in Australian crime fiction (77).[4]

My analysis of ecologemes in Outback noir in this chapter is, due to my choice of texts, necessarily limited to white understandings of environment in general and to white readings of the outback in particular, with all its historical and cultural baggage. Furthermore, as a white European scholar myself, I am in no position to analyse texts from a non-white perspective, and that limits my analysis to land and landscape as well as of any ecological entity to white understandings and readings of them. Those differ considerably from Indigenous understandings of Country, which is so much more than just land. In the words of Margo Neale, series editor of First Knowledges, "Country holds information, innovations, stories and secrets – from medicine, engineering, ecology and astronomy to social mores and how to live. . . . It is the wellspring from which all knowledge [i.e., sciences, humanities, and ancestral knowledge] originates" (1–2).

Outback Noir and Its Ecologemes

Despite locations being paramount in general and regional crime fiction, the Australian variant was slow to embrace the unique Australian landscape – but when it did, it did with full force. Stephen Knight has identified two strands of how place is being handled in Australian crime fiction, one being "a complete lack of any reference to place" and one he calls "an essentially touristic version", in which we find larger-than-life landscapes which enact their own revenge in the form of particularly Australian catastrophes, both environmental (Knight lists "flood, fire, avalanche") and animalistic (as, e.g., by means of poisonous snakes or man-eating saltwater crocodiles) (*Continent of Mystery* 144). This second strand is touristic in the sense that it revels in descriptions of environments alien to non-Australians and thus allows for the voyeuristic gaze of the tourist. The Australian landscape thus becomes a tourist attraction and has, in recent years, morphed into a unique selling point, especially in the case of Outback noir.

What, then, is the outback, and what does it denote in a white imaginary that it has come to signify a particularly Australian regional variant of crime fiction? In *The Australian National Dictionary* we find two definitions which capture a geographical and a national imaginary meaning of the term: "Sparsely inhabited country that is remote from a major centre of population" ('Outback' s.v. C 1), and, in the form of the special compound "great (Australian) outback", it stands for "the outback, esp. as perceived in a romanticised literary depiction of life there" (s.v. C 2). The national allegory of the outback here is defined in relation to humans instead of as a landscape in its own environmental right. At the same time, the outback is associated with endless plains and a blazing sun, the kind of sublime Bill Ashcroft calls "horizonal sublime", "the terrible horizon of trackless desert" (Ashcroft 142) – and as such seems rather ill-suited to be considered noir at all.

However, Evan McHugh, although writing about true crime, has a point when arguing that it is the remoteness and isolation, both of and in the outback, which are especially menacing and present distinctive challenges for detection. For Greg Dolgopolov, the term Outback noir is a particularly apt expression exactly because of its oxymoronic nature: Despite its "inversion of the traditional noir visual iconography of the nocturnal urban scene", its focus on the rural and on the horizonal sublime is just a veneer for "a dangerous, twisted noirish world hiding in the blind spot" (7). What is more, the blazing sun of the outback does everything but expose both crime and troubled past of the noir characters (Dolgopolov 9).

The outback of Outback noir, as we can see, is not limited to the geographical definition in terms of settlement structures and its remoteness to urban centres but goes beyond and features other ecological realities of the Australian environment. Yet whereas Knight considers, e.g., floods and fires "force[s] of vengeance" and reads them in postcolonial terms as aiding and abetting in revenging colonial land taking ('Crimes Domestic' 20), I would like to tap into econarratology and add a distinctly ecocritical edge to this by analysing them as ecologemes which are particular to Australian climate change realities. Following Erin James (23–24), I understand econarratology as being concerned with the site-specific representations of the environment and how those translate into narratological patterns. In her discussion of *The Dry* and *Scrublands*, Rachel Fetherston puts this into practice when she argues that these novels use the generic conventions of crime fiction "to engage with current ecological crises in a specifically Australian, rural context" (4). She reads the agency of what she refers to as "nonhuman phenomena" (2) in the tradition of Knight's observation of the land itself bringing about justice (Knight, 'Crimes Domestic' 20; Fetherston 10–11), but through an additional ecocritical lens. Therefore, she points out, if we are talking about place in this context, we need to include "environmental perspectives and understandings of nature" (Fetherston 3). I would like to add a complementary reading to this by positioning Fetherston's non-human as ecologemes and as a new set of narratological category somewhere between character and setting. In the case of Outback noir, this means reading ecologemes for their narratological influence and as necessary elements of genre analysis, instead of mere parts of the setting in the sense that they have come to supplement, even replace, defining characteristics of the genre while ensuring generic stability. Ecologemes in Australian crime fiction are more than just markers of place but are endowed with an agency of their own. At the same time, they are not quite characters but presences cloaking the characters and silently conversing with them. They are not restricted to creating a certain atmosphere, that of an inverse noir, but they are lived realities. As such, they fulfil a vital function both within and without the storyworld: Outside the storyworld, they help readers to identify Outback noir and to experience a multi-dimensional imaginary of the outback from the comfortable safety of their armchairs. Within the storyworld, they are mixed up in the detective work, either by covering the crime in glaring daylight and under layers of fine red dust or by taking the detective figure by the hand and leading them through the vastness of the outback.

Most obviously, the outback in Outback noir novels invites a geocritical reading in terms of ecologemes of place, the "sweeping plains" and "far horizons" of Mackeller's poem (Mackellar 10, 13). It does not need a close reading of the four novels to instantly realise the extent to which they are embedded in a certain spatial structure. The blurb introduces *The Dry* and *Scrublands* as set "in the small country town of Kiewarra" and "a country town", respectively, and speaks of "the remote border of their vast cattle properties . . . [in] the outback" and "an isolated part of Australia" in the case of *The Lost Man* as well as of "the desolate outback town of Finnigans Gap" in *Opal Country* (itself a spatially telling title). Less overtly, it is ecologemes of climate, "of droughts and flooding rains" (Mackellar 12), which are a unique characteristic of Outback noir. In fact, landscape and ecological reality are closely connected: "Drought", writes C.A. Cranston, "is a prevalent image in literature and in actuality, so prevalent that the landscape imagology of Australia . . . is likely to be that of the desert biome" (171). At this point, Outback noir both deviates from and harks back to the parent of regional crime fiction, Nordic noir. Whereas in Nordic noir the readers are supposed to feel the chilling cold of a Northern climate sweeping the beginnings of the Arctic tundra, in Outback noir they cannot escape "the midday heat, ferocious and furnace-dry" and riverbeds which have turned into "a mosaic of cracked clay, baked and going to dust" (*SL* 5). Although ecologemes of weather and climate often take the form of universally known entities and concepts, they are still place-specific due to the ecological realities of climate change which are experienced differently all over the world. In Australia, it is El Niño Southern Oscillation events (consisting of El Niño and La Niña) which are a key factor when it comes to the climate of (Eastern) Australia, droughts being the result of an El Niño episode and floods that of an occurrence of La Niña (Garden 35). However, in the four novels I have chosen for this chapter, it is El Niño with its heatwaves, droughts, and frequent fires which takes centre stage, described as "a high-pressure system suspended above eastern Australia like a spiteful god, banishing clouds and forbidding moisture" (*SL* 394).

The drought is most explicitly represented through references which are particularly Australian, i.e., references to vegetation and, to a certain degree, to the dryness of the earth, all of which can be considered ecologemes. On the other hand, the realities of climate change are also represented through more universal references, such as the sun, the sky, or heat, elements which nevertheless have a unique quality in an Australian context and thus also fit my definition of ecologemes. Each of the four novels can be read for a different set of ecologemes: It is vegetation together with the sun and the sky (which, combined, equal a deadly fire hazard) in *The Dry*; the sky and a red earth in *The Lost Man*; the heat, mulga scrub, and bushfires in *Scrublands*; and a drying salt lake together with, again, the heat, in *Opal Country*.

Of those ecologemes, it is the vegetation which is most particularly Australian and not found in other parts of the world. Originally a signifier of national identity via place, the bush in Outback noir no longer only works as a means to distinguish rural life from that in urban areas (cf. 'Bush' s.v. A 3) but has come to draw attention to the effects of climate change through an emphasis on the changes in vegetation. In *The Dry*, this shift of emphasis is established by focusing on how the drought has changed the colour palette of nature: "Outside [the car window], a thin wire fence protecting yellow scrub flashed past. Beyond, the fields were beige and brown" (*TD* 95). Brown and yellow are the predominant colours when it comes to descriptions of the bushland surrounding Kiewarra, so much so that it even extends to the pictures of schoolchildren on display where, mise-en-abyme-like, "[i]n every attempt at landscape, the fields were coloured brown" (*TD* 104). Those colours of the drought stand out precisely because they are not overly focused on, but mentioned almost in passing, as a wearily tolerated new reality. To get the full picture, however, the reader needs to have some background knowledge of the Riverina region in which Kiewarra is located. This region in the south-west of New South Wales is primarily an agricultural region with a number of regional centres where droughts threaten the very existence of its inhabitants, something the murder and frequent allusions to in the novel initially seem to attest to.

In a very similar way, *Scrublands* also plays with the idea of scrubs in the national imaginary as opposed to the actual vegetation of scrubs. According to the Australian National Dictionary, scrub is both "a wide range of generally low and apparently stunted forms of vegetation, often thick, impenetrable, and frequently growing in poor soil" ('Scrub' s.v. 1) and, as the Scrub, "the country as opposed to the town" ('Scrub' s.v. 3). In this capacity, it is linked to the bush. Unlike the landscape of *The Dry*, which is described by means of colours, the representation of vegetation in *Scrublands* goes a step further and already assesses the brownish and yellowish tinges of the grass as symptoms of a general fatal sickness caused by the drought: "The very country Martin is driving through looks sick: anaemic trees, spindly shrubs and, between them, more dirt than grass" (*SL* 76). Where the vegetation is not dying, it is already dead, "killed off by the sun and lack of water" (*SL* 33). Again, it is helpful to locate fictional Riversend (already a telling name) on a map to fully grasp the impact of climate change brought across in the novel through its use of ecologemes. Riversend is part of the Murray region, named for Australia's longest river, the Murray River. Accordingly, a riverbed which is nothing but "cracked bare mud" (*SL* 26) and "parched" (*SL* 125) is the representation of a lived reality of El Niño.

The very dryness of both river and scrub, both of which are also found in *The Dry*, have to be read alongside the oppressive heat and its origins, the sun and the cloudless sky. Although heat, sun, and sky are universal phenomena, they nevertheless are ecological markers of the particularly Australian, Outback, variety of crime fiction, for before the summer of 2022, hardly anyone in the Northern parts of Europe would have fully understood the meaning of "day after day of burning blue sky" (*TD* 1). They are described in hellish terms in *Scrublands*, where the sky is drained of its blue and is of a blinding white to mark "[a]nother hellish day in the 'town of death'" (*SL* 178), where it is "hot as Hades" (*SL* 43) and "the sun beats down" (*SL* 326). Those ecologemes indicate the reality of climate change in Australia and thus are not quirks used to create a corporate identity for Outback noir or to set it apart from other regional varieties of crime fiction but constitute a narratological category of its own right.

This is due to the rather unconventional generic history of crime fiction in Australia, which for rather long lacked recognisable references to a particularly Australian setting (Knight, *Continent of Mystery* 144). Knight attributes this to two key factors, publishing demands dictated by English publishing houses on the one hand and the repression of colonial dispossession on the other hand (144). That we now find ecological references to signify Australianness ties in with what Nicole Watson has argued about the normalcy paradigm of Whiteness (76–77): Instead of taking up debates of sovereignty or Native Title, contemporary Australian crime fiction written by non-Indigenous authors withdraws to representations of climate and literary imaginations of spatial structures as signposts of a national crime fiction identity. Furthermore, the ecologemes found in contemporary crime fiction have a hand in the detection, as in their ubiquity they both partner with the crime and the investigator.

In both *The Dry* and *Scrublands*, the lack of water together with the heat, "ferocious and furnace-dry" (*SL* 1), help to bring destruction in the form of bushfires. Bushfires have long been ingredients of Australian crime fiction as examples of how nature avenges the crime and punishes the criminal (Knight, 'Crimes Domestic' 20), and although in *The Dry* the bushfire fulfils exactly this purpose, the bushfire in *Scrublands* is of a different kind. For one, it does not mark the grand finale but unearths yet another murder. Additionally, instead of bringing justice and restoring order, it is an agent of destruction, fuelled by dry scrub. "In bushfires [as opposed to house fires]", Constable Haus-Jones explains to the detective figure, disgraced journalist Martin Scarsden,

> It's heat, pure and simple. The fire front will generate temperatures of hundreds of degrees. If it catches you in the open, it'll cook you alive. . . . You hear stories of people sheltering in swimming pools, farm dams, water tanks. Won't help. Water keeps 'em from burning, but the air is super-heated so they can't breathe – burns their lungs out from the inside.
>
> (SL *95)*

They also are especially vicious, as they might flare up time and time again: "the mulga scrub could smoulder for weeks, even months", and "the roots could burn away underground with little sign of it above ground. The only thing that would put the fire out once and for all would be soaking rain and plenty of it" (*SL* 115). Bushfires are both marked and unmarked ecological entities here, as on the one hand they are exceptional in their nature (as compared to other fires), but on the other hand, they have become commonplace, a constant threat, due to the combination of heat and dry vegetation (*SL* 76). Their sole purpose in Australian crime fiction is no longer to bring the criminal down, but to emphasise the consequences of climate change reality in Australia; in this case, perturbances in the El Niño Southern Oscillation weather patterns.

Those patterns, and the change in climate they stand for, are much more explicitly referred to in both *The Lost Man* and *Opal Country*. It is especially the change from La Niña to El Niño events, as exemplified in flooding riverbeds and lakes in one year and devastating droughts in the next, which are mentioned in those novels. Much of *The Lost Man* dwells upon the isolation of the outback and its horizonal sublime, where the sky in its emptiness and limitless mirrors the vastness of the land, recreating the imagery of an idealised outback setting. The particularly Australian flavour of the (literally) Outback noir in this novel is accomplished by drawing the readers' attention to the ubiquitous red dust which is a constant companion – it fills every crack and crease and "settle[s] on every surface" (*TLM* 21). Commonly referred to as the red centre, Australia's interior is more often than not painted in terms of red soil, and red seems to be the predominant colour the outback is associated with both inside and outside Australia. This ties in with understanding the outback in general, and the red centre in particular, as an environment which is inhospitable, even hostile, to people – something which also features in *The Lost Man,* because the titular victim had died in a 45-degree Celsius heat in the open country and far away from any homestead, let alone town (*TLM* 1). The recurrent reminders of the red dust throughout the novel have to be read together with its descriptions of the sky, which "loom[s] empty and large" (*TLM* 3), a "monstrous" (*TLM* 4) and "cloudless" (*TLM* 39) menace which allows "the blazing sun" (*TLM* 355) to scorch the earth. Here, the interaction of sky, sun, heat, and soil in *The Lost Man* mirror the actual processes of chemical weathering at work in some parts of Australia, as it is the hot and dry climate which causes the iron found in rocks to oxygenate and thus to turn red.

All of these elements I consider ecologemes, as through the ways they are characterised in the novel they lose their universal character and become genuine environmental markers for Australia. As for their impact on crime fiction, they pervade the entire course of the crime fiction story and accompany the double narrative of crime and detection (cf. Bennett 239). They connect both narratives through their presence and are the constant which links the present events (the detection) to those of the past (the crime). The fact that those ecologemes are used to depict the Australian Outback as hell on earth, though, is a purely Western point of view. For Australia's Aboriginal People living in the Outback (itself a category invented by Australia's settler society), the red centre of Australia is far from inhospitable, but nourishing and sacred Country, and its red earth is even represented in the Australian Aboriginal Flag. Due to the ever-increasing temperatures in Central Australia as a consequence of climate change, however, Central Australian Aboriginal People nevertheless fear for their survival as well as for that of their culture. Thus, the ecologemes of sky and earth are taking on additional meanings which reflect the realities of climate change in Australia. Furthermore, the changes between climate extremes are indicated right at the beginning of *The Lost Man*: "The ground was typically sandy and sparse for eleven months of the year and hidden under murky floodwater for the rest" (*TLM* 1). From the beginning, the novel uses the ecologemes of "droughts and flooding rains" to mark its Australianness, but at the same time, these ecologemes also speak of climate change (Mackellar 12). In Australian crime fiction, they do not simply create atmospheric density but collaborate with both crime and detective in subliminal ways. Stephen Knight has already shown that land and the agents it uses to exact revenge "are not merely useful, and even simplistic, plot devices", but "that there are evaluative capacities in the land itself" (*Continent of Mystery* 144), an idea I would like to broaden to incorporate representations of climate change in contemporary Australian crime fiction. Ecologemes "become disturbingly involved", to borrow from Knight (144); they are insinuating presences and constant reminders of the noirish quality that lies beneath the coat of red sand and behind the impenetrable wall of heat.

We find a similar representation of climate extremes in *Opal Country*, in which the reader is invited to witness the water evaporating from Lake Kalingra in the vicinity of the opal town of Finnigans Gap, "seven or eight hundred kilometres inland" (*OC* 12), and La Niña is explicitly mentioned. The timeframe of *Opal Country*, e.g., is measured in terms of the change from El Niño patterns to La Niña events, a change which occurs in the time elapsed between the events in *Scrublands* and those of *Opal Country*. This is contemplated by Detective Sergeant Ivan Lucic, who featured as a minor character in *Scrublands* and is the main protagonist of *Opal Country*, when he remembers the last time he went "this far inland", i.e., to the scrublands of Riversend: "Back then, the earth was bleached, barren and bone-dry, in the grip of a terrible drought" (*OC* 16) whereas now, "the rainclouds have returned, sweeping in week after week" (*OC* 16) because "some mighty switch out in the Pacific has altered its orientation" (*OC* 16). Here, the extreme weather phenomena are directly linked to climatic conditions and the rapid change from rainfall to drought is something the novel returns to time and time again in its description of the ways in which Lake Kalingra keeps disappearing. That Lake Kalingra is an indicator for climate change becomes clear in passing, when one of the various murder suspects explains that it is the "[f]irst time it's had any water in it for years", even for more than a decade (*OC* 57). It is a salt lake, and the average temperature is more than once estimated to be over 50 degrees Celsius (*OC* 57, 367) with the main character likening it to "the surface of a hostile planet" (*OC* 57), even that "of an alien sun" (*OC* 367). Both in its watery and dry condition, Lake Kalingra is the symbol of climate extremes and as such epitomises the effects of climate change. It condenses the frequent references to changing and varying climate patterns across Australia we find in the novel (cf. *OC* 20, 405) into a single ecological entity. At the same time, it is both witness and accomplice to murder and fraud (*OC* 481, 366). The representation of the heat, constantly estimated to be in the 40s (degrees Celsius) in Finnigans Gap and which has followed the heavy rains in *Opal Country*, is that of a brutal force concentrating its troops, almost of a beastly life form of its own, "ferocious" (*OC* 29) and "howling" (*OC* 405). It is characterised through a language of violence and its coexistence with the miners is framed as a constant battle (cf. *OC* 111, 241, 366).

There are a number of other ecologemes to be found in *Opal Country* and which follow the same patterns as in the other novels I have discussed in this chapter. The heat, e.g., is again complemented by a certain variety of sky and sun, all of them working together to stifle and suffocate. We find the same kind of brown vegetation that was so prevalent in *The Dry*, only now, we watch its slow death, its turning from green to brown parallel to the disappearance of Lake Kalingra. And then there is the earth again. Like in *The Lost Man*, it is described as "red soil" (*OC* 57), but there is also the titular dirt, a reference which gets lost by publishing the novel as *Opal Country* instead of as *Treasure and Dirt* outside Australia. Dirt, in fact, is particularly Australian (mining) vernacular and refers to the kind of debris in which opal is found (cf. 'Opal' s.v. 1). The treasure of course is opal, the most valuable of which, black opal, is found only in Finnigans Gap (and its real-life equivalent, Lightning Ridge). Although opal is by no means singular to Australia, it is nevertheless typical of Australia, as is mining itself. At this point, through its use of ecologemes as a new generic element of crime fiction, the novel also touches upon environmental crimes by centring on an elaborate con involving a rare-earth mine. Mining itself is a highly contested issue in Australia, with financial interests regularly clashing with those of the Traditional Owners and Custodians, sometimes causing irreparable loss, as in May 2020, when mining giant Rio Tinto destroyed Juukan Gorge, a sacred site which was also witness to and had preserved thousands of years of environmental changes. By linking Australia's mining tradition to one of the world's most sought-after resources, the novel moves beyond the national to address global environmental concerns. What is left out, though, are any references to the traditional owners of the site where the Kalingra mine is located and thus to mining as a colonial project.

A Narratological In-Between

The ecologemes of Outback noir I have discussed in this chapter paint the picture of a sunburnt country replete with references to anthropogenic climate change instead of singular and non-recurrent natural disasters. As site-specific ecological entities, they also change the narratological patterns of the genre. In Outback noir, ecologemes often coincide with the national imaginary, as in the case of the bush or the outback, shifting them from mere tokens of an imagined national identity to signifiers of climate change realities. Outback noir is a particularly apt genre to narrate this twofold scope of ecologemes, as it subverts romanticised escapist fantasies – you cannot flee the dark undercurrents of the outback, neither the life-threatening realities of climate change. In the case of Outback noir, ecologemes are still embedded in particularly Australian understandings of the rural and based on an Australian version of pastoralism and thus work as recognisably Australian signposts. Reading them from an ecocritical perspective, however, allows us to see beyond their status as national icons and unique selling points and lets us understand them as entities speaking of climate change in a particularly Australian context. Reading them with an econarratological lens helps us to scrutinise the generic conventions of crime fiction in general, as they feature as a new narratological category that is neither setting nor character. Hovering uneasily between the two, they are constant presences shadowing the characters who actively contemplate them and engage in dialogue with them, and vice versa. Outback noir, which, as a regional crime fiction variety, depends on displaying a uniquely Australian flavour. At the same time, it is deeply involved in global and national histories of crime fiction. Therefore, the use of ecologemes as markers of both genre and Australianness as well as of climate change realities in Outback noir positions them as new and emerging nodal points in a network of crime fiction.

Notes

1. By Arrangement with the Licensor, The Estate of Dorothea Mackellar, c/- Curtis Brown (Aust) Pty Ltd.
2. The Australian original was published under the title *Treasure and Dirt* by Allen and Unwin in 2021. For its 2022 publication in Europe, the British publisher Wildfire changed the title to *Opal Country*, which is the edition I am using in this chapter.
3. In the Australian context, *The Australian National Dictionary* defines a larrikin as a "person who acts with apparently careless disregard for social or political conventions; a person who is unsophisticated but likeable and good-hearted" ('Larrikin' s.v. 2.a), a definition which captures both the anti-authoritarian stance of the larrikin and the rather positive connotations it carries in an Australian context.
4. Recent exceptions to this written by non-Blak authors are Peter Papathanasiou's *The Stoning* and Emma Viskic's Caleb Zelic series (*Resurrection Bay*, *And Fire Came Down*, *Darkness for Light*, and, most recently, *Those Who Perish*), which do feature strong and prominent Indigenous characters and also touch on Indigenous issues. Crime fiction written by Blak authors differs from its non-Blak counterpart in that it focuses on Indigenous characters, their identity in Australian society, and their experiences in a climate of racism (Watson 79–80).

Bibliography

Anderson, Jean. "Strategies for Strangeness: Crime Fiction, Translation and the Mediation of 'National' Cultures". *The Translator*, vol. 22, no. 2, 2016, pp. 221–231, doi:10.1080/13556509.2016.1184881.

Ashcroft, Bill. "The Horizonal Sublime". *Antipodes*, vol. 19, no. 2, 2005, pp. 141–151.

Ashman, Nathan. "Hard-Boiled Ecologies: Ross Macdonald's Environmental Crime Fiction". *Green Letters*, vol. 22, no. 1, 2018, pp. 43–54, doi:10.1080/14688417.2018.1431139.

Ballard, Michel. *Versus: La Version Réfléchie*. Ophrys, 2003.

Bennett, Donna. "The Detective Story: Towards a Definition of Genre". *PTL*, vol. 4, 1979, pp. 233–266.

Cranston, C. A. "Reconstructing Representations: 'Australia' as Ecocritical Andragogy". *Ecocritical Concerns and the Australian Continent*, edited by Beate Neumeier and Helen Tiffin, Lexington Books, 2020, pp. 163–189.

Dolgopolov, Greg. "Ghosting in the Outback Noir". *Coolabah*, no. 29, 2021, pp. 4–16, doi:10.1344/co2021294-16.

Doolan, Emma D. *Hinterland Gothic: Reading and Writing Australia's East Coast Hinterlands as Gothic Spaces*. Queensland University of Technology, 2018.

Erdmann, Eva. "Nationality International: Detective Fiction in the Late Twentieth Century". *Investigating Identities: Questions of Identity in Contemporary International Crime Fiction*, edited by Marieke Krajenbrink and Kate M. Quinn, Rodopi, 2009, pp. 11–26.

Fetherston, Rachel. "'Little Difference between a Carcass and a Corpse': Ecological Crises, the Nonhuman and Settler-Colonial Culpability in Australian Crime Fiction". *JASAL*, vol. 21, no. 2, 2021, pp. 1–17, https://openjournals.library.sydney.edu.au/JASAL/article/view/14914/13575.

Forrester, Kim. "'The Lost Man' by Jane Harper". *Reading Matters*, 26 Dec. 2020, readingmattersblog.com/2018/12/02/the-lost-man-by-jane-harper/.

Garden, Don. "Colonial History, Economy and Culture: The Influence of South-Eastern Australian Weather and Climate". *Nature and Environment in Australia*, edited by Beate Neumeier et al., WVT, 2018, pp. 33–47.

Gifford, Terry. "Pastoral, Anti-Pastoral, and Post-Pastoral". *The Cambridge Companion to Literature and the Environment*, edited by Louise Westling, Cambridge UP, 2014, pp. 17–30.

Hammer, Chris. *Scrublands*. Wildfire, 2019.

———. "The New Class of Australian Crime Writers". *CrimeReads*, 22 Jan. 2019, crimereads.com/the-new-class-of-australian-crime-writers/.

———. *Opal Country*. Wildfire, 2022.

Hansen, Kim Toft, and Anne Marit Waade. *Locating Nordic Noir*. Springer, 2017.

Harper, Jane. *The Dry*. Abacus, 2017.

———. *The Lost Man*. Little, Brown, 2019.

Hayward, Keith J., and Steve Hall. "Through Scandinavia, Darkly: A Criminological Critique of Nordic Noir". *The British Journal of Criminology*, vol. 61, no. 1, 2020, pp. 1–21, doi:10.1093/bjc/azaa044.

Hedberg, Andreas. "The Knife in the Lemon: Nordic Noir and the Glocalization of Crime Fiction". *Crime Fiction as World Literature*, edited by Louise Nilsson et al., Bloomsbury Academic, 2017, pp. 13–22.

Hill, Annette, and Susan Turnbull. "Nordic Noir". *Oxford Research Encyclopedia of Criminology and Criminal Justice*. Oxford UP, 26 Apr. 2017.

Hughes-d'Aeth, Tony. "Landscape (After Mabo)". *The Routledge Companion to Australian Literature*, edited by Jessica Gildersleeve, Routledge, 2021, pp. 292–303.

James, Erin. *The Storyworld Accord: Econarratology and Postcolonial Narratives*. U Nebraska P, 2015.

Kim, Julie H. "Introduction: National Identity and International Crime Fiction in the Age of Populism and Globalization". *Crime Fiction and National Identities in the Global Age: Critical Essays*, edited by Julie H. Kim, McFarland & Company, 2020, pp. 1–11.

King, Stewart. "Place". *The Routledge Companion to Crime Fiction*, edited by Janice M. Allan et al., Routledge, 2020, pp. 211–218.

Knight, Stephen. *Continent of Mystery: A Thematic History of Australian Crime Fiction*. Melbourne UP, 1997.

———. "Crimes Domestic and Crimes Colonial: The Role of Crime Fiction in Developing Postcolonial Consciousness". *Postcolonial Postmortems: Crime Fiction from a Transcultural Perspective*, edited by Christine Matzke and Susanne Mühleisen, Rodopi, 2006, pp. 17–33.

———. *Australian Crime Fiction: A 200-Year History*. McFarland & Company, 2018.

Lungu-Badea, Georgiana. "Remarques sur le Concept de Culturème". *Translationes*, vol. 1, no. 1, 2009, pp. 15–78, doi:10.2478/tran-2014-0003.

Mackellar, Dorothea. "My Country". *Macquarie PEN Anthology of Australian Literature*, edited by Nicholas Jose, Allen & Unwin, 2009, pp. 388–389.

Moore, Bruce, ed. "Bush". *The Australian National Dictionary: Australian Words and Origins*. 2nd ed., vol. 1: A–L, Oxford UP, 2016.

———. "Larrikin". *The Australian National Dictionary: Australian Words and Origins*. 2nd ed., vol. 1: A–L, Oxford UP, 2016.

———. "Opal". *The Australian National Dictionary: Australian Words and Origins*. 2nd ed., vol. 2: M–Z, Oxford UP, 2016.

———. "Outback". *The Australian National Dictionary: Australian Words and Origins*. 2nd ed., vol. 2: M–Z, Oxford UP, 2016.

———. "Scrub". *The Australian National Dictionary: Australian Words and Origins*. 2nd ed., vol. 2: M–Z, Oxford UP, 2016.

Neale, Margo. "First Knowledges: An Introduction". *Songlines: The Power and Promise*, edited by Lynne Kelly and Margo Neale, Thames & Hudson, 2020, pp. 1–5.

Nestingen, Andrew, and Paula Arvas. *Scandinavian Crime Fiction*. U Wales P, 2011.

Nilsson, Louise. "Mediating the North in Crime Fiction". *Journal of World Literature*, vol. 1, no. 4, 2016, pp. 538–554, doi:10.1163/24056480-00104007.

Pezzotti, Barbara. "Transnationality". *The Routledge Companion to Crime Fiction*, edited by Janice M. Allan et al., Routledge, 2020, pp. 94–101.

Puxan-Oliva, Marta. "Crime Fiction and the Environment". *The Routledge Companion to Crime Fiction*, edited by Janice M. Allan et al., Routledge, 2020, pp. 362–370.

Reed, Sarah M. A. *Translating Cultural Identity: French Translations of Australian Crime Fiction*. Peter Lang, 2019.

Rolls, Alistair, et al. "Translating National Allegories: The Case of Crime Fiction". *The Translator*, vol. 22, no. 2, Jan. 2016, pp. 135–143, doi:10.4324/9781315161778.

Schaffer, Kay. *Women and the Bush: Forces of Desire in the Australian Cultural Tradition*. Cambridge UP, 1988.

Shaw, Janice. "Australian Crime Fiction: Such is Life for Hard-Boiled Larrikins". *Crime Fiction and National Identities in the Global Age: Critical Essays*, edited by Julie H. Kim, McFarland & Company, 2020, pp. 35–56.

Stadler, Jane, et al. *Imagined Landscapes: Geovisualizing Australian Spatial Narratives*. Indiana UP, 2016.

Stougaard-Nielsen, Jakob. *Scandinavian Crime Fiction*. Bloomsbury Academic, 2017.

Venuti, Lawrence. *The Translator's Invisibility: A History of Translation*. 2nd ed., Routledge, 2008.

Vermeer, Hans J., and Heidrun Witte. *Mögen Sie Zistrosen? Scenes & Frames & Channels im Translatorischen Handeln*. Groos, 1990.

Watson, Nicole. "Deadly Detectives: How Aboriginal Australian Writers Are Re-Creating Crime Fiction". *AlterNative: An International Journal of Indigenous Peoples*, vol. 15, no. 1, 2018, pp. 75–81, doi:10.1177/1177180118818187.

PART IV

Criminality and Justice

21

REVISING CRIME IN FICTION

An Environmental Invitation

Marta Puxan-Oliva

The global environmental crisis, and efforts to redress it, have cast doubt on the conventional definition of crime at the highest level, legally, politically, and culturally. The fact that campaigns such as "Stop Ecocide" and the "International Parliamentary Alliance for the Recognition of Ecocide" are arguing for "ecocide" to be recognised as an international crime showcases the fact that environmental harm inflicted on a daily basis, and the overlap of bodies and institutions in charge of regulating it in a coordinated range of scales, have undoubtedly produced a deep questioning of the concept of "crime" itself. In the past few years, even a new field in criminology has emerged, green criminology, which directly targets this problem and proposes a revision of environmental crime. In the process, the boundaries between what does (and what does not) constitute environmental crime have become significantly blurred. Just to pick one simple example, we would certainly consider illegal fishing a major environmental crime but would not so easily define massive bottom trawling as a "crime", even though it is scientifically documented as irreversibly devastating to the seabed and its ecosystems. Since many trawling practices are permitted, this globally used fishing technique might simply not qualify as criminal practice. And yet, if its obvious economic profit is disregarded, it is not hard to imagine this fishing technique being banned in the not-too-distant future. Examples are endless, including gas emissions, techniques such as fracking or mountaintop removal mining, low-frequency active sonar in submarine surveillance, and toxic and plastic dumping. In the light of crime fiction as a genre and crime in fiction in a broader sense, the current discussion over redefining crime prompts us to revise how this notion works in fiction and how the idea of "crime" is changing today.

This chapter argues that crime fiction is strongly contributing to reveal the ambivalences and maladjustments that lie at the root of our working interpretations of crime. Crime fiction undertakes this task through its recent interest in environmental crime and its formal elaboration, thereof, by means of embedded narratives that question the weak boundaries that today separate some spectacular, bloody crimes from other less visible ones, particularly those that operate on other, not necessarily human scales. In this, environmental crime is a cornerstone. I will review the distinctions between environmental harm and crime proposed by green criminologists, will consider how crime fiction expresses this problem, and will address how criminality has permeated several narratives that, despite not explicitly adhering to crime fiction as a genre, also have environmental crime at their core. Finally, I will delve into Juan Villoro's Mexican novel *Arrecife* (*The Reef*, 2012) to illustrate the problem of defining environmental criminality. Within this, I claim that literature strongly participates in the current redefinition of crime and harm whilst stimulating new formal challenges.

DOI: 10.4324/9781003091912-26

Environmental Criminality in the Spotlight: Redefining Crime

What is legally defined as "crime" is a social and cultural convention that changes through different political and legal systems over time. While this is probably widely accepted, attachment to, and concern for, already legally established crimes often shortcuts discussion over what we consider crime. However, when new sensibilities and contexts emerge, the blurring contours of legally established crimes are shaken, as has occurred in relation to environmental crime. Environmental harm is hard to assess as it often has long-term consequences, is accumulative, barely visible, and tends to inflict ecosystemic harm rather than direct harm to human beings. All these features condition awkward accommodation within a legal system that is mostly based on readily identifiable and traceable harms inflicted on humans. Oftentimes, environmental harm is difficult to detect, trace, and prosecute since it is a slow kind of violence, to use Rob Nixon's expression. In addition, harm that qualifies as environmental crime is regulated through different legal bodies, such as Administrative Law versus Criminal Law (De la Cuesta), and at local, national, and international scales that need to agree in various points so as to find a difficult, fragile collective agreement. This is due to the perception that environmental crime is a global concern that needs to be agreed upon in international treaties in order to palliate the effects of a global environmental crisis. These layers make the definition of environmental crime an extremely complex matter. Nevertheless, the peculiarities of environmental harm and crime demand new reflection and revision with regard to how environmental criminality is being assessed today. A new field has emerged from this preoccupation: green criminology. Though varied in their approaches, which range from studying already established environmental crimes, such as poaching, to clamorous harms such as plastic waste disposal, green criminologists examine environmental harm from a sociolegal point of view. As Avi Brisman and Nigel South define it:

> Green criminology refers to the study of environmental crimes and harms affecting human and non-human life, ecosystems and the biosphere. More specifically, green criminology explores and analyzes: the causes, consequences and prevalence of environmental crime and harm, the responses to and prevention of environmental crime and harm by the legal system (civil, criminal, regulatory) and by nongovernmental entities and social movements, as well as the meaning and mediated representations of environmental crime and harm.
>
> *(Brisman and South 1)*

A substantial body of this scholarship undertakes the task of revising the idea of crime with a broader view since, as Potter points out, "[w]hen we strip away the value positions to the bare science that underpins green criminology, we are left with an important observation: a green perspective can contribute greatly to our general understanding of crime" (Potter 125).

At the core of this revision is an expansion of the concept of crime beyond legal boundaries, by focusing on inflicted harm and on the consequences that practices have rather than on the actions themselves and their agents. As Morelle-Hungría explains:

> Environmental harm is considered a crime within the standpoint of critical criminology, both as cause of the crime and as a factor to take into account in the prevention of risk situations, in this ecocriminology can be a uniting point between crime and environmental harm.
>
> *(Morelle-Hungría 11)*[1]

Focusing on harm so as to weigh up what should be considered criminal practices involves the central concern for the consequences to the environment and to human beings. As Morelle-Hungría explains, placing harm at the centre helps us identify the extent to which some practices can impact on the environment and human beings in the long run and, as such, can help our legislation prevent irreparable damage. That is, a flexible understanding of crime that has harm at the basis, he argues, would help us review our current established crimes and include some practices that are today legal yet extremely harmful for the environment.[2] Rob White discusses this along similar lines:

> If ecological (and social and economic) welfare is to be maximised, then there is a need to expand notions of what actually constitutes environmental crime. Harm, as conceived by critical green criminologists, for example, demands more encompassing definitions than that offered by mainstream law and traditional criminology. This is because some of the most ecologically destructive activities, such as clear-felling of old-growth forests, is quite legal, while more benign practices, such as growing of hemp (an extremely strong fibre), is criminalised. Green criminology provides an umbrella under which to theorise and critique both illegal environmental harms (that is, environmental harms currently defined as unlawful and therefore punishable) and legal environmental harms (that is, environmental harms currently condoned as lawful but which are nevertheless socially and ecologically harmful). How harm is conceptualised is thus partly shaped by how the legal – illegal divide is construed within specific research and analysis.
>
> *(22)*

The simultaneous focus on harm to both the environment and human beings also favours taking an ecocentric perspective, which understands that we are part of an ecosystem that needs to be addressed as a whole, not only in relation to the effects that practices have on human beings. The ecosystemic perspective of green criminology takes sustainability as a measure of environmental criminality.[3] These challenging perspectives to criminology and to criminal and administrative law push forward the sharp boundaries of crime, adopt an ecosystemic perspective, and invite understanding of the deep interconnections within the ecosystem that make sustainability a clear future priority. It also forces us to address the fact that, as White reminds us, "[w]hat constitutes environmental crime is, therefore, contentious and ambiguous" (19–20), a contested and ambivalent conceptualisation that is now constituent to how we understand environmental criminality and whose possibilities and shortcomings we need to revise.

Crime Fiction and the Interrogation of Environmental Crime

Recent scholarship has addressed crime fiction and the environment in several aspects, such as the foregrounding of ecological degradation (e.g. Ashman, Jordan), of Indigenous reservations and the environment (e.g. Bandyopadhyay, Dechême, and Di Gregorio), of plants and animals (e.g. Carroll, McLauchlan), and of the global dimension of environmental damage (e.g. King; Hollister; Farrier; Puxan-Oliva).[4] Criminology and crime fiction share clear interests that require further exploration. As Matthew Levay observes:

> Both are interested in what counts as a crime and who counts as a criminal as well as the conditions that produce these designations, and both investigate the problem of how to prevent acts of crime through institutional, individual, or other means.
>
> *(273)*

As Levay traces in his chapter "Crime Fiction and Criminology", developments in criminology have historically gone hand in hand with crime fiction's changing views and focus. Indeed, recent treatment of environmental criminality in crime fiction can be read alongside the redefinition proposals of green criminology. In particular, crime fiction has strongly suggested the contested association of environmental harm with crime, which constitutes the fundamental argument for a number of green criminologists. As I have pointed out elsewhere (Puxan-Oliva, 2020), crime fiction often addresses the criminality of environmental harm through the creation of 'double stories', where the typical and more localised crime plot is placed *alongside* the larger frame of environmental damage, which ultimately resonates with the target crime. Through this narrative strategy, the genre's central feature of a crime prepares the reader to apprehend an environmental criminality that goes beyond the specific murder and which mostly ties together the main crime with the consequences of indirect environmental victims. On many occasions, environmental harm is related to economic business whose profit derives from activities that transform the environments and affect ecosystems as a whole; on other occasions, the environmental crime goes unpunished because there is no justice system that recognises such acts as criminal.

This is clear in crime fiction novels like Roberto Ampuero's *El alemán de Atacama* (1996), Ricardo Piglia's *Blanco nocturno* (2010), and Donna Leon's *Earthly Remains* (2017), which trigger reconsideration of environmental harm as criminal. Ampuero's *El alemán de Atacama* features the murder of a German activist who proposes a change to the management of aquifers in the Atacama Desert, which barely supply an Indigenous village with water due to increasing "ecotourism". The murder was committed by a company that illegally buried toxic waste from European chemical experiments in the desert, a practice that inevitably poisoned the already fragile freshwater wells.[5] Piglia's *Blanco nocturno* tells the story of the transformation of the Argentinian rural world through the story of the industrialisation of the countryside. Piglia's novel focuses less on the environment, but it does emphasise how capitalist transformations bring about corruptive practices that ultimately lead to murder. Studied by Aina Vidal-Pérez in this collection (Chapter 32), Leon's *Earthly Remains* presents a murder in a Venetian lagoon where the population of bees is also mysteriously dying. Commissario Brunetti uncovers that the murder is due to the discovery of industrial toxic waste disposal in the lagoon, which is in fact responsible for the ecosystemic death of species. Ampuero's and Leon's novels directly expose their environmental concerns, placing toxic waste dumping clearly in the realm of environmental criminality.

What these three crime fictions have in common is that the environmental situations they portray inflict harm that is explicitly tied to the murders around which the crime novel revolves. This is a neat inversion: while the murder crime is clearly identified in the narrative, it directly refers to the environmental condition, which ultimately invites the reader to interrogate the underlying environmental harm, which is not explicitly exposed as criminal. In this sense, criminality invites a rethinking of environmentally damaging actions not only as preconditions for already established crimes but as criminal in themselves. In the case of Ampuero's and Leon's novels, toxic waste disposal is technically illegal, yet both narratives point to toxic dumping as a broader unresolved problem that is not always properly regulated. Neither prosecution nor punishment for this practice feature as part of the novels' central plots, as attention is directed more towards the individuals responsible for the murders and towards the exploits of the investigating detectives. That is, explicit legal actions in the novels target the murderers, but not the problem of toxic waste disposal as an environmental crime. These crime-focused novels establish new paths for writing using double criminal stories: one crime for which it is easy to trace and find the individuals responsible; another, underlying one, which leaves environmental crime as a highly ambivalent concept.

Environmental Crime Beyond Genre

In its recent engagement with the environment, crime fiction has opened the path to consider other novels that are interested in environmental crime, and which turn to this core component of crime fiction. A growing body of contemporary works that would not necessarily fit into the genre of crime writing are currently questioning environmental harm and inviting us to think of it as criminal. Environmental criminality is suggested in forms that range from novels that use the afore-described pattern of a double crime plot, through some that directly treat environmental harm as crime, to others that reveal environmental harm at a major scale, clearly inviting us to conceive it in global criminal terms. In novels like Wajdi Mouawad's *Anima* (2012), Juan Villoro's *Arrecife* (2012), and Rafael Chirbes' *En la orilla* (2013), a murder tied to a special environmental circumstance drives the plot.[6] In Mouawad's *Anima*, the questioning of environmental harm is indirectly driven by the historical removal of Native Americans from their own territories and their enclosure in reservations, leading to drug and weapon trafficking. This disastrous transformation of their way of life is forced by an integration process that destroys the communities' ecosystemic way of living. Written from a highly polyphonic perspective of multiple animal narrators, the novel adopts a strong ecosystem perspective that decentres the criminal story. As we will see, Villoro's *Arrecife* features a narcotraffic murder in an enormous hotel complex in a coral reef in the Mexican gulf that has destroyed the coast. In a similar vein to *Arrecife*, Chirbes' *En la orilla* develops around the degraded environmental and socio-economic circumstances produced by the boom in construction and tourism that along the Spanish Mediterranean coasts in the 1960s, leading to major debts that result in a character's murder in the coastal wetlands.

Other novels such as Luis Sepúlveda's *Mundo del fin del mundo* (1989) target environmental crime directly. This novel deals with Japanese illegal whale hunting in the Chilean Patagonia, where an unregistered ship claims that whales are captured for scientific purposes. While there is no actual human murder in this novel, criminal whale hunting serves to denounce the specific illegal practice as much as to place it within an ambivalent framework that combines it with other legal and illegal environmental harms such as deforestation, with the aim of upscaling it to planetary damage. Another case is John Grisham's *Gray Mountain* (2014), which centres on the difficult legal prosecution of mountaintop carbon mining. The mining process itself inflicts severe damage upon the people working on, and living around, the extraction site, due to the technique's release of toxic gases that ultimately become linked to a range of health problems, including various forms of respiratory disease. The novel mostly focuses on environmental victims but legally denounces environmental harm and its complicated regulation due to the increasing profit and long-term consequences of these nonetheless authorised practices. Sepúlveda's and Grishman's novels directly target environmental damage as a crime using the cases of illegal whale hunting and mountaintop mining, but they scale them to their contested regulation and prosecution.

Finally, novels such as Nathaniel Philbrick's *In the Heart of the Sea* (2000), Wu Ming-Yi's *The Man with the Compound Eyes* (2011), and John Lanchester's *The Wall* (2019) use the environmentally damaging circumstances as background to the main plot, which tends to focus on the characters' experience of a direct or indirect environmental crisis. Philbrick's *In the Heart of the Sea* focuses on the survival story of the *Essex*'s sailors who suffered the rare attack of the sperm whale that inspired Melville's *Moby Dick*. The novel explores the dreadful voyage of the remaining seamen across the Pacific, but it also lays out the massive nineteenth-century whale hunting industry as partly explaining the whale's "revenge". Wu Ming-Yi's *The Man with the Compound Eyes* features a tsunami that is caused by an enormous plastic island that crashes against the Taiwanese coast, devastating a whole village. While exposing the catastrophic consequences of environmental plastic waste dumping and linking it with the sudden phenomenon of a tsunami induced by climate change, the novel nonetheless focuses on the love story of two characters rather than on criminality. In keeping environmental damage as an existing circumstance rather than incentivising a denouncing tone, the novel manages to downplay the great catastrophic long-term effects of plastic dumping and climate change, revealing the slowness of this kind of harm in its controlled narrative tone and rhythm. Lanchester's science fiction novel, *The Wall*, presents a post-apocalyptic scenario after the Change, a sudden modification of the Earth's condition, where seawater levels have risen, and the United Kingdom has survived. The surviving country seeks protection against the millions of sea refugees who try to reach land with a robust wall guarded by an entire population under the grip of a totalitarian regime. The environmental catastrophe remains as a generational memory, focuses on the harm inflicted on the environmental victims, and addresses the diffuse culpability of a collective crime with no individual victim or individual responsibility, which David Farrier describes as characteristic of Anthropocene noir fiction. *The Wall* works as an Anthropocene fiction that emphasises environmental harm from a mostly anthropocentric point of view, the proportions of which signal the importance of preventing environmental damage so that what are today speculations about the future consequences might be prevented. These novels lay bare the magnitude of environmental harm, while keeping it silent in the novel, since focus does not initially refer to environmental damage as such, but to stories in which environmental harm indirectly causes a catastrophe and dire post-catastrophic lives. In this sense, these narratives emphasise the invisibility and long-term effects of environmental harm that green criminologists identify as the place where we should act and legislate to prevent major consequences.

As this literature shows, environmental harm and crime have become a growing interest for contemporary fiction, instigating a hybridisation of genre that blurs the boundaries of what we understand as crime fiction. As Hollister suggests, ecological awareness stimulates a perspective akin to noir, in the sense that criminality is examined more broadly, without a dead body or the need for resolution, via a foregrounding of setting that invites us to take a wider perspective that goes beyond the genre to pay attention to fictions of ecological violence that take on a global perspective. Not only does Hollister's observation invite us to consider other non-fiction novels in the relations between criminality and ecology, but it also drives our attention to the *problem* of this relationship in and of itself, as seen and adopted by a larger corpus of varied literary works. In order to display the specific problem of ambiguity and the difficult apprehension of environmental criminality, the novels reviewed here borrow multiple narrative strategies, especially from crime fiction and science fiction (Minler; Heise), to be able to better locate the discussion on harm and crime and to reach the global scale (Farrier; Trexler) that the prevention of future planetary damage requires. All of all these examples plot strategies that contribute to the exploration of an underlying criminal story, which leaves environmental crime as an ambivalent force underpinning the main story.

Juan Villoro's *Arrecife*: The Ambiguity of Environmental Criminality

The Mexican novel *Arrecife* (2012) by Juan Villoro is a good example of the bewildering definition of environmental criminality. The novel would probably not qualify as crime fiction, but at most a "narconovela" or "narcoficción" (Adriansen and Kunz). In fact, focus is not on the central murder, probably not even on narcotraffic, but rather on the scenario in which the plot takes place; a massive hotel in the fictional Mexican Maya area of Kukulkán that rises in pyramidal form in a peninsula upon the second largest coral reef in the world. The narrator, a tourist resort musician, ex drug addict and ex-player in the rock band Los Extraditables, Antonio Gándara, tells the story of a double murder in La Pirámide hotel. The first victim, a diver called Ginger Oldenville, turns up dead in the hotel's aquarium with a harpoon in his back and a hammock rope knotted to his penis. The hotel manager, Mario Müller, the narrator's childhood best friend, is quick to label the death a homosexual suicide, as he attempts to conceal the hotel's connection to the Cruci/Ficción organisation. This story is problematised, however, by the discovery of a second body, that of another expert diver, Roger Bacon, on nearby shores. Like Müller, Bacon is also found with a pairing rope knot tied to his penis.

As half-hearted investigation develops and the narrator gradually comes to know, from different sources, Roger Bacon had discovered a cocaine cargo and told Ginger, who was determined to report it. As readers learn later, the US hotel's major shareholder, Michael Peterson – usually called Gringo

Peterson or just Gringo – had organised the murder through his London-based finance manager, James Mallett, and the head of security, Leopoldo Támez. The deep reasons for the crime turn out to be more personal than expected, since, as Peterson confesses to a complicit narrator, he saw in Ginger's willingness to betray La Pirámide's enterprise an excuse to punish him for deserting the elite Coast Guardianship as an expert diver when he was sent to the Persian Gulf. As Peterson's own child had drown in a pool, he could not forego Ginger's refusal to save lives despite his specialised diving training. Furthermore, Peterson considered this a suitable crime, since it would simply lead the hotel to bankruptcy, which would be the ideal condition to pursue and even increase international money laundering. All the agents involved, from the police officers to the multiple individuals who run and work at La Pirámide had reasons to harm Ginger. In one way or another, all the characters benefited from the narcotraffic network, for which reason the crime is mostly under-investigated and certainly left unpunished at the end of the novel. As Mario explains:

> "If you ask around", he said, "you'll learn that the leading candidate for governor, the head of the opposition, the Secretary of Tourism, the whores, the bishop – all of them wanted it to happen. That's where the heat was coming from. When the rocks are glowing red and air is practically roasting your face, it's not so hard to come together for a sacrifice. Just ask the Maya: the best people are always the ones to die."
>
> *(223)*[7]

The discovery of the murderer's causes and effects drives the plot, but the unravelling occurs informally through the narrator's partial learning of information and his well-founded suspicions. In this way, the novel only partially reveals the whole story in a narrative strategy that Andrew Pepper observes in other crime fiction narcotraffic novels such as the Mexican Juan Pablo Villalobos' *Fiesta en la madriguera* (*Down the Rabbit Hole*, 2010). In this case, this partial revelation of the murders and the narcotraffic business strengthens the novel's attention to the grotesque enterprise of La Pirámide, which has been built in order to manage this dirty money effectively. La Pirámide is both a producer of happy adventurous tourists as well as a factory of criminals on the fringes of society. Even more ironically, Mario Müller's benevolent actions towards lost people – such as the illegal US migrant to México, yoga instructor Sandra, or the narrator himself who, lost in drug consumption, was rescued by his friend and assigned the job of producing ambient music – turn La Pirámide into a hotel residence that recalls a sanatorium or a rehabilitation centre for marginalised people involved in crimes and drug consumption.

In *Arrecife*, attention is driven from crime to place. As in many "crimate" novels – to use Stewart King's term – making sense of place becomes especially central. In *Arrecife*, this place is a tourist resort that was state planned and authorised, since "It was the President's decision to put Kukulcán on the map" (54).[8] This decision to transform the local area at the risk of devastating its ecosystems was motivated by the claim to offer jobs to poor local communities, and at the expense of the fishermen who were removed from those shores:

> The biological reserve was transformed into a golf course, money flooded the area, and suddenly there were jobs for people who used to suck mango seeds to curb their hunger. Monoliths of glass and steel came to dominate the coast.
>
> *(54)*[9]

Highly connected and internationally promoted, the resort's hosting of the Miss Universe pageant and (ironically) a world convention on climate change had contributed to a successful façade. However, forty years of aggressive construction and coastal use had resulted in environmental devastation and unsustainable local economics. In spite of La Pirámide's drug business, the once prosperous tourist enclave had undergone a process of heavy, rapid degradation, since

> [W]hat had once been tolerably dirty development now became excessively dirty. Diseases arrived, and so did pollution and the guerrilla and rainfall every three days and the drug trafficking and beheadings. The countless churches that popped up in the region offered little consolation.
>
> *(54)*[10]

This strong decline is mostly due to the environmental harm entailed by the construction of the town, which not only affects the reef's sustainability but directly destroys the tourism business it was built for

> Practically all of the hotels in Kukulcán were vacant. They rose up along the shore like vertical mausoleums, circled by seagulls and ravaged by plants and rats.
>
> The cruise ships no longer pulled into the pier where an enormous statute of Sebastian, our sainted sculptor, dominated the landscape. In the distance we could see the ships passing by. Only their trash made it to the beach. In the evenings, children and old people wearing rags would emerge to sort through it. I'd seen them make off with spoons, plastic bags (their contents mysterious), and bits of soggy food.
>
> The coastal region had entered a period of hardship. The tourist enclave hadn't heeded the warnings about building on the sand: the wind beat mercilessly against the façades and then rushed out to sea, taking the beach along with it. Every day a slow boat from Santo Domingo would bring sand to fill the cavities along the shore. The coast-line was slowly devouring itself.
>
> Also, the oil rigs and city plumbing had contaminated the water, which was now threatening the second-largest coral reef in the world. [Only La Pirámide survived with some success, thanks to the risky temptations created by Mario Müller.] Things at The Pyramid weren't so bad, especially in contrast to the thirty-story buildings of the hotel's competitors, where the only signs of life.
>
> *(35)*[11]

Construction right on the sand becomes fatal to beach conditions, oil drilling and sewage dumping heavily pollutes the waters, and the weather becomes more aggressive as a result of climate change. This devastation does not only affect the coast and seawater, but also the local population, who are left with much less profitable activities, such as local men burning vegetation in order to grow corn or little girls selling coconut soap painted pink. As Tony reflects: "The present-day Maya had burned the forest to plant corn. Their best option was narcotrafficking; after, that, The Pyramid; and after that, burning trees" (60).[12]

Paul Goldberg argues that the association of narcotraffic with environmental degradation, often based on the destruction of the traditional agriculture, is common in Mexican narcotraffic novels, such as Leonidas Alfaro's *Tierra blanca* (1996), Gerardo Cornejo's *Juan Justino Judicial* (1996), and Elmer Mendoza's *El amante de Janis Joplin* (2001). *Arrecife* does not resort to the pastoral but focuses on the combination of mass tourism and drug trafficking in the Mexican Maya area. In fact, Golderg recalls that in 2012, the same year *Arrecife* was published, a report documented "the destruction of habitat in Guatemala's Maya's Biosphere Reserve", detailing that

> [A]lthough the forest has been under threat by logging and oil drilling for years, destruction of the Western half of the Reserve has increased during the past decade as drug traffickers clear land to build airstrips and roads in order to transport goods northward.
>
> *(33)*

Whether observed from a pastoral point of view, like the aforementioned novels, or from a global capital perspective, narcotraffic and environmental harm are intimately linked.

To tackle the economic crisis brought about by environmental degradation of this area, Mario Müller engages with a new kind of tourist resort that has managed to keep the tourist clientele that other hotels in the tourist enclave of Kukulcán have been unable to retain over the years. As the narrator Tony explains:

> Der Meister had been Peterson's solution for avoiding total disaster. The reef's days were numbered. Under constant rainfall the hotels had been closing one by one, or operating at 10 percent occupancy. The area was moving toward decline, but Mario had found a solution: he was peddling a tropics amped on adrenaline, complete with venomous spiders, and excursions that created the illusion of miracle survival – necessitating raucous after-parties.
>
> *(35)*[13]

Mario's solution to massification and continuous destruction of the environment is to offer a unique tourist experience based on danger and fear in a form of "tanatourism", a kind of tourism that "searches for real or symbolic encounters with death and violence" and which "is driven by the desire to visit places that are inherently associated to death" (Adriaensen, 109). In México, tanatourism associates with narcotraffic or with experiences such as crossing the US border in what is called "turismo morbo" ("morbid tournism"), "turismo negro" ("dark tourism"), or "narcoturismo" ("narco-tourism") (Adriaensen 110). Basically, La Pirámide offers a catalogue of fear experiences such as kidnapping, guerrilla attacks, dangerous animals, and convenient, terrifying performances of Mayan mythology in situ. The farce even includes playful Mayan interpretations of landscape transformation. When Mario and Tony are out with several clients on an exciting Mayan excursion, the slow environmental change is perceptible:

> "The Earth is titled!" Mario shouted, mimicking a slanted wingspan.
>
> The land did in fact slope down toward the sea at an almost imperceptible angle. When it rained, the water washed away the soil that had been loosened by all the tree-felling. The mud clouded the reef, and this prevented the coral from receiving adequate sunlight. We had successfully added deforestation to the list of the region's misfortunes.
>
> "Now this is real Maya apocalypse!" Mario shouted. He. Coughed from the strain.
>
> *(58)*[14]

Mario uses the slow environmental violence whose ecosystemic effects result in coral death in the reef to display a cosmic Mayan catastrophe. Even if Mario Müller's adventure hotel model has proved highly effective for several years, it does not prove so when risk and fear are not only faked, but murders are real and two actual corpses lie in La Pirámide. While not fully investigated, the double murder is, certainly, a dangerous advert. The crime brings the owner and managers face to face with deciding their hotel's future. Gringo's ultimate option is to go bankrupt, since as Mario explains to Tony:

> Abandoned hotels are a fantastic racket. Have you seen all those empty buildings lining the beach? They're inhabited by rats and badgers; seagulls build their nests in the attics. But officially there are no vacancies. It's the best form of money laundering. I learn so much from the British. They're the ones who invented these so-called *offshore* paradises in their former colonies. Some figures from the *Financial Times*: Ten per cent of all money laundering happens in London. Hotels that go broke are the perfect places for simulating investments and setting up ghost accounts.
>
> *(106)*[15]

However, James Mallet believes the hotel still has many real tourism options and proposes to convert La Pirámide into a cultural and ecological tourist project, since these areas have enormous, underexploited potential, as he argues to Tony Gándara:

> Atrium donates to NGO's and has its own non-profit organization. . . . "*We are global players*", he added in English. "We're interested in your country, Tony. La Pirámide is the ideal setting for our new Project. . . ."
>
> "Let's call it 'sustainable colonialism' shall we? Or 'soft imperialism'. The whole scheme – it's not only safer than Mario Müller's, it's also more marketable. We could open a children's center with a Harry Potter Pavilion! The only thing more commercial than culture is ecology. . . ."
>
> "Guilt", he said while he licked the rice paper, "is a lucrative business. The people who pollute the most are the same ones who invest the most in organic food and ecotourism." [La Pirámide still has great potential: ecology and culture.]
>
> *(184–5)*[16]

The new model would profit from social guilt and be based on cultural and ecological consumption; the drug business would remain intact.[17] The final outcome of La Pirámide is unknown, as it is the final resolution of the murder crimes. In fact, these are examples of a poignant way in which criminality is treated in this novel. While the murders and narcotraffic are clearly identifiable crimes, *Arrecife* powerfully engages in blurring the contours and nature of criminality. Criminality is diffused in the multiple ways in which the novel plays with facts and imagination and around the practice of performed fake incidents, mostly related to crime.

La Pirámide bases its survival on a mixture of "provocation, seduction and violence" (Martha Elia Arizmendi Domínguez and Gerardo Meza García, 37), which operates through the performance of fake criminal episodes that ignite fear in the customers who, after extreme experiences such as professional actors' kidnappings of tourists – which sometimes last over three days – end up with clients enjoying a cocktail on the beach or by the pool, joyfully revisiting their exciting experience. Tony witnesses one of these kidnappings without even knowing it is fake, as a result of which he is knocked down with a violent Chinese hold that leaves him in the infirmary, though unharmed. On another occasion, during a Mayan excursion, some actors fake a guerrilla assault, and Tony himself is assigned to act as a guerrilla member.

However, violent episodes are not only limited to La Pirámide's contracted events. The security chief Leopoldo Támez is attacked by two armed men, an aggression that we later learn has been paid for by Támez himself in order to regain his colleagues' trust and appear as a victim rather than Ginger's and Roger's potential murderer. Similarly playing with performance and fake criminality, Mario had also set up the story of a gay suicide in a strategy that involved inverting the order of the murders and tying hammock rope knots to the victim's penises:

> "You don't get what I'm saying", Ríos said. "There was a set up. Bacon didn't die in saltwater and his body wasn't discovered after Ginger's. But someone really wanted us to see it that way. Why? So we'd focus on the knot tied around the guy's dick. So we'd make the connection to Cruci/Fiction and the other knot, which we already had. Everything happened in reverse. Bacon was investigating the same thing Ginger was. Killing one diver made it necessary to kill the other."
>
> *(124)*[18]

These criminal performances at the basis of La Pirámide's tourism project get confused with the real criminality that is taking place: the two murders and the narcotraffic crime, which expands to surrounding criminal victims, such as the girls hosted by a nearby refugee institution. Therefore, fake criminality complicates the limits of what is true and what is false, and what is harm and what is entertainment, what is violence and what is pleasure.

Further, the blurring between performance and reality extends to fact and imagination in another key aspect of the novel. They get mixed up deep in the character's life: Tony has strong memory lapses that are filled with Mario's memories of their friendship. As Margherita Cannavacciuolo observes, these memories are not always facts clearly remembered by Mario to reconfigure a past which has been forgotten. Rather, their truth is often uncertain: "Mario and Antonio seem to constitute a memory diptych where, in Mario's embodying of the memory that his friend lacks, the already blurred limits between memory and imagination completely dissolve" (Cannavacciuolo 146).[19] Tony does not remember many episodes in his life, but Mario provides him with memories that might very plausibly be fake but are convenient for Mario's future plotting. *Arrecife* is, in many ways, a novel about rebirth and rehabilitation, about finding new identities that can make our own bearable. However, even this rehabilitation is uncertainly crafted over a restored memory. In blurring the lines between imagination and fact, the novel invites performative reinterpretation of actual reality. As seen with the reinterpretation of the Mayan apocalypse, environmental degradation is invited to be read as part of the scene, as lying in the uncertain terrain of dubious criminality, between what we see and what we want to see. The limits of criminality and harm are strongly questioned and even subverted in this novel. *Arrecife* points to environmental harm as interpretable both as criminal and as a product of our militant interpretation. Undoubtedly, were the project of La Pirámide oriented towards ecotourism, the environmental harm would be completely reinvented. The novel leaves the reader with no criminal resolution, let alone legal prosecution: neither the murders nor the narcotraffic, nor even the environmental destruction of the coastal area are solved in the novel. This lack of resolution, combined with the play between performance and reality, undoubtedly leaves criminality as the most ambiguous and fragile area.

A growing corpus of contemporary novels, in several traditions and from diverse genres, interested in the environmental crisis has enabled a revision of environmental criminality, highlighting its ambivalent location in our current imagination and legislation. In addressing environmental harm in conjunction with criminal elements, novels such as the aforementioned contribute to push forward the legal delimitations of environmental crime so as to invite reflection on the underlying status of environmental damage and the visible consequences it has on sustainability and human beings. Leading literary discussions on crime and the environment, crime fiction is developing interesting plot, time, and space narrative strategies that effectively link environmental harm to crime, such as double plotting, thereby extending our conception of crime to potential, less visible and less resolved criminality. Environmental criminality is undertaken by a relevant number of other contemporary novels that frequently borrow formal conventions of the crime novel in order to tackle environmental criminality through various narrative strategies, from downplaying an otherwise evident degradation of the environment; using legal prosecution as the basis of the plot so as to directly indicate environmental crime; and a polyphony of animal narrators to force an ecosystem understanding of the story; to deep ambiguity in structural inversions such as the confusion created by performed and real crime in Juan Villoro's *Arrecife*. The polemical invitation of green criminologists to revise environmental harm and crime demands serious consideration, which should at least generate public acknowledgement that environmental crime cannot be satisfactorily resolved as legally established today, and that great ambiguity lies at the heart of its current legal form. As literary critics we should be duly attentive to the challenging task that contemporary fiction has decidedly undertaken in the interrogation of environmental criminality, which certainly deserves, from our part, further exploration.

Acknowledgements

I am grateful to Aina Vidal-Pérez, for our discussion and her suggestions, and to Steward King for his revision and observations on a chapter draft.

Funding

This chapter was possible thanks to the grant Programa Ramón y Cajal (RYC2019–028186/AEI/10.13039/501100011033) in a contract funded by the Ministerio de Ciencia e Innovación, the Agencia Estatal de Investigación, the Universitat de les Illes Balears and the European Social Fund ("The European Social Fund Invests in your future").

Notes

1 My translation of the original: "El daño ambiental es considerado como un crimen dentro de los posicionamientos críticos criminológicos, tanto como causa del crimen como factor a tener en cuenta en la prevención de situaciones de riesgo, ante ello, la ecocriminología puede ser un nexo de unión entre la concepción de crimen y daño social".

2 Morelle-Hungría identifies classifications of harms that include for example White's (2008) green harm, as referring to biodiversity in general and other questions such as climate change; brown harm, as referred to polluting activities; and white harm for pathogens and biotechnology; and distinctions between *primary green crimes*, which are directly harmful for the environment and biodiversity, and *secondary green crimes*, which are "consequence of a negligent or illegal corporate or state activity" (4).

3 In addition to the ecosystemic approach, green criminologists also adopt "anthropocentric" or "biocentric" perspectives (see Birsman and South; Morelle-Hungría).

4 One of the most relevant publications in this relationship is the special issue in *Green Letters*, "Crime Fiction and Ecology", edited by Jo Lindsay Walton and Samantha Walton and published in 2018.

5 See Marún and Canepa for this novel. I have developed the analysis of this case in Puxan-Oliva.

6 A rich analysis of animal narrative voices in Mouawad's *Anima* can be found in Patron and Weber, while an ecocritical analysis of Chirbes' *En la Orilla* is discussed by Vidal-Pérez.

7 All quotes of the novel are included in the English published translation in the text, and in the original Spanish in the endnotes. In the original: "-Si preguntas por ahí sabrás que el candidato a gobernador, el líder de la oposición, el secretario de turismo, las putas y el señor obispo querían que pasara eso. El calor existe para eso. Cuando las piedras arden y el aire te raspa la cara, no es difícil ponerse de acuerdo para hacer un sacrificio. Pregúntale a los mayas: el mejor debe morir". (218)

8 In the original: "Kukulkán fue puesto en el mapa por decisión presidencial" (61).

9 In the original: "La reserva biológica se transformó en campo de golf, el dinero inundó la zona, hubo trabajo para gente que mataba el hambre chupando huesos de mango, los monolitos de cristal y acero dominaron la costa" (61).

10 In the original: "[l]uego el tolerable desarrollo sucio se volvió demasiado sucio. Llegaron los virus, la contaminación, la guerrilla, las lluvias cada tercer día, el narcotráfico, las decapitaciones. Las numerosas sectas que prosperaban en la zona ofrecían poco consuelo" (61).

11 In the original: "Casi todos los hoteles de Kukulcán estaban vacíos. Se alzaban por la costa como mausoleos verticales, orbitados de gaviotas, invadidos de plantas y de ratas.

Los cruceros ya no se detenían en el embarcadero donde se alzaba una inmensa escultura de Sebastián (una geométrica estrella de mar, color azul cobalto). Veíamos a lo lejos los barcos que seguían de largo. Su basura llegaba a la costa. Al atardecer, niños y ancianos vestidos de andrajos aguardaban desperdicios. Los había visto rescatar cucharas, bolsas de plástico de contenido incierto, comer restos de comida mojada.

El litoral había entrado en una fase de agonía. La ciudad turística no atendió las advertencias sobre los riesgos de construir sobre la arena: el viento golpeaba en las fachadas sin escape alguno, y regresaba rumbo al mar, llevándose la playa. Todos los días, un barco lento llegaba de Santo Domingo con arena para llenar los huecos en la orilla. La costa se devoraba lentamente a sí misma.

Las plataformas petroleras y los drenajes habían contaminado el agua, poniendo en riesgo el segundo arrecife de coral más grande del mundo. Sólo la Pirámide sobrevivía con éxito, gracias a las arriesgadas tentaciones creadas por Mario Müller.

La situación no era mala, sobre todo en comparación con los edificios de treinta pisos donde la única señal de vida era el repentino cortocircuito de un aparato eléctrico" (42).

The sentence in brackets in the text is missing in the translation, so I added it to the English quote.

12 In the original: "Los nuevos mayas quemaban la foresta para sembrar maíz. Su mejor alternativa era el narcotráfico; la segunda, la Pirámide; la tercera, quemar plantas" (66).

13 In the original: "*Der Meister* había sido su solución ante la catástrofe. Los días del arrecife estaban contados. Bajo una lluvia incesante, los hoteles habían cerrado uno tras otro, o trabajaban con una ocupación del diez por ciento. La zona estaba destinada al deterioro, pero Mario encontró una solución: un trópico con adrenalina, arañas venenosas, excursiones que creaban la ilusión de sobrevivir de milagro y la necesidad de celebrar en forma tempestuosa". (42–3).

14 In the original: "-¡La tierra está inclinada! – gritó Mario, haciendo un ademán oblicuo.

El terreno se deslizaba hacia el mar en un declive apenas perceptible. Con las lluvias, el agua arrastraba la tierra aflojada por la tala de árboles. El lodo oscurecía el arrecife, impidiendo que los corales recibieran la luz del sol. A las desgracias de la zona debíamos agregar la deforestación.

-¡Es el verdadero apocalipsis maya! – el esfuerzo lo hizo toser" (65).

15 In the original: "Los hoteles abandonados son un espléndido negocio. ¿Has visto los edificios de la Costera? Ahí viven ratas, tejones, las gaviotas hacen nidos en las azoteas, pero oficialmente están llenos. Es la mejor forma de lavar dinero. He aprendido mucho con los ingleses. Ellos inventaron los paraísos *offshore* en sus antiguas colonias. Te doy un dato del *Financial Times*: el 10% de todo el blanqueo de dinero se hace desde Londres. Los hoteles quebrados son perfectos para simular inversiones y llevar una contabilidad fantasma" (109).

16 In the original: "Atrium ha hecho donativos a ONG, tiene su fundación de Obra Social . . . *We are global players*. Nos interesa tu país, Tony. La Pirámide es el escenario ideal para un nuevo proyecto. (. . .)

– Llámalo 'colonialismo sustentable' o 'imperialismo sutil'. El plan no es sólo más pacífico que el de Mario Müller; es más rentable. ¡Podemos abrir un proyecto infantil, un pabellón dedicado a Harry Potter! Lo único más comercial que la cultura es la ecología. (. . .)

– La culpa es un supernegocio – dijo mientras ensalivaba el papel de arroz – . Los que más contaminan son los que más invierten en comida orgánica y ecoturismo. La Pirámide aún tiene mucho potencial: ecología y cultura" (182).

The sentence in brackets in the text is missing in the translation, so I added it to the English quote.

17 For a helpful survey of the relations between cultural and ecological tourism and their great economic possibilities see Picornell Belenguer and Martínez Tejero.

18 In the original: "No me ha entendido: hubo un montaje. Bacon no murió en agua salada ni fue descubierto después. Alguien quiso que las cosas fueran vistas de ese modo. ¿Para qué? Para que nos fijáramos en el nudo en el pene, para que lo asociáramos con Cruci/Ficción y con el otro nudo, que ya teníamos. Todo ocurrió al revés. Bacon estaba investigando lo mismo que Ginger. Matar al primer buzo obligaba a matar al segundo" (126).

19 My translation. In the original: "Mario y Antonio parecen constituir un díptico memorial donde, al encarnar Mario la parte de la memoria que le falta al amigo, los límites ya borrosos entre memoria e imaginación se disuelven del todo" (Cannavacciuolo 146).

Bibliography

Adriaensen, Brigitte. "Turisteando en Narcolandia: la comodificación de la violencia en *Arrecife* de Juan Villoro." *Narcoficciones en México y Colombia*, edited by Brigitte Adriaensen and Marco Kunz, Iberoamericana-Vervuert, 2016, pp. 109–124.

Adriaensen, Brigitte, and Marco Kunz, editors. *Narcoficciones en México y Colombia*. Iberoamericana-Vervuert, 2016.

Ampuero, Roberto. *El alemán de Atacama*. Random House De Bolsillo, 2013.

Arizmendi Domínguez, Martha Elia, and Gerardo Meza García. "Espacio narrativo en *Arrecife* de Juan Villoro." *Un recuento de los días transcurridos. La escritura de Juan Villoro*, edited by Martha Elia Arizmendi Domínguez, Universidad Autónoma del Estado de México, 2019, pp. 31–40.

Ashman, Nathan. "Hard-Boiled Ecologies: Ross Macdonald's Environmental Crime Fiction." *Green Letters*, vol. 22, no. 1, 2018, pp. 43–54, doi:10.1080/14688417.2018.1431139.

Bandyopadhyay, Nibedita. "'The Green Sleuth': An Analysis of the Environmentalism in the Selected Detective Fictions of Sunil Gangopadhyay." *Green Letters*, vol. 22, no. 1, 2018, pp. 67–77, doi:10.1080/14688417.2018.1431142.

Bishop, Katherine E. "'When 'Tis Night, Death Is Green': Vegetal Time in Nineteenth-Century Econoir." *Green Letters*, vol. 22, no. 1, 2017, pp. 7–19, doi:10.1080/14688417.2017.1413990.
Brisman, Avi, and Nigel South. "Green Criminology and Environmental Crimes and Harms." *Sociology Compass*, vol. 12, 2018, pp. 1–12, doi:10.1111/soc4.12650.
Canepa, Gina. "El desierto como basurero en *El alemán de Atacama* de Roberto Ampuero: ecoliteratura en los tiempos de globalización." *Polis: Revista Latinoamericana*, no. 17, 2007.
Cannavacciuolo, Margherita. "Repoblar el recuerdo. Fracturas temporales en *Arrecife* de Juan Villoro." *Ressegna iberistica*, no. 99–100, 2013, pp. 141–156.
Carroll, Alicia. "'Leaves and Berries': Agatha Christie and the Herbal Revival." *Green Letters*, vol. 22, no. 1, 2018, pp. 20–30, doi:10.1080/14688417.2018.1438303.
Chirbes, Rafael. *En la orilla*. Anagrama, 2013.
Dechêne, Antoine, and Luca Di Gregorio. "'Knock it off with the Mystical Horseshit:' Detection in the American West." *Green Letters*, vol. 22, no. 1, 2018, pp. 55–66, doi:10.1080/14688417.2018.1431140.
De la Cuesta, José Luis. "El derecho al ambiente: su protección por el derecho penal." *E-Revue Internationale de Droit Penal*, 2017, pp. 1–20.
Farrier, David. "Animal Detectives and 'Anthropocene noir' in Chloe Hooper's *A Child's Book of True Crime*." *Textual Practice*, vol. 32, no. 5, 2018, pp. 875–893, doi:10.1080/0950236X.2016.1275756.
Goldberg, Paul. "Narco-Pastoral: Drug Trafficking, Ecology, and the Trope of the Noble Campesino in Three Mexican Narconovelas." *ISLE: Interdisciplinary Studies in Literature and Environment*, vol. 23, no. 1, 2016, pp. 30–50, doi:10.1093/isle/isw012.
Grisham, John. *Gray Mountain*. Doubleday, 2014.
Heise, Ursula K. "Science Fiction and the Time Scales of the Anthropocene." *ELH*, vol. 86, no. 2, 2019, pp. 275–304.
Hollister, Lucas. "The Green and the Black: Ecological Awareness and the Darkness of Noir." *PMLA/ Publications of the Modern Language Association of America*, vol. 134, no. 5, 2019, pp. 1012–1027, doi:10.1632/pmla.2019.134.5.1012.
Jordan, Peter. "Carl Hiaasen's Environmental Thrillers: Crime Fiction in Search of Green Peace." *Studies in Popular Culture*, vol. 13, no. 1, 1990, pp. 61–71.
King, Stewart. "*Crimate* Fiction and the Environmental Imagination of Place." *The Journal of Popular Culture*, vol. 54, no. 6, 2021, pp. 1235–1253, doi:10.1111/jpcu.13083.
Lanchester, John. *The Wall*. Norton, 2019.
Leon, Donna. *Earthly Remains*. Atlantic Monthly Press, 2017.
Levay, Matthew. "Crime Fiction and Criminology." *The Routledge Companion to Crime Fiction*, edited by Janice Allan et al., Routledge, 2020, pp. 273–281.
Marún, Gioconda. "Una denuncia ecológica: *El alemán de Atacama* de Roberto Ampuero." *Lain American Detective Fiction: New Readings*, edited by Shelley Godsland and Jacqueline Collins, Manchester Metropolitan UP, 2004, pp. 112–121.
McLauchlan, Laura. "Sadness and the Noir of Urban Hedgehog Conservation." *Green Letters*, vol. 22, no. 1, 2017, pp. 101–112, doi:10.1080/14688417.2017.1413991.
Milner, Andrew, et al. "Ice, Fire and Flood." *Thesis Eleven*, vol. 131, no. 1, 2015, pp. 12–27, doi:10.1177/ 0725513615592993.
Morelle Hungría, Esteban. "Ecocriminología, la necesaria visión ecosistémica en el siglo XXI." *Revista Electrónica de Criminología*, vol. 03–02, 2020, pp. 1–14.
Mouawad, Wajdi. *Anima: Roman*. Lémeac/Actes du Sud, 2012.
Nixon, Rob. *Slow Violence and the Environmentalism of the Poor*. Harvard UP, 2011.
Patron, Sylvie. "Narrations d'*Anima*. Un récit non naturel?" *Langues d'Anima. Écriture et histoire contemporaine dans l'œuvre de Wajdi Mouawad*, edited by Claire Badiou-Monferran and Laurence Denooz, Garnier, 2016, pp. 41–61.
Pepper, Andrew. "Crime Fiction and Narcotics." *The Routledge Companion to Crime Fiction*, edited by Janice Allan et al., Routledge, 2020, pp. 371–378.
Philbrick, Nathaniel. *In the Heart of the Sea: The Epic True Story that Inspired Moby Dick*. Viking Press, 2000.
Picornell Belenguer, Mercè, and Cristina Martínez Tejero. "From Pleasant Difference to Ecological Concern. Cultural Imaginaries of Tourism in Contemporary Spain." *Companion to Iberian Environmental Cultural Studies*, edited by Luis I. Prádanos. Tamesis, 2022, pp. 195–206.
Piglia, Ricardo. *Blanco nocturno*. Anagrama, 2010.

Potter, Gary R. "Justifying 'Green' Criminology: Values and 'Taking Sides' in an Ecologically Informed Social Science." *The Values of Criminology and Criminal Justice*, edited by M. D. Cowburn, Polity Press, 2013, pp. 125–140.
Puxan-Oliva, Marta. "Crime Fiction and Environment." *The Routledge Companion to Crime Fiction*, edited by Janice Allan et al., Routledge, 2020, pp. 362–370.
Sepúlveda, Luis. *Mundo del fin del mundo*. Tusquets, 1994.
Vidal-Pérez, Aina. "La piscina global. El Mediterráneo de Rafael Chirbes desde el *spatial turn* y la ecocrítica." *452°F*, no. 21, 2019, pp. 73–91.
Villoro, Juan. *Arrecife*. Anagrama, 2012.
———. *The Reef*. Translated by Yvette Siegert. George Braziller Publishers, 2017.
Walton, Jo Lindsay, and Samantha Walton. "Introduction to Green Letters: Crime Fiction and Ecology." *Green Letters*, vol. 22, no. 1, 2018, pp. 2–6, doi:10.1080/14688417.2018.1484628.
Weber, Julien. "Composer Avec Les Animaux Dans Anima De Wajdi Mouawad." *L'Esprit Créateur*, vol. 57, no. 1, 2017, pp. 99–110, doi:10.1353/esp.2017.0008.
White, Rob. "The Conceptual Contours of Green Criminology." *Emerging Issues in Green Criminology: Exploring Power, Justice and Harm*, edited by Reece Walter et al., Palgrave MacMillan, 2013, pp. 17–33.
Wu, Ming-Yi. *The Man with the Compound Eyes*. Harvill Secker, 2013.

22
CRIMINAL VIOLENCES
The Continuum of Settler Colonialism and Climate Crisis in Recent Indigenous Fiction

Rebecca Tillett

Recent works by Indigenous North American authors are redefining what constitutes 'crime' fiction through an examination of the criminal violences of the settler-colonial state and the impact of those violences upon Indigenous individuals, communities, and lands. These violences, driven by greed and marked by white privilege and entitlement, are evident in a range of related federal policies that have resulted both in the genocide of Indigenous peoples and in the ecocide of the natural world. As a result, these violences also extend our understandings of environmental racism. Kyle Powys Whyte (Potowatomi) has argued that for Indigenous peoples "climate change and colonialism are interrelated" because both "start . . . with greed" ("Climate Change" 9). Indeed, the extreme "anthropogenic environmental changes" that accompanied colonialism in North America saw such a "rapi[d] disrupt[ion]" of Indigenous peoples that, Whyte comments, Indigenous peoples in the twenty-first century "often understand their vulnerability to climate change as an intensification of colonially-induced environmental changes" ("Indigenous Climate Change Studies" 154). Whyte argues, therefore, that the contemporary climate crisis is both "unprecedentedly old" and profoundly colonial ("Climate Change" 9). Recent twenty-first century Indigenous fiction traces the roots of this unprecedentedly old climate crisis in the insatiable behaviours and ideologies that drive capitalist settler-colonial systems. In these fictional texts, the slow and criminal violences imposed upon the environment are directly related to the equally entrenched and lengthy slow and criminal violences imposed upon Indigenous peoples by settler-colonial systems, a series of ideologically driven "delayed destruction[s]" that are similarly "attritional", "incremental and accretive" (Nixon 10). This chapter explores how twenty-first century Indigenous fiction acts to expose the inextricable links between the violences of settler colonialism and the violences of climate crisis in the Anthropocene era and to make visible the continuum of criminal violences in North America.

This continuum can be measurably dated to the violences enacted by early European settlers in the Americas, and this dating acts to challenge some of the ongoing arguments surrounding our conceptualisation of the Anthropocene. While the International Working Group on the Anthropocene declared in 2016 that the era of human-caused climate crisis should be dated from the mid-twentieth century, even as late as 1964, the timing – and indeed the term itself – is much contended (Carrington, 'The Anthropocene epoch'). Quite apart from the fact that the 'universality' of the term 'Anthropocene' fails to address highly unequal power relations between 'developed' and 'developing' nations and their concomitant climate responsibilities, this dating also fails to recognise the inextricable links between the criminally violent and insatiable behaviours and ideologies driving settler colonialism

DOI: 10.4324/9781003091912-27

and racialised capitalism and their related human and climate outcomes. Accordingly, many critics argue, the term 'Capitalocene' is more accurate (Moore, 2016). Moreover, as Simon Lewis and Mark Maslin have argued, the Anthropocene should, more correctly, be dated from 1610, a moment that marks not only the "Columbian Exchange" of plants and animals between Europe and the Americas that irreversibly transformed the ecosystems of both, but also the 'Orbis spike' – a notable and measurable "dip in atmospheric CO_2" in the Americas caused by the permanent reduction in numbers of Indigenous peoples through the violences enacted by settler-colonisers (175). As Heather Davis and Zoe Todd (Métis) argue, not only is "the foundational and epistemic violence of European colonialism" evident within the Anthropocene, but the Anthropocene itself is an "*extension and enactment of colonial logic*" that forced a "radical transformation of the biosphere" of the Americas (769, original emphasis). In this context, we can reframe our conceptualisations of the Anthropocene, and of environmental racism, to effect what Erik Kojola and David N. Pellow call "a deeper engagement with histories and ongoing practices of domination that devalue life and lead to premature death for marginalized peoples and nonhuman species" (30). Above all, we can trace and expose the continuum between the criminal violences of capitalist driven settler colonialism and profit-driven climate crisis by considering the interlinked ideologies of colonial and environmental North American histories. As Davis and Todd argue, these histories have "always [been] about changing the land, transforming the earth itself, including the creatures, the plants, the soil composition and the atmosphere. . . . moving and unearthing rocks and minerals" because "all of these acts were intimately tied to the *project of erasure* that is the imperative of settler colonialism" (770, emphasis added).

The writers under consideration in this chapter – Louise Erdrich (Turtle Mountain Chippewa), Cherie Dimaline (Métis), and Waubgeshig Rice (Anishinaabe Wasauksing First Nation) – all explore the twenty-first century impact of this lengthy continuum of related criminal violences on Indigenous peoples and on Indigenous lands. In her 2012 novel, *The Round House*, Erdrich reveals the historical trajectories of settler-colonial ideologies that have enabled and facilitated an epidemic of sexual violence against Indigenous women and girls. Engaging directly with the ongoing Missing and Murdered Indigenous Women and Girls (MMIWG)[1] crisis, Erdrich traces vitally important historical connections between white American conceptions of land, law, and gender that have resulted in such extensive, even genocidal, sexual and physical violence against almost incomprehensible numbers of Indigenous women and girls. Erdrich's novel deliberately acts to consider this kind of criminal violence within the contexts of both American law and tribal law: when culturally crucial Indigenous land is both desacralised and set outside of settler-colonial state law – effectively made 'law-less' – the direct result is that criminal violences are unpunishable and go unpunished. Similar contentious settler-colonial histories are explored in Cherie Dimaline's 2017 young-adult novel *The Marrow Thieves*, which traces how climate crisis is directly related to genocidal settler-colonial practices in a speculative yet imminent future. Here, the majority white North American population has lost its ability to dream, and Indigenous peoples are rounded up to have their dreams 'harvested' in former Indian school buildings. Dimaline taps into the criminally violent histories of the settler-colonial Indian education system exposed by the recent Canadian Truth and Reconciliation Commission (2008–2015), which resulted in the undocumented deaths of thousands of Indigenous children, many still lying in unmarked graves. By situating these futuristic 'dream factories' on historical sites of genocide, Dimaline's story points suggestively to the equally criminally violent and damaging notion of 'extraction' and extractive ideologies, and – crucially – considers how Indigenous peoples continue to resist the consumption of their bodies, cultures, and lands by the settler-colonial state. My final focus, on Waubgeshig Rice's 2018 novel *Moon of the Crusted Snow*, assesses how Rice creates a further imminent and probable future to expose the direct relationships between twenty-first century climate crisis and genocidal settler-colonial practices. Rice's novel, which casts a white interloper within an Anishinaabe tribal community as the traditional Windigo, a dangerously contagious cannibal monster, further explores not only how the voraciousness of capitalism drives climate crisis, but also how the ideologies of settler-colonial cultures consume both land/environment and Indigenous peoples. Rice's novel offers a cautionary demonstration of how the insatiable desire for ever-greater 'profits' must be combatted as an imperative for our very survival in an era of climate crisis. Most importantly, Rice demonstrates that powerful, enduring, and all-consuming Windigo ideologies and policies can and should be successfully resisted.

Significantly, Erdrich, Dimaline, and Rice all present traditional Indigenous values and worldviews as pertinent and valuable counterpoints to the continuum of settler-colonial and capitalist criminal climate violences. Whyte comments that "Indigenous intellectual traditions are rooted in philosophies that work to understand how the actions of human societies are entwined with environmental change", noting that "one aspect of these traditions concerns political philosophies of diplomacy for peoples who share ecosystems" ("Climate Change" 8). All three writers and texts discussed here explore how these political philosophies work in practice, exposing and rejecting the established hierarchies of significance that have been forcibly enacted by the mutually supportive systems of patriarchy, colonialism, and capitalism and that directly result in criminally violent and fatally damaging profit-driven climate crises that impact on Indigenous peoples across the globe. While Erdrich, Dimaline, and Rice all interrogate the criminal violences of broken imperial world systems, they also all demonstrate how we might see and live in very different, more sustainable ways.

Displacing Settler-Colonial Legal Violences: Louise Erdrich's *The Round House*

Louise Erdrich's *The Round House* (2012) tells the story of the violent rape and attempted murder of an Ojibwe woman, Geraldine Coutts, in a text that explores the racialised nature of the law in the United States, where 'ownership' of 'property' is valued above the well-being of either human or environmental bodies. Told from the perspective of Geraldine's 13-year-old son Joe, Erdrich's story exposes the damage that Geraldine's experience inflicts upon her: it is an act of such profound criminal violence that it "nearly sever[s]" Geraldine's "spirit from her body" (*Round House* 7, 45). Sarah Deer (Creek) argues that sexual violence is "one of the most devastating threats to contemporary indigenous culture[s]" because it actively contributes to the ongoing traumas of "colonization and cultural domination" ("Toward and Indigenous Jurisprudence of Rape" 121). In this context, Erdrich traces the impact of Geraldine's rape upon her immediate family – Joe and her husband, the tribal judge Bazil – but also upon the wider reservation community as a whole. Moreover, Rebecca Macklin suggests that the overwhelming impact of violent crimes of this nature for colonised communities is actively demonstrated through the novel's consideration of Ojibwe "kinship", "relationality", and intersectionality (12). The crime is, therefore, situated within a continuum of settler-colonial and legal criminal violences; this is evident in the novel's afterword, which positions the text as a direct response to a growing public awareness and discussion of the levels of sexual violence experienced by contemporary Indigenous women in North America. In 2016, a National Institute of Justice study found that more than half of Indigenous women in the United States have been sexually assaulted, with over a third experiencing rape, at a rate nearly 2.5 times higher than the white national average (Center for Public Integrity, 2018). In addition, there is a widespread crisis of Missing and Murdered Indigenous Women and Girls (MMIWG)[2] in both the United States and Canada. The Canadian National Inquiry's final report of 2019 collated the personal testimonies of the families and communities of those missing and murdered and noted that such criminal violences are not just "a national emergency" that "has been centuries in the making" but quite simply an act of "genocide" ("Reclaiming Power and Place" 9–10). It is to America's great shame that it has so far failed to investigate MMIWG in any meaningful way.

Erdrich's novel clearly demonstrates the ways in which such violence is made yet more devastating by the fact that tribal lands have jurisdictional challenges that make the prosecution of sexual offences, and the pursuit of justice for Indigenous survivors, almost impossible. In this context, these sexual violences mirror those to which Indigenous lands are also subject: they are both physical (colonising and conquering) and legal (subject to imposed Euro-American criminal and property laws). Geraldine's terror thus derives in part from the uncertainty that her attacker will be brought to justice: in order to successfully prosecute the crime, it must be clear *who* committed it – whether the perpetrator was "Indian or non-Indian" (*Round House* 12) – and *where* it was committed in order to decide which police group – state, local, or tribal – has jurisdiction. Geraldine is forced to repeatedly voice her trauma in statements to three different police representatives because, as Eric Cheyfitz and Shari Huhndorf note, the Major Crimes Act of 1885 ensures that "Indian tribes do not have the inherent jurisdiction to try and to punish non-Indians" (268). As Erdrich commented in her subsequent 2013 *New York Times* article "Rape on the Reservation", despite the fact that "more than 80 percent of sex crimes on reservations are committed by non-Indian men", there is "a gap in the law" which ensures that they "are immune from prosecution by tribal courts" ("Rape on the Reservation"). And there is open public knowledge of this fact within the novel: Geraldine's white rapist, Linden Lark, openly brags that "I won't get caught. . . . I know as much law as a judge", and he also deliberately blindfolds Geraldine so she is unable to say with any accuracy where the rape took place (161). This failure to identify or prosecute the perpetrator ensures that Geraldine's safety and well-being – and that of the community more widely – remains persistently threatened: a constant reminder of the reach and impact of settler-colonial criminal violences. While Lark could be "barred from the reservation . . . there was really no way that could be enforced", and Lark continues to enter the reservation freely to play golf (*Round House* 266). As Joe comments, the jurisdictional complexities and the impotence of tribal justice is "a statement of our toothless sovereignty" (142).

Erdrich is careful to trace how ongoing sexual violences enacted against Indigenous women are both grounded in and enabled by complex histories of systemic and violent settler-colonial racism, dispossession, and exploitation. While Lark bluntly declares that he is "one of those people who just hates Indians generally", his claim is undeniably disingenuous in its attempts to elide complex national and familial histories that emerge from the systemic racism of the settler-colonial state (161). Colonial histories are physically present in the bodies of the Indigenous dead who lie in the tribal cemetery and represent what Joe calls the "gut kick of our history", including the three tribal men lynched in 1897 by local white inhabitants of Hoopdance (100, 99). This actual historical event,[3] the subject of Erdrich's 2008 novel *The Plague of Doves*, draws together the histories of systemic colonial racism with those of individual families: Joe's grandfather, Mooshum, is the sole survivor of the lynching, while Lark's "great-uncle was in the lynching party" (211). Moreover, Lark's parents are identified as "petty thieves" due to their two former legal cases that sought to use legal loopholes to defraud the tribe (54). For Bazil, as tribal judge, these histories expose the complex origins of Lark's racial "contempt" and demonstrate that he "thinks he can get away with it. Like his uncle" (211).

These complex histories are equally evident in the figure of Mayla Wolfskin, whose disappearance and murder by Lark is directly linked to Geraldine's own rape: the teenage Mayla is physically present at the site of Geraldine's rape and, in her role as "tribal enrollment specialist", Geraldine holds the written evidence of the father of Mayla's child, her former employer Governor Curtis Yeltow (*Round House* 3).[4] The 'relationship' between Governor Yeltow and the teenage Mayla further exposes the ongoing exploitation of Indigenous women's bodies through unequal racialised and gendered power relations. Lark's attempt to prevent political scandal by erasing the tribal record deliberately singles Geraldine out for attack: it is her job "to know everybody's secrets", and her files detail the children "of incest, molestation, rape", and of "BIA superintendents, police, and priests" (149). Geraldine's files are not simply a record of enrolled tribal members but a detailed history of systemic settler-colonial dispossession, exploitation, and criminal violences. Lark's attack on Geraldine attempts not only to eradicate those problematic histories, but to overwrite and silence the voices and experiences of Indigenous women.

The site of Geraldine's attack demonstrates the intrinsic connections between the violences enacted against Indigenous women and those enacted against Indigenous lands, and Geraldine's violated body is described in terms that echo the act of colonial settlement: she is both "discovered, and invaded" (*Round House* 112). The attack takes place in the traditional tribal Round House that is the heart of culture and ceremony for the Indigenous community, a crucially significant site of Indigenous political resistance and cultural "survivance", constituting what Gerald Vizenor (White Earth Anishinaabe) identifies as the "active sense of [Indigenous] presence" and "the continuance of [cultural] stories" (1). So, while the Round House is where prohibited Indigenous cultural ceremonies could be held disguised as Christian prayer meetings,[5] it is also a "storied space" where, Silvia Martínez-Falquina comments, "acts of re-storying" can overwrite the narratives of settler-colonial violences (153). The Round House's structure itself is born of story, made on the instruction of Old Buffalo Woman who gave her life to save the tribal community in the times of scarcity following the forced relocations of the nineteenth century. Old Buffalo Woman's instructions are explicit: "the Round House will be my body, the poles my ribs, the fire my heart. It will be the body of your mother and it must be respected the same way" (214). In this context, the violation of Geraldine is a violation of the central cultural meaning of the Round House: a violation of tribal land, of tribal identity, of tribal sovereignty – "a breach of sacred etiquette" (*Round House* 52).

Importantly, the Round House retains its animate power,[6] alerting Joe to the location of Geraldine's attack by uttering "a grieving cry" that "cried out . . . in my mother's anguished voice" (59, 60). Since Geraldine's rape cannot be prosecuted – it is a crime committed by a white man on tribal land – the Round House is, within settler-colonial constructions, effectively 'law-less', a space where criminal violences are unpunishable and go unpunished. Yet its profound animacy acts to displace settler-colonial legal violences and to replace them with Indigenous cultural understandings of justice that are intimately tied to the land. Within this Indigenous legal framework, Lark is also transformed, identified as Windigo, the insatiable cannibal monster. This has a powerful three-fold effect. First, it identifies Lark's acts of criminal violence as both monstrous and directly reflective of the criminal violences of the settler-colonial state. Second, it identifies Lark's worldview as dangerously contagious since "Windigos are not born, they are made" (Kimmerer, *Braiding Sweetgrass*, 194). And third, it actively re-stories Lark both within an Indigenous frame of reference and within Indigenous forms of justice. It is through this re-storying that Erdrich forces her reader to recognise not only ongoing Indigenous resistance to entrenched settler-colonial criminal violences, but also the continuance of Indigenous sovereignty. The final sentence is, tellingly, "We just kept going" (317).

Unsettling Settler-Colonial Histories: Cherie Dimaline's *The Marrow Thieves*

Further criminal violences are traced within Cherie Dimaline's 2017 novel *The Marrow Thieves*, which compares the genocidal histories of the settler-colonial system with the ecocidal processes of capitalist extractive economies. Set sometime after 2047, Dimaline's novel depicts a world that has been "broken" by the capitalist ideology of "too much taking for too damn long". It is told from the perspective of the Indigenous teenager Frenchie, one of a disparate group of Indigenous individuals who have fled the remaining cities to avoid the state-sanctioned detention and slaughter of all Indigenous peoples (Dimaline 87). As "the shapes of countries changed forever", as oil pipelines "spewed bile", and city skylines "looked like a ruined mouth of rotted teeth", "half the population was lost" to disaster and disease while the remaining population are forced to "wor[k] longer hours" to compensate (87, 26). This intensification of capitalist production causes a "plague" of "dreamlessness" to sweep the remaining settler-colonial population, leading not only to suicide but to a lack of productivity (53). In this "world . . . suddenly gone mad", ensuring productivity and maintaining profit is all important, and dreams become yet another 'product' to be consumed (47). Accordingly, the capitalist settler-colonial system creates a new industry, where the only remaining dreams – those of Indigenous communities still resisting the imposition of capitalist 'values' – can be extracted and redistributed to the settler-colonial society. Beginning with "volunteers", this new industry soon turns to "prisoners" before sending out "Recruiters" to round up entire Indigenous communities (87). Juxtaposing Indigenous bodies that have their dreams fatally "siphon[ed] . . . right out of our bones" with Indigenous lands that are simultaneously annexed, "commoditized. . . . filled with water companies and wealthy corporate investors", Dimaline demonstrates how both have always been identified by the settler-colonial state as consumable resources (88). Dimaline's young-adult story thus has two specific aims: to expose the continuum of criminal violences that connect the genocidal histories of the settler-colonial state with the environmental racism practised by twenty-first century capitalist processes and to demonstrate the power of Indigenous cultures and worldviews and the sovereignty of Indigenous Nations.

In the first case, Dimaline's speculative fiction rests on genocidal truths: the criminally violent histories of the settler-colonial Indigenous education system. The Indian residential schools not only aimed to force assimilation by banning Indigenous languages and cultural practices, converting students to Christianity, and imposing English names, they also created brutal regimes and living conditions that resulted in the undocumented deaths of thousands of Indigenous children. In Canada alone, more than 150,000 Indigenous children were sent to the schools resulting in approximately 6,000 deaths, while in the United States the figures can still only be estimated (*Guardian*, "US to investigate 'unspoken traumas'"). The extent of the state genocidal practices in Canada was exposed by the 2008–2015 Truth and Reconciliation Commission (TRC), whose collected testimonies of the decades of state-sanctioned emotional, physical, and sexual abuses of Indigenous children provoked an international outcry.[7] In June 2021, the discovery of 215 children's remains at the Kamloops School, British Columbia, and 751 remains at the Marieval School, Saskatchewan, led to the United Nations calling for Canada to urgently "perform an exhaustive investigation into uncovering the remains of residential school children buried in unmarked graves across the country" (Newlove). It is notable that America has also responded: in June 2021 the first Indigenous US Interior Secretary Deb Haaland, who had previously spoken of the trauma undergone by her Laguna Pueblo grandparents "stolen from their families as children" to be sent to residential schools (Haaland), announced an investigation to "uncover the truth about the loss of human life and the lasting consequences" of US Indian boarding schools (*Guardian*, "'US to investigate 'unspoken traumas'"). The responses of both countries, therefore, suggest an unspoken acceptance that there are potentially thousands more Indigenous children's bodies lying in unmarked graves.

Dimaline's novel engages with these criminal violences: as the new processes of harvesting dreams become more efficient, the state "needed too many bodies", and the Indigenous community leader Miigwan (Miig) bluntly comments that the settler-colonial authorities "turned to history" for "how to best position the culling" (89). At this point in the novel, with the erasure and absorption of Indigenous bodies now state and corporate policy, "new residential schools started growing up like poisonous mushrooms" (89). Dimaline's siting of these new state "death camps" (Dimaline 81) on historical sites of genocide also draws our attention to the 'agents' who expose those historical connections, what Daniel Heath Justice names the "racist collusion" of authorities and institutions (138). Alongside the "Recruiters" who argue that the sacrifice of Indigenous lives is "for the good of the nation" and the scientists who create the fatal extraction process, we also see the "Headmistress" and the "Cardinals", names that evoke the central historical roles played by Indian agents, truant officers, social workers, residential school administrators, the Christian church, and local clergy in the punishment and erasure of Indigenous bodies (Dimaline 150, 172). We are shown the violent process of 'extraction', where all that remains of Indigenous bodies are vials of bone marrow, the "crop" that has been "harvest[ed]" (26, 143). As Miig is horrified to discover when he hijacks a 'delivery' truck, the labelling of the "frosted test tubes" that contain the "thick, viscous fluid" reflects historical settler-colonial treatment of Indigenous peoples; there are no names for those who have 'donated', they are identified only by number, age, gender, and tribe: "46522Y. 64 year-old female. Metis" (144). And Miig's horror is compounded by his despair that "66542G, 41-year-old male. Euro-Cree" is all that remains of his missing husband Isaac (144). This "erasure of the individual" is, Megan Cannella argues, "the core objective of all colonialist efforts" (115); and it becomes impossible for Dimaline's reader to separate the impacts of these fictional state-sanctioned policies and processes upon Indigenous communities from the actual historical traumas caused by settler-colonial criminal violences. As Patrizia Zanella argues, the text enacts "a heightening of already-present forms of [criminal] violence" (182). It is notable that both fictional and actual settler-colonial genocidal practices are, to use Achille Mbembe's term, necropolitical: settler colonialism is itself a form of necropower, where nation-state retains the right to determine which social and/or cultural groups should live or die within their borders. Moreover, the cultural impact of those historical state-sanctioned attempts to destroy Indigenous cultures actively creates what Mbembe terms "*death-worlds*, new and unique forms of social existence in which . . . [entire] populations are subjected to conditions of life conferring upon them the status of *living dead*" (39–40). In the novel, the slow violences that extract Indigenous bone marrow – the murders that are "real slow" – are equated with the capitalist processes and ideologies whose slow violences are resulting in climate crisis: the fatal "miscalculation of infallibility" at the centre of the capitalist worldview (Dimaline 81, 87). As Dimaline's extractive processes demonstrate, capitalism works in collusion with the settler-colonial state to disproportionately affect Indigenous communities in the twenty-first century.

Dimaline counters this horror story with an expression of Indigenous cultures and worldviews and the sovereignty of Indigenous Nations. The text is interspersed with a series of "coming to stories" (how each individual joins Miig's group) and chapters of "Story", where Miig details the cultural knowledges that are crucial to the group's survival, including resistance to the histories of settler-colonial genocide and capitalist ecocide (Dimaline 1, 21). The telling of these stories is essential; it is, Moritz Ingwersen argues, "an enactment of sovereign agency" ("Reclaiming Fossil Ghosts"). This is evident in Miig's theft of the marrow vials to sing "each of them home", releasing the remains into the last waters unaffected by the poisons of capitalist industrial processes; but it is especially evident in the actions of the elder Minerva[8] who sacrifices herself to save Miig's group (Dimaline 145). Revered as an Indigenous language speaker, Minerva is a receptacle of memory with essential cultural knowledge to pass on. It is Minerva's presence within the extractive system that destroys it and, crucially, it is destroyed by song, by ceremony, by language: "every dream Minerva had ever dreamed was in the language. It was her gift, her secret, her plan" (172). Confronted with a forbidden language that carries Indigenous ceremony, by Minerva's "blood memory, her teachings, her ancestors", the extractive mechanics fail (172). While Minerva's song gains power from its call to wider histories as it "echoed through her relatives' bones, rattling them in the ground under the school itself. . . . morphing her singular voice to many", it also gains power from its call to ceremony; by "changing her heartbeat to drum", she becomes "the weapon that could bring them all down" (173, 206). This emphasis upon the importance of community and ceremony is echoed by the larger Indigenous community-in-hiding that Miig's and Frenchie's small group join. This diverse community organises to reclaim Indigenous ceremonies and languages: they "piec[e] together the few words and images each of us carried" and "start a youth council" to "pas[s] on the teachings right away" (214). It is no accident that one of the first words that Frenchie learns to write is "family", or that he writes it on "a creamy curl of birch bark" because the Indigenous understanding of family here includes all living things (214). While Minerva doesn't survive her rescue, at the close of the text the community gain back Miig's missing husband Isaac, who acts as a further repository of Indigenous cultural knowledge because he too "dream[s] in Cree" (228). As the community realise that "as long as there are dreamers left, there will never be want for a dream", they can finally follow Minerva's last instruction: "kiiwen . . . go home"

(231, 210). This sense of 'home' – of the intricate relationships between peoples and place and of the essential experiential knowledges that come from *living* those relationships – enables Dimaline to counter the "transgenerational narratives of displacement and abuse" with Indigenous "hope and endurance" (Cannella 120). By countering the criminal violences of the setter colonial state and of extractive capitalist consumption, the community can "start healing the land" because "when we heal our land, we are healed also" (Dimaline 193).

Dis-locating Windigo Ideologies: Waubgeshig Rice's *Moon of the Crusted Snow*

The criminal violences that link genocide and ecocide are also the focus of Waubgeshig Rice's (Anishinaabe Wasauksing First Nation) 2018 novel *Moon of the Crusted Snow*, set on isolated Indigenous lands in northern Ontario and told from the perspective of the young father Evan Whitesky. Set in a not-too-distant future, in an Anishinaabe tribal community cut off by capitalist driven climate crisis, the introduction of a settler-colonial interloper – a traditional cannibal Windigo – into the community explores not only how the voraciousness of capitalism drives climate crisis, but also how the ideologies of settler-colonial cultures consume both land/environment and Indigenous peoples through an "insatiable hunger for Indigenous flesh, cultures, and lands" (Léna Remy-Kovach). As the community loses all power and food supplies, they are forced to look to traditional cultural hunting practices and cultural values that eschew 'selfishness' because "everyone's survival depends on cooperation" (Rice 95). Notably, the interloper Justin Scott's arrival threatens community bonds that have been made fragile by centuries of settler-colonial oppression, where entire generations have been made dependent upon fossil fuels and settler-colonial capitalism, and as a result, very "few hunted anymore" (120). While it is precisely this history of criminal violences that makes the community especially vulnerable to the Windigo ideologies of Scott, it is also that history that makes them resilient. As the elder Aileen comments, although there is no word for apocalypse in Ojibwe, the community has nonetheless experienced apocalypse "over and over", from the removal and deforestation that "ended . . . our world", to the removal of "our children" that ended "our world . . . again"; and yet, crucially, "we're still here" (149, 150).

Told originally when winter starvation was a very real possibility, traditional Windigo stories acted to strengthen the social taboos against the kind of greed that would waste restricted resources. The Windigo's contagious nature acts to expose and renounce the selfishness that Basil Johnston (Anishinaabe) identifies "as the worst human shortcoming", while validating a communal sense of obligation (222). Traditional Windigo stories convey a sense of moral and ethical horror at individual actions that reject a sense of obligation to the group in favour of fleeting personal gain. In the twenty-first century, the Windigo retains its storied significance not simply, Robin Wall Kimmerer (Potowatomi) argues, as "the embodied nightmare of greed" but as the logical end result of the criminal violences of settler-colonial capitalist consumption which "insatiably devours the earth's resources" ("Greed does not have to define our relationship to Land"). As Alicia Elliott (Tuscarora) notes, the moral and ethical power of these reimagined Windigo tales confront the "monstrous reality" of the consumption of Indigenous lands and bodies precisely because Indigenous writers write from the experience of "know[ing] what it's like to live in a world where the horror never stops" ("The Rise of Indigenous Horror"). Rice's novel demands that its community – and its readers – look at the damage caused to Indigenous bodies and the natural world by the intertwined criminal violences of settler-colonial and capitalist ideologies. Importantly, the dangers of acting through selfishness and greed – as Windigo – are juxtaposed with examples of traditional Anishinaabe values in story and dreams, particularly via the figures of Evan's father, Dan, and the elder Aileen. The twin effect is to expose the ways in which the processes of settler-colonial capitalism actively create death-worlds for Indigenous peoples and to demonstrate how traditional Anihsinaabe stories perform crucial cultural 'storywork'.

Settler-colonial death-worlds rely on the elimination of Indigenous cultural knowledges that are essential to survival outside of capitalist infrastructures. Rice clearly shows how communal pride in the fact that "survival has always been an integral part of their culture" sits uneasily alongside how wide swathes of the community are so reliant upon capitalist commodities that they are resistant to the demands to conserve fuel and food that contradict settler-colonial ideologies and capitalist lifestyles (Rice 48). As Indigenous values are replaced with those of settler-colonial capitalism, it is little wonder that while Evan "had been out on the land learning real survival skills [hunting] with his father", his younger brother Cam "had chosen to stay behind, learning simulated ones in video games" (Rice 34). Rice shows us both the community leadership's preparedness for just such an emergency *and* the ways in which Windigo ideology has already infiltrated the community via the violent imposition of settler-colonial capitalist values and products. While the tribal council has conserved a "massive cache of non-perishable food[s]" over "nearly two decades" (Rice 96, 73), it is notable that food shortages are nonetheless accompanied by panic buying, demonstrating exactly the same "rampantly individualistic" actions witnessed in the settler-colonial city to the south (Martens 199). At the local grocery store, "owned by the biggest . . . chain in the country" with "a monopoly" on reservations, the store staff "plea[d] with people to be mindful of the needs of others", while the store manager is delighted to discover that crisis is "good for business!" (Rice 59, 61, 62). The arrival of the Windigo Scott adds a further injection of rampant individualism: "invading the [Indigenous] . . . space", Scott demands the "unrivalled" "hospitality of the Ojibwe" irrespective of shortages (100, 102). It becomes clear that Scott's worldviews and ideologies work to undermine traditional Indigenous values; his hunting philosophy directly contradicts Anishinaabe directives not to "take more than you need", and the very Indigenous values which allow Scott to entrance to the community in the first place – the recognition by the tribal chief Terry Meegis that "we can't just turn him away . . . it's not our way" – are subsequently undermined by Scott's own killing of another starving white interloper demanding aid (125, 109, 139). Scott's provision of restricted commodities – "cigarettes and free-flowing booze" – to vulnerable community members, and his parties that waste valuable resources such as heat and electricity, further threaten community values (107, 125, 131).

Scott becomes an embodiment of the settler-colonial capitalist state: vulnerable community members "fal[l] under" his "spell" because he "promises . . . easier living;" and the situation escalates as it becomes clear that food has gone missing from the communal cache, with some community members "suspected of hoarding supplies from others especially the elderly" (Rice 151, 175). Evan's suspicions, that "Scott was intimidating people into handing over their food or trading it for the few drops of contraband booze that remained", demonstrates the historic pressures enacted by the settler-colonial state on Indigenous communities (175). Yet it is Scott's identification as Windigo that exposes the cannibalistic nature of voracious consumption. As members of the Indigenous community begin to die "through sickness, mishap, violence, or by their own hands", crucially these deaths all centre on Scott as a physical manifestation of the "lingering misery" caused by the criminal violences of the settler-colonial state that make "the community's tragedies . . . suffocating" (151, 153, 196). Preying on the "palpable" desperation caused by the threat of imminent starvation, Scott brags of an unidentified 'solution' as readers get their first glimpse of his "dark" and "cavernous" mouth "behind his big teeth" (182). It is in Evan's subsequent warning dream that Scott becomes fully Windigo: in the garage where the 22 Indigenous bodies are now missing, Evan is instead confronted by a "feral odor" of "the beast that Scott had become", whose "emaciated torso", "long incisors", and "blacked out" eyes mark the incurable and insatiable hunger of the Windigo (187). Notably, Scott's justification for cannibalising and cooking "our relatives" is rooted in his identification of the Indigenous community as "freeloaders" taking "handouts" who can only make a valuable contribution through their transformation into a profitable consumable in death (203, 201, 202).

Crucially, it is via Evan's dream and via traditional storywork that the community regains the ability to finally ward off Windigo settler-colonial capitalism. As Jo-ann Archibald et al. argue, Indigenous storywork is a powerful "decolonizing methodology" that reveals the "meaningful education[al]" nature of Indigenous stories (xi, 1). For Archibald et al., Indigenous storytelling is "mindfulness in action", and this can be detected in the traditional stories told by Evan's father, Dan, to Evan's own children, which offer crucial intergenerational information on how to avoid infection by settler-colonial Windigo ideologies through the maintenance of Indigenous cultural values (12). Dan's story, of Nanabush and the geese,[9] is included in full to emphasise its significance and details how Nanabush had been too "lazy" to hunt for winter food (Rice 170). Although Nanabush goes on to "trick" and kill 30 migrating geese, the goose he is cooking burns and the rest are stolen while he sleeps because his "diiyosh" – his "bum" – fails to wake him in time. Nanabush is so furious that he punishes his "diiyosh" by putting it into the fire and burning all the skin off, causing himself very real pain (173). The educational nature of the story is immediately evident to Evan's children: "always be ready for winter" and "don't be greedy" (174). And these are the values that open the novel also, as Evan hunts for moose and demonstrates the accompanying ceremonies of reciprocity: "as he took from the earth, he gave back" in recognition of moose's sacrifice, and Evan plans "to give a lot of the meat away" to those in need (5, 6). This commitment to the well-being of the community as a whole acts as a much needed corrective to the kinds of individualism being promoted by the Windigo Scott. While Aileen works to pass on her knowledge of the "old medicine ways" that will be essential if there are no "new supplies . . . from the hospital down south", it is notable that the children are now "learning their language earlier and better than their parents had", a fact that directly confronts the settler-colonial histories of state-sanctioned "forced and often violent assimilation at church-run residential schools" (128, 53). While these residential schools "imprisoned stolen children" who were "punished" for "speaking Ojibwe", it is notable that the children of Rice's Indigenous community are both dis-locating Windigo ideologies and relocating traditional Anishinaabe cultural knowledges (147, 128).

In the Nanabush storytelling session, Evan's son Maaingan asks an essential question: "why would . . . [Nanabush] want to hurt himself?" (Rice 174). It is a question that reverberates through the entire text and exposes the cultural importance of Nanabush himself: as Leanne Betasamosake Simpson (Mississauga Nishnaabeg) argues, Nanabush "stories the land with a sharp criticality necessary for moving through the realm of the colonized into the dreamed reality of the decolonized, and for navigating the lived reality of having to engage with both at the same time" (17). It is not surprising, therefore that it is a question that readers return to repeatedly, especially in the context of the Indigenous community members who act against the communal interest embodied in traditional Anishinaabe values. Rice's novel offers a cautionary demonstration of how the insatiable desire for ever-greater 'profits' must be combatted as an imperative for our very survival in an era of climate crisis, demonstrating that powerful and all-consuming Windigo ideologies and policies can and should be successfully resisted, because why would we want to hurt ourselves? Most importantly, as Bussière notes, Rice "fram[es] the apocalypse as a politically mobilizing and agency-creating mechanism that allows Indigenous peoples to build on traditional knowledge about how to survive using the resources available and succeed in a new world" (56–57). At the close of the text, the Indigenous community "decid[e] to take control of their own destiny": as they leave the lands to which they had been historically removed in favour of 'going home' they "didn't look back" (Rice 212, 213). The challenge that Rice poses to settler colonialism's conceptualisation of its own permanence is evident on the novel's fly leaf, which bluntly states that "as one society collapses, another is reborn" (213). It is a particularly pertinent challenge in the Anthropocene/Capitalocene.

Relocating Indigenous Sovereignties: Conclusions

All three writers discussed here refuse settler-colonial formulations that equate Indigenous bodies with Indigenous lands as resources to be conquered and consumed and actively redefine our understandings of both crime fiction and environmental racism. Instead, they foreground Indigenous understandings of the relationships between human bodies and the natural world. This shift of focus deliberately challenges the mutually supportive systems of patriarchy, colonialism, and capitalism that directly result in criminally violent and fatally damaging profit-driven climate crises that adversely impact on Indigenous peoples. As Davis and Todd comment, "the Anthropocene is not a new event, but is rather the continuation of practices of dispossession and genocide, coupled with a literal transformation of the environment, that have been at work for the last five hundred years" (761). Even more importantly, the writers discussed here demonstrate how we all might see and live in very different, more sustainable ways that recognise and reject the fatal kinds of "Windigo thinking" that saturates settler-colonial capitalism (Kimmerer, *Braiding Sweetgrass*, 309). This Windigo thinking, Kimmerer argues, embraces "a systemic policy of sanctioned greed" that causes fatal harm to Indigenous lands and bodies interpreted as consumable resources (*Braiding Sweetgrass* 308).

And this shift of focus onto Indigenous understandings of our place within the natural world is equally evident in the recent work of Indigenous theorists such as Glen Coulthard (Yellow Knives Dene) and Leanne Betasamosake Simpson (Mississauga Nishnaabeg), who are committed to a decolonisation of academic methodologies that commit similar theoretical violences. Coulthard's and Simpson's work engages with Whyte's argument that opened this chapter, regarding the kinds of "political philosophies of diplomacy" that are necessary "for peoples who share ecosystems" ('Climate Change' 8). Both Coulthard and Simpson therefore argue that contemporary Indigenous writing should be read through a lens of "grounded normativity", which Coulthard defines as a "place-based foundation for Indigenous decolonial thought and practice" that is "deeply informed by what the land *as a system of reciprocal relations and obligations* can teach us about living our lives in relation to one another and the natural world in nondominating and nonexploitative terms" (*Red Skin, White Masks* 14). Simpson's argument in favour of 'land as pedagogy' reflects the kinds of experiential storytelling that all three novels engage in; and she states that

> A "theory" in its simplest form is an explanation of a phenomenon. . . . "Theory" is generated and regenerated continually through embodied practice and within each family, community and generation of people. "Theory" isn't just an intellectual pursuit – it is woven within kinetics, spiritual presence and emotion, it is contextual and relational.
>
> *(7)*

All three writers engage in just this embodied, contextual, and relational practice, demonstrating the kind of experiential understanding of the world that is surely an essential requirement in an era of profound climate crisis. Moreover, this embodied practice actively relocates Indigenous sovereignties to promote sustainable ways of seeing and living in the world. As Whyte comments, it is an approach that is essential "in the face of [our] collective survival" in the Anthropocene/Capitalocene ("Indigenous Climate Change Studies" 156, 157).

Notes

1. MMIWG has more recently begun to be expanded – to MMIWG2S – to reflect the experiences of Two-Spirit (those observing Indigenous gender variant community roles) and LGBTQIA+ persons in this context.
2. The Canadian National Inquiry's final report, 2019, extends this to include two-spirit, lesbian, gay, bisexual, transgendered, queer, questioning, intersex, or asexual victims.
3. For details, see Peter G. Beidler, 2013.
4. See Julie Tharp, 2014, 35, for details of Jancita Eagle Deer, upon whom Mayla's story is based.
5. Indigenous religious cultural practices were outlawed from the later nineteenth century until the passing of the 1978 American Indian Religious Freedom Act.

6. See Macklin (2020) for a detailed discussion of animacy within the text.
7. See the TRC website, Truth and Reconciliation Commission of Canada (rcaanc-cirnac.gc.ca).
8. Minerva's name does, of course, also deliberately recall the Roman goddess of wisdom.
9. This is the same Nanabush to whom Old Buffalo Woman gives the instructions for the building of the Round House in Erdrich's novel.

Bibliography

Archibald Q'um Q'um Xiiem, et al. *Decolonizing Research: Indigenous Storywork as Methodology*. Zed Books, 2019.

Beidler, Peter G. *Murdering Indians: A Documentary History of the 1897 Killings that Inspired Louise Erdrich's the Plague of Doves*. Jefferson, McFarland and Co, 2013.

Bussière, Kirsten. "Beginning at the End: Indigenous Survivance in *Moon of the Crusted Snow*." *Foundation*, vol. 49, no. 136, 2020, pp. 47–58.

Cannella, Megan E. "Dreams in a Time of Dystopic Colonialism Cherie Dimaline's *The Marrow Thieves* and Louise Erdrich's *Future Home of the Living God*." *Displaced: Literature of Indigeneity, Migration, and Trauma*, edited by Kate Rose, Routledge, 2020, pp. 111–120.

Carrington, Damian. "The Anthropocene Epoch: Scientists Declare Dawn of Human-Influenced Age." *The Guardian*, Guardian News and Media, 29 Aug. 2016, www.theguardian.com/environment/2016/aug/29/declare-anthropocene-epoch-experts-urge-geological-congress-human-impact-earth?CMP=Share_AndroidApp_Other.

Center for Public Integrity, "Murdered and Missing Native American Women Challenge Police and Courts", 27 Aug. 2018, https://publicintegrity.org/politics/murdered-and-missing-native-american-women-challenge-police-and-courts/.

Cheyfitz, Eric, and Shari M. Huhndorf. "Genocide By Other Means: US Federal Indian Law and Violence Against Native Women in Louise Erdrich's *The Round House*." *New Directions in Law and Literature*, edited by Elizabeth S. Anker and Bernadette Meyler, Oxford UP, 2017, pp. 264–278.

Coulthard, Glen. *Red Skin, White Masks: Rejecting the Colonial Politics of Recognition*. U Minnesota P, 2014.

Davis, Heather, and Zoe Todd. "On the Importance of a Date, or Decolonizing the Anthropocene." *ACME: An International Journal for Critical Geographies*, vol. 16, no. 4, 2017, pp. 761–780.

Deer, Sarah. "Toward an Indigenous Jurisprudence of Rape." *Kansas Journal of Law and Public Policy*, vol. 14, 2004, pp. 121–154.

———. "Decolonizing Rape Law: A Native Feminist Synthesis of Safety and Sovereignty." *Wicazo Sa Review*, vol. 24, no. 2, 2009, pp. 149–167, doi:10.1353/wic.0.0037.

———. *The Beginning and End of Rape: Confronting Sexual Violence in Native America*. U Minnesota P, 2015.

Dimaline, Cherie. *The Marrow Thieves*. Cormorant Books, 2017.

Elliott, Alicia. "The Rise of Indigenous Horror: How a Fiction Genre is Confronting a Monstrous Reality | CBC Arts." *CBCnews*, CBC/Radio Canada, 17 Oct. 2019, www.cbc.ca/arts/the-rise-of-indigenous-horror-how-a-fiction-genre-is-confronting-a-monstrous-reality-1.5323428.

Erdrich, Louise. *The Round House*. Harper Collins, 2012.

———. "Rape on the Reservation." *The New York Times*, 27 Feb. 2013, www.nytimes.com/2013/02/27/opinion/native-americans-and-the-violence-against-women-act.html.

Guardian Staff. "US to Investigate 'Unspoken Traumas' of Native American Boarding Schools." *The Guardian*, Guardian News and Media, 22 June 2021, www.theguardian.com/us-news/2021/jun/22/us-investigation-native-american-boarding-schools.

Haaland, Deb. "Opinion | Deb Haaland: My Grandparents Were Stolen from Their Families as Children. We Must Learn about This History." *The Washington Post*, WP Company, 25 June 2021, www.washingtonpost.com/opinions/2021/06/11/deb-haaland-indigenous-boarding-schools/.

Ingwersen, Moritz. "Reclaiming Fossil Ghosts: Indigenous Resistance to Resource Extraction in Works by Warren Carious, Cherie Dimaline, and Nathan Adler." *Canadian Literature*, vol. 240, 2020, pp. 59–76.

Johnston, Basil. *The Manitous: The Spiritual World of the Ojibway*. Minnesota Historical Society Press, 2002.

Justice, Daniel Heath. *Why Indigenous Literatures Matter*. Wilfrid Laurier UP, 2018.

Kimmerer, Robin Wall. *Braiding Sweetgrass: Indigenous Wisdom, Scientific Knowledge and the Teachings of Plants*. Milkweed Editions, 2013.

———. "Robin Wall Kimmerer: Greed Does Not Have to Define Our Relationship to Land." *Literary Hub*, 1 June 2020, lithub.com/robin-wall-kimmerer-greed-does-not-have-to-define-our-relationship-to-land/.

Kojola, Erik, and David N. Pellow. "New Directions in Environmental Justice Studies: Examining the State and Violence." *Environmental Politics*, vol. 30, no. 1–2, 2020, pp. 100–118, doi:10.1080/09644016.2020.1836898.
Lewis, Simon L., and Mark A. Maslin. "Defining the Anthropocene." *Nature*, vol. 519, no. 12, 2015, pp. 171–180.
Macklin, Rebecca. "Natural Violence, Unnatural Bodies: Negotiating the Boundaries of the Human in MMIWG Narratives." *Interventions International Journal of Postcolonial Studies*, vol. 23, no. 4, 2020, pp. 1–17.
Martens, Reuben. "Petromelancholia and the Energopolitical Violence of Settler Colonialism in Waubgeshig Rice's Moon of the Crusted Snow." *American Imago*, vol. 77, no. 1, 2020, pp. 193–211, doi:10.1353/aim.2020.0010.
Martínez-Falquina, Silvia. "Grieving Places, Sovereign Places: Storied Space in Louise Erdrich's the Round House." *Roczniki Humanistyczne*, vol. 68, no. 11, 2020, pp. 145–160, doi:10.18290/rh206811-9.
Mbembe, Achille. "Necropolitics." *Public Culture*, vol. 15, no. 1, 2003, pp. 11–40.
Moore, Jason W., editor. *Anthropocene or Capitalocene? Nature, History, and the Crisis of Capitalism*. PM Press, 2016.
———. "Introduction." *Anthropocene or Capitalocene? Nature, History, and the Crisis of Capitalism*, edited by Jason W. Moore, PM Press, 2016, pp. 1–11.
Newlove, Nigel, and Citynews Staff. "UN Calls on Canada to Search for Remains of Missing Residential School Children." *Vancouver.citynews.ca*, City News, 2 June 2021, vancouver.citynews.ca/2021/06/02/un-calls-on-canada-to-search-for-remains-of-missing-residential-school-children/.
Nixon, Rob. *Slow Violence and the Environmentalism of the Poor*. Harvard UP, 2013.
"Reclaiming Power and Place: The Final Report of the National Inquiry into Missing and Murdered Indigenous Women and Girls." Vol. 1a, 2019, www.mmiwg-ffada.ca/final-report/.
Remy-Kovach, Léna. "Insatiable Hunger for Indigenous Flesh, Cultures, and Lands: Colonialism as a Ravenous Monster in *Monkey Beach* and *Kiss of the Fur Queen*." *Parlour*, no. 3, 2018, pp. 1–7.
Rice, Waubgeshig. *Moon of the Crusted Snow*. ECW Press, 2018.
Simpson, Leanne Betasamosake. "Land as Pedogogy: Nishnaabeg Intelligence and Rebellious Transformation." *Decolonization: Indigeneity, Education & Society*, vol. 3, no. 3, 2014, pp. 1–25.
Tharp, Julie. "Erdrich's Crusade: Sexual Violence in *the Round House*." *Studies in American Indian Literatures*, vol. 26, no. 3, 2014, p. 25, doi:10.5250/studamerindilite.26.3.0025.
Vizenor, Gerald. *Survivance: Narratives of Native Resistance*. U Nebraska P, 2008.
Whyte, Kyle Powys. "Indigenous Climate Change Studies: Indigenizing Futures, Decolonizing the Anthropocene." *English Language Notes*, vol. 55, no. 1–2, 2017, pp. 153–162, doi:10.1215/00138282-55.1-2.153.
———. "Climate Change: An Unprecedentedly Old Catastrophe." *Grafting*, vol. 1, 2018, pp. 8–9.
Zanella, Patrizia. "Witnessing Story and Creating Kinship in a New Era of Residential Schools: Cherie Dimaline's the Marrow Thieves." *Studies in American Indian Literatures*, vol. 32, no. 3–4, 2020, pp. 176–200, doi:10.1353/ail.2020.0023.

23

ENVIRONMENTAL RACISM AND POST-KATRINA CRIME FICTION

Ruth Hawthorn

Hurricane Katrina and "The City That Care Forgot"

Hurricane Katrina and its aftermath confirmed the secondary meaning of New Orleans' reputation as "The City That Care Forgot". Although ostensibly a natural disaster, the storm's devastating impact was a result of government neglect both before and after Katrina hit. The Bush administration's defunding of wetland conservation and "gutting of 'disaster mitigation'" programs, the inadequate levees and the glacially slow response of federal agencies in organising rescue and aid meant that whole areas of the city were devastated, particularly the lower Ninth Ward which was predominantly home to Black residents (Klein). Jeremy Levitt and Matthew Whitaker write: "As the world watched the coverage of Hurricane Katrina and its desolation in horror, America's racial inequality and shocking levels of poverty, vulnerability, and dislocation, particularly among African Americans, were laid bare before the world" (3). As thousands of stranded citizens took refuge in the Superdome and the Convention Centre, the sensationalised (and often scandalously inaccurate) media accounts of looting and lawlessness, scathingly critiqued in Spike Lee's epic documentary series *When the Levees Broke* (2006), were racially inflected, working to marginalise New Orleans residents, particularly through framing those who were unable to leave as responsible for their fate and labelling citizens displaced by the storm as 'refugees' (a term usually reserved for displaced citizens from another nation). Diane Negra places this othering of New Orleans in the context of neo-facist nationalist myths, arguing that "Given the cultural proclivity in the United States to trade in competitive strength/weakness dialectics, it is not surprising that a strong impulse is to expel faltering cities as damaged parts of the national body" (10).

While 9/11, the other epochal national catastrophe of the early twenty-first century, enabled the deployment of chauvinistic narratives of America's unity in the face of an external enemy, Katrina laid bare the nation's deeply entrenched societal inequality, exposing America's fault lines through the proliferation of "incidents and images decisively outside the bounds of what is conceived as the American 'way of life'" (Negra 1). At the same time that impoverished African American citizens were being both scapegoated for staying behind and demonised as violent thugs and thieves in media coverage of the storm-ravaged city, armed vigilante groups in the white suburb of Algiers Point were openly celebrating "the opportunity to hunt black people", under the auspices of defending their neighbourhood (Thompson). These violent crimes, including murder, went uninvestigated until years

DOI: 10.4324/9781003091912-28

later, only after a piece on the killings was co-published by *The Nation* and the investigative journalism outlet *ProPublica* in 2008 (Thompson).

These divisions were further exacerbated by the uneven restoration efforts which, accelerated by what Naomi Klein has described as "disaster capitalism", consisted of "orchestrated raids on the public sphere" by private contractors, many of whom had also been employed in the US occupation of Iraq (6). Just days after the storm hit, one Republican congressman, Richard Baker, openly gloated to lobbyists that Katrina offered an opportunity for social cleansing: "We finally cleaned up public housing in New Orleans. We couldn't do it, but God did" (Babington). The public housing projects demolished in the wake of the hurricane accounted for "more than half of all of the conventional public housing" (Bullard, *Black Metropolis* 189). Klein summarises the appalling inequality of post-Katrina New Orleans, "where two very different kinds of gated communities emerged from the rubble":

> On the one hand were the so-called FEMA-villes: desolate, out-of-the-way trailer camps for low-income evacuees, built by Bechtel or Fluor subcontractors, administered by private security companies who patrolled the gravel lots, restricted visitors, kept journalists out and treated survivors like criminals. On the other hand were the gated communities built in the wealthy areas of the city, such as Audubon and the Garden District, bubbles of functionality that seemed to have seceded from the state altogether. Within weeks of the storm, residents there had water and powerful emergency generators. . . . Between the two kinds of privatized sovereign states was the New Orleans version of the Red Zone, where the murder rate soared and neighbourhoods like the storied Lower Ninth Ward descended into a post-apocalyptic no-man's land.
>
> *(421)*

Klein's evocative description of the divided city is borne out by a May 2008 progress report which stresses that these stark geographic disparities of post-Katrina New Orleans were racialised: "On nearly every indicator, the storm impact and recovery experience for black households is significantly different than for whites, even after examining these issues by income levels" (Alfred 12). In a similar vein, Robert Bullard and Beverly Wright observe that improved flood fortifications can be mapped along racial lines, "closely correspond[ing] with the race of neighbourhoods, black neighbourhoods such as the Ninth Ward, Gentilly, and New Orleans East receiving little, if any, increased flood protection" ("Race, Place" 38). The environmental injustices faced by the Black citizens of New Orleans are clearly not purely the result of a singular, catastrophic, "Act of God" weather event,[1] but rather attest to the damage of environmental racism which Bullard defines as "any policy, practice, or directive that differentially affects or disadvantages (whether intended or unintended) individuals, groups, or communities based on race or color" ("Environmental Racism" 103). Such environmental discrimination underlines that fact that some communities are, politically, more easily expendable than others and therefore left more vulnerable to the extreme weather events caused by an increasingly unstable climate; it is a form of what Rob Nixon has termed "slow violence . . . a violence that occurs gradually and out of sight, a violence of delayed destruction" (2). He argues that catastrophes such as Katrina are the result of systemic political neglect of particular communities:

> Discrimination predates disaster: failures to maintain protective infrastructures, failures at pre-emergency hazard mitigation, failures to maintain infrastructure, failures to organize evacuation plans for those who lack private transport, all of which make the poor and racial minorities disproportionately vulnerable to catastrophe.
>
> *(59)*

In New Orleans, the areas which were most vulnerable to the storms' impact were shaped by environmentally racist neoliberal policies. Carlton Waterhouse observes that Katrina's uneven impact

> typifies the experience of African Americans, Latinos, Asians, Native Americans, and poor whites in facing environmental and ecological hazards nationwide. Federal, state, and local governments routinely provide lower protection standards, fewer resources, and slower responses to the risk faced by people of color and poor whites.
>
> *(156)*

This context of long-standing, racialised environmental disparities in New Orleans underpins my ecocritical perspective in this chapter. Ron Eyerman argues that Hurricane Katrina, although it also caused both individual and collective traumas, can be further understood as a cultural trauma that fundamentally challenges America's sense of national identity. He asserts, however, that "the hurricane in itself did not cause a cultural trauma; what did was the failure of those charged with collective responsibility, the upholders of the covenant, to act accordingly" (7). Environmental issues are, at root, social issues, and I will be exploring how two post-Katrina crime novels – James Lee Burke's *The Tin Roof Blowdown* (2007) and Sara Gran's *City of the Dead* (2010) – reflect and respond to the kinds of cultural narratives which allow certain communities to be seen as less worthy of care in moments of ecological crisis.

Post-Katrina Fictions

Arin Keeble observes that crime narratives have dominated fictional representations of Katrina and its aftermath, contrasting this with the body of "literary" fiction emerging from 9/11 and questioning whether that prestige distinction maps onto the broader cultural response to these epochal catastrophes:

> Given the hierarchies of literary prestige, it is tempting to draw conclusions about Katrina being the purview of crime fiction rather than literary fiction while 9/11 novels have been "literary." Indeed, this division seems to invite extrapolation: Does the value that America places on New York compared to New Orleans and the Gulf Coast somehow relate to the kinds of narratives that have approached the respective events? More pointedly, is the stark disparity attached to these very different American cultural centres linked to the enduring value discrepancy between literary and genre fiction?
>
> *(6)*

Keeble's study acknowledges and questions but ultimately upholds this literary hierarchy, as the texts which he selects as his Katrina "cycle" are those that "for a range of reasons – some potentially questionable – have been seen as literary, 'complex' or cultural" (7). He talks in broad terms about the genre fiction he excludes, echoing critical commonplaces which posit detective fiction as a formally (and often ideologically) conservative genre:

> Another way of understanding the prevalence of crime narratives of Katrina might be the narrative frameworks they provide for a subject that seems so complex and far-reaching. Crime narratives can manage this unwieldy subject with its well defined narrative frameworks.
>
> *(6)*

The crime genre, in this view, is a form of comforting escapism which works to neutralise disturbing subjects through rigid narrative patterning, particularly an emphasis on solution and closure.

Some popular detective narratives do, as Keeble suggests, work to contain the subject of Katrina within familiar narrative structures. Lyndsay Steenberg argues that in most of the forensic crime television shows dealing with Katrina, the storm and its aftermath are de-politicised as "the Katrina event is slotted into preexisting narrative formats where spectators can watch it being 'solved' and, ultimately, dismiss the event as merely one act of violence among many others from which we can learn and move on" (38). A similar pattern can be observed in other bestselling crime fiction series. In Erica Spindler's *Last Known Victim* (2009), the storm-ravaged city is the backdrop for a fairly conventional serial killer police procedural where the investigation leads to the main detective also uncovering what happened to her husband (who went missing during the hurricane). Nevada Barr's *Burn* (2010) sees her beleaguered sleuth, Anna Pigeon (more usually found protecting endangered species and habitats), caught up in a case of child sex trafficking while visiting New Orleans National Park. The publisher's blurb maps the series' wilderness discourse onto the inhabitants of the post-Katrina city, claiming "it will take all of Anna's skills learned in the untamed outdoors to navigate the urban jungle . . . and to rescue the most vulnerable of creatures from the most savage of animals", but it does not share the ecological concerns of the rest of her series (back cover). The storm's impact does not move beyond the incidental in these texts, the plots of which revolve around unrelated crimes. In Spindler's novel, a detective works out that kidnapped victims are being kept in the devasted Ninth Ward due to the smell of mould on an escapee's clothes, while in Barr's text there are brief, uncritical mentions of the city's uneven recovery: "Katrina might have brought much of New Orleans to its knees, but it had done wonders for the French Quarter" (35). These novels fit Nahem Yousaf's formulation of post-Katrina detective narratives where "the city's breakdown is relieved only by the resilience of its sleuthing protagonists who bear witness to the city's devastation as they battle its crime" (555).

Neither of these texts could be described as environmental crime fiction which, as Samantha Walton and Jo Lindsay Walton suggest, grapples more explicitly with moral complexities and tends to eschew such neat narrative resolutions, after all:

> What kind of justice can be enacted when antagonistic actors and agencies are no longer the "evil geniuses" of classic detective fiction, but distant and oppressive corporate powers, outside regular theorisations of responsibility and the clutches of justice, or even the agencies of justice themselves?
>
> *(3)*

Jamin Creed Rowan contends that the hard-boiled genre more broadly works to expose "extractivist infrastructures" through its attentiveness to urban localities:

> As hard-boiled detectives navigate the city's infrastructure, they push back against the kind of infrastructural amnesia upon which those with power depend to acquire control and to cover up the environmental, social, and economic crimes that they commit to maintain their power.
>
> *(393)*

This vein of ecocritical scholarship challenges the rather vague descriptors Keeble uses to elevate his Katrina canon as such ecologically attuned detective texts can, quite uncontroversially, be described as "complex" and "cultural".

James Lee Burke's much-lauded *Tin Roof Blowdown* (2007) and Sarah Gran's *City of the Dead* (2011) engage explicitly with the kinds of ethical tensions Walton and Walton delineate, evoking the context of environmental racism, corruption and media spectacle that shape the broader cultural significance of Katrina. This chapter explores the ways in which these texts show how Katrina tests the "well defined narrative framework" Keeble attributes to crime fiction, rather than being comfortingly contained within the familiar pattern of disruption to resolution. They favour instead the partial and conflicted resolutions of noir which Deborah Bird Rose has described as the apposite form for addressing our contemporary ecological crises: "under the weight of burdens from the past carried into the future as inescapable fate, there is a sensibility of discontent and anxiety, disillusionment and loss of confidence in the possibility of effective agency" (215). The more pessimistic cadences of noir (or hard-boiled) texts, with their self-destructive and morally ambiguous protagonists match the complexities of contemporary ecological crises. Detectives in noir narratives are incapable of achieving the kinds of satisfying solutions found in Golden Age texts, as guilt is more diffuse and pervasive, frequently stretching to implicate the investigative figures themselves, rather than being attributable to a single criminal who is reassuringly apprehended by the text's close. Noir texts unsettle the binaries of victim/criminal, hero/villain and this deconstruction, as Bird Rose argues, reflects our fraught ecological entanglements:

> We are all tangled up in political, cultural and economic systems we know to be destructive – of course we are guilty. We participate in actions that damage the earth – yes, in some sense we are criminals. We seek to understand the coercive bonds that entangle us in systems of damage, and so – yes, we are detectives trying to figure out what is going on, really. And we are now suffering, and in future will suffer more deeply, the effects of all this damage – we are victims too.
>
> *(215)*

Katrina was an event which laid bare these kinds of uncomfortable complicities and shifting subject positions; as Anna Hartnell argues, it "questioned all of the divisions that 9/11 so comfortingly consolidated: between past and present, American and non-American, the innocent and the enemy" (4). The two texts considered here work to highlight the systems of damage which shape the storm's impact. While Burke's novel vociferously critiques the Bush administration's inadequate response to the crisis and explicitly draws attention to America's systemic racism, these polemic passages are undermined by the novel's central plotlines, one of which focuses on the fate of a Black gang rapist as he seeks redemption in the flooded city. Burke's text intermingles fact and fiction, but his particular framing of the breakdown in law and order following the storm actually works to reinforce the media sensationalism that he claims to counter, conceiving of his writing as "giv[ing] voice to those who have none" (Dwyer). In contrast to Burke's uneasy rehearsal of clichés, I argue that Gran's text offers a more nuanced response to the disaster. In a novel which is both hallucinatory and elegiac, Gran confronts the intersecting injustices of New Orleans, unsettling formulaic framings of heroism and criminality. Both texts end in a wishful fantasy of justice, laying bare the limits of the detective's efficacy against the myriad forms of slow violence Katrina entailed.

The Tin Roof Blowdown

Burke's novel begins with his detective Dave Robicheaux recounting a recurring PTSD dream about his service in Vietnam. On waking from these nightmares of botched rescue and "images of brown water", his mantra includes the reassurance that the worst has past, and he "will never again have to witness the wide-scale suffering of innocent civilians, nor the betrayal and abandonment of our countrymen when they need us most" (2). Following Katrina, however, he realises that such comfort is no longer available:

> That was before a storm with greater impact than the bomb blast that struck Hiroshima peeled the face off southern Louisiana. That was before one of the most beautiful cities in the Western Hemisphere was killed three times, and not just by the forces of nature.
>
> *(2)*

From the outset, then, Burke frames Katrina's impact as national trauma, compounded by political choice and institutional neglect, linking the contemporary disaster to an earlier example of catastrophic government failure. The Vietnam War and Hurricane Katrina are events which challenge America's exceptionalist narratives and both have been credited as sounding the death knell for the triumphalist post-WWII narrative of the American Century. David Schmitz argues that the United States' humiliation in Vietnam "marked the end of the postwar era and visions of American omnipotence" (111) while Hartnell sees Katrina's impact as "the repudiation of the ideals that animated the mirage of the 'American Century'" (1). There are clear parallels between Robicheaux's nightmare images of drowning, chaos and abandonment in Vietnam and the apocalyptic scenes which emerged from New Orleans in the wake of Katrina; in drawing this lineage, Burke makes a point about the dangers of cultural amnesia (a recurring theme in his work), suggesting that veterans' traumatic Vietnam memories "are the stuff of history now and are not considered of interest to those who are determined to recreate them" (2).[2] He conceives of crime writers as telling "sociological stories" that work to counter the politically expedient repression of national failures and argues the best of these texts are "written from the ground up" (Masciotra).

Burke sees the writer's duty to "tell the truth about the period he lives in and to expose those who exploit their fellow men", and in *Tin Roof Blowdown* the particular crimes Robicheaux investigates are set against a backdrop of endemic violence, corruption and profiteering in the wake of the storm (Wroe). As he joins the rescue effort with his New Iberia colleagues in the immediate aftermath, Robicheaux observes the devastated city becoming "food for every kind of jackal in the book" with members of law enforcement taking part in the looting: "a rogue group of NOPD cops had actually set up a thieves headquarters" (121, 38). Later, as the rebuilding efforts are underway, a wearily dismayed Robicheaux describes the FEMA contracts in a passage of explicit social critique which would not be out of place in Klein's *Shock Doctrine*:

> Staggering sums of money were given to insider corporations who subcontracted the jobs to small outfits that used only nonunion labor. A $500 million contract for debris removal was given to a company in Miami that did not own a single truck, then the work was subcontracted to people who actually load debris and haul it away. Emergency roof repairs, what are called "blue roof jobs", involved little more than tacking down rolls of blue felt on plywood. FEMA provided the felt free. Insider contractors got the jobs for one hundred dollars a square foot and paid the subs two dollars a square foot. In the meantime, fifty thousand nonunion workers were brought into the city, most of them from the Caribbean, and were paid an average of eight to nine dollars an hour to do the work.
> *(148)*

This carving up of the city for massive profit on the backs of migrant workers on minimum wages speaks to the "distant and oppressive corporate powers, outside regular theorisations of responsibility and the clutches of justice" which Walton and Walton identify as a central feature of ecological crime fiction (3). The cataloguing of wrongs which fall beyond the realm of anything Robicheaux can reasonably be expected to address in his law enforcement position presents "the destruction of New Orleans [as] an ongoing national tragedy and probably an American watershed in the history of political cynicism" (Burke 148).

Maria Hebert-Leiter argues that, in Burke's novel, "Conflating local crimes with the crime of abandonment allows for a rewriting of Katrina chaos that allows for a type of fictional vengeance against the federal government and its failure to recognize the level of disaster that occurred" (105). The text is punctuated by these digressive moments of righteous ire, directed at both the entirely unprosecutable "crime of abandonment" at the hands of Bush's government and the "global experience of avarice and venality" of corporate elites who are benefitting from the city's tragedy (148). However, this backdrop of political critique sits uneasily against the framing of the particular crimes

Robicheaux is investigating, as Burke's mix of fact and fiction uncomfortably traffics in the racist tropes which dominated the wall-to-wall media coverage in the days following Katrina, stigmatising the storm's survivors. Referring to the sensationalised reporting of post-Katrina New Orleans, Nasem Yousaf observes the interplay between news and crime fiction:

> Regardless of accuracy, post-Katrina stories took on the texture of crime genre clichés, and such images, in turn, find their way back into novels . . . Post-Katrina fictions that focus on murder, rape and looting, such as *The Tin Roof Blowdown*, must inevitably also negotiate the racializing as well as the criminalizing of the city.
>
> *(559–560)*

We can see this cross-fertilisation from tabloid-style reporting when Robicheaux meets his friend and ex-partner, Clete Purcell, outside the Convention Centre. Clete relates the chaotic conditions inside: "Street rats are shooting guns in there and raping anybody they want" (36). Reports of widespread gun crime and rape within the Superdome and Convention Centre dominated the news cycle in the days following the storm. Although, as is frequently the case in disaster zones, there was a rise in sexual assaults following Katrina, it is now also well established that many of the more lurid stories of violence were exaggerated or fabricated, eagerly amplified by a press excited by what Carl Smith describes as the "primordial appeal of the story of a city in chaos and people running loose" (Carr).[3] Not only does Clete's account remain uncontested in the novel, but the plot revolves around the actions of Bertrand Melacon and his brother Eddy who, tapping into the nightmarish vein of Southern fears of miscegenation, "had a thing about young white girls" (66). These two Black serial rapists, who live by a self-interested expediency, turn their attention to looting rich White neighbourhoods during the storm along with a couple of their associates. In the process, they also manage to kill a priest who was attempting to rescue his parishioners from the church attic and steal his boat, unwittingly rip off a gangster, taking his seemingly cursed blood diamonds, and – stretching narrative credulity to breaking point – coincidentally turn up at the house of Thelma Baylor, their rape victim from two years earlier, where one of their group is shot to death, and Eddy is paralysed. Their actions are framed as a kind of racialised vengeance: "Eddy and Bertrand saw the storm as a gift from God. White people in New Orleans had been making money on the black man's back for three hundred years. It was time for some payback" (58). This narrative focus on Bertrand and his brother seeking personal profit from the chaos echoes the media reports presenting the city's Black populace as out of control violent criminals, contributing to a situation in which "Emergency personnel became afraid of the very people they were supposed to save" (79).

Bertrand's depiction is quite clearly couched in well-worn racial stereotypes. Robicheaux observes:

> Bertrand Melancon seemed to personify for Clete what he hated most in the clientele he dealt with on a daily basis. They were raised by their grandmothers and didn't have a clue who their fathers were. They got turned out in jail and thought of sexual roles in terms of prey and predator.
>
> *(76)*

In his review of the novel, Gary Younge voices his frustration with this plot strand:

> With its huge creole community, pigmentocracy, black middle class and Cajun influence, this part of Louisiana was richer than the sum of these banalities before the storm rolled in. So any fictional, crime-based portrayal of Katrina that has its core narrative driven by black rapists and looters has little hope of catching the drama of the moment.

Indeed, this narrative choice seems to run entirely counter to the detective-hero (and mouthpiece for Burke's politics) Robicheaux's frequent expressions of disgust at the widespread racism he encounters, not least among colleagues laughing about lynchings: "These guys not only made me ashamed I was a police officer, they made me ashamed I was a white man" (90). The flashbacks of the brothers' attack on Thelma Baylor are harrowingly graphic and we also see the ongoing impact of her trauma in the present (8, 66, 154–155). *Tin Roof*'s byzantine plots are linked together, then, through the figure of a Black rapist in ways which uncomfortably reinforce racial tensions and anxieties stoked by an unrestrained media. As Steenberg observes:

> Repeated reference to (the threat of) rape in the aftermath of Katrina became a particular talking point across news coverage. Frequently this crime was connected to African American men, suggesting a longer American ideological association first, with the suffering female body as standing in for the traumatized national body and second, of the body of the African American man as an alien force threatening the national/female body.
>
> *(33)*

Given this particular element of Katrina coverage and the broader context Louisiana's long history of lynchings, it is disturbing that the text comes close to exculpating the shooting of looters. White vigilantism is by no means confined to Louisiana's past, as Thompson's account armed militia defending Algiers Point from "thieves, outlaws or, as one member put it, anyone who simply 'didn't belong'" attests. His investigative article aims to redress the balance of media coverage which had exacerbated already fraught racial fault lines: "Immediately after the storm, the media portrayed African-Americans as looters and thugs . . . Now it's clear that some of the most serious crimes committed during that time were the work of gun-toting white males" (Thompson).

There are significant parallels between the events Thompson reports and those which take place in Burke's novel. In Thompson's account, however, it is suggested that the gunmen "defending" Algiers Point fire indiscriminately at Black people, as their White enclave, which suffered relatively minor storm damage, became an evacuation point for flood victims. We find a similar militia in Burke's novel, led by the Baylors' neighbour, Tom Claggart, who uses his knowledge of Thelma's assault to attempt to involve her father, Otis Baylor, in defending the area from "the Snoop Dogg fan club" (53). Here, however, Claggart's alarmist rhetoric is justified, as not only are the Black interlopers there to steal what they can, it turns out that, in a frankly overwhelming coincidence, they are in fact the very men who attacked Thelma. As the details of the case emerge, Robicheaux sympathises with Baylor, who he wrongly suspects of pulling the trigger: "I know what happened to your daughter. If my daughter were attacked by degenerates and sadists, I'd be tempted to hand out rough justice too. In fact, any father who does not have those feelings is not a father" (90). Given the fact that similar real-life shootings went uninvestigated for years, until they were exposed by Thompson's reporting, Robicheaux's lengthy investigation into the shooting at the Baylors' house takes the text quite far from Burke's professed commitment to laying bare uncomfortable social truths.

Burke's aim in his characterisation of Bertrand seems to be to at least partially redeem him, explaining his violent proclivities as a result of his background: "Dave knows he grew up in a world which was morally insane and full of abuse" (Wroe). We are told that he "could not remember a time when he was not afraid" and that "Bertrand woke each morning with a nameless fear that was like a hungry animal eating a hole through his stomach" (220). After the murder of Father LeBlanc, Bertrand is haunted by the voices of those he left to drown in the church attic. This guilt propels his clumsy attempts to make amends with Thelma and her family by offering them the stolen diamonds, leading Robicheaux, who acts as the young man's confessor, to conclude that

> After his brother was shot, I think Bertrand tried to become the kind of person he could have been if he'd had a better shake when he was a kid . . . Bertrand was able to perform a couple of noble deeds before he disappeared. That's more than we expect from most men who started off life as he did.
>
> *(372–373).*

This epilogue finds Robicheaux accepting his powerlessness and moving beyond his rage at the fate of New Orleans: "The failure to repair the levees before Katrina and the abandonment of tens of thousands of people to their fate in the aftermath have causes that I'll let others sort out" (369). Bertrand's quiet exit from the text – he takes a boat and sails "Way on out yonder" – allows Robicheaux to end the case through a fantasy of salvation for the young man which contrasts with the frustrations of more powerful criminal players emerging unscathed and unrepentant, "whacking golf balls with over-the-hill television celebrities at a Lafayette country club" (370). This final image, with "the ruined City of New Orleans" receding into the background conjures a peace for Bertrand which he never experienced in life, recasting the earlier scene where he sees strange lights in the flood water after pushing Father LeBlanc to his death. Here, the lights become communion wafers and Bertrand is joined with the troubled priest through their shared imperfections:

> The luminosity that radiates from them lies in the very fact that they have been rejected and broken. But in a way he cannot understand, Bertrand knows that somehow all of them are safe now, including himself, inside a pewter vessel that is as big as the hand of God.
>
> *(373)*

While the ending seems to be framed as the calm following the storm, it is a closure which rests on the detective's comforting faith and is explicitly introduced as wishful thinking: "I have a fantasy of about Betrand Melacon and my old friend Father Jude LeBlanc" (373). It serves a belated invocation of protection for two who the detective was unable to save.

City of The Dead

Sara Gran's much lesser-known Claire DeWitt series takes place in a world which the author describes as,

> just a little to the left of reality – it's not an alternate universe, but it's slightly out of register with the mundane world. One facet of this world is that detectives are more prominent than they are in our real, more boring world.
>
> *(Dos Santos)*

City of the Dead is the first in the series and sees Gran's eccentric private detective return to New Orleans to take on a missing person case. DeWitt is hired by Leon, the nephew of Assistant District Attorney, Vic Willing, who sat out the hurricane in his French Quarter apartment only to go missing during the chaos that followed. Willing's status as a member of New Orleans elite meant that, unlike the thousands of other missing citizens, "even the cops had looked for Vic. They had found nothing" (6). While the plot, like Burke's, relies heavily on coincidence, this is not so glaring or unexpected within Gran's text where the detective's methods range from the traditional – finger printing, crime scene searches and subject interviews – to the significantly more outlandish – premonitory dreams, the I Ching and sharing joints laced with embalming fluid with persons of interest. This slightly quirky narrative world may seem ill-suited to the weighty subject matter of Katrina, but the novel encompasses many of the same issues as Burke's work, and the frequently hallucinatory narration evocatively depicts the strangeness of the shattered cityscape the detective must interpret:

> The damage didn't end. It seemed like it should be over, and then on the next block it was worse: buildings missing walls, houses pushed by the force of the water into other houses, cars on top of cars, blocks of houses half collapsed, boats on sidewalks, parking lots of cars covered with chalky white dust the dirty water left. It had been more than a year since the storm. But on some blocks it was as though nothing had happened since then, literally nothing, not even a breeze or rainfall or a bird or even a breath.
>
> *(23)*

This passage, from DeWitt's first survey of New Orleans in the course of her investigation, presents a traumatised city which seems both eerily arrested in time while also continuing to experience damage, fitting Caruth's understanding of trauma as "unclaimed experience", which remains beyond the grasp of narrative articulation, which "defies even as it claims, our understanding" (5) Metaphors of drowning permeate the depictions of the storm's survivors as they remain stuck in the storm's wake, despite the passing of time. Trauma is the text's central theme, occurring on individual, collective and cultural levels, and the text's resolution lies in the excavation of repressed memories at the edge of the flood waters.

It emerges that Vic Willing was a serial abuser of the boys he came into contact with through his work in the DA's office and on mentoring schemes for at risk youths. He was shot by one of his victims, Terrell, while defying his nephew's expectations and taking part in the rescue effort: "swimming around helping people, getting dirty – I don't really see him doing that" (7). Gran's depiction of those who Leon awkwardly terms "you know, Superdome people" contrasts sharply with Burke's avaricious Melacons (4). While DeWitt's main suspect, Andray, is capable of violence, he repeatedly risks his life to protect his younger and less hardened friend. Unlike *Burn* or *Last Known Victim*, however, Katrina is not just the backdrop here, with opportunistic criminals using the chaos to cover their crimes. The murder occurs *because* of the storm. Negra argues that

> Where 9/11 speedily and durably installed a set of themes and archetypes of cultural heroism that helped to inscribe a narrative of worthy self-sacrifice and barbaric terrorism, the events associated with Katrina, although they did indeed generate a range of heroic actions, did not lead to the lionisation of particular individuals or groups. Explanations for this are surely rooted in if not limited to the fact that in the latter instance those most conspicuously impacted were the black urban poor rather than white professionals.
>
> *(16)*

The shooting of Vic Willing, which occurs at the water's edge, reflects the implausibility of traditional conceptions of heroism in the midst of "garbage and shit" strewn flood waters (257). In stark contrast to Bertrand, Terrell goes to the water with the intention of going out on the rescue boats, distinguishing himself from his friends: "I'm not sittin' around doin' nothing when kids dyin' out there. People stranded on the rooftops and shit" (257). When he reaches the water, the nightmarish scenes challenge his Hollywood visions of heroism: "People was looking through the dead people looking for their kids and shit. It was like in church, when they talk about hell? Like being hot and dead people all over and shit?" (257) Nevertheless, he moves forward to pitch in but notices Willing "Just a few feet away . . . getting out of a boat with this kid, this boy – little kid about twelve. Real dark kid, all wet, shaking" (258). The sight of the powerless child in his abuser's arms reactivates Terrell's trauma, as he fears Willing will "just take that boy, and, you know", so he shoots (258).

Terrell's fantasy of self-sacrifice is derailed first by the abject scenes of suffering flood victims, then by the unexpected confrontation with his individual traumatic past as he poignantly recounts

> That night . . . It was like – like before that happened? When I was going down to the water on my way there? Walking through the city? It was like – like I was going to be a hero or some shit. Like I was going to be savin' people. I saw this, like a picture of myself, like a picture in my head. Like a movie, you know? And I was in a boat. Just like he was. And I was saving all these boys, from the water. And it seemed *real* . . . it felt really good, you know? Like I'd already done it. Like the movie, the thing I was seeing in my head, me saving all those kids, I'd already done it. So I don't understand how . . . .
>
> *(258)*

Narratives of "worthy self-sacrifice" are not possible, it seems, amidst the "fucking mess" of the storm's aftermath (257). In the wake of Katrina, Willing undertakes a self-imposed penance for his years of predation on the very communities now stranded by the storm, rescuing people and animals, "taking all the ones everybody else left behind . . . boatload after boatload" (244). This narrative of redemption is cut short as he is recast as abuser when Terrell misreads his act of lifting the boy from the flood waters. The temporal disjunctions in Terrell's account, imagining his future heroism as a done deal before being abruptly reimmersed in his past experience of abuse by the unexpected appearance of Willing, echoes the earlier description of the wrecked city, suspended in time, where the "damage didn't end":

> Like it was happening all over again. Like it was *still happening*. Like it never stopped. Like everything I'd done since then, everything I am, all my friends, my brothers – it was all just gone and it was just, just that fucked-up time forever, that tiny little fucked-up time *forever*. Like I was in that room with him over and over again, and I wasn't getting out.
>
> *(258)*

Terrell's personal trauma collides with the larger collective trauma of the storm's incomprehensible damage, arresting Willing's attempt at self-redemption and short-circuiting Terrell's own hero-narrative before it even begins.

As excavator of this memory, DeWitt's role as detective merges with that of the therapist, helping Terrell recover the pieces of his fractured self-image: "There will never be any shortage of floods . . . There will always be people who need to be rescued. And there will never, ever be enough people to save them all" (261). The future she promises him is notably not that of a better world but rather the possibility of helping in a world that is irredeemably damaged. DeWitt entrusts Terrell's rehabilitation to federal law enforcement. Given the chaos of the post-Katrina justice system portrayed earlier in the novel, this reliance on official channels of justice seems as fantastic as Burke's luminescent communion wafers, reiterating DeWitt's assertion that "New Orleans is no town for happy endings" (229).

Conclusion

In the weeks following Hurricane Katrina, George W Bush made frequent addresses to the nation. These speeches were an effort in damage limitation, in the face of rising anger at his administration's delayed response to the disaster. On 10 September 2005, he proclaimed: "Today, America is confronting another disaster that has caused destruction and loss of life. This time the devastation resulted not from the malice of evil men, but from the fury of water and wind" (BBC). Two days later, in what seems to be an attempt to address the growing charges of institutional racism, he claimed "the storm didn't discriminate and neither will the recovery effort. When those Coast Guard choppers were pulling people off roofs, they didn't check the colour of a person's skin. They wanted to save lives" (BBC). Bush attempted to regain control of the narrative of Katrina, but both of these proclamations, as we have seen, were demonstrably false. While the storm may not have discriminated, its impact was not felt equally, and the environmental racism underpinning disparities in flood protections meant some neighbourhoods emerged relatively unscathed while others were decimated. As Ted Steinberg argues:

> By recruiting an angry God or chaotic nature to their cause, those in power have been able to rationalize the economic choices that help to explain why the poor and people of color – who have largely borne the brunt of these disasters – tend to wind up in harm's way. The official response to natural disaster is profoundly dysfunctional in the sense that it has both contributed to a continuing cycle of death and destruction and also normalized the injustices of class and race. *(xiii)*

Both Burke's and Gran's texts work against these kinds of attempts to depoliticise Katrina as a tragic but unavoidable weather event. Their depictions of crimes in the wrecked city highlight the context of environmental racism, mapping the geographic disparities of New Orleans along racial fault lines. Rather than using the recovering city as an exotic backdrop for crimes which could occur anywhere (as other crime writers have done) these novels interrogate the storm's meanings, making a complicated contribution to "the public articulations of collective pain and suffering" which Eyerman argues are necessitated by cultural trauma (8).

Notes

1. There are multiple critical accounts outlining why the emphasis on ecological disasters as "natural" is politically expedient, glossing over the manmade elements of these crises. See, for example, Steinberg (particularly the original Preface and Preface to the 2nd edition) and Gordon.
2. Haunting is a key theme in Burke's work, and Robicheux is visited by the ghost of a Confederate soldier in *In the Electric Mist with Confederate Dead* (1993). In a recent interview, Burke makes clear that he believes his beloved Louisiana landscape is inhabited by spectres of America's violent past, persisting in order to remind present inhabitants of their ecological responsibilities: "I believe the spirits of Confederate and Union soldiers are still in those marshes and among the hummocks and the flooded woods and the miles and miles of wetlands that are among the most beautiful in the world. I think these spirits are there to warn us about our role as stewards of the earth. The southern coast of the state is literally washing away. Some brave souls, even corporate ones, are trying to save it. It's a daunting task, particularly in our present political environment" (Dwyer).
3. A 2006 report into sexual violence in during Hurricanes Katrina and Rita found a total of 47 cases, which the authors describe as a "remarkably high rate of prevalence" (National Sexual Violence Resource Centre, 3)

Bibliography

Alfred, Dana. Progress for Some, Hope and Hardship for Many. Louisiana Family Recovery Corps, May 2008.

Babington, Charles. "Some GOP Legislators Hit Jarring Notes in Addressing Katrina." The Washington Post, WP Company, 10 Sept. 2005, www.washingtonpost.com/archive/politics/2005/09/10/some-gop-legislators-hit-jarring-notes-in-addressing-katrina/685fa514-1b19-4893-934b-9200d1f4d608/.

Barr, Nevada. Burn. St Martins Press, 2010.

Bird Rose, Deborah."Anthropocene Noir." *Arena Journal*, no. 41/42, 2013/2014, pp. 206–219.

Bullard, Robert D. "The Threat of Environmental Racism." *Law and the Environment: A Multidisciplinary Reader*, edited by Robert V. Percival and Dorothy C. Alevizatos, Temple UP, 1997, pp. 103–108.

———. *The Black Metropolis in the Twenty-First Century Race, Power, and Politics of Place*. Rowman & Littlefield Publishers, 2007.

Bullard, Robert D., and Beverly Wright. "Race, Place and the Environment in Post-Katrina New Orleans." *Race, Place and Environmental Justice after Hurricane Katrina: Struggles to Reclaim, Rebuild and Revitalize New Orleans and the Gulf Coast*, edited by Robert Bullard and Beverly Wright, Westview Press, 2009, pp. 19–47.
Burke, James Lee. *The Tin Roof Blowdown*. Orion Books, 2007.
Carr, David. "More Horrible than Truth: News Reports." *The New York Times*, 19 Sept. 2005, www.nytimes.com/2005/09/19/business/media/more-horrible-than-truth-news-reports.html?searchResultPosition=1.
Caruth, Cathy. *Unclaimed Experience: Trauma, Narrative, and History*. 20th anniversary ed. Johns Hopkins UP, 2016.
Dossantos, Alexandra. "Sara Gran Interviewed – Strand Magazine." *Strand Mag*, 19 Oct. 2018, strandmag.com/sara-gran-interviewed/.
Dwyer, Murphy. "James Lee Burke on Art, Addiction, and Making Movies." *CrimeReads*, 24 Jan. 2019, https://crimereads.com/james-lee-burkes-louisiana/.
Eyerman, Ron. *Is this America?: Katrina as Cultural Trauma*. U Texas P, 2015.
Gordon, Ruth. "Katrina, Race, Refugees and Images of the Third World." *Hurricane Katrina: America's Unnatural Disaster*. *Hurricane Katrina: America's Unnatural Disaster*, edited by Jeremy Levitt and Matthew Whitaker, U Nebraska P, 2009, pp. 226–254.
Gran, Sara. *City of the Dead*. Faber, 2011.
Hartnell, Anna. *After Katrina: Race, Neoliberalism, and the End of the American Century*. State U New York P, 2018.
"In Quotes: Bush on Katrina." *BBC News*, BBC, 16 Sept. 2005, news.bbc.co.uk/1/hi/world/americas/4252906.stm#:~:text=This%20time%20the%20devastation%20resulted,will%20be%20stronger%20for%20it.&text=My%20impression%20of%20New%20Orleans,neither%20will%20the%20recovery%20effort.rring-notes-in-addressing-katrina/685fa514-1b19-4893-934b-9200d1f4d608/.
Keeble, Arin. *Narratives of Hurricane Katrina in Context: Literature, Film and Television*. Palgrave, 2019.
Klein, Naomi. *The Shock Doctrine: The Rise of Disaster Capitalism*. Penguin, 2007.
Levitt, Jeremy, and Matthew Whitaker. "'Truth Crushed to Earth Will Rise Again': Katrina and its Aftermath." *Hurricane Katrina: America's Unnatural Disaster*, edited by Jeremy Levitt and Matthew Whitaker, U Nebraska P, 2009, pp. 1–22.
Masciotra, David. "James Lee Burke on Art, Fascism, and the Hijacking of American Christianity." *CrimeReads*, 11 Aug. 2020, crimereads.com/james-lee-burke-on-art-fascism-and-the-hijacking-of-american-christianity/.
Murphy, Dwyer. "James Lee Burke on Art, Addiction, and Making Movies." *CrimeReads*, 24 Jan. 2019, https://crimereads.com/james-lee-burkes-louisiana/.
National Sexual Violence Resource Centre, *Hurricanes Katrina/Rita and Sexual Violence*, www.nsvrc.org/sites/default/files/2012-03/Publications_NSVRC_Reports_Report-on-Database-of-Sexual-Violence-Prevalence-and-Incidence-Related-to-Hurricane-Katrina-and-Rita.pdf.
Negra, Diane. "Introduction: Old and New Media after Katrina." *Old and New Media after Katrina*, edited by Diane Negra, Palgrave Macmillan, 2010, pp. 1–21.
Steenberg, Lindsay. "Uncovering the Bones: Forensic Approaches to Hurricane Katrina on Crime Television." *Old and New Media after Katrina*, edited by Diane Negra, Palgrave Macmillan, 2010, pp. 23–40.
Steinberg, Ted. *Acts of God: The Unnatural History of Natural Disaster in America*. 2nd ed., Oxford UP, 2006.
Thompson, A. C. "Post-Katrina, White Vigilantes Shot African-Americans with Impunity." *ProPublica*, 19 Dec. 2008, www.propublica.org/article/post-katrina-white-vigilantes-shot-african-americans-with-impunity. Accessed 18 July 2022. Picador, 2008.
Walton, Jo Lindsay, and Samantha Walton. "Introduction to Green Letters: Crime Fiction and Ecology." *Green Letters*, vol. 22, no. 1, 2018, pp. 2–6, doi:10.1080/14688417.2018.1484628.
Waterhouse, Carlton. "Failed Plans and Planned Failures: The Lower Ninth Ward, Hurricane Katrina, and the Continuing Story of Environmental Injustice." *Hurricane Katrina: America's Unnatural Disaster*, edited by Jeremy Levitt and Matthew Whitaker, U Nebraska P, 2009, pp. 156–181.
Wroe, Nicholas. "A Life in Writing: James Lee Burke, Bestselling Chronicler of the Good and Evil in Modern America." *The Guardian*, Guardian News and Media, 17 Nov. 2007, www.theguardian.com/books/2007/nov/17/featuresreviews.guardianreview14.
Younge, Gary. "Review: The Tin Roof Blowdown by James Lee Burke." *The Guardian*, Guardian News and Media, 1 Dec. 2007, www.theguardian.com/books/2007/dec/01/featuresreviews.guardianreview12.
Yousaf, Nahem. "Regeneration through Genre: Romancing Katrina in Crime Fiction from *Tubby Meets Katrina* to *K-Ville*." *Journal of American Studies*, vol. 44, no. 3, 2010, pp. 553–571.

24

SEEKING ENVIRONMENTAL JUSTICE

Muti in Southern African Crime Fiction

Felicity Hand

Medicine murders, or *dipheko* or *muti*, are a ritual still practised in parts of Southern Africa. Young children, usually prepubescent girls, are abducted, and their organs are removed while they are still alive to be used in traditional potions, considered to enhance the power – political, economic or sexual – of the recipient. In her novel, *The Screaming of the Innocent* (2003), Motswana writer, Unity Dow, now a minister in the government of Botswana, offers a powerful denunciation of the perpetuation of these practices and the silent connivance of members of the population. Likewise, the South African authors known together as Michael Stanley explore the world of witch doctors and traditional medicine in their crime novel *Deadly Harvest* (2015). In this chapter, I analyse these novels from an environmental justice perspective as the two female detectives – one a professional, the other a young woman who finds herself forced into the role of amateur sleuth – fight against ignorance, poverty, superstition and sexism in a society where upper class, male dominance has guaranteed a certain degree of tolerance towards ritual murder. I first outline my main theoretical approach: environmental justice covers a large field and has moved in subtle ways since its early days, in many aspects moving towards and clearly including ecofeminism. I have chosen to explore these novels from the perspective of environmental justice as it prioritises concepts like basic civil liberties and social, gender and class equality. Although these two novels have little in common apart from the subject matter – one is written by a leading human rights activist and politician while the other is the work of two white South African retired male university professors as part of the Detective Kubu series – I contend that both show the deeply rooted patriarchal attitudes to women and to the poor which are often perpetuated by the very victims of these cultural traditions.

Environmental Justice in Southern Africa

The two novels to be discussed in this chapter are from and about Southern Africa, and any analysis of an African text must take on board how race, gender and class inequality intersect alongside environmental injustice. Environmental justice as an extension of social justice clearly privileges the human over the non-human as it prioritises concepts like self-determination, basic civil liberties, social equality and access to natural resources among other demands of disadvantaged peoples. It is vital to understand the origin of these social differences and how they are connected, as Schlosberg and Collins (361) point out:

> The environmental justice movement has never been about equity alone; environmental justice has always focused on how injustice is constructed – why those already exposed to other forms of disadvantage are also subject to environmental bads.

DOI: 10.4324/9781003091912-29

Environmental justice demands that nobody should be discriminated against on the basis of their nationality, race, class or gender, and "it relates to how humans relate to each other as well as how humans interact with the environment" (Bassey). Placing humans as opposed to flora and fauna at the centre of a web of social, economic, political and environmental relationships may appear to be anthropocentric, but environmental justice concerns itself with relational injustices among human beings as well as their relationship with nature. In contrast, ecological justice seeks to distribute justice to nature as a whole, disavowing any anthropocentric bias. Nevertheless, and despite the importance of decentring the human subject as regards the natural world, in this chapter I draw on environmental justice theories that foreground silenced or marginal perspectives in human collectives: the disenfranchised, the poor, the disabled and those whose health, livelihood and general well-being is of little concern to the elites as is the case of the victims of *muti* in the two novels (Caminero-Santangelo 32).

Both *The Screaming of the Innocent* and *Deadly Harvest* expose what Belinda Dodson calls the "structural injustices of material inequality" (84) which include public and private social relations, related discourses and widespread belief systems, all of which generate these social injustices. Caminero-Santangelo claims that social justice and representations of nature, power and privilege feature largely in African writing, not just those texts that would be classified under the heading of environmental or ecocritical literature. He argues that economic inequality and the political rights of the lower classes are closely intertwined with environmental policies, therefore issues of oppression and liberation take centre stage (7). From this standpoint, class and gender inequalities form the backbone of texts that speak out against environmental injustices, as the human animal is part – not centre – of a web of interconnected species. I am aware that recent African ecocritical scholarship laments the lack of "interlinking of human and nonhuman lives in African societies represented in literary works" (Iheka 2), but I second Anthony Vital's argument that there is a need for African ecocriticism provided it encourages societies to construct "more equitable, sustainable, and healthy ways of inhabiting their place":

> Ecocriticism, if it is to pose African questions and find African answers, will need to be rooted in local (regional, national) concern for social life and its natural environment. It will need, too, to work from an understanding of the complexity of African pasts, taking into account the variety in African responses to currents of modernity that reached Africa from Europe initially, but that now influence Africa from multiple centers, European, American, and now Asian, in the present form of the globalizing economy.
>
> *(88)*

Likewise, Chengyi Coral Wu notes that "Anglo-American ecocriticism has developed so narrowly that it has ignored ecocriticism that developed outside of Anglo-American contexts" (149). She goes on to claim that while African scholars have no axe to grind with ecocritical theories, they may query the "values and assumptions that Anglo-American ecocriticism tends to endorse and treat as universal" (151). Wu argues that African literature focuses on neocolonial – that is Western – answers to environmental issues and the exploitation of African natural resources by multinational companies. Certainly, the agenda for African ecocriticism must be different, if only because of the negative impact of the continent's colonial past but also because the emphasis is on the creation of awareness, which leads on to agency and the possibility of resistance. In my view this is what encourages the connection between environmental justice and the principles of ecofeminism. Dodson (2002), drawing on Warren (1997), claims that there exists a clear overlap between the two methodologies, as both seek to explore how women, people of colour, the underclass and the non-human natural world are exploited. Ecofeminism emerged in the West as a result of the pacifist, feminist and ecology movements of the late 1970s and the early 1980s. It traced a clear connection between patriarchal domination and violence against marginalised peoples, especially women and nature. One of its main tenets was that the worldwide struggle to preserve the planet with all its living beings goes hand in hand with the liberation of women.

Ecofeminism, therefore, draws attention to sexual and environmental justice and traces connections between human and non-human environments from feminist viewpoints. Ecofeminist critics claim that sexist and classist discourses of nature have been instrumental in the formation of master narratives of domination. It is true that early ecofeminism was often criticised for essentialising women as being more suited to understanding and leading in environmental issues than men (Johnson and Johnson 1994, 107). Great importance was also attached to women as reproducers of the human species while failing to contemplate a woman's decision to reject motherhood. Women were identified as being too closely related to nature in opposition to men, who were deemed to be the creators of culture. This meant that women were often positioned as passive and their bodies a site of oppression without choice or freedom. Certainly, Unity Dow's women characters refute this notion of female passivity as will be seen later.

For the early ecofeminists, the root cause of women's domination was seen to be patriarchy, its institutions and its values – based on capitalism – that have reduced them to an inferior status. Natural surroundings have been destroyed, the earth resources mercilessly exploited and flora and fauna killed in order to satisfy the greed of capitalist accumulation. In time, the term ecological feminism has become more inclusive as it seeks to include other marginalised people such as former colonised subjects, usually ethnic minorities and the poor. In Val Plumwood's terminology, humans participate in "a network of related oppressions", and all systems of oppression operate in a system of domination of a minority over a majority (1994, 79). These interlocking, interrelated dualisms – nature vs culture, active vs passive, reason vs emotion to cite just a few – operate together in order to justify oppressions. Differing types of exploitation – race, class and gender – mutually reinforce one another as they overlap to form a complex map of patriarchal domination. Ecological feminism, as it tends to be called nowadays, responds to all kinds of patriarchal domination including the disregard for the environment and all living creatures, although scholars working in this area may still have to struggle to have their work acknowledged both as feminist and as ecological. Greta Gaard reminds us:

> In the early 1990s, despite feminist leadership in the antitoxics and antinuclear movements and women's leadership in the burgeoning environmental justice movement, feminist scholars still conceived of social justice, interspecies ethics, and environmental concerns as separate.
>
> *(2011, 33)*

Attitudes to non-human animals can also reflect attitudes to minority groups, weaker subjects or anyone regarded as the other. In this respect, Plumwood points out the close link between male chauvinist attitudes towards women and towards non-human animals thus establishing a species hierarchy. The boundary signalling the frontier between human and non-human animals has been shown to be fuzzy,[1] and ecological feminism questions the criteria for separating the human species from the rest of the living community, as any nature-humanity dualism is a consequence of a patriarchal ideology. It is true that the human species sets itself apart from non-human animals because of our capacity to think and reason but this does not make those creatures that do not possess these qualities any less valuable. This assumption of innate superiority as regards non-human animals is a political stance and can easily be extended to any creature, human or otherwise, that is deemed inferior, as I will suggest in my discussion of the two novels.

Plumwood's "dualistic conception of human identity" (1993, 34) needs to be probed as it reveals both subordination and dependency. The strong are only identified as strong because there exist weak 'others', but strength need not be defined in physical terms only. The objectification of the weaker 'other' entails the belief that he or she has no purpose of his/her own, which demands consideration. The purpose of the weaker other is defined in terms of the master's purposes (Plumwood 53). The master does not recognise the other as his (or her) fellow human and fails to recognise their desires or needs, which allows him to impose his own purpose. The weaker other has thus no intrinsic value; it is merely useful. However, this also means that the master depends on the other as s/he provides him with a needed resource, so in fact the dependency is stronger on the part of the master, and the more dependent he becomes on the resource, the more he will disparage his/her humanity. This kind of reading obviously brings to mind the master-slave relationship but, as I will show in my reading of the two novels, it is applicable to the neocolonial context of Botswana and the traditional beliefs surrounding *dipheko*. My reading of two African novels from an environmental justice perspective is thus an attempt to test these theories with two crime fiction novels that, first and foremost, aim to stir consciousness among their readers. Moreover, these African texts extend Western theories and complement what is understood as environmental justice and ecological feminism.

Ritual Murder: Deviancy or Tradition?

Ritual murder is a controversial term in itself. Many experts in witchcraft violence argue that the killing of the victim in these cases is not ritualised as there are no specific symbolic actions that have to be carried out, nor prescribed words that need to be uttered (Minnaar 2003, 87). Anthropologist Charlanne Burke states that:

> *Dipheko* is technically not witchcraft since it is the process of procuring human body parts through murder to make into medicine to do witchcraft.[. . .] witnesses' accounts attest either to being bewitched and thus forced to help in *dipheko* killings, or fearing future harm through witchcraft should they refuse to help.
>
> *(Burke 2000, 205)*

The term preferred by scholars is "medicine murder" or *muti*, as it is called in Southern Africa because the victim's body parts are used to make magic potions and used by the individual for self-serving motives. However, for the purposes of this chapter I am using the term "ritual" in the sense of a prescribed code of behaviour or a clearly pathological symptom of obsessive-compulsive neurosis. For the practitioners of *dipheko*, as it is specifically called in Botswana, even calling this action a murder is to elevate it above what Harriet Ngubane calls "just an act of wanton wickedness" (191). She argues that it is not *deliberately* evil or harmful towards the victim herself since this killing is a means to an end, but this end does not justify the means since respect for human life is simply disregarded. Unlike *muti* murders, a human sacrifice, such as a ritual sacrifice to a spirit, may be regarded as an act of worship or an act of atonement and must be understood as designed to serve the greater needs of the community as a whole. Ritual murders, on the other hand, manifest a complete disregard for human life, which is what leads me to read the novels from both an environmental justice and an ecological feminist standpoint.

Dow's novel was inspired by a real-life event that occurred in 1994. A 14-year-old schoolgirl named Segametsi Mogomotsi was found murdered on 6 November 1994 in Mochudi, Botswana. She went missing the previous day, and her body was found naked and mutilated in an open space the next morning. The *dipheko* sparked protests by the students at school where she attended, as well as among the citizens of Mochudi. The protests led to riots in neighbouring Gaborone, prompting the government of Botswana to call in London's Scotland Yard. To date, no one has been formally charged with the murder. Unity Dow wrote her novel, *The Screaming of the Innocent* (2003), loosely based on these events, and the murder of Mogomotsi also inspired the South African writers who publish under the name Michael Stanley in their novel *Deadly Harvest*, to be discussed more fully later.[2] According to Burke, Segametsi Mogomotsi was picked, not because she was a clever girl and a good student, which she was, but because she was poor, and they thought no one would ask awkward questions if she disappeared (209). This underlines the disregard for marginalised people, and, I argue, reinforces the argument that the impoverished, especially if they are female, hover disturbingly on the species boundary, as the poor are depicted and classified as animal.

Burke goes on to claim that the actions of the businessmen-killers reveal a fixation on material wealth. Unhappily, at the same time, their motivations in capturing Segametsi parallel the mother's dreams for her daughter growing up to support her financially (Burke, 209). That the frontier between humans and non-humans is fixed is, as Huggan and Tiffin point out, a "fiction" (135). Cynthia Willett queries whether one species should serve as the yardstick for deciding the moral worth of any other species (Willlett, 8). In *Deadly Harvest*, the fate of an albino, a *leswafe*, is determined because he is not considered quite human. The two men who abduct him have no qualms about delivering him up to a witch doctor who uses his body parts with no consideration of the man's human rights. Defining or deciding just who or what qualifies as being an equal human being is based, according to the Australian philosopher Peter Singer, on "the moral principle of equal consideration of interests rather than on the possession of some characteristic" (2015: 237). In the two novels under consideration, both female children and albinos are clearly dismissed by elite members of society for lacking any "equal consideration of interests", as they are ruthlessly murdered for the instrumental purpose of providing increased social, political and sexual power through the so-called medicinal use of their physical organs.

Dipheko, in fact, normally consists of the abduction of young children, usually prepubescent girls, and the removal of their organs while they are still alive to be used in traditional potions that are considered to enhance the power – political, economic or sexual – of the recipient. The victim has to be alive when the body parts are removed as – in the words of those who participate in the murder: "she has to see and scream to release the power of the *dipheko*" (SI 211).

***The Screaming of the Innocent*: Challenging the Impunity of the Urban Male Elite**

Unity Dow, former high court judge and now Minister of Education and Skills Development, has written a powerful denunciation of the perpetuation of these practices and the silent connivance of members of the population. In *The Screaming of the Innocent*, the police resolve that the missing girl, Neo Kakang, has been killed by wild animals, a story which never really convinces her mother and the other villagers in Gaphala, and proceed to close the case. The allegation that wild animals have caused the death of the child highlights the main stumbling block in any interspecies understanding owing to the inherent bias that humans have towards our own species. As Willet argues: "Every time a particular human skill cannot be found in nonhumans, it gets used to assert human superiority" (2014: 47). Singer reminds us that the human animal often looks to other animals for justification of his or her acts. In the case of killing other animals for food, humans may feel vindicated, but they forget that non-human animals do not have a choice, whereas the human animal can reflect on the morality of killing for food (2015: 224). In the context of *muti*, they can or should reflect on the morality of killing for purely egomaniacal and dissolute reasons.

Five years after the crime, Amantle Bokaa, a national service worker at a remote health centre stumbles upon hidden evidence – Neo's abandoned clothes – and opens the case again. Amantle sets out to unearth what really happened to Neo and why the police closed the case when the evidence clearly points to the girl having been a victim of a ritual killing. In the process, her investigation unearths a cover-up that implicates the rich and powerful. Neo's clothes, stiff with blood after 5 years, become a silent plea for someone to clarify her disappearance. In her role as unexpected amateur detective, Amantle, with the help of her lawyer friend, Boitumelo, forces a showdown between the poor, uneducated villagers and the powerful Botswana elite. The novel does not bring the guilty to justice, but Dow's novel does highlight a society which is made up of a network of oppression that plagues women and the rural communities. She suggests that such a society can only exist provided these deeply rooted patriarchal attitudes are perpetuated by the very victims of these cultural traditions. Grass roots environmental justice movements can do little to short circuit the access to power and resources that remain in the hands of the powerful elites, who hinder the separation of environmental policies from social justice laws that would really empower the poor and the underclasses.

The novel opens with the three main participants of the killing being introduced ironically, as "three pillars of the community" (22). Mr Disanka is a "good husband, father, lover, businessman and community leader" (9). Head Man Bokae is an unlikeable character, known to resort to witchdoctors to get his own way, and Deputy Headmaster Sebaki, who, we are told, "had engineered, through witchcraft, the headmaster's fatal car crash" (19). The three men meet, Mr. Disanka being the leader, and agree to hunt what they call "a hairless lamb", after which the novel then fast forwards to the present, five years later, and is focalised through the character Amantle. Through her, Dow relates how children are taught from an early age to fear the men out looking for "hairless lambs".

> These men speak Setswana and can therefore trick a child into stopping and listening to them while they're pretending to be kind or to be simply looking for directions. They prefer to capture darkish children because they find they can use them to make the best *dipheko*, which means 'traditional strengthening medicines'. Those men can *lobela* you *dintsi*, that is, kill you so effectively, efficiently and secretly that not even flies will find out you've died. Those men are after body parts, especially young breasts, anuses and brains. They're known by many frightening names, including *Bo-Rakoko*: 'The Brain Men'. Because their powers are so strong, it's impossible for them to get caught.
>
> *(59)*

I have quoted this passage at length as it reveals Dow's desire to transmit how deeply ingrained these beliefs are in Botswanan society and, what is even more important, how firmly people believe in the impunity of the perpetrators of this practice. I argue that one of the motives for writing this novel is the need for people to rid themselves of these fears and speak out against such abuses, bearing in mind that children from rural areas, not from the wealthier urban centres, tend to be the prey of the so-called Brain Men. The child Amantle grows up but ceases to doubt the veracity of these tales when she notices how "a school child did occasionally disappear, so it seemed that ritual murders mightn't belong to the realm of myths after all" (59). It cannot be overlooked that the child whose body parts are needed for the medicine ceases to be human as she is described as a "hairless lamb". In fact, the animal imagery present throughout Dow's novel highlights the animalisation of the poor, rural, female victims and how speciesism is another form of oppression that is linked with, and reinforced by, other oppressive structures (Gaard 2011, 38). As commented on earlier, the unfortunate Tanzanian citizen, Mabulo Owido, falls victim to the blurred boundary between humans and non-humans (DH 169–176). In Dow's novel, Mr. Disanka's victim is likened to a vulnerable, artless animal, unaware of the lurking carnivore: "She was an impala that wasn't noticing the poacher for what he was, an impala that was mistaking the poacher for a game warden" (7).

After the discovery of the mysteriously displaced bloodied clothes, there is a flashback to the report of the missing child, Neo. This episode has a twofold purpose. On the one hand, it becomes clear that the local police are terrified about *dipheko* and do not want to get involved for fear they or their families should suffer any consequences, such is the power that people believe the practitioners of *dipheko* hold. On the other hand, it highlights one of the main issues of the novel, the immense gap between the rich and the poor, the rural communities and the urban elite and the perpetuation of patriarchal attitudes. The fact that Dow's characters rebel against this hierarchy is indicative of the changes that Dow foresees needing to take place in Botswana if these obscene practices are ever to be eradicated. Annie Gagiano argues that the socially entrenched practice of *dipheko* is maintained by mainly male elitist groupings who are safeguarded by a widespread fear of their activities. For her, the whole novel "functions as a powerful, harrowing arraignment of the wicked as well as of the cowed and apathetic members of the community" (Gagiano 2006, 54). However, Dow creates strong characters that do not give in easily to the pressure of hierarchy. Neo's mother is not convinced that the bloodied clothes – a vital piece of forensic evidence – have simply disappeared from police custody. "She hadn't been fooled by either of the officers: a dead daughter gives people strength. She'd never have thought she could challenge a man of the law" (64). Amantle, nominated by the villagers as their spokesperson, refuses to be bullied by the police. "She has power – more than she first realises – because she has the confidence afforded her by formal education, and because she has already learnt how to organise herself and others in opposition" (Lenta 2004, 43). The clear differentiation between rural communities and the urban elite highlights the deepening class divides that pose a threat to a real environmental justice in contemporary Botswana.

The truth is that the state's representatives of justice and security have no interest in tracking down those guilty of the killing – all they want to achieve is to pacify the troublesome villagers with a show of commitment to justice, because violent suppression of their anger would be counter-productive. Mrs. Molapo, the director of the Tirelo Sechaba (Botswana National Service), the only woman present at a crucially important meeting – because it would have been unwise to leave her out – "puts her finger on the spot" (Gagiano 2006, 56):

> It's true that very few ritual-murder cases are actually solved. Children, especially girls, disappear or die under very mysterious circumstances, and no-one's called to account. And when the villagers demand answers, we, the government, are the very ones who try to shut them up. *(146)*

Amantle and her two lawyer friends join forces with the resentful villagers and Neo's devastated mother in order to take on the bureaucratic, patriarchal structures that work to conceal abuses and maintain entrenched inequalities. The denouement of the novel throws up a ghastly irony: The Minister for Safety and Security turns out to be the chief instigator of Neo's horrendous murder. Not only is he not accused of the crime, but he manages to recuperate the bloodstained evidence in a pretence to reopen the murder investigation.

Deadly Harvest: Questioning the Species Boundary

In their 2015 novel, *Deadly Harvest*, Stanley's character, Assistant Superintendent David Bengu, popularly known as "Kubu" (hippopotamus), on account of his size and who features in many of their books, is joined by a new recruit in the Botswana Criminal Investigation Department, Samantha Khama. As the CID's first female detective, she works hard to establish her credentials in the male-dominated police force and uncover the culprits behind a series of ritual murders. Khama finds herself fighting against ignorance, poverty, superstition and sexism in a society where upper class, male dominance has guaranteed a certain degree of tolerance towards ritual murder. Moreover, the HIV and AIDS epidemic in the country is also part of the backdrop to the ritual murder cases. Despite laudable efforts by the government, Botswana has one of the highest infection rates in the world.[3] Thus, there is a collusion between the fear of contamination and a somewhat unhealthy respect towards the witch doctors as the belief in at least some of their powers is very widespread.

The novel opens with the disappearance of 10-year-old Lesego Betse. After 4 months, no corpse has been found, and the now cold case has been assigned to Khama, who declares that she wants to make a difference for women, "[t]o give crimes against them the same attention as the police give crimes against men" (16). To start off her investigation, Khama consults with a European professor of anthropology who has specialised in traditional healing. She shrugs off the explanations as "complete nonsense" (34) but later realises that the white man is well informed: "These are my people he's talking about, she thought. I'm ashamed for them" (35). The anthropologist explains that:

> Potions for luck usually involved animals thought of as fortunate for a special reason: the scaly anteater – safe from attack with its armour-plating; the klipspringer – escapes easily by jumping between rocks on hooves that seem to hold like Velcro (35).

The link between animals used for *muti* and the next stage up – the use of live human children – is made early on in the novel. While Samantha Khama is working on the cold case, another little girl goes missing, and her distraught father eventually goes mad in his fruitless attempts to find her. The girl's mother had died of AIDS leaving Witness Maleng to bring up Tombi on his own. In his desperation, and as the police do not seem to be moving, he visits a traditional healer who provides an enigmatic solution to his woes:

> You must seek a man. A man who was nothing and is now everything. A man no one knew and now all know. A man who was weak and now is powerful. That is where you must look. That is where you will find out about her.
>
> *(76)*

Influenced by the healer's words, Maleng believes that a rising star in Botswana politics, Bill Marumo, must be his daughter's murderer, and he breaks into his house and kills him. Maleng feels no remorse: "Marumo had got what he deserved – raping young women, using *muti*, and who knows what else. Bastard" (147). Through this misguided father – because in fact Marumo was just another pawn in the witch doctor's power and not the actual murderer of Tombi – the first serious challenge to patriarchy is made. Samantha Khama denounces the social and gender gap between the rich and powerful, men like Bill Marumo, and the poor and female, like Maleng's daughter, Tombi. She accuses the institutions of shelving cases of *muti* – like the real-life case of Segametsi Mogomotsi – "because high-up *men* in Botswana were involved . . . Justice for some, a blind eye for others" (17; emphasis in original). These deepening class divides put any real environmental justice on hold.

In a parallel scene, the nephew of the deputy commissioner of police, Joshua Gobey, visits a witch doctor in order to achieve "real power" (83). The witch doctor, who turns out to be a highly respectable member of society, dons an animal disguise to enhance the mystery surrounding him. Joshua's reaction is indicative of the generalised fear that witch doctors or traditional healers incite in the population:

> At the side of the table, with the light somewhat behind it, sat something large. The face had sunken eyes and a baboon snout with exposed teeth. The torso was bare and strong, a leopard skin wrapped around the loins. The baboon head is a mask, Joshua thought. And what right does he have to leopard skin, the mark of royalty? He swallowed. The most powerful witch doctors were said to be shape-changers, becoming baboons or hyenas at will to do their evil work in the night. He felt an urge to run, but stood his ground.
>
> *(82)*

Stanley's witch doctor, unlike Dow's, actually becomes an animal when seeing his clients – or, perhaps, one should say victims as men like Joshua who desire power over others in fact succumb to the dictates of the person who has the power to grant it to them. The animalisation of the witch doctor – he dons the guise of a baboon and a leopard – blurs the border between humans and non-humans. Within the non-human group, leopards and apes rank high in comparison with the lambs of Dow's novel, which underlines the previously mentioned fuzziness between species boundaries. Moreover, the aggressive animal identities that the witch doctor assumes suggests that it is acceptable to treat people as if they were animals (Huggan and Tiffin 135). Like the policemen in *The Screaming of the Innocent*, Kubu's subordinates are unwilling to admit to the fear that witch doctors conjure up in their minds. Samantha is adamant that "[t]hey're all phonies, yet they're such a big part of our culture. People actually believe in them. In this day and age". Her boss, Kubu takes a more practical view: "As long as people believe in them, they'll be around – whether we think of them as charlatans or not" (139).

A connection between the two missing girls is soon made when a gourd containing *muti* from Lesego's body parts is recovered from the dead man's home. A further development occurs when an albino man from Tanzania is reported missing. A *leswafe*, as they are called in Botswana, has "great power" (83) according to the witch doctor visited by Joshua Gobey, and even Assistant Superintendent Kubu is aware that "witch doctors sought albinos for making their most powerful potions" (222). The two underlings who are hired to abduct Mabulo Owido believe that albinos have "powerful spirits" (171). The unfortunate man has his organs removed while he is still alive, like the little girls. The choice of victims is completely arbitrary because they are merely a class of interchangeable items which can be used as resources to satisfy the master's needs. Owido is not important as a person; he is objectified, as any albino would have equally served the witch doctor's purpose. Likewise, Tombi and Lesego, like Dow's Neo Kakang, are not seen as individual girls with their own wishes, ambitions and personalities. Indeed, any young girl with similar characteristics would have sufficed.

Khama and Kubu conduct a laborious investigation but obtain satisfactory results in the end. Using people's fears of the alleged power of albinos, Kubu obtains a confession from one of the men who abducted Owido. This fear of the other is what kindles the blind faith in the magical powers supposedly commanded by the witch doctor and by the rich and powerful men who are his clients. The tentacles of the witch doctor reach into high places. The deputy commissioner of police, who has used *muti* in the past, officially dies of emphysema but "[i]t was the witch doctor's curse that killed him" (252). Kubu is convinced that the deputy commissioner's attempt to unmask his own nephew's dealing with *muti* sealed his fate. In this way, Stanley does not pass judgement on what their character Samantha sees as primitive superstition. In fact, she is portrayed somewhat unfavourably in the novel as being too intense and too belligerent, whereas Kubu is depicted as a modern, caring Motswana man, nothing like the monolithic image Samantha has of the male sex: "Who cared that a little girl was murdered for body parts, when the reputation of *men* had to be protected" (17; emphasis in original). Dow's women, apart from being much more numerous, take on a multiplicity of complex roles. Stanley's women, especially their main character, are shallower, more vacuous, and less convincing. Kubu stands out as the model to imitate and Stanley take pains to show him in his private capacity as loving son, husband, and father of daughters, one biological and the other adopted after her mother died of AIDS.

While all types of exploitation mutually determine and support one another, for example race, class and gender, patriarchal domination can only be understood when exploring specific practices of masculine identity in concrete social and cultural contexts (Plumwood 12). Stanley, like Dow, singles out the arrogant neocolonial elite who have enriched themselves both by accumulating enormous wealth and by the imposition of an authoritarian political culture, which in turn has led to radical inequalities among the population.[4]

Environmental Justice Awareness Through Crime Fiction

Ironically, it is the men who live a more urban, Westernised lifestyle that use *muti* medicine in the knowledge that they can buy the silence of the authorities and forestall the investigations into their victims' disappearances. Dow's Botswana is one where the rich can easily exercise such powers because of the alienation between rural and urban people. According to Margaret Lenta, this alienation is not "ignorance or detachment: the rich and urban prey on the poor and rural, and to that extent they need them" (Lenta 2004, 40). Dow's searing presentation of such "profoundly corrupt male leaders brings the powerful down, symbolically, to a level where even rural, uneducated people are able to see through their shiny façades to the rottenness underneath" (Gagiano, 56). The police, sharing the murderers' belief in their magic powers, dare not oppose them although they can hardly be ignorant of who is behind the ritual murders. Likewise, the novel suggests that Disanka's wife and mother, each in her own way, are aware of Disanka's involvement in the recent case being broadcast over the radio. The grandmother's reassurance that *dipheko* is an act of violence enacted by the rich on the poor, and her firm belief that without her son's intervention the crime will remain unsolved, point to what Lenta describes as a "mixture of understanding and wilful ignorance which most people use to distance themselves from painful facts" (Lenta, 44).

> "What's with this child?" Mma-Disanka [the grandmother] asks the other four family members who are present, including Lesego. "You've been acting very strangely recently. You're either angry or scared. What are you afraid of? You think someone will come and hurt you like they did that little girl? No one would dare, Sego. Your father's too powerful for anyone to hurt any of you. These poor children always come from a poor family . . . But really, Rra-Lesego [Mr Disanka], you must go over there to help with the search. And those poor villagers will need people like you to make the police move. You'll see; nothing will happen – nothing!"
>
> *(190–191).*

Lesego, elder daughter of Disanka, has been acting strangely ever since she unwittingly saw her father and his accomplices bring home bloodstained plastic bags which may easily have contained human body parts. Her father's realisation that his beloved daughter has uncovered his most evil desires is a just, if feeble, punishment for his crimes. Lesego cuts all ties with her family, and so Disanka pays a high price for his illicit urges. However, the reader can feel no sympathy for this predator, unlike perhaps Rra-Naso, an unwilling partner in crime. Possibly the most moving scene in the whole novel – and certainly the most gruesome – is the confession by this old man who had befriended Neo's mother and had given her great comfort in her loss all these 5 years. It turns out that he had been the local accomplice in the search for a suitable hairless lamb. Rra-Naso had been cornered into finding a suitable child for the three so-called pillars of the community in order to carry out their "harvesting".

> He promised me five goats if I found him a hairless lamb, a child with no sins yet. He said he wanted a girl who hadn't yet had her first period and who hadn't been with a man yet. I'm a poor man, a weak man. And I didn't want to do it, but he came to me many, many times; many nights he came. He brought food. I was afraid to say no to his food; *you can't say no to people like that*. . . . I was afraid of him. He looked at Naso, my child, when he was leaving, and I was afraid. . . . I was afraid – very afraid – when I saw him looking at my daughter.
>
> *(209; emphasis mine)*

Neo's screaming has haunted Rra-Naso for so long that he finally commits suicide but not before revealing the identity of the fourth man in the conspiracy: the Minister for Safety and Security. Unfortunately, the death of the vital witness, together with the discovery that the only evidence the villagers had was once again in the hands of the enemy, means that the Neo Kakang murder will forever be a cold case. However, Dow's powerful denunciation can work towards raising awareness about such matters and bringing about change in people's attitudes towards the lesser powerful members of society.

Deadly Harvest conforms more closely to the formula of a detective novel with its emphasis on police procedure and the endless interviews and laborious investigations that ensue. Moreover, its multiple plot lines converge neatly to reach narrative closure (Scaggs 2005, 94). Stanley's dedication is to Alice Mogwe and Unity Dow "who fight the battles we just write about" (vii).[5] Environmental justice denounces a high-class male-dominated social system that casts women, low-class men (and the environment) as victims of patriarchal acts of exploitation, but neither women nor men are a unified construct and race, class, age, region and ethnicity need to be taken into consideration. Young girls in both novels are treated worse than animals, and the albino Owido does not conform to hegemonic African masculinity. Thus, any call for an ecological masculinity demands that the values pervading environmental movements need to be directed towards expressions of care not just for nature but for men's compatriots and ultimately for their own selfhood (Hultman & Pulé, 2018: 51–52).

Morati Disanka or Kudu's daughter Tumi are never going to be victims of *muti* as their social class protects them from this kind of violence. Mr. Disanka in *The Screaming of the Innocent* and Dr. Pilane in *Deadly Harvest* fail to perceive the other – the little girls and the albino man – as an independent other. They attribute their victims with possessing the qualities that they, the masters, desire, need and lack. Environmental justice forms the backbone of the two novels I have discussed, as they illustrate the basic issues behind the movement of what is often called brown ecology, that is the issues that affect the poor, regardless of race, class or gender. The consumers of muti medicine all live in "a rich area" (DH 91); Mr. Disanka's elder daughter "has her own bedroom – in her village, an almost unheard-of luxury" (SI 183), while the victims Lesego Betse, Tombi Maleng and Neo Kakang all hail from humble families living in rural, makeshift, cramped dwellings, which highlights the growing gap between social classes. The police are reluctant to deploy resources to search for the missing girls as they can profit little from stirring up scandal when powerful elites are behind the abductions, as "ritual murders aren't committed by poor people" (SI 171). As Samantha cries out in despair: "If the president had a daughter, and she disappeared, may be things would change" (DH 132).

Conclusion: Justice for Humanity

The perpetuation of *muti* murders with the concomitant objectification of the more defenceless members of society highlights the continuing power of patriarchal values. Education, in the case of *dipheko*, does not seem to solve the issue as the kind of people, usually men, who take part in this kind of practice tend to originate from the social elite. These men have become wealthy and prosperous because of their class position which allows them to use less educated and poorer people to catch their victims, as is the case in both novels. The lethal combination of power and knowledge – formal education and traditional witchcraft beliefs – makes these men practically invincible. Ordinary people may resent what they see as "poverty, corruption, and lack of care and respect" (Burke, 211) but too frequently fail to speak out against their lack of real power. I contend that Dow's and Stanley's novels move inexorably in the right direction, showing how the deeply rooted patriarchal attitudes to women and to the poor are often perpetuated by the very victims of these cultural traditions precisely *because* of these entrenched power disparities. In *The Screaming of the Innocent*, Amantle and her friends therefore "pose very serious ethical questions about culture and challenge the encrustations of power, tradition and dogma that bedevil their society" (Kalua 2009, 54).

Thus, despite the eventual foreclosure of justice, Dow's novel still functions as a powerful call to action. A failure to grasp this awareness can only perpetuate the inequalities that exist between the rich and the poor, the urban and the rural and, above all, between male impunity and privilege and female disempowerment. Likewise, *Deadly Harvest* highlights the coerced complicity between the powerful elites and the less powerful peoples of society. Stanley's novel, like Dow's, calls for a return to the basic principles of human society through allegiance to a Setswanan concept: *botho*, translated as respect. It refers to a profound interconnectedness among people – I am because you are – which is not so far away from the principles of environmental justice, where the web of connectedness is not one of dominance but one of equality and freedom.

Acknowledgment

Research for this article has been funded by the project Rhizomatic Communities: Myths of Belonging in the Indian Ocean World, Reference: PGC2018-095648-B-I00. Ministerio de Ciencia, Innovación y Universidades/ Fondo Europeo de Desarrollo Regional (FEDER).

Notes

1. See Huggan & Tiffin 2010, 134–139 for further discussion on the question of species boundaries.
2. To a lesser extent Alexander McCall Smith (1998) also draws on this case.
3. See the United Nations Development Programme, www.bw.undp.org/content/botswana/en/home/sustainable-development-goals/goal-3-good-health-and-well-being.html Accessed February 12th 2022.
4. See for example Good (2008) and for a critique of Good's view, Samatar (2009).
5. Alice Mogwe is the director of Ditshwanelo [rights], a human rights organisation in Botswana.

Bibliography

Bassey, Nnimmo. "Environmental Justice and the Right to Life and Dignity." *The Guardian Nigeria News – Nigeria and World News*, 6 Apr. 2021, guardian.ng/opinion/environmental-justice-and-the-right-to-life-and-dignity/.

Burke, Charlanne. "They Cut Segametsi into Parts: Ritual Murder, Youth, and the Politics of Knowledge in Botswana." *Anthropological Quarterly*, vol. 73, no. 4, Oct. 2000, pp. 204–214, doi:10.1353/anq.2000.0009.

Caminero-Santangelo, Byron. *Different Shades of Green. African Literature, Environmental Justice, and Political Ecology*. U Virginia P, 2014.

Dodson, Belinda. "Searching for a Common Agenda. Ecofeminism and Environmental Justice." *Environmental Justice in South Africa*, edited by David A. McDonald, Ohio UP, 2002, pp. 81–108.

Dow, Unity. *The Screaming of the Innocent*. Spinifex Press, 2002.

Gaard, Greta. "Ecofeminism Revisited: Rejecting Essentialism and Re-Placing Species in a Material Feminist Environmentalism." *Feminist Formations*, vol. 23, no. 2, Summer 2011, pp. 26–53, doi:10.1353/ff.2011.0017.

Gaard, Greta, Simon C. Estok, and Serpil Oppermann, editors. *International Perspectives in Feminist Ecocriticism*. Routledge, 2013.

Gagiano, Annie. "Entering the Oppressor's Mind: A Strategy of Writing in Bessie Head's *a Question of Power*, Yvonne Vera's *the Stone Virgins* and Unity Dow's *the Screaming of the Innocent*." *The Journal of Commonwealth Literature*, vol. 41, no. 2, 2006, pp. 43–60, doi:10.1177/0021989406065771.

Good, Kenneth. *Diamonds, Dispossession and Democracy in Botswana*. Boydell & Brewer, 2008.

Huggan, Graham, and Helen Tiffin. *Postcolonial Ecocriticism. Literature, Animals, Environment*. Routledge, 2010.

Hultman, Martin, and Paul M. Pulé. *Ecological Masculinities. Theoretical Foundations and Practical Guidance*. Routledge, 2018.

Hund, John, editor. *Witchcraft Violence and the Law in South Africa*. Protea Book House, 2003.

Iheka, Cajetan. *Naturalizing Africa. Ecological Violence, Agency, and Postcolonial Resistance in African Literature*. Cambridge UP, 2018.

Johnson, David Kenneth, and Kathleen R. Johnson. "The Limits of Partiality. Ecofeminism, Animal Rights, and Environmental Concern." *Ecological Feminism*, edited by Karen J. Warren, Routledge, 1994, pp. 106–119.

Kalua, Fetson. 2009. 'Identities in Transition.' *Scrutiny2*, vol. 14, no. 2, 2009, pp. 48–58, doi:10.1080/18125440903462133.

Lenta, Margaret. "Postcolonialism in an Anti-colonial State: Unity Dow and Modern Botswana." *Kunapipi*, vol. 26, no. 2, 2004, pp. 34–46.

Minnaar, Anthony. "Legislative and Legal Challenges to Combating Witch Purging and *muti* Murder in South Africa." *Witchcraft Violence and the Law in South Africa*, edited by John Hund, Protea Book House, 2003, pp. 73–92.

McCall Smith, Alexander. *The No. 1 Ladies' Detective Agency*. Birlinn, 1998.

Moolla, F. Fiona, editor. *Natures of Africa. Ecocriticism and Animal Studies in Contemporary Cultural Forms*. Wits UP, 2016.

Ngubane, Harriet. "The Predicament of the Sinister Healer: Some Observations on Ritual Murder and the Professional Role of the Inyanga." *The Professionalisation of African Medicine*, edited by G. Chavunduka and M. Last, Manchester UP, 1986, pp. 189–204.

Plumwood, Val. *Feminism and the Mastery of Nature*. Routledge, 1993.

Samatar, Abdi Ismail. "Diamonds, Dispossession and Democracy in Botswana." *African Studies Review*, vol. 52, no. 3, 2009, pp. 192–194, doi:10.1353/arw.0.0254.

Scaggs, John. *Crime Fiction*. Routledge, 2005.

Schlosberg, David, and Lisette B. Collins. "From Environmental to Climate Justice: Climate Change and the Discourse of Environmental Justice." *WIREs Climate Change*, vol. 5, no. 3, 2014, pp. 359–374, doi:10.1002/wcc.275.

Singer, Peter. *Animal Liberation*. The Bodley Head, 2015 [1995 3rd edition].

Stanley, Michael. *Deadly Harvest*. Orenda Books, 2015.

Vital, Anthony. "Toward an African Ecocriticism: Postcolonialism, Ecology and Life & Times of Michael K." *Research in African Literatures*, vol. 39, no. 1, Spring 2008, pp. 87–121, doi:10.1353/ral.2008.0005.

Warren, Karen J., editor. *Ecofeminism: Women, Culture, Nature*. Indiana UP, 1997.

Willett, Cynthia. *Interspecies Ethics*. Columbia UP, 2014.

Wu, Chengyi Coral. "Towards an Ecocriticism in Africa: Literary Aesthetics in African Environmental Literature." *Natures of Africa. Ecocriticism and Animal Studies in Contemporary Cultural Forms*, edited by F. Fiona Moolla. Wits UP, 2016, pp. 141–165.

25

A FORM OF WILD JUSTICE

Carl Hiaasen's Deployment of Carnivalesque Environmental Ethics and Moral Technology

Anna Kirsch

Carl Hiaasen is a prolific environmental crime writer specialising in outrageous and darkly ironic satires aimed at highlighting the breakage in human-environmental relations. In her handbook, *Twentieth-Century Crime Fiction*, Lee Horsley identifies Hiaasen, and his forerunner John D. MacDonald, as two "important contributors to the literary campaign against the destruction of the physical world, using the satiric potential of crime fiction to raise serious moral questions about human interactions with nature" (175). Throughout his novels, Hiaasen uses carnivalesque performativity as a mode of environmentalist critique, encouraging readers to place environmental protection over human life by aligning the environment with characters who embody nature, such as Skink: the roadkill-eating ex-governor of Florida. The carnivalesque becomes a narrative tool through which Hiaasen influences his reader's ecological moral decision-making, infusing the texts with a gleeful sense of schadenfreudic delight in eco-radical violence and an empathetic connection to characters who act to defend it, such as Skink. The humour in Hiaasen's writing comes from the incongruity produced by the inversion of commonly held moral and ethical codes which demand we place human interests as the centre of concern.

In a period spanning three decades, Skink transverses the opening and closing of an ecological carnivalesque where nature presents a satirical challenge to human authority. Since his appearance in Hiaasen's second novel, *Double Whammy*, in 1987, Skink has appeared in seven novels: *Native Tongue* (1991); *Stormy Weather* (1995); *Sick Puppy* (1998); *Skinny Dip* (2004); *Star Island* (2010); *Skink: No Surrender* (2014); and *Squeeze Me* (2020). In each of his appearances Skink has acted as a catalyst for carnivalesque eco-radical activism with each novel representing a different stage of Skink's psychological journey from oddball Floridian to eco-radical folk hero.

Hiaasen's Cultural Ecology

Hiaasen's writing style embodies the essential idea of the environmental movement: that everything is interconnected, and human action can influence everything else. While Hiaasen's settings are uniquely Floridian, he confronts the environmental issues of economic growth and human activity within a complex global ecosystem. Hiaasen stages different facets of environmental degradation in each of his novels, allowing nature to avenge itself through his narrative framings that direct the reader into moral judgements. His positivity towards eco-radicalism is unique not only in the crime genre, but to an extent, popular culture in general. This positive narrative framing of radical environmentalism is indicative of a form of what the moral psychologist Mark Alfano calls "moral technology". In *Character as*

 DOI: 10.4324/9781003091912-30

Moral Fiction, Alfano argues that moral technology is a process of guiding what we are and do from an ethical point of view (9). Neo-Aristotelian philosopher Rachel Haliburton expands on Alfano's work, describing detective fiction as essentially an extended version of a philosophical trolly problem which can serve as a powerful tool for those "engaged in the task of moral education" (34). Hiaasen uses a constellation of rhetorical devices (especially tone and perspective) as moral technology to encourage his readers to engage with environmental injustice within a moral framework.

Hiaasen's style of satire embodies a Rabelaisian commitment to the carnivalesque grotesque, as his environmental villains frequently find messy deaths at the hands of the natural environment they sought to exploit and control. Indeed, all of Hiaasen's novels contain spectacles where nature, usually in the form of an animal agent, reduces humanity to the material level. Hiaasen offers his readers a carnavalesquely grotesque world where a dolphin sexually assaulting a developer's thug (*Native Tongue*), or a rich socialite making her way through the guts of a python (*Squeeze Me*), are experienced as environmental justice. These surrealist qualities of the carnivalesque reorientate the reader's sympathetic responses and liberates them from traditional social rules and expectations, realigning, in the process, their perception of justice to not only include the non-human, but to prioritise the non-human.

The carnivalesque is, rightfully speaking, a collection of performative phenomena also classified under a variety of terms such as humour, comedy, parody, satire, and the grotesque. In a literary context, the term carnivalesque was developed by the Russian formalist Mikhail Bakhtin who used it to refer to "the varied popular-festive life of the Middle Ages and the Renaissance" (218). Despite the historical temporality of Bakhtin's formulation of the carnivalesque and its role as a state-licensed form of misrule, the term has become an often-used interpretive device routinely appropriated by those seeking to deconstruct the commodified spectacle of many contemporary staged events and themed attractions (Ravenscroft and Gilchrist 35). Surprisingly, despite becoming a popular theoretical frame for modern festivals and spectacles, Bakhtin's study of the medieval and early modern carnivalesque in *Rabelais and His World* remains one of the few to attempt a vigorous definition of the carnivalesque as a phenomenon.

As a literary theorist and philosopher working under Stalinist censorship, Bakhtin's view of state officialdom was shaped by a very specific set of cultural expectations, where state structures were blatantly authoritarian. Indeed, the authoritarian nature of the Russian government likely contributed to one of the deepest contradictions in Bakhtin's understanding of the carnivalesque. While Bakhtin ostensibly positions the carnival as a period of misrule and mischief, it is still ultimately sanctioned by the power of the state, calling its deeper political function into question. Nevertheless, Bakhtin's formulation became popular with American comedy theorists in the 1960s as a tool for engaging with American counterculture. Because the carnivalesque is a mode of experience where social norms are overturned and ignored (even if only for a short while), it creates an anarchic, discursive space for activists of all stripes, including radical eco-activists. For example, the sociologist Paul Joosse emphasises the role the carnivalesque played in the promotional materials for the Earth Liberation Front or ELF, which were especially designed to appeal to teenage radicals, specifically teenage boys ("Leaderless Resistance and Ideological Inclusion" 360). In Hiaasen's fiction, he uses the surreal qualities of the carnivalesque to normalise not only acts of eco-radicalism, but also murder.

While much has been written about Hiaasen's hard-boiled and satiric credentials, less critical attention has been paid to his ties with the tragic tradition, specifically the kinship Hiaasen's approach to environmental vengeance can claim with revenge tragedies. Hiaasen's most iconic reoccurring character, the roadkill-eating ex-governor Skink, mimics the traditional malcontent of revenge tragedies where violence is justified because "the homicide is close to the top of the pyramid of power and consequently it is not possible to rely on the authorities to ensure justice is carried out" (Ascari 22). In his pursuit of ecological justice and punishment, Skink cannot seek redress for environmental wrongs through the legal system because those responsible for environmental degradation are frequently at the top of the pyramid of power, leaving limited legal options for pursuing environmental crimes/criminals.

To this day, there has not been a better summation of the complex attitude readers have towards revenge and its moral implications than the one presented by Francis Bacon, one of the Renaissance's leading figures in natural philosophy and scientific methodology. In his essay "Of Revenge", Bacon writes of revenge as "a kind of wild justice; which the more man's nature runs to, the more ought law to weed it out" (347). Bacon also notes that "the most tolerable sort of revenge is for those wrongs which there is no law to remedy" (348). For Bacon, revenge represents a suspension of rationality and justice, yet revenge is occasionally the only way to punish those who are unreachable through legal means. Bacon sticks to the conventional human versus nature binary, where human rationality is pitted against non-human irrationality and where the term "wild" is used pejoratively to denote the animalistic passions unleashed by the act of revenge. In contrast to Bacon, who associates wildness with a form of savage irrationality, Hiaasen's novels propose a new form of wild justice which celebrates the wild. Drawing on the long tradition of the carnivalesque, Hiaasen deploys narrative tone and dramatic irony to reduce the emotional impact of human suffering, all the while dispensing grotesque punishments against those who he perceives to have sinned against the environment. In his development of an ecologically conscious narrative form driven by a mode of green morality, Hiaasen presents a series of eco-revenge fantasies which imbue saving the environment with a sense of moral imperative. His use of carnivalesque satire, combined with his commitment to grotesque yet appropriate ecological punishments, demonstrates a streak of moral puritanism. For Hiaasen, the greatest sin is to be blind to the beauty of nature. In an interview, Hiaasen remarks,

> I think you could take a cross section of all those and sort of put them in a single category of people who just don't care. They don't care about violating a place. And I think that Florida's attraction when I grew up is that it's such an overwhelming place to be that it becomes a character in my novels, and when that character is violated by somebody – whether it's a developer or a lobbyist or a tourist throwing a carton of hamburgers out the window they all become a target to me because I just see no greater sin than to come to a place that's so beautiful and trash it out, you know, and so bad things happen to those people in my books. (Byrne)

In Hiaasen's work, South Florida is more than a place; it becomes a character functioning simultaneously as both victim and avenger. In his comprehensive critical monograph, *Carl Hiaasen: Sunshine State Satirist*, David Geherin examines Hiaasen's entire body of work, from his journalism for the *Miami Herald* to his literary influences. Geherin's work on Hiaasen is particularly concerned with a sense of place, a quality which has been taken largely for granted in crime fiction studies. Even while critics recognise that every crime novel has a physical locality, the process of reading with conscious attention to "ambiance, location, history and place memory" is an acquired skill (S. Walton 115). According to Stewart King, who argues for a growing global readerly consciousness based on a sense of translocality, Hiaasen is part of a growing body of environmental crime fiction where place can be read as "both localised and cosmopolitan" (217). Hiaasen's specific mode of actively environmentalist crime fiction creates a sense of place, not just in the physical sense, but in the sense of emplacement within an ecology.

Hiaasen began his writing career in the 1980s, just as Florida replaced Southern California in the American consciousness as a lotus land intimately linked with economic opportunity and easy living (Willett 76). In Florida, Hiaasen found the perfect place to explore a landscape fully committed to conspicuous pleasure and consumption where he could explore the psychogeography of consumerism and test "the moral consequences of living the sunniest and emptiest of American dreams" (Van Dover and Jebb 326). Hiaasen witnessed first-hand the consequences of those dreams as he watched many of his childhood haunts be transformed into a concrete jungle (Geherin, *Scene of the Crime* 110–11).

Indeed, Hiaasen's early years inspired his column for the *Miami Herald*, where he raged against the sleazy development of Florida funded by the tourist industry and its creation as an artificial paradise.

Hiaasen's personal experience of the slow and creeping process of urban development is what – David Geherin postulates – fuelled Hiaasen's turn towards the Juvenalian mode of satire, a blunt and angry form of discourse directed at the folly of human behaviour. Geherin contrasts this with the Horatian mode, which he considers gentler in tone and often couched as fatherly advice (*Carl Hiaasen: Sunshine State Satirist* 175). Geherin's classification of Hiaasen's humour is supported by Hiaasen, who has described his anger at the obliteration of the natural environment as his inspiration for writing (Silet 11). Hiaasen channels his natural anger towards environmental destruction into a surreal jumble of exuberant violence, adapting and advancing the work of his hard-boiled predecessors.

One of Hiaasen's greatest influences within the crime genre is the prolific hard-boiled writer John D. MacDonald, who wrote around seventy-eight novels as well as short stories for the same pulp magazines that shaped the careers of the Californian touchstone authors such as Raymond Chandler and Dashiell Hammett (Van Dover and Jebb, 331). MacDonald is best described as a proto-environmentalist crime writer because he used his detective to bitterly rage against overpopulation and capitalist development during a period when the environmental movement was still at the margins of the cultural consciousness. For example, MacDonald's 1962 novel, *A Flash of Green*, was considered to have such a compelling environmental message that, when it was adapted for the screen in 1984, a group of developers took action to stop its release (Horsley 176–7). MacDonald's success in addressing environmental issues and educating readers on environmentally destructive development was an inspiration to Hiaasen regarding what was possible within the mode of crime fiction.

In "Carl Hiaasen's Environmental Thrillers: Crime Fiction in Search of Green Peace", Peter Jordan skips over MacDonald's influence to directly connect Hiaasen to Raymond Chandler. Jordan uses Chandler as a touchstone to classify Hiaasen's peculiar brand of biting carnivalesque and to justify his place in a coalescing cannon of crime writers (Jordan 62). The similarity between Hiaasen's Florida and Chandler's Southern California is also highlighted by David M. Parker, who comments on California's and Florida's strange ability to attract "an unusual number of eccentric people" (Parker 308). However, despite attempts to link him to a specific genre model, Hiaasen is innovative in his use of absurdist logic and refuses to present a traditional mystery, or even a plot governed by cause and effect.

Hiaasen's combination of realism and absurdity has been praised by Tony Hillerman, an award-winning crime fiction writer in his own right, who described Hiaasen's first novel, *Tourist Season* (1986), as "another remarkable example of what talented writers are doing these days with the mystery novel" (Hillerman). It is not the complexity of a mystery that beguiles Hiaasen's readers, but the simplicity of his characterisations and the psychological pleasure of justice served. Hiaasen has a distinct voice within the crime genre; his narrative structure is, on analysis, predicated on a variety of formulaic tropes, with many of his characters experiencing grotesque endings at the hands of various non-human embodiments of environmental vengeance (both animal and object).

This violent object interaction is best described by Alan Gibbs as "cartoon realism" (76). However, despite aptly describing Hiaasen's authorial tone, Gibbs is one of the few critics who sees Hiaasen's tone of grotesque surrealism in a negative sense, maintaining that it compromises Hiaasen's representation of gender. While Gibbs is correct in his analysis of Hiaasen's simplistic treatment of gender dynamics, his critique overlooks how Hiaasen centres environmental preservation as a virtue ethic. The simplicity of Hiaasen's narratives allows his use of moral technology to shine, with his heroes acting as "moral cartographers who map out the features of an ethical landscape for readers" (Haliburton 83). Indeed, Skink acts as more than a moral tour guide; he is also judge, jury, and executioner. In all his novels, especially the Skink texts, Hiaasen plays with revenge fantasies against those oblivious to the fragile beauty of nature, often presenting murder as justifiable, or even desirable.

The Invention of an Icon: *Double Whammy* and the Roadkill-Eating Ex-Governor

Hiaasen creates Skink in his second novel *Double Whammy*. From inception, Skink bursts onto the page already a legend, or at least something of an unwashed phenomenon. As a character, Skink is the physical embodiment of many of Hiaasen's fantasies. Indeed, Hiaasen has described Skink as a renegade character who endears himself to readers because he "says and does things that lots of us wish we could get away with" (Burke). Initially, Skink functions as an assistant for Hiaasen's detective figure, R.J. Decker, a former journalist, and neophyte private eye. However, it is Skink's biocentric morality that dominates *Double Whammy*.

In a long section of exposition, Hiaasen informs his readers that Skink, legally known as Clinton Tyree, was the governor of Florida in the mid-1970s before he mentally cracked under the pressure of being "a completely honest man" in a state where corruption runs rampant on both sides of the political aisle. (*Double Whammy* 92). Hiaasen describes how, after his election, Skink shocked everyone by refusing a development bribe and reporting the incident to the FBI, a move which sent one of the largest land development firms in the Southeastern United States "belly up like a dead mudfish" (*Double Whammy* 93). After pressure was brought to bear on Skink by various other developmental interests, he vanished, leaving behind a resignation note citing disturbing moral and philosophical conflicts. Skink, in his apparent madness, has entered the first stage of the carnivalesque. By breaking the bonds of cultural normality and hegemony, he is able to view the world with fresh and anarchic eyes.

In *Double Whammy*, Hiaasen introduces the reader to the high-stakes world of bass-fishing tournaments with a plot involving a cheating conspiracy and a string of related murders. The duo of Decker and Skink soon discover that there are more things fishy in the state of Florida than cheating at fishing tournaments, when Decker's employer, Dennis Gault, a rich playboy with family connections to the sugar industry, frames Decker for the murder of a prominent bass-fishing celebrity Dickie Lockhart. Lockhart is found in a grotesque manner in a fish tank with a fancy fishing lure in his mouth, the eponymous Double Whammy. It is up to Skink to clear Decker's name and confront the various environmental villains of the novel through a series of shambolic spectacles designed to distribute retributive justice.

Upon Lockhart's death, the world of professional fishing and television evangelism collide and combine in unpredictable ways. Hiaasen attacks the spread of urban and commercial development via his characterisation of Richard Weeb, a fraudulent televangelist, and the head of a sleazy condo scheme. Weeb is the mastermind behind combining a fishing tournament to honour Lockhart with the marketing of his development, Lunker Lakes. It is up to Skink to devise an elaborate plan to save the Everglades from Weeb while simultaneously punishing Dennis Gault for his complicity in Lockhart's murder. Skink begins his mission by infiltrating Weeb's televised faith healing session posing as a blind man willing to fake a miraculous recovery. Skink waits for the cameras to roll before yelling "Squeeze My Lemon Baby" at a shocked gathering of Evangelicals, who are only too happy to escape the Hindenburg that is Lunker Lakes (*Double Whammy* 302). While Skink is merciful and does not kill Weeb, the exposure of his fraud and subsequent arrest is almost crueller. After dealing with Weeb, Skink discloses the location of his enormous tame lunker bass Queenie to Gault's minions, with the intention of setting up a natural confrontation between man and fish.

While it is unclear whether Skink intended to kill Gault, the results speak for themselves. Gualt's death is outrageously grotesque as Queenie pulls him overboard. In a passage characteristic of Hiaasen's careful employment of branding and consumerism, Hiaasen puts Gault through the equivalent of a meat grinder when he is pulled into the propeller of his "brand-new turbo model SST" (*Double Whammy* 306). Skink declares that all he did to Dennis Gault was to arrange a natural confrontation, one no different from any other. Yet Skink's abdication of responsibility rings hollow. He understands the premeditative nature of the trap set for Gault and clearly expresses remorse for his actions when they affect Queenie, if not for Gault's death.

In the novel, Skink characterises the dike separating Lunker Lakes from the Everglades as the moral seam of the universe "with evil on the one side, good on the other" (*Double Whammy* 281). Hiaasen uses the image of the dike to instil a physical sense of the metaphysical qualities of good and evil and the thin line of justice dividing them. However, there is a physical component to the good and evil dichotomy of the dike. Skink learns to his chagrin at the denouement of *Double Whammy* that the dike separating Lunker Lakes from the Everglades was the literal container of toxic sludge hostile to all aquatic life. In Hiaasen's moral ecology, it is the innocent fish that Skink endangers in his pursuit of ecological justice that weighs most heavily on his conscience. Skink's failure to read the land correctly in *Double Whammy* sends him towards his future role as an itinerant eco-avenger and environmental malcontent.

In *Double Whammy*, Skink is still rooted within a community, albeit on the outskirts. Skink is living off the grid, but not in the same way as he does in later novels. However, after learning that he placed his pet bass in mortal danger by unknowingly exposing her to water containing lethal phosphate levels, Skink enters a period of self-imposed flagellation. The last view the reader has of Skink is of his bare bottom mooning Decker as he dives into the poisoned waters of Lunker Lakes to save his faithful companion. As Skink fades from view, he and his fish seem to merge with the water, and soon all Decker can hear is the gentle sounds of something swimming out into the Everglades.

Mentoring in the Eco-Radical Tradition and the Opening of Carnivalesque Activism: Skink's Glory Days in *Native Tongue, Stormy Weather*, and *Sick Puppy*

Skink is, fundamentally, an everyman eco-avenger with a core of deeply held ecological principles. From his first appearance in *Double Whammy*, Skink functions as a grizzled guide to environmental radicalism. However, it takes more than one appearance to establish a pattern. After *Double Whammy*, Skink subsequently features in *Native Tongue*, *Stormy Weather*, and *Sick Puppy*. In these three novels, Hiaasen establishes Skink as a mentor for eco-radical activists as they embark on their own ecological journeys through Florida's sleaze, instilling in his chosen pupils an essential environmental sensibility.

As with all Hiaasen's plots, *Native Tongue* is composed of a series of interconnected, episodic interactions. Hiaasen starts the novel in *medias res* where a tourist family in a bright red Chrysler LeBaron convertible have what they believe to be a rat tossed into their car, a rat which is then shot by a friendly state trooper. It is later revealed that the animal tossed in the car is half of a pair of Mango Voles, which in typical Florida fashion, is actually a cosmetically altered pine vole or *Microtus pitymys* stolen from a Disney knock-off theme park at the behest of Molly McNamara, the founder of the radical environmental group the Mothers of Wilderness. In a serendipitous incident of synchronicity, McNamara is an old friend of Skink's from his gubernatorial campaign who was inspired into activism by Skink's elaborately plotted symbolic violence in service of ecological vengeance.

Like the opening of Pandora's box, the primary function of the failed vole heist is to unleash chaos. Afterwards, Joe Winder, a disenfranchised former journalist rotting in his current position as a publicity writer for the Amazing Kingdom of Thrills, finds himself drawn into a series of escalating events surrounding the cover-up of the vole fraud by his tasteless ex-con employer Francis X. Kingbury, or Frankie King. In *Native Tongue*, Hiaasen reveals what Skink has been doing since his disappearance into the moral seam of the universe in *Double Whammy*. Since his last appearance, Skink has escalated his violence to include shooting at airplanes and rental cars in what can only be described as a doomed effort to stem the tide of tourists who are financing the destruction of Florida's fragile environment.

Skink requires an ecological apprentice and appears from the swamp to save Winder not just physically, but also from the soul-destroying reality of his job. Winder's first view of the irascible Skink is when Skink saves him from thugs sent by his ex-employer. It is a memorable first impression, with Winder's confused mind seeing a "silvery beard of biblical proportions. Mismatched eyes: one as green as mountain pines, the other brown and dead. Above that, a halo of pink flowers" (*Native Tongue* 108). Once again, Skink functions as an ecological saviour rather than as a primary protagonist. Under Skink's influence, Winder becomes increasingly militant in opposition to his ex-employer's development plans. Skink initiates this escalation by, in a perfect illustration of the principle of Chekhov's gun, gifting Winder a firearm. Once armed, Winder attacks Kingbury's development site, holding workers hostage and forcing them to inflict property damage (*Native Tongue* 250). However, this initial attack only proves to be a minor wrench in the works of the ever-devouring engine of growth in Florida. Even Winder's second attack on the construction site with an RPG becomes a referendum on monkey-wrenching techniques. Hiaasen uses the carnivalesque surrealism of the explosion as a form of performance activism already popular among eco-radical activists, but this use also underscores how these actions are losing their ability to shock or change cultural attitudes towards development. While the spectacle of a forty-foot flame shooting into the sky and great globs of cement raining down does make a radical political statement, the repetitive nature of human-object violence can lose meaning (*Native Tongue* 412–13). As Winder's love interest, Carrie, delicately points out, there is a certain lack of imagination and permanence in specific sabotage techniques, to the point that Winder could be blowing up bulldozers for the rest of his life.

Hiaasen embellishes upon traditional sabotage techniques when he has Skink and Winder infiltrate the theme park during a staged pageant involving a sanitised version of Florida history, where the "plundering, genocide, defoliation and gang rape that typified the peninsula's past" is removed and repackaged as feel-good fun (*Native Tongue* 279). When the controlled pageantry of the marketing department of the Amazing Kingdom of Thrills and the wild unpredictability of Skink's carnivalesque eco-activism meet, the result is, quite literally, explosive. Skink's sabotage ends the carnival of consumerism, but the pageant's closure grants licence to Skink's crowning as an eco-radical Lord of Misrule. This metaphorical crowning lends legitimacy to Skink's later actions.

Before burning down the amusement park, Skink presides over one of the more morally ambivalent incidents of Hiaasen's carnivalesque brand of wild justice involving the rape and subsequent drowning of Pedro Luz, a corrupt security guard for the developer Kingbury, by a dolphin. There is a certain sense of satisfaction in the Cetacean revenge Skink facilitates on Luz because Luz had previously killed the scientist who discovered the vole fraud by feeding him to an Orca whale, killing both in the process. While the incident with the dolphin ought to be thoroughly reviled, Hiaasen directs the reader emotionally by instilling a sense of poetic justice into Skink's vigilante actions. This incident with the dolphin is a literal illustration of how Hiaasen turns the Baconian interpretation of wild justice as dangerously uncontrollable on its head, by openly celebrating the idea of nature (the wild) revenging itself on the humans who seek to control it.

Skink's next appearance is in Hiaasen's hurricane novel, *Stormy Weather*, where he makes a climactic entrance strapped to a bridge anticipating a hurricane with the fierce intensity of a biblical preacher. However, much to Skink's disappointment, the aftermath of the hurricane is more bathos than pathos, as it fails to cleanse the face of Florida from the unsightly growth that is human development. Worse, the hurricane unleashes new problems, as a variety of opportunistic grifters stream into wrecked communities eager to extract their pound of flesh from the survivors. In Hiaasen's words:

> It hadn't been the cataclysmic purgative he had hoped for and prophesied. Ideally a hurricane should drive people out, not bring people in. The high number of new arrivals to South Florida was merely depressing; the moral caliber of the fortune-seekers was appalling- low- life

hackers, slick-talking scammers and cold-blooded opportunists, not to mention pure gangsters and thugs.

(*Stormy Weather* 239)

Skink, with such disappointing examples of humanity before him, is soon in the grips of apocalyptic withdrawal, latching onto an ideal target for his brand of ecological re-education in the form of Max Lamb, a junior accountant at a New York advertising firm on his honeymoon with his wife Bonnie. Max comes to Skink's attention while he is dragging his increasingly distressed wife around the wreckage of other people's lives in a dark form of disaster tourism. Skink captures Max for enforced mental reconditioning when Max unadvisedly chases after an escaped monkey who has pilfered his expensive camera. When Max asks why he is being kidnapped, Skink chastises him for failing to recognise a hurricane as a holy event and for "desecrating the environment" with his opportunistic filming (*Stormy Weather* 75). Skink's solution for Max's lack of ecological reverence is to attach a shock collar to Max's neck and begin quizzing him on Florida's history, punishing him whenever he gets an answer wrong. Skink's attempts to re-educate Max do not actually succeed, but they indirectly lead to an adorably creepy love story.

In *Stormy Weather*, the hurricane becomes "a Swift-like satiric devise" washing all of Hiaasen's characters together (Horsley 181). The hurricane sets up a meet-cute scenario between Max's wife Bonnie, and Augustine Herrera, the owner of the escaped monkey who beset her husband and lured him away from civilisation. Augustine is a slightly mad millionaire seriously disturbed by environmental degradation. While Augustine wishes to strike out in defence of nature, he has no trustworthy knowledge of heavy explosives. Rather than act on his occasional desire for destruction, Augustine drifts through life donating to acceptable environmental causes such as the Sierra Club or the Nature Conservancy (*Stormy Weather* 26). When the unacknowledged lovers meet Skink, they find an environmental savant ready to induct both of them into a new form of environmental activism. Indeed, *Stormy Weather* closes with both Bonnie and Skink tying themselves to a bridge in anticipation of another hurricane.

In *Sick Puppy*, a new eco-radical activist, Twilly Spree, makes his appearance. Twilly is an independently wealthy and mentally unstable eco-enthusiast who encounters Palmer Stoat, a powerful and well-connected political fixer, when he sees him littering on the highway. Initially, wishing to punish Stoat for his littering, Twilly is shocked to find Stoat is organising a deal with the government to build a bridge that will allow for a major island development project. Twilly's violent threats against the development project spur the developers to hatch an ill-advised plan to send one crazy eco-radical after another.

For the first and only time in Hiaasen's novels, Skink is depicted as emotionally vulnerable and hauntingly human with a personal life beyond his status as an urban legend. The developers blackmail Skink with his brother's well-being in exchange for his services in the disposal of Twilly. Unsurprisingly, the blackmail attempt on Skink backfires, and Skink soon wriggles his way out of the developers' control, repaying their scheme by pulling the current governor's pants down and etching the word SHAME into his buttocks with a buzzard's beak (*Sick Puppy* 417). In Twilly, Skink finds more than a protege, he finds an individual whose mind and moral compass are the equivalent of his own. Indeed, after Stoat is killed by a charging rhinoceros he had paid to hunt, the closure of *Sick Puppy* mimics the generic conventions of the Hollywood western with Twilly and Skink riding into the sunset together in pursuit of yet one more ecologically oblivious litterbug.

Hiaasen's deployment of the carnivalesque in *Sick Puppy* has been questioned by critics such as the noted ecocritic, Richard Kerridge who uses the term "political resignation" to capture the perfunctory performance of environmental vengeance in the novel (87). However, Kerridge is perhaps overeager in his application of the term resignation. Indeed, *Sick Puppy* contains some of Hiaasen's most explosive moments of eco-radical protest. The novel does, however, signal a transitional mode in Hiaasen's approach to the carnivalesque, which becomes more prominent in his later fiction. In *Sick Puppy*, Hiaasen's increasing focus on the political appropriation of carnivalesque performativity comes to the fore. Rather than being part of the carnival atmosphere, Hiaasen begins to use his characters' eco-radical acts of violence to initiate a sense of closure to a carnival of greed. While Skink's approach to eco-radical activism has not changed, the world's attitude towards eco-radicalism has.

An illustrative instance of the political appropriation of the type of carnivalesque protest usually associated with eco-radicals occurs in *Sick Puppy*, when Hiaasen describes how resistance to the development of Toad Island – or Shearwater island, as the developers have rebranded it – came from a small group of "embittered landlords masquerading as environmentalists", who used performance activisms, such as staging a Thoreau-quoting petition, as a way to extract more money for their property (*Sick Puppy* 53). Hiaasen's description of avaricious property owners using performative activism signals the beginning of a new and darker chapter in the carnivalesque performance and environmental activism, one where the roles of government officials and eco-radical activists are inverted.

Nature's Carnivalesque Revenge as Closure and Nostalgia: The Bakhtin Tradition in *Skinny Dip*, *Star Island*, *Skink No Surrender*, and *Squeeze Me*

Hiaasen entices his readers with a combination of natural disgust and schadenfreudic delight in the inversion of traditional power structures. In his Skink novels, Hiaasen uses the appearance of his eco-radical protagonist as a signal of the commencement of a festive time of eco-radical misrule. However, while Hiaasen's later Skink novels – *Skinny Dip*, *Star Island*, *Skink: No Surrender*, and *Squeeze Me* – all still construct Skink's actions as an eco-radical antidote to a world gone mad, they also all involve increasingly carnivalesque institutional structures and politicians that make the delivery of ecological justice through eco-radical misrule all the more complicated. At first glance, *Skinny Dip* is barely worthy of inclusion in a discussion on Skink as he only appears briefly at the end of the novel to claim those individuals' needing punishment for their trespasses against nature. However, *Skinny Dip* marks a dramatic shift in Skink's technical function as an ecological moralist. In *Skinny Dip*, Hiaasen uses Skink to install order, in the sense of punishing the murderer, rather than as a catalyst for rebellion. Instead of signalling the beginning of a period of eco-radical misrule, Skink's appearance signals the closure of a different kind of misrule, that of human acquisitiveness.

There are two vastly different crimes and criminals in *Skinny Dip*. The first crime is an attempted murder, while the second is the systematic poisoning of the Florida Everglades. *Skinny Dip* starts when Chaz Perrone, a biologist so incompetent that he is unfamiliar with the direction the Gulf Stream flows, pushes his wife Joey overboard while on a cruise. Perrone's misunderstanding of his environment leads to a failure of his murder plot, as Joey finds a bale of marijuana and floats back to Florida, bent on revenge. Perrone's attempted murder reveals his involvement with Red Hammernut, a wealthy agribusiness owner who placed Perrone into the South Florida Water Management District to falsify water testing near his farm. At the end of *Skinny Dip*, Hiaasen punishes the witless would-be wife-killer Chaz Perone by placing his fate in Skink's capable hands. Skink's references to the famous Tennyson quote about nature being "red in tooth and claw" – usually associated with the idea that evolution is a violent competition – leaves little doubt of his intentions (*Skinny Dip* 496). In *Skinny Dip*, Skink becomes an embodiment of ecological order committed to enforcing natural rhythms through Darwinian confrontation. For Hiaasen, Hammernut's and Perrone's crimes against the environment are sins equivalent to murder, or perhaps even greater than murder. Hiaasen writes:

> The murder of the Everglades, as perpetrated by Red Hammernut and others, is insidiously subtle and undramatic. Unlike more telegenic forms of pollution, the fertilizers pouring by the ton from the sugarcane fields and vegetable farms of southern Florida do not produce stinking

> tides of dead fish or gruesome panoramas of rotting animal corpses. Instead, the phosphates and other agricultural contaminants work invisibly to destroy a mat of algae known as periphyton, the slimy brown muck that underlies the river of grass and is the most essential nutrient. As the periphyton begins to die, the small fish that feed and nest there move away. Next to go are the egrets and herons, the bluegills and largemouth bass, and so on up the food chain.
> (*Skinny Dip* 468)

In this passage, Hiaasen establishes his message that the true crime in his novels is environmental degradation. Hiaasen slows his crime caper to a narrative crawl forcing the reader to see environmental degradation as premeditated murder. The anthropomorphism of nature is Hiaasen's cleverest, and most morally dangerous, contribution to the crime fiction genre. By giving characters like Skink a certain moral latitude with their approach to ecological justice, Hiaasen initiates a form of green moralism which justifies a certain level of violence when used in defence of the environment.

In *Star Island*, Hiaasen delivers the expected murder of his developer almost as an afterthought focusing most of its attention on the antics of Cherry Pye, a vapid and drug-filled, Lindsay Lohan-esque celebrity and her almost consensual kidnapping by a crazed paparazzo. While *Star Island*'s critique of America's fascination with celebrity is a subject worthy of Hiaasen's satirical attention, the lack of punishment of any kind for the quite frankly spoiled Cherry leaves *Star Island* without any clear unifying moral direction, besides the side plot where Skink fulfils his prime directive as an eco-avenger. Skink is placed in direct conflict with the developer Sebago, who has illegally recruited some itinerant, machete-bearing crackheads to shear twenty wild acres of red mangroves to provide a premium view of the Atlantic from the future townhouses that he and his investors plan to erect (*Star Island* 79). In typical fashion, Skink deploys a painful and bizarre form of punishment, forcibly attaching a sea urchin to an intimate place on Sebago's person (*Star Island* 264). As with all his acts of weird and bizarre corporeal punishment, Skink acts in an extra-legal capacity as judge, jury, and executioner of environmental justice. However, instead of Skink murdering the unfortunate Sebago for his greed, Sebago is killed by a hitman hired by a disgruntled investor. Skink's willingness to allow Sebago to live distinguishes *Star Island* from other Skink novels, making him a more socially palatable kind of hero.

Hiaasen continues Skink's publicity makeover by cross branding the ex-governor into one of his young-adult novels, *Skink: No Surrender*. The novel is narrated from the perspective of a young teenager, Richard Sloan, who embarks on a cross-country tour with Skink to save his cousin from a paedophile. In his able summary of the novel, David Geherin damns *Skink: No Surrender* with faint praise, complaining that Hiaasen's distinct form of environmental rage is "largely pushed to the background" in favour of a plot following online predation (*Carl Hiaasen: Sunshine State Satirist* 160). Geherin's complaint is arguable, as Hiaasen frames environmental degradation within a criminal framework by having Skink's target be a paedophile who also takes pot-shots at local wildlife. For Skink, anyone who attacks animals is "a mentally defective evolutionary mistake who deserves everything he is about to do to them" (*Skink: No Surrender* 196). Skink's retribution is framed in an understanding that those willing to transgress against animals may well transgress against their fellow humans.

In *Skink: No Surrender*, Skink acquires folkloric status and takes on superhuman qualities, including a seeming physical invulnerability to death, as testified to by the various (fictional) Wikipedia pages Richard consults. Richard's first meeting with the ex-governor occurs when he encounters a straw sticking out of the sand and discovers a buried Skink lying in wait for a turtle poacher (*Skink: No Surrender* 5–7). Skink emerges from the ground in a grotesquely dirty birth to take on the roles of guide and mentor. The themes of death and rebirth are strong in *Skink: No Surrender*, which ends with yet another rebirth in the form of the discovery of an Ivory-Billed Woodpecker, a species thought to be extinct.

Skink's next adventure in *Squeeze Me* is a madcap serpentine romp set among the landed gentry of Palm Beach during the early stages of the coronavirus pandemic. *Squeeze Me* opens with the disappearance of a prominent high-society matron, a fierce supporter of the president, and the discovery of a monster python with a suspiciously large bulge. The pinnacle of *Squeeze Me* is Skink's cunning plan to introduce a collection of LSD tripping pythons to various prominent tourist locations throughout Palm Beach, including the fictionalised ground of Trump's Mar-a-Lago, which Hiaasen thinly disguises under the alias of Casa Bellicosa (*Squeeze Me* 7). In a suitably carnivalesque fashion, *Squeeze Me* closes with the president wearing an African mask due to a tanning bed accident involving a monstrously large python.

Hiaasen refuses to name Trump in *Squeeze Me*, instead offering a thinly veiled caricature of the president known only by his secret service code 'Mastodon'. Neither Hiaasen's caricature nor the factual President Trump offers the conventional strait-laced authority figure prerequisite for eco-radical derision. Indeed, *Squeeze Me* has a distinct sense of nostalgia for a time when the government represented a conventional straight man. Hiaasen deliberately emphasises the parallels between Skink's attitude towards overpopulation and earlier literary eco-radicals, such as George Hayduke in Edward Abbey's 1975 novel *The Monkey Wrench Gang*, by giving the most telling reference to the coronavirus pandemic as a global environmental reality. Skink muses that he can never go back to Key West because he "might end up rooting for the goddamn virus" (*Squeeze Me* 334). Skink uses George Hayduke as an alias to conjure a nostalgic return to debates surrounding misanthropy and the radical environmental movement and pays tribute to early eco-radical writings.

Unfortunately, Skink's eco-radical deployment of the carnivalesque loses potency in the face of the refusal of politicians to assume the staid and stuffy elegance of political statecraft. In Hiaasen's writing, Skink's eco-radicalism is increasingly forced to take on the role of moral closure in the face of a society unmoored from what Bakhtin termed "official strict forms of social relations" (208). Indeed, while Bakhtin saw the carnivalesque as signalling a licensed time of misrule and a social safety valve for protest against governmental structures, politicians are increasingly turning to carnivalesque performance as a political strategy. The turn towards political carnivalism is not new, but it became nationally visible in the United States during the 2016 presidential election.

As a presidential candidate, Donald Trump turned his berserk behaviour into an asset, claiming it set him apart from other politicians. Trump's campaign strategy traded Trump's viability as a comic figure exploiting a "public need for revelry and breaks with political social norms" (Symons 186). After his election, Trump consistently failed to embody the solemnity expected in higher office, and political satirists and comedians were faced with the difficulty of satirising a public persona that was already ridiculous. Trump used the chaos he created as an alternative leadership strategy successfully confounding those who sought to criticise him. In their article "Berserk!: Anger and the Charismatic Populism of Donald Trump" sociologists Paul Joosse and Dominik Zelinsky make the connection between acts of violence and carnivalesque protest by aligning Trump's trenchant politics with what they term "berserk-charisma", which is "acquired through a self-abandoning, bloodthirsty rage" (12). Joosse and Zelinsky emphasise the violent nature of carnivalesque politics exploring in detail how Trump, and his supporters, employed carnivalesque violence during the attempted insurrection of 6 January 2021.

In *Squeeze Me*, Hiaasen responds to Trump's carnival politics with an optimistically depressed future for the deployment of carnivalesque acts in direct-action environmentalism. While Hiaasen does not abandon the simmering rage and active defence of the natural world that marked his previous work, he is far less optimistic in his depiction of the impact acts of radical environmentalism will have in creating long-term cultural change. Hiaasen leaves a new generation of radical environmentalists in a world where they can only hope to stem the ongoing tide of the assault against the environment. While Hiaasen does give the reader hope for the future in the form of the irrepressible animal wrangler Angie Armstrong, he does not set her up as an eco-radical replacement for Skink as he does Skink's previous proteges such as Twilly Spree in *Sick Puppy*.

In *Squeeze Me*, Hiaasen demonstrates all the qualities of political resignation which Richard Kerridge claims to detect in *Sick Puppy*. While Hiaasen describes Angie as envious of Skink's "capacious anger and high torque after a lifetime of crushingly predictable futility", he is clear that Angie is unable and unwilling to replace Skink (*Squeeze Me* 270). Indeed, Angie has learned the hard way that Skink's brand of ecological punishment has unintended consequences. In the past, Angie had attempted environmental vengeance by feeding part of a poacher to a tame alligator named Lola, a decision which earned Angie prison time and led to the alligator being killed (*Squeeze Me* 35). In *Squeeze Me*, Skink retreats into a nostalgic eco-radical past, and there is no clear successor ready and willing to assume the mantle of eco-avenger as there was in *Sick Puppy*.

In *Squeeze Me*, eco-vengeance runs against the reality of an equally carnivalesque political status quo. When radical activists and the representatives of political institutions both seek to incite a carnivalesque atmosphere, the carnivalesque loses its secondary function as social commentary and there is no inversion of social power structures. In *Squeeze Me*, Skink seeks to recapture the sense of eco-radical activism as a form of mischievous play. His plan to attack the president, and other Palm Beach big-wigs, with psycho-drugged pythons is meant, as with Skink's role in earlier novels, to incite eco-radical misrule, but the scheme fails to recapture the same sense of carnivalesque revelry. *Squeeze Me* feels like a swan song for Skink who teeters, figuratively speaking, between being a relic of an eco-radical past and an actor in an as yet uncharted eco-radical future.

Conclusion: A Future for the Carnivalesque in Environmental Dialogues

Hiaasen represents an articulation of the environment's place within popular culture, offering a snapshot of the average citizen's comprehension of ecological worries. While Hiaasen does not offer a constructive way beyond the Anthropocene moment, he creates narratives flexible enough that readers can interpret the ethical frameworks surrounding them. Hiaasen's environmentally conscious crime fiction provides a familiar narrative space for the discussion of sensitive topics allowing ecological issues to emerge organically from the narrative. As Mary Evans argues in *The Imagination of Evil: Detective Fiction and the Modern World*, crime fiction is an essentially moral genre "about morality: its limits, its meaning and its value" (Evans 2). In his writing, Hiaasen addresses the moral problems green criminology faces when it fails to identify a clear perpetrator behind the crime of environmental degradation. Hiaasen dispenses a kind of moral justice where the reader is assured closure will come, and the villain will suffer, frequently in painful and ironic ways. Hiaasen's writing fits within a larger emerging subgenre of overtly environmental crime fiction, which has emerged in response to ever-shifting political, national, ideological, and social contexts.

Increasingly, crime fiction places the detective as a figure who may be "called upon to bear witness, diagnose, organise, protest, persuade, suffer, mourn, and act" in a world that is becoming increasingly unthinkable (Walton and Walton 3). Hiaasen responds to this challenge for environmental action with novels which blend darkly ironic revenge fantasies with the thesis that nature is the victim of man's rapacious consumption of natural resources. Hiaasen does not offer this as a viable example for the future of environmental justice, but he does offer it as a satisfying fantasy version of environmental justice in the present. Further, Hiaasen fails to address the question of whether we should make moral environmental judgements and just what an environmental moral judgement in the real world would look like.

Instead, Hiaasen gives us a fictional world where nature punishes the greedy and those who appreciate nature are rewarded. Hiaasen's brand of humour highlights the incongruity of human expectation by deploying characters such as Skink to create "a breakage in the bond" of everyday experience (Critchley 41). Hiaasen's deployment of the carnivalesque grotesque in the disposal of his ecological villains is part of his unique authorial voice which occupies a space between the serious and the satiric. With unbridled energy, Hiaasen's work celebrates the human ability to change and adapt, creating an ecological carnivalesque marked by its rapacious inversion of social norms. In Hiaasen's world, Nature, with a capital letter, will have vengeance against those who seek to exploit it.

Bibliography

Abbey, Edward. *The Monkey Wrench Gang*. Harper Perennial Modern Classics, 2006.
Alfano, Mark. *Character as Moral Fiction*. Cambridge UP, 2013.
Ascari, Maurizio. *A Counter-History of Crime Fiction: Supernatural, Gothic and Sensational*. Palgrave Macmillan, 2007.
Bacon, Francis. "Of Revenge." (1625) *Francis Bacon: The Major Works*, edited by Brian Vickers, Oxford UP, 2002, pp. 347–348.
Bakhtin, Mikhail. *Rabelais and His World*. Translated by Helene Iswolsky, Indiana UP, 1984.
Buell, Lawrence. *The Future of Environmental Criticism: Crisis and Literary Imagination*. Blackwell, 2005.
Burke, Monte. "Carl Hiaasen on Writing, Ben Carson, Fishing and the Enduring Appeal of Skink." *Forbes*, 9 Nov. 2015, www.forbes.com/sites/monteburke/2015/11/09/carl-hiaasen-on-writing-ben-carson-fishing-and-the-enduring-appeal-of-skink/?sh=11c25b6d5a3c.
Byrne, Jennifer. "Interview with Carl Hiaasen." *Foreign Correspondent*, 16 May 2001.
Critchley, Simon. *On Humor*. Routledge, 2002.
Evans, Mary. *The Imagination of Evil: Detective Fiction and the Modern World*. Continuum, 2009.
Geherin, David. *Scene of the Crime: The Importance of Place in Crime and Mystery Fiction*. McFarland, 2008.
———. *Carl Hiaasen: Sunshine State Satirist*. McFarland, 2018.
Gibbs, Alan. "Gender Politics and Morality in Carl Hiaasen's Crime Novels." *Writing America into the Twenty-first Century: Essays on the American Novel*, edited by Elizabeth Boyle and Anne-Marie Evans, Cambridge Scholars Publishing, 2010, pp. 76–91.
Haliburton, Rachel. *The Ethical Detective: Moral Philosophy and Detective Fiction*. Lexington Books, 2018.
Hiaasen, Carl. *Double Whammy*. Pan Books, 1990.
———. *Native Tongue*. Pan Books, 1992.
———. *Stormy Weather*. Warner Books, 1995.
———. *Sick Puppy*. Pan Books, 1999.
———. *Star Island*. Sphere, 2011.
———. *Skink No Surrender*. Indigo, 2014.
———. *Skinny Dip*. Warner Books, 2014.
———. *Squeeze Me*. Sphere, 2020.
Hillerman, Tony. "Terrorists and Beauty Queens." *The New York Times*, 16 Mar. 1986, archive.nytimes.com/www.nytimes.com/books/97/11/16/home/hiaasen-tourist.html.
Horsley, Lee. *Twentieth-Century Crime Fiction*. Oxford UP, 2005.
Joosse, Paul. "Leaderless Resistance and Ideological Inclusion: The Case of the Earth Liberation Front." *Terrorism and Political Violence*, vol. 19, no. 3, 2007, pp. 351–368.
Joosse, Paul, and Dominik Zelinsky. "*Berserk!*: Anger and the Charismatic Populism of Donald Trump." *Critical Sociology*, vol. 48, no. 6, 2022, pp. 1073–1087.
Jordan, Peter. "Carl Hiaasen's Environmental Thrillers: Crime Fiction in Search of Green Peace." *Studies in Popular Culture*, vol. 13, no. 1, 1990, pp. 61–71.
Kerridge, Richard. "Narratives of Resignation: Environmentalism in Recent Fiction." *The Environmental Tradition in English Literature*, edited by John Parham, Ashgate, 2002.
King, Stewart. "Place." *The Routledge Companion to Crime Fiction*, edited by Janice Allan, Jesper Gulddal, Stewart King, and Andrew Pepper, Routledge, 2020, pp. 211–218.
Parker, David M. "Is South Florida the New Southern California?: Carl Hiaasen's Dystopian Paradise." *The Florida Historical Quarterly*, vol. 90. no. 3, 2012, pp. 306–323.
Ravenscroft, Neil, and Paul Gilchrist. "Spaces of Transgression: Governance, Discipline and Reworking the Carnivalesque." *Leisure Studies*, vol. 28. no. 1, 2009, pp. 35–49.
Silet, Charles L. P. "Sun, Sand and Tirades: An Interview with Carl Hiaasen." *Armchair Detective*, vol. 29, no. 1, 1996, pp. 9–18.
Symons, Alex. "Trump and Satire: America's Carnivalesque President and His War on Television Comedians." *Trump's Media War*, edited by Catherine Happer, Andrew Hoskins, and William Merrin, Routledge, 2019, pp. 183–197.
Van Dover, J. K., and John F. Jebb. *Isn't Justice Always Unfair?: The Detective in Southern Literature*. Bowling Green State U Popular P, 1996.
Walton, Jo Lindsay, and Samantha Walton. "Introduction to Green Letters: Crime Fiction and Ecology." *Green Letters: Studies in Ecocriticism*, vol. 22, no. 1, 2018, pp. 2–6.
Walton, Samantha. "Studies in Green: Teaching Ecological Crime Fiction." *Teaching Crime Fiction*, edited by Charlotte Beyer, Palgrave Macmillan, 2018, pp. 115–130.

26

ENVIRONMENTAL CONCERNS IN CARL HIAASEN'S CRIME FICTION

David Geherin

No one writes funnier crime novels than Carl Hiaasen, whose absurdist sense of humour has produced books filled with outrageously twisted characters and comic plots. But like some of the great satirists he has been compared to – Jonathan Swift, Mark Twain, Evelyn Waugh, and Joseph Heller – he understands how effective humour can be as a weapon to expose human behaviour and evil actions. Philip Roth once described satire as "moral rage transformed into comic art" (119). In Hiaasen's case, the moral outrage that fuels his comic writing is a response to the ongoing assault on the environment.

With its tropical climate, 1300 miles of oceanfront, scenic waterways, verdant woodlands, and teeming wildlife, Florida would seem to be a paradise. Instead, it's a place that can break a nature lover's heart, for the natural beauty that is the state's pride is also the reason it has attracted millions who have moved there to enjoy that beauty, resulting in its current perilous state. For a writer who is as concerned about the environment as Carl Hiaasen, Florida provides constant inspiration. For over four decades, first as an investigative reporter, then a newspaper columnist, and most importantly a bestselling novelist, he has been a voice celebrating the state's natural beauty while imploring his readers to fight to save as much of it as possible before it's too late.

Hiaasen's love affair with nature began at an early age. In an essay entitled "Last of the Falling Tide", he recalls how the stories his father and grandfather told about the Florida Keys created an image in his young mind of

> a mystical, Oz-like destination: a string of rough-cut jewels, trailing like a broken necklace from Florida's southernmost flank – the water, a dozen shades of blue and boiling with porpoises and gamefish; the infinite churning sky, streaked by pink spoonbills and gawky pelicans and elegant ospreys. This I had to see for myself.
>
> *(71)*

What he saw when he made his first trip there at age 6 with his father in 1959 didn't disappoint: "Standing at the brim of those velvet horizons, gulping the sharp salty air, I understood what my father and grandfather meant. This was an honest-to-God wilderness, as pure and unspoiled and accessible as a boy could imagine" (72). But the essay also sounds an ominous warning. "Over centuries the Keys had survived droughts, flood, and the most ferocious of hurricanes. What was there to fear from man" (73)?

DOI: 10.4324/9781003091912-31

Hiaasen, a third-generation Floridian who was born in Fort Lauderdale in 1953 and grew up in Plantation, a small town a few miles west of there, would soon get an unsettling answer to that question. At the time, Plantation was a rural place at the edge of the Everglades. Every day after school, young Carl and his buddies would ride their bikes to a place that seemed to him as exotic as the Serengeti, where they would fish and catch snakes. But this childhood paradise wouldn't last forever. One day the dirt path they rode their bikes on was paved over and turned into an eight-lane highway lined with shopping malls. When the bulldozers first arrived, the trio yanked the survey stakes out of the ground: "We didn't know what else to do. We were little and the bulldozers were big" (Stevenson xv). This episode fuelled his anger and, as he grew older, he discovered a more effective way of fighting back against such an obscene onslaught: he could write about it and become a fierce advocate for preserving what remains.

Hiaasen's Environmental Journalism and the Influence of John D. MacDonald

Hiaasen decided he wanted to be a journalist after getting his first typewriter at age 6. In high school, he created a satirical newsletter and later wrote for his college newspaper, first at Emory University and later at the University of Florida. After graduating in 1974 with a degree in journalism, he was hired for his first newspaper job at *Cocoa Today*. Two years later, he joined the *Miami Herald*, where he worked as an investigative reporter. Much of his journalism reflected his environmental concerns. One of his proudest accomplishments was "North Key Largo: The Last Stand", a series of investigative articles he co-wrote with Brian Duffy, a fellow *Miami Herald* reporter in 1982, about a planned development of condominiums and hotels projected to house as many as 45,000 residents on 12,000 acres of unspoilt wilderness in North Key Largo, 50 miles south of Miami. The environmental consequences of such a development would be devastating, as the last pristine stretch of unspoilt land in the Keys would be bulldozed over, posing a threat to the nearby John Pennekamp Coral Reef State Park, the second largest reef system in the world. The series resulted in a state investigation that led to the shutdown of Port Bougainville, the largest project, and the purchase of much of the property by the state of Florida for preservation. Hiaasen's background in investigative journalism and environmental activism helped establish his credentials as a columnist who could speak with authority on subjects of his own choosing. But he was no longer restricted to a journalistic style of writing and could express his opinions in the satirical style he first used in his high school newsletter and college columns, one he would later also put to good use in his fiction.

One of the strongest formative influences on Hiaasen as an environmental novelist was another popular Florida crime novelist, John D. MacDonald, who is best known as the author of a series of 21 novels, each with a colour in the title, about a "salvage consultant" named Travis McGee that appeared between 1964 and 1985. McGee lived on a houseboat called "The Busted Flush", moored at Slip F-18, Bahia Mar in Fort Lauderdale, not far from where Hiaasen grew up. Hiaasen says he felt a deep personal connection to McGee in that the two of them roamed the same beaches, fished the same flats, "and avoided the same tourists" ("Introduction" viii).

MacDonald employed McGee as a spokesperson for his views on many subjects, including the environment. For example, as he drives along a Florida highway in *The Turquoise Lament* (1973), McGee describes a

> Tacky wilderness of franchised food, car dealerships, boat dealerships, trailer dealerships, motels, auction houses, real estate agencies, rent-anything emporiums, used cars, used trailers, used campers, used boats. Had I not seen a boat for sale every few hundred yards, I would never have known I was within five hundred miles of salt water.
>
> *(186)*

Seeing toxic smoke being spewed into the air from the Borden phosphate and fertiliser plant in Bradenton even prompts McGee to make a direct appeal to the reader: "Anybody can walk into any brokerage office and be told where to look to find a complete list of the names of the directors and where they live. Drop the fellows a line, huh" (185)?

In an introduction Hiaasen wrote in 1994 to new paperback editions of the McGee series, what he said about MacDonald also serves as an appropriate introduction to what he aimed to do in his own fiction:

> MacDonald wanted his readers to do much more than see Florida. He seemed to want them to care about it as deeply as he did; celebrate it, marvel at it, laugh about it, grieve for it, and even fight for it.
>
> *(x)*

The Satiric Vision of Hiaasen's Environmental Crime Fiction

Hiaasen's first three novels – *Powder Burn* (1981), *Trap Line* (1982), and *Death in China* (1984) – are conventional crime thrillers co-authored with *Miami Herald* colleague Bill Montalbano. Stylistically, they lack the comic voice and environmental concerns found in his regular columns, but they taught him a great deal about storytelling and the publishing business. He felt it was now time to "cut loose and have some fun" (Silet). Following the example of John D. MacDonald, who showed him that one could write about the destruction of a beautiful place like Florida and do it in the context of a page-turning story, he decided to write a crime novel. This would provide the framework that would enable him to combine his environmental concerns with the absurdist humour and satiric sting of his newspaper columns. His bad guys wouldn't be serial killers or armed robbers; in his view, the worst kind of criminals are those who conspire to kill the environment, and his novels would allow him to use his fertile imagination to create fitting punishments for the bad guys.

Nature often plays a significant role in fiction. It can serve as a decorative background to the action or, as in the case of Tony Hillerman's Navajo novels, be so beautifully described and integrated into the action that one can't imagine the books being set anywhere else. It can also be used for symbolic purposes. For example, the action in Ernest Hemingway's "Big Two-Hearted River" moves from a burnt-out town to a beautiful river which symbolises the journey of Nick Adams, who seeks the restorative power in the quiet beauty of a Michigan river to help him deal with a wartime experience that has left him as burnt out as the town he passes through on his way to the river.

In Hiaasen's crime fiction, Florida is much more than a symbol or a scenic background to the zany action. It is the central character and the driving force that inspires the fiction, and its destruction is, as Peter Jordan noted, "a sin in its own right, not just a symptom or effect of the general moral corruption" (64). Florida is portrayed both as a victim of man's greed and carelessness as well as the battleground in an ongoing war between the forces of good – those committed to preserving as much of it as possible – and evil – those land developers and their many enablers for whom profit trumps preservation. Hiaasen considers what happened to his beloved state to be nothing less than a crime:

> They've taken everything that was perfect about Florida and destroyed it. They've straightened the rivers, bulldozed the beaches, drained the Everglades. That to me is immoral. . . . I have no problem saying that most developers have the moral footing of drug dealers. It is the exact same thing.
>
> *(Booth)*

Tourist Season (1986) opens with a pair of troubling mysteries: someone has been murdering people, including the head of the Miami Chamber of Commerce, who was killed by a toy plastic alligator that was shoved down his throat; and popular *Miami Sun* columnist Skip Wiley has gone missing, possibly a victim of foul play. The two mysteries are solved when it is revealed that Wiley is the person responsible for the killings. This also includes the chilling death of an elderly widow who is tossed into crocodile infested waters and a trio of golfers who are killed when a putted ball dropping into the cup sets off a bomb blast. After years spent using his column to sound the alarm about the threat to Florida due to overpopulation and overdevelopment, Wiley decides to act. Calling himself El Fuego, leader of *Las Noches de Diciembre*, he and a trio of cohorts – a Seminole Indian, a former Miami Dolphins fullback, and an Ivy League-educated Cuban American devoted to anti-Castro activities – launch a campaign of terrorism aimed at scaring people into leaving the state instead of arriving in record numbers. Wiley's hope is to empty out the state of as much of its population as he can.[1]

Tourist Season is the most radical of Hiaasen's environmental crime novels in that the main character doesn't just attack tourism in his columns, he murders tourists in order to get his point across. Wiley's heart is in the right place, and his diagnosis of Florida's problem is on target:

> Where there was water, we drained it. Where there were trees, we sawed them down. The scrub we simply burned. The bulldozer was God's machine, so we fed it. Malignantly, progress gnawed its way inland from both coasts, stampeding nature.
>
> *(329)*

In the second half of the novel, Wiley's terrorist actions become less violent – dropping several hundred snakes from a helicopter onto a boat filled with travel writers who have come to Florida to promote the state as a tourist destination – and are much closer to the comic spirit that will characterise the rest of Hiaasen's novels.

In *Double Whammy* (1987), Miami private eye R.J. Decker is hired by a professional fisherman to investigate cheating on the highly competitive bass-fishing circuit. He decides he could use some help from someone with more fishing expertise than he has and is directed to a local guide who lives in an isolated cabin by a lake. The man's name is Skink. Skink is an imposing figure who stands 6-foot-6, wears Marine-style boots and a luminous yellow rain suit, and has his braided silver-gray hair hanging down beneath a flowered shower cap he wears on his head. After an attack by a trio of teenage thugs playing a game of what they call "bum-bashing", he has also lost an eye, which he replaces with a glass eye taken from a stuffed barn owl. He lives alone in the woods, survives on fresh roadkill, and amuses himself by firing a pistol at jetliners filled with tourists and potential time-share customers as they begin landing at the Fort Lauderdale airport. Who is this crazy guy?

We get a radically different perspective on Skink when we learn about his past. His real name is Clinton Tyree, and he is a former All-American college football player and a decorated Vietnam veteran. He's also an ex-governor of Florida. A rarity in Florida politics, an honest politician, he warned that the state was on the brink of an environmental cataclysm. The only way to save it was to halt its breakneck growth. This immediately set him at odds with the bankers, builders, highway contractors, etc. who were only interested in making money. Unlike most other elected officials, he refused to sell his vote to the special interests. After casting the lone dissenting vote in a proposal to sell a coastal wildlife preserve to a development corporation, he walked away from Tallahassee in disgust and disappeared into the safety and solitude of the woods.

Hiaasen didn't intend to write a series of novels built around a single protagonist, but in Skink he found a character he could use on a recurring basis to liven things up: "He's the sort of character I wish existed in real life", Hiaasen says, "and it's one of the great joys of the novel, being able to turn him loose and have him kick some butt" (Pleasants 268). He features in many of Hiaasen's novels and even gets his name in the title of one of Hiaasen's young-adult books. He mainly serves as an avenging angel who devises ingenious ways to punish those who defile, deface, and destroy the environment. Hiaasen calls him the moral centre of his novels. As Skink suggests to a like-minded friend: "Somebody's got to be angry or nothing gets fixed. That's what we were put here for, to stay pissed off", words that also apply to Hiaasen himself (*Sick* 304). Most of Skink's actions are harmless: he tosses a dead baby manatee on the stage during the Miss Florida pageant to remind the public of the consequences of rampant waterfront development, and he dresses one land developer in a diaper after attaching a spiky sea urchin to his scrotum and then ties him to a poisonwood tree. Some, however, like burning down an entire amusement park, might go too far for every reader's taste.

Double Whammy also features the first in a long line of Hiaasen villains who share a common evil: they lack all respect for the environment and gleefully destroy huge chunks of it in pursuit of profit. Charles Weeb is a phony TV evangelist (he isn't an ordained minister and doesn't have a church) and a condominium developer who intends to build 29,000 units on land at the edge of the Everglades. Characters like him are blind to nature's beauty; he dismisses the Everglades as being like "the fucking Sahara . . . except with muck" (121). The bass-fishing competition is to be held on an artificial lake on Weeb's property that has become poisoned with lethal levels of phosphate that kill all the 2000 bass he had trucked in for the event. At one end of the lake there is an old earthen dike that "separated the lush watery Florida Everglades from concrete civilization" (339). That barrier, which Skink calls "the moral seam of the universe . . . Evil on the one side, good on the other", becomes a symbol of the battle between those competing forces for the survival of what's left of Florida's natural environment (353).

Native Tongue (1991) is another prime example of Hiaasen's skill in combining elements of the crime novel with an environmental theme, all of it filtered through his comic imagination and sense of the absurd. The novel features a kidnapping, a pair of murders, and an attempted murder. What's unusual about the story is that the kidnapping victims were the last surviving pair of rare blue-tongued voles stolen from the Amazing Kingdom of Thrills, a large amusement park on North Key Largo. One of the murder victims was Dr. Will Koocher, the park's veterinarian, and the target of the attempted murder was Joe Winder, the park's PR man. Hiaasen's comic ingenuity transforms everything. It turns out the rare voles, like everything else in the park, are fakes; they are common pine voles whose tongues have been dyed blue. The body of Will Koocher, who was about to go public about the fraud, ends up in the pool containing Orky the Killer Whale, who chokes to death when he tries to eat Koocher's dead body. And Winder's life is saved by the timely intervention of an odd-looking stranger with one eye and what appears to be a halo of flowers on his head. Readers of *Double Whammy* will immediately recognise the return of Skink.

Francis X. Kingsbury, owner of the Amazing Kingdom of Thrills, is one of Hiaasen's great comic villains. His real name is Frankie "The Ferret" King, who several years earlier was arrested in New York on racketeering charges. After ratting out his co-conspirators, several of whom were members of gangster John Gotti's crime organisation, he entered the government's witness protection program. He changed his named to Francis X. Kingsbury and moved to Miami, where he proved to be a wizard at selling Florida real estate before deciding to become a developer himself. His first project was the Amazing Kingdom of Thrills, designed as South Florida's competition to Orlando's Disney World. Not only are his blue-tongued voles fake, but he also defrauded the U.S. government out of the $200,000 in grant money he received under the Endangered Species Act to save them. He also tries to pass off Orky the Killer Whale's dead body as fresh tuna and attempts to sell it to South Korea. But when the kidnapping of the voles and the death of Orky bring unwanted attention to the park, he orders Pedro Luz, the park's head of security, to kill Will Koocher and Joe Winder. As bad as all these crimes are, however, the worst in Hiaasen's estimation is what Kingsbury is doing to the property adjacent to his park.

Not only has Kingsbury paved over 2000 acres of Florida land for his Amazing Kingdom of Thrills, his next project, the Falcon Trace Golf and Country Club, will do even more damage. Joe

Winder witnesses the latest butchery to nature with his own eyes when, on the way to his favourite fishing spot, he's shocked to see that the trees he normally passes through have all been cut down. Hearing the sound of a high school band ironically playing "America the Beautiful", he approaches the stage where the local mayor is telling the crowd what a lucky day this is for the Florida Keys for this is the future site of Falcon Trace, which will contain 202 luxury waterfront homesites surrounding an 18-hole golf course. Disgusted, Winder heads for the water, fishing rod in hand, only to find the water murky and the fish all gone. He nearly trips on the branch of a red mangrove that has been ripped out and dumped in the water. Doing that is illegal, "but who besides the fish would ever know" (125)?

Native Tongue illustrates both how Hiaasen's reporting influences his fiction and how his fiction differs from his journalism. Falcon Trace is based on Port Bougainville, the planned resort that was the subject of his newspaper series, "North Key Largo: The Last Stand". In his journalism, he was bound by the facts; in his fiction, he is able to use his imagination to go beyond facts and invent his own endings. Falcon Trace is doomed after Skink, acting as nature's avenger once again, burns Kingsbury's park to the ground. Kingsbury's past catches up with him when a hitman hired by the Gotti gang is dispatched to bump him off. And Pedro Luz is romanced to death by an amorous bottle-nosed dolphin in the same pool where he tossed Will Koocher to his death.

In *Strip Tease* (1993), Hiaasen educates his readers about the sugar industry's threat to the environment by highlighting the comic antics of a U.S. congressman who falls in love with a nude dancer at the Eager Beaver strip club in Fort Lauderdale. The story begins when drunken congressman David Dilbeck jumps on stage and attacks another inebriated patron who is trying to hug the dancer. Unfortunately, the scene is captured on film, and Dilbeck soon becomes the target of blackmail. Dilbeck may be a comic buffoon, but there isn't anything the least bit funny about what he represents. Nothing can be allowed to endanger his re-election to Congress so he can continue voting to use taxpayer money to pay lucrative subsidies to the Florida sugar farmers and to overlook the environmental damage the industry is guilty of. A behind-the-scenes fixer named Malcolm "Moldy" Moldowsky will do whatever it takes, including murder, to protect that vote for his Big Sugar clients. Though Dilbeck is unaware of the murders, he shows no concern for the part he plays in Big Sugar's crime against nature. He loses no sleep over the far-reaching impact of cane growers "flushing billions of gallons of waste into the Everglades. Dilbeck didn't understand what the fuss was all about. In truth, he didn't much care for the Everglades; it was torpid, swampy, crawling with bugs" (320).

Sick Puppy (1999) features a quartet of bad guys who together represent another of the ongoing assaults on Florida's environment: Robert Clapley, whose land development company is planning a new resort community on Shearwater Island that will level hundreds of acres and displace untold numbers of animals to make way for luxury homesites, 16-story condominium towers, and a pair of championship golf courses; project manager Karl Krimmler, whose mission in life is to obliterate as much of the landscape as possible: "In nature Krimmler saw neither art nor mystery, only bureaucratic obstacles. . . . Nothing fogged him in gloom so much as the sight of earth-moving machinery sitting idle" (194); Palmer Stoat, a powerful lobbyist, problem fixer, and deal broker for whom no cause was too abhorrent: "he'd work for anybody and anything, if the price was right" (125); and Dick Artemus, the current Florida governor, who gladly does the bidding of people like Clapley and Stoat without any regard to the environmental consequences as long as the generous political contributions keep coming in.

The good guy is Twilley Spree, an eco-avenger like Skip Wiley and Skink, who has serious anger-management issues. A five-million-dollar inheritance has given him the freedom to follow his passion for nature which, like Hiaasen's, was triggered by a childhood experience. At age 14, what he saw when he returned to Marco Island, where he had lived when he was younger, enraged him. The pristine beach where he used to collect seashells has been turned into a "concrete picket of towering hotels and high-rise condominiums" that cast "tombstone shadows across the sand" (24). Since then, he has devoted his life to fighting for nature and he has a very creative imagination when it comes to doling out punishments to those who defile it. For example, he follows one politician, who has been receiving illegal campaign contributions from a cattle ranch that has been flushing raw manure into an estuary, into a bathroom where he lectures him on the immorality of water pollution. Then he pees all over his shoes and asks him, "How do you like it" (27)?

Spree first encounters Stoat when he sees him tossing some empty fast food cartons out of his Range Rover. He follows him for over a hundred miles to his home, then returns in a rented garbage truck and dumps four tons of smelly trash into his wife's BMW convertible. When he witnesses a second incident of Stoat's littering, he deposits 3000 dung beetles into his Land Rover. Stoat, however, proves to be a slow learner, so Spree changes tactics and kidnaps his dog Boodle. Learning that Stoat is behind the planned Shearwater Island project gives him an even greater incentive to act. He sends him an ear and a paw taken from a dead dog that resembles Boodle to pressure him to halt the Shearwater project. In response, Clapley hires a hitman to remove the troublesome meddler.

Skink is drawn into the proceedings when Governor Artemus threatens to evict his troubled brother from the remote state lighthouse where he has been living in order to force him to find Spree before he can further endanger the Shearwater project. Skink locates Spree and later prevents Clapley's hitman from killing him. Then he returns to the governor's mansion to punish Artemus for his role in backing the Shearwater project: he pulls Artemus's pants down and carves the word SHAME in large letters on his bare buttocks with one of the buzzard beaks attached to the end of his braided beard.

Sick Puppy ends with a typical example of Hiaasen's revenge on his environmental criminals. Stoat and Clapley visit a private pay-to-hunt game preserve to shoot a decrepit, overweight, and nearly blind rhino purchased from a theme park in Argentina for the "hunt". What follows confirms Skink's faith that "Nature eventually settles all scores, sets all things straight" (326). While Skink, Spree, and Boodle watch the hunters from a distance, the dog slips out of his leash and heads towards the unfamiliar animal scent he has just whiffed. The docile rhino remains impervious to the dog's barks until he playfully bites its tail, which sends the terrified animal lumbering off towards Clapley and Stoat, killing both of them. The deaths of the two men provide a happy ending as the Shearwater Island project goes bankrupt, allowing Spree to purchase the property and deed every parcel of land for preservation.

On the opening page of *Skinny Dip* (2004), Joey Perrone is struggling to answer a puzzling question: why did her husband Chaz just toss her off the cruise ship where they were celebrating their second anniversary? A former collegiate swimmer, she's strong enough to make it close enough to land, where she is rescued by Mick Shanahan, who previously appeared in *Skin Tight*. With his help, she sets out to seek revenge against her no-good husband and find out why in the world he wanted to kill her. The answer exposes another serious crime against nature. Chaz Perrone is a biologist who loathes the outdoors: "Nothing about nature awed, soothed or humbled him – not the solitude or the mythic vastness or the primordial ebb and flow. To Chaz, it was all hot, buggy, funky-smelling and treacherous" (75). Despite this, he obtains a PhD in wetlands ecology and manages to get a job with a Florida state agency monitoring the levels of phosphorus in the Everglades. Like many of Hiaasen's villains, he's an idiot and a fraud. For example, his plan to kill his wife by tossing her off the ship depends on the Golf Stream taking her body south. He's likely the only marine biologist around who doesn't know that the Gulf Stream flows north, which is what makes it possible for his wife to swim safely to land in Florida.

Over the years the original Everglades have sadly been developed, converted to agriculture, or otherwise debauched to such a degree that only 10% of it still remains. As Hiaasen put it in one of his columns;

> Nature's plumbing has been rejiggered so that farms and cattle ranches can tap onto the Everglades at will, use the water, then dump it back as waste. Florida's famous river of grass is being used not only as a fountain, but as a toilet.
>
> *(Paradise 61)*

Fortunately, in the 1990s, the U.S. Congress and the Florida Legislature allotted $8 billion to restore a natural and unpolluted flow to the Everglades.

Red Hammernut is the millionaire owner of 13,000 acres of vegetables he irrigates with unlimited water from the Everglades, then flushes more fertiliser back into the Everglades than the state's largest cattle ranch and sugarcane grower combined. He's looking to find a scientist "who loves money more'n saw grass and mud fish" (322) to falsify the toxic phosphorus levels (322). Chaz Perrone is the perfect candidate. Despite his disgust with the Everglades, which he deems a "streaming shithole", he dutifully collects water samples, then secretly dumps them and records toxic levels so low that the local Sierra Club awards a plaque to Red Hammernut for his clean-up efforts.

Joey learns that the reason her husband tried to kill her was because he mistakenly thought she had discovered his fraudulent water-sample reports. Justice, however, awaits both Chaz and his boss. Red Hammernut decides Chaz has to be killed in order to keep his fraud a secret and so he can continue polluting unchecked. Tool, the hitman he hires to carry out the job, instead allows Chaz to escape into the Everglades. When Hammernut berates him for doing that, Tool impales him with one of the highway fatality crosses he collects to plant in his garden. Meanwhile, Chaz encounters a strange-looking one-eyed man as he's trying to elude the alligators in the Everglades. Readers of Hiaasen's previous novels will recognise the stranger as Skink, who leads Chaz deep into the Everglades. He reminds him of Tennyson's famous line, "Nature, red in tooth and claw", but Hiaasen leaves it up to the reader to imagine how nature will take its revenge this time.

Boyd Shreave in *Nature Girl* (2006) is a Texas-based telemarketer who tries to sell wooded acres in north-central Florida to residents living in the southern part of the state. He's a familiar type in that like many of Hiaasen's characters; he neither loves nature nor appreciates its beauty. Hiaasen engineers a way of getting him to Florida where he is given an opportunity to see unspoilt nature with his own eyes. Sadly, he is unmoved:

> Not being the spiritual sort, Boyd Shreave saw no divine hand in the unbroken wilderness that lay before him; no grand design in the jungled labyrinth of creeks and islets. Such unspoiled vistas inspired in Shreave not a nanosecond of introspection; when it came to raw nature, he remained staunchly incurious and devoid of awe. He would much rather have been back in Fort Worth, watching *American Idol*, swilling beer and gorging himself on microwave burritos. *(279)*

The punishment Hiaasen metes out to him is fittingly ludicrous: when he tries to shoo away a bald eagle, the bird dumps a prodigious load of excrement all over him. Hiaasen's crime novels have been widely praised for their bizarre characters, outlandish action, and riotously hilarious plots, but he is far more than a mere entertainer. "Anyone can make jokes", Hiaasen says. "I want to make jokes with an edge, so that when people are laughing, they are thinking" (Kogan). Like the best satirists, he knows that humour can be an effective way of drawing the reader's attention to serious matters. His is an art of comic exaggeration and in his fiction, he follows in the footsteps of Flannery O'Connor, who once explained why she employed exaggerated characters and action: "To the hard of hearing you shout and to the almost blind you draw large and starling figures" (113). He knows you must first get the reader's attention before you can communicate your message.

While Hiaasen has been called "the Mark Twain of the crime novel" (Hillerman) and "Florida's resident conscience and America's finest satirical novelist" (Adams), some critics have faulted his novels for allowing humour to obscure the environmental issues. Richard Kerridge complained that using absurd characters as representative of the evils Hiaasen is satirising ultimately diminishes his critique, leaving the impression that "environmental problems could be blamed only on the most grotesque excesses of consumer desire" (91). Larry Byrne has also faulted Hiaasen's comic treatment of environmental themes especially when compared to the more direct approach of his columns, which he says are more likely to urge the reader to action: "It is safe to say that most of the millions of readers who cannot put a Hiaasen novel down have little sympathy for, and thus easily ignore his ecological agenda" (184).

Such criticism overlooks several important points: 1) Hiaasen's crime novels are an extension of his mission, not a substitute for it. He continued to write his newspaper column during his fiction-writing career, so he never abdicated his job as an advocate for environmental causes; 2) His novels reached countless millions of new readers, many of whom had no access to his columns; 3) He recognises that a novelist's "one and only responsibility it to entertain. . . . Any writer who thinks otherwise has been misinformed" (Brunet). In his view, the task of the satirical novelist is to find ways of using humour to open the eyes of his readers to a problem, not necessarily to propose any solution to it. Those who wish that Hiaasen's novels contained a greater call to action might be more sympathetic to the novels he began writing for young readers in 2002, where he makes a strong pitch for a productive partnership between children and adults in achieving their environmental goals.

Hiaasen's Young-Adult Ecofiction

When an editor first approached Hiaasen with the idea of writing a novel for kids, his initial reaction was, "Are you out of your freakin' mind. Have you not read any of my grown-up novels" ("Book-Expo")? But after thinking more about it, he felt that if he toned down the action, eliminated the sex and violence, and cleaned up the language, he could still keep the humour while writing about crimes that would be more appropriate for a younger age group. More importantly, he could reach a new audience with some of the same environmental concerns found in his adult novels.

Hoot (2002) begins with a pair of mysteries: 1) Who is the barefoot boy running full-tilt past the bus as it approaches a stop on its way to Trace Middle School?; and 2) Who has been vandalising the construction site of the latest Mother Paula's All-American Pancake House by pulling up surveyor's stakes, placing small alligators into the portable toilets, and removing the seats from all the earth-moving machines? The answer to both questions, we learn, is a young boy who goes by the name of Mullet Fingers because of his skill in catching the slippery fish with his bare hands. He is in many ways a young version of Skink in that he too lives alone in the woods having run away from an unloving mother who sent him off to military school on the day she married her new husband. He is also a budding eco-terrorist. Shocked to discover that several underground dens occupied by pint-sized burrowing owls will be destroyed once construction begins on the new pancake house, he does what he can to delay that from happening.

The central character in the novel is Roy Eberhardt, the new kid at Trace Middle School who has just moved to Florida from Montana. He is curious about the young boy he saw running past the bus stop and sets out to find out who he is. They soon become friends, and Mullet Fingers brings Roy to a favourite secret place that is teeming with wildlife. Roy is dazzled by the "wondrous quiet" far removed from "the honking and hammering of civilization" just 20 minutes from his own backyard (176). When Mullet Fingers tells him what is about to happen to the owls, he becomes an ally in his crusade to save them: "It wasn't just about the owls", he comes to realise, "it was about everything – all the birds and animals, all the wild places that were in danger of being wiped out" (205). He does some snooping around at City Hall and discovers that the file containing the required Environmental Impact Statement about the presence of burrowing owls on the construction site is missing. We later learn that the file was illegally removed by a city councilman who, like many elected officials in Florida, was paid a bribe so that construction could proceed. Just as the bulldozers are to be fired up at the ground-breaking ceremony, Mullet Fingers squeezes into one of the owl burrows and shouts, "You bury those birds . . . you gotta bury me, too" (267). At that point Roy, his skateboarding buddies, Mullet Fingers' stepsister Beatrice, and her soccer teammates, all gather together to form a human barricade around him and begin singing "This Land is Your Land". Even the actress who plays Mother Paula in the TV commercials joins the group. The resultant publicity forces the Mother Paula company to abandon plans for its newest location and promise that the site will become a permanent sanctuary for burrowing owls.

The crime in *Flush* (2005) is also environmental: someone is poisoning the water at a beach where kids swim and loggerhead turtles lay their eggs. Young Noah Underwood suspects the guilty culprit to be the owner of a gambling boat docked at a nearby marina. With the help of his kid sister Abbey (named after environmental novelist Edward Abbey), they launch Operation Royal Flush: they pool their money to purchase dozens of one-ounce bottles of fuchsia food-colouring gel and then slip aboard the ship to flush the contents down the boat's toilets. Once they have determined that the boat is indeed the source of the toxic water, they take matters into their own hands to stop it. "It's our duty to clean up after the brainless morons", Noah says. The smart humans "owe it to every other living creature not to let the dumb humans wreck the whole planet" (139). The focus in Hiaasen's adult novels is primarily on the bad guys, and the reader gets satisfaction from watching the comic ways they are punished, often by nature. In his young-adult books, the focus shifts to the good guys, and instead of watching the bad guys being punished, these readers get to cheer on the young characters in their heroic efforts to right what they see as a wrong.

Hiaasen didn't start out with the idea of writing any more young-adult novels after *Hoot*, but the critical acclaim the book received (it was named a Newbery Honor Book) and the positive response he received from young readers, changed his mind: "Writing for kids was probably the smartest thing I ever did for my psyche", he says. "You get this tremendous response to the environmental message" (Gilson).

> Kids are born with a natural curiosity about nature and real compassion for wildlife. . . . I won't reach them all, but all it takes is one out of a hundred, one who becomes a purist and goes to the county commission meetings and hollers about some horrible development project.
>
> *(Burke)*

Since 2002, he has alternated between novels for adults and those for kids, distinguishing between them by using two-word titles for the grown-up novels, single-word titles for the others.

In his young-adult novels, Hiaasen creates environmental situations that younger readers can easily identify with. Instead of an abstract idea like "Save the Rain Forest", he depicts actual battles they can win, like figuring out a way to save the owls or stop the dumping of raw sewage that is polluting the beach where they swim. His books for kids have been singled out for praise by several critics in the field for introducing environmental themes in a genre that often relies on fantasies and superheroes. David Aitchison noted that while the young-adult book market is saturated with "self-centred, escapist, and morbid fiction" (156), Hiaasen's novels for young readers recuperate adolescence "as a time for taking responsibility for one's local community, the biophysical environment, and the world at large" (142). Elsewhere, Alexandra Panos has applauded his books for offering young readers an opportunity to "think about how to act and behave in the face of environmental challenges" and for serving as "an argument against prioritising capitalism over the environment and humanity" (6).

Conclusion: Hiaasen's Eulogy to a Paradise Lost

For over four decades, both in his columns and in fifteen crime novels and six books for young readers, Carl Hiaasen has been delivering a heartfelt eulogy to a paradise lost or, to use the words of the title of a collection of his journalism, a "paradise screwed" by man. But he's realistic in knowing the limits to what he can accomplish:

> If you go into journalism or you go into any kind of writing thinking you're going to change the world or change the course of events, you're out of your mind, you're gonna end up

> disappointed and crushed. If you go into it thinking "OK, maybe someone who thinks the same way will know after reading this that they're not alone", if you just go into it thinking you're going to give voice to people who maybe don't have any sense of a voice, that's a good thing. That's a little more realistic.
>
> *(Salustri)*

There's no scientific way of measuring the effectiveness of his novels in advancing the cause of environmental activism, but Hiaasen can take some satisfaction in knowing his message is being heard. "It is comforting, even emboldening to read Hiaasen's work", wrote Monte Burke,

> and realise that we are not alone, that it is perfectly okay to be mad and exasperated about these dark forces that seem to have left us defenseless. Hiaasen gives us a voice, and for this reason he is one of the most important writers of the last half century. He is our howler, Southern in focus but American – even global – in reach
>
> *(Burke)*

But it is the hundreds of letters he has received from his young readers that have given him something else: hope that it's not too late to do what is necessary to slow down the continuing assaults on the environment. "I'm optimistic about the future generations of citizen-readers", Hiaasen says. "They're smart, they're not easily fooled, and they really, truly care about preserving the last wild places in the country" (Gulli).

Note

1. Though Skip Wiley and his band of eco-terrorists may remind readers of Edward Abbey's 1975 novel, *The Monkey Wrench Gang*, about a similar quartet of environmental activists in the American Southwest, Hiaasen confessed he had not read that novel until after he wrote *Tourist Season*. But he would pay tribute to Abbey in several of his later novels: on his driver's license, Skink, his most popular character, uses the name of George Hayduke, the leader of the Monkey Wrench Gang; and a father in one of his young adult novels names his daughter Abbey in honour of his favourite writer.

Bibliography

Adams, Tim. "Scourge of the Scumbags." *The Guardian*, Guardian News and Media, 1 July 2000, www.theguardian.com/books/2000/jul/02/crimebooks.features.

Aitchison, David. "Little Saboteurs, Puerile Politics: The Child, the Childlike, and the Principled Life in Carl Hiaasen's Ecotage Novels for Young Adults." *Children's Literature Association Quarterly*, vol. 40, no. 2, 2015, pp. 141–160, doi:10.1353/chq.2015.0018.

Booth, William. "He Came from the Swamp." *The Washington Post*, WP Company, 4 Mar. 1992, www.washingtonpost.com/archive/lifestyle/1992/03/04/he-came-from-the-swamp/4a12de69-bfae-4d75-a8b3-f3bc8278e455/.

Brunet, Rob. "Rob Brunet Interviews Carl Hiaasen." *The Thrill Begins*. undated. Web. 8 Aug. 2017.

Burke, Monte. "Carl Hiaasen: The Last Great Howler." *Garden & Gun*, 2013, gardenandgun.com/feature/carl-hiaasen-the-last-great-howler/.

Byrne, Larry. "Fighting the Good Fight on All Fronts: Carl Hiaasen and the Rhetoric of Ecocriticism." *Florida Studies: Proceedings of the 2008 Annual Meeting of the Florida College English Association*, edited by Claudia Slate and April Van Camp, Cambridge Scholars Publishing, 2009, pp. 181–189.

Geherin, David. *Carl Hiaasen: Sunshine State Satirist*. McFarland, 2019.

Gilson, Nancy. "Q&A: Carl Hiaasen: Fla. Novelist Crazy about Home State." *The Columbus Dispatch*, 25 Oct. 2015, eu.dispatch.com/story/entertainment/books/2015/10/24/q-carl-hiaasen-fla-novelist/23749579007/.

Gulli, Andrew F. "Interview with Carl Hiaasen." *Strand Magazine*, 6 July 2013. Web. 2 Mar. 2017.

Hiaasen, Carl. "North Key Largo: The Last Stand." *Paradise Screwed: Selected Columns of Carl Hiaasen*, edited by Diane Stevenson, UP of Florida, 2009, pp. 25–27.

———. *Tourist Season*. Warner Books, 1987.
———. *Native Tongue*. Knopf, 1991.
———. *Strip Tease*. Knopf, 1993.
———. "Last of the Falling Tide." *Heart of the Land: Essays on Last Great Places*, edited by Joseph Barbato and Lisa Weinerman, Pantheon, 1994, pp. 71–77.
———. "Introduction." *The Quick Red Fox*, edited by John D. MacDonald, Fawcett, 1995, pp. vii–xi.
———. *Sick Puppy*. Knopf, 2000.
———. *Paradise Screwed: Selected Columns of Carl Hiaasen*. Edited by Diane Stevenson, Putnam, 2001.
———. *Hoot*. Knopf, 2002.
———. *Skinny Dip*. Knopf, 2004.
———. *Double Whammy*. Grand Central Publishing, 2005.
———. *Nature Girl*. Knopf, 2006.
———. *Flush*. Knopf, 2007.
———. "BookExpo America 2014: Carl Hiaasen." *YouTube*, 30 May 2014. Web. 15 Feb. 2021.
Hillerman, Tony. Quoted on Dust Jacket of *Native Tongue*. Knopf, 1991.
Jordan, Peter. "Carl Hiaasen's Environmental Thrillers: Crime Fiction in Search of Green Peace." *Studies in Popular Culture*, vol. 13, no. 1, 1990, pp. 61–71.
Kerridge, Richard. "Narratives of Resignation: Environmentalism in Recent Fiction." *The Environmental Tradition in English Literature*, edited by John Parham, Ashgate, 2002, pp. 87–89.
Kogan, Rick. "On the Water with Author Carl Hiaasen." *Chicago Tribune*, 24 Sept. 1995, www.chicagotribune.com/news/ct-xpm-1995-09-24-9509240075-story.html.
MacDonald, John D. *The Turquoise Lament*. Fawcett Crest, 1996.
O'Connor, Flannery. *Conversations with Flannery O'Connor*. Edited by Rosemary M. Magee, UP of Mississippi, 1987.
Panos, Alexandra. "Beyond Sanctioned Activism in Carl Hiaasen's *Flush*: Sacrifice Zones in Realistic Fiction." *Journal of Children's Literature*, vol. 43, no. 1, Spring 2017, pp. 6–15.
Pleasants, Julian M. "Interview with Carl Hiaasen." *Orange Journalism: Voices from Florida Newspapers*. UP of Florida, 2003, pp. 245–276.
Roth, Philip. *Reading Myself and Others*. Farrar, Straus and Giroux, 1975.
Salustri, Cathy. "The Man Who Invented Skink." *Great Florida Road Trip*, 30 Sept. 2020, greatfloridaroadtrip.com/the-man-who-invented-skink/.
Stevenson, Diane. "Introduction." *Kick Ass. Selected Columns of Carl Hiaasen*, edited by Diane Stevenson, UP of Florida, 1999, pp. xiv–xxiv.

27

NEW ENERGY, OLD CRIME

Forms of Individual and Collective Responsibility in Nordic Crime Series

Leonardo Nolé

The Nordic countries famously pride themselves on their progressive energy policies. Denmark, Finland, Iceland, Norway, and Sweden have shared a joint platform to cooperate in energy research since the 1970s,[1] and they plan to become almost fossil free by 2050 through a huge commitment to renewable resources (Sovacool 570). This widespread interest in energy policies and their social and environmental consequences is mirrored in recent Nordic TV productions, such as Norwegian *Okkupert/Occupied*, Danish *Bedrag/Follow the Money* and *Borgen – Riget, Magten, og Æren/Borgen – Power & Glory*, Finnish *Karppi/Deadwind*, and Icelandic *Ófærð/Trapped*. These narratives often take ecological and energy themes as a background or a dramaturgical starting point, but they still depict accurate dynamics of the international energy system (Bruhn 66). As a response to *Norden*'s "eco-exceptionalist self-perception", many of these narratives choose the form of crime fiction and link their discussion of contemporary energy culture to corrupted politics, international scams, and murderous interpersonal relationships, creating a less triumphalist picture than one would expect (Mrozewicz 85). The examination and questioning of shared values and relevant social issues is a well-established practice in Nordic public television productions. Modelled on the example of the Danish dramas produced by DR, these fictional narratives are famously characterised by "ethical and social layers besides an entertaining plot" (Redvall, "Double Storytelling" 34). This critical perspective allows for the representation of the Nordic countries as a "globalized space that actively participates in the global flows of labor, of an economic and historical system that has produced social, cultural and environmental inequalities" (Jensen and Loftsdóttir 2). As a consequence, as Linda Rugg (614) and Julia Leyda (87) have claimed, these texts embody a peculiar form of the so-called "Scandinavian guilt" – defined in the earlier examples as the awareness of the impact that energy systems and policies have on both society and the environment. In certain narratives, this feeling of guilt is "covered up, repressed, and redirected in order to maintain an image of individual and/or national coherence and innocence", while in others it is "foregrounded, debated . . . in order to promote social change" (Oxfeldt 1).

In this chapter, I will position two Nordic crime series, *Karppi* and *Bedrag*, as proof of these broader, oppositional tendencies to explore the limitations and the possibilities of crime fiction in the representation of environmental crimes. On the one hand, I will show that *Karppi*'s reliance on its main detective, combined with the need to deliver a successful investigation, leads to the displacement of ecocrimes and to a focus on a single murder and its very personal motives. In this way, the series ensures Finland's absolution and misrepresents the crimes against the environment. On the other

DOI: 10.4324/9781003091912-32

hand, I will point out that *Bedrag*, a nuanced noir with several protagonists and an unsatisfactory ending, offers a more complex definition of crime, which calls into question the social and environmental values of an entire system. Thanks to its focus on a diverse set of criminals, the series develops a form of collective guilt that challenges the behaviour of both characters and viewers. Despite their differences, I will argue that the two shows successfully represent the national and international scale of the energy business and its entanglement with global capital. *Karppi*'s and *Bedrag*'s discussion of renewable energy – and wind power in particular – underline the resistance to change exhibited by the present energy system while warning of the violent outcomes that inevitably result from a capitalist exploitation of new forms of energy.

Before moving to my textual analysis, allow me a brief note on methodology. My discussion will be limited to the first season of both TV series, even if multiple seasons are now available, with more likely be released by the time this chapter is published. Far from being a constraint, this choice will allow me to look at two comparable narrative objects produced by public service broadcasters, each designed to expose a national audience to similar controversial themes in a shared socio-historical context. It is certainly true that, as the seasons have evolved, some of the characteristics of the series have changed, building more complex systems that would require greater critical space to explore. However, while keeping in mind Jason Mittell's discussion of "complex TV", I hope to be able to justify, in what follows, that the first seasons of *Karppi* and *Bedrag* provide definite narrative arcs that deserve critical attention in themselves (18). In *Karppi*, the more traditional structure of the investigation certainly helps to deliver some form of closure, while in *Bedrag*, the choice for a more open ending not only provides the hook to season two, but complicates fundamental questions related to the representation of global and systemic crimes. I also believe that a comparative analysis of the two texts can be seen as a privileged way to investigate Nordic popular narratives as, in Andrew Nestingen's words, a "forum for urgent cultural political debates" (198) where one can follow the "transformation of collective self-understanding" (14).

The Tradition of a Reassuring Success

Karppi is a Finnish drama created by Rike Jokela for YLE, Finland's public broadcasting company. The first season aired in Finland at the beginning of 2018, followed by a second season in 2020 and a third one in 2021. *Karppi* is an orthodox crime drama that quotes influential classics of the genre, from *Twin Peaks* to Nordic noir precursors like *Forbrydelsen/The Killing* and *Bron-Broen/The Bridge*. The opening credits immediately point to traditional elements of crime fiction, thanks to the use of greyscale, slow motion, and the focus on police equipment and operations. The title is itself revelatory: the English version chosen for its international release on Netflix, *Deadwind*, suggests a link between wind energy and crime, whereas the original version follows an established tradition and comes from the name of the main detective, Sofia Karppi (Pihla Viitala). In addition to the representation of a female police detective who is also a single mother, which is one of the hallmarks of recent crime fictions produced in *Norden* (Povlsen 74), *Karppi*'s location, colour, and atmospheric choices reveal the influence of more famous Nordic noirs. According to Kim Toft Hansen and Anne Marit Waade, who have followed its transformation into a commercial brand, the genre is characterised by an "enigmatic city and landscape melancholy, suspense-filled enigmatic plot developments, a visually commodified style, conscious or unconscious intertextual associations with influential romanticized imagery" (89). *Karppi* is set primarily in Helsinki, represented on a par with other Nordic capitals as an inscrutable city capable of nurturing intriguing mysteries. The specific Nordic climate – with its very long nights, incessant autumn rain and winter snowfall, and its eerie sea of ice that always seems on the verge of cracking – is also a recurring element that helps set a distinctive "melancholic mood" (297). *Karppi*'s locations confirm Nordic noir's predilection for purely local elements. Helsinki is in fact constantly exposed to the viewers' gaze through establishing shots with aerial views of the most recognisable spots, like the shoreline and the city's cathedral.

The crime that Sofia and her fellow detective Sakari Nurmi (Lauri Tilkanen) are asked to investigate is also of a traditional kind, since it develops from the discovery of a dead woman, Anna Bergdahl (Pamela Tola). As other ecocrime fictions, *Karppi* begins "with human suffering, a human victim, which . . . slowly unfolds into the detection of environmental crime" (Rugg 610). Despite the hints for a more personal motive behind the murder – like the fact that the victim was buried in her wedding dress and with white callas in her hands – the investigation starts by linking the crime to the crime scene. The body is found on the construction site of a new residential area supplied by an innovative wind turbine park owned by Tempo, the company where the victim was working at the time of her death. For most of the show, the detectives and other main characters seem to connect this project to the murder, opening the narrative to ecological and energy questions. In the first minutes of episode one, for instance, viewers are shown a promotional video released by Tempo to advertise the technical innovations of their bladeless wind turbines built with an extremely durable material. Episode three is mostly devoted to a TV debate between Tempo's owner Alex Hoikkala (Tommi Korpela) and a politician openly against the company's proposal and in favour of nuclear power. This is an important moment in the series because the debate is presented as a discussion of wind power's potential and Finland's likely decision to shut down its nuclear power plants in the near future. Despite the controversial and divisive issues at stake, both opponents win parts of the debate, and their arguments seem to present equally convincing reasons for wind and nuclear power. The visual narrative fosters this same idea by the means of a television studio perfectly divided in two symmetrical parts. By staging a realistic TV debate, *Karppi* raises urgent political questions, like the unequal access to energy resources and the challenges of a just transition towards new forms of energy. Although Alex strenuously defends it, his project – based on self-generated renewable energy – is overshadowed by the suspicion of elitism: ordinary citizens seem worried that this luxury residential area would steal clean energy from poorer people.

Interestingly, the time spent by the narrative discussing competing directions in the Finnish energy system does not lead to a clear conclusion; nuclear and wind power are ultimately presented as equally viable possibilities. This impasse speaks to the complexity of the contemporary public debate around energy futures and highlights two of its main characteristics. First, it proves that our energy system is invested in the sustainment of the status quo and the validation of current power dynamics (Goodbody and Smith 14). Second, it shows that this same trend can sometimes hide itself behind the marketing rhetoric of technological innovation and futurity. According to Imre Szeman and Dominic Boyer, "much of the contemporary discussion about energy in relation to the environment imagines energy as an input into modern social and material processes that does not alter their character or nature very much" (3). In his analysis of today's official narratives that discuss energy futures, Szeman strongly criticises the discourse he defines as "techno-utopianism", precisely because it promises to solve the scarcity of energy resources with "scientific innovations that are in perfect synchrony with the operations of the capitalist economy" ("System Failure" 102). In *Karppi*'s TV debate, the arguments of both opponents highlight the widespread misconception that the future of energy is merely a technological problem; their speeches are built on a "defensive" language that insists on "changes in input in order to preserve global capitalism and its systems of property and profit" (Szeman and Boyer 3). As the narrative makes clear, neither opponent can win over the other because their differences are only superficial, when in fact both share the same capitalist-based approach to the present and future of energy.

The series further complicates its discussion of the relationship between energy and the capitalist system with the introduction of a new suspect, a German eco-activist (August Wittgenstein). By investigating him, the narrative can dwell on the European and international scale of Tempo's business and reveal the wrongdoings of a German partner working on the same project. This is when ecocrimes come to light, as it turns out that the foreign company is using Tempo's material to illegally store nuclear waste in disused mines all over Germany, contaminating water and soil and causing harm to human bodies. But this is also the moment when the main narrative unfolds in a different direction, with Karppi explicitly dismissing this lead. The last episode provides a more traditional solution to the mystery than opened the narrative. The detectives discover that Anna's death had nothing to do with her job at Tempo, since it was linked to a sexual assault that took place in the village in the Finnish countryside where she grew up. The murderer is in fact an old acquaintance, himself conventionally a victim of violence within his own family. Thanks to this sudden personal turn, the story of the main crime becomes a very private story, diverting attention from the social and environmental context the show previously framed. As a visual proof of this shift, domestic spaces play a significant role in *Karppi*'s last episode. Anna's murder is revealed to have taken place in the basement of her own apartment, and the killer is finally arrested in what remains of his old family house.

By following this more traditional motive behind the crime, the plot moves towards a more conservative ending. In his discussion of classic detective fiction, Franco Moretti argues that "in finding one solution that is valid for all – detective fiction does not permit alternative readings – society posits its unity and declares itself innocent" (144). Having to deal in the end with only one victim and one perpetrator, the police in *Karppi* can ensure a satisfactory solution and a general restoration of the social order which entails the success of Finland's national system on various grounds, together with its innocence. First of all, Finland is shown as a place where justice can still be delivered. The city and national police are represented as incredibly efficient, and Karppi and Nurmi are always supported by their department, even when they behave outside the rules. With minor exceptions, Finland's political system is also portrayed as transparent and committed to the common good. Besides the debate on future energy strategies broadcasted on national TV, other important public discussions are live streamed and watched by ordinary citizens, like the vote of Helsinki's city council on Tempo's construction permit. More importantly for my argument, Finland is represented as alien to ecological crimes. Even if the beginning of the show invites the viewers to be suspicious of Tempo and its owners, in the end their name is cleared. Finnish characters are held accountable for infidelity, personal violence, and abuse, but almost none of them are directly linked to environmental crimes and their social consequences. Even besides the various levels of personal responsibility, *Karppi*'s ideological narrative suggests that ecocrimes only happen outside of Finland's borders. In episode eleven, Karppi and Nurmi secretly travel to Hamburg where they witness the illegal storage of nuclear waste. From a visual point of view, the location is characterised by warmer colours, which are completely different from Finland's whitish/greyish palette.

In this way, *Karppi* demonstrates the limitations of detective fiction in representing environmental crimes which require large spatial and temporal scales and push us to rethink the role of perpetrator and victim beyond individual – and often even human – categories. Bounded by the starring role assigned to the detective, her limited geographic jurisdiction, and the need to solve the mystery in a short amount of time, the series classifies the environmental crimes as a red herring and shifts the focus to a murder with a single culprit and very local implications. This individual, limited perspective responds to the need to find a satisfactory solution that re-establishes the status quo and reassures viewers of their innocence, thus preventing the show from exploring the social consequences of ecocrimes and questioning personal and collective entanglements.

When the Criminal Is the System

Bedrag is a financial crime drama created by Jeppe Gjervig Gram for DR, the Danish Broadcasting Corporation. The first of its three seasons aired in Denmark at the beginning of 2016 and in the United

Kingdom (BBC) shortly thereafter. *Bedrag* moulds some of the *topoi* of detective fiction to accommodate a more complex narrative structure. The main detective, Mads Justesen (Thomas Bo Larsen), abdicates the predominant role that is traditionally reserved to this character and shares the stage with three other protagonists, whose ethical and psychological intricacies are equally scrutinised throughout the episodes: Alexander Søndergren (Nikolaj Lie Kaas), the CEO of the Danish company Energreen, which presents itself as a key player in wind power business; Claudia Moreno (Natalie Madueño), head of legal department at Energreen; and Nicky (Esben Smed Jensen), a mechanic that occasionally steals cars to make extra money for his family. Through these four perspectives, the series shows "a deep investment in realism, in its depiction of social institutions . . . of family life . . . and of the underprivileged as well as the affluent class" (Engelstad 34). The main characters belong to various socio-economic statuses, from the upper class (Alexander and Claudia) to the middle class (Mads) and the working class (Nicky). Thanks to this wide spectrum, it is society at large that materialises as the primary object of the detective's and the viewers' investigation.

One of the most generative sequences in the series starts with a focus on the foreign workers employed by Energreen and the funeral wake they organise for a fellow colleague who died on the job. The sequence's soundtrack, Lucinda Williams's "Born to Be Loved", offers a social and political commentary to the visual narrative – "you weren't born to be abandoned, you weren't born to be forsaken, you weren't born to be mistreated, you weren't born to be a slave". Thanks to effective editing, the narrative seems to extend the accusation to the entire society, since the images of the wake are juxtaposed with images of Mads, Claudia, Nicky, and their families. While the foreign workers are mourning a premature death that demonstrates their dispensability, insecurity, and powerlessness, the Danish characters and their children are all in safe domestic environments. Moreover, the details of the home interiors reveal the differences even within Danish citizens (Hansen and Waade 182), hinting to the critique that Nordic noirs make of the recent developments to the welfare state. "The success of the welfare system has led to complacency", Audun Engelstad writes,

> and the risk of losing sight of the founding values society rests upon is real. Big corporations trample over small enterprises, politicians cater to their own interests or give favors to friends, and the police hold the lower class and immigrants in low esteem.
>
> *(35)*

The manifold nature of the crime and the multiplication of the criminals in *Bedrag* are instrumental in constructing this complex picture of contemporary Danish society and the welfare state. As the title indicates, this is the story of a financial fraud. Energreen and its CEO publicly present themselves as working with wind power to ensure a better future for Denmark and the whole world, but the development of the plot will reveal their interest in exploiting the growth of the renewable energy market and the large public funding for their personal economic interest, no matter the social and environmental cost of their choices. Energreen's CEO is often shown in the visual narrative behind the glass of his luxury office or behind the screen of a television or computer: he is exposed and distant at the same time, always ready to show off to the public to cover up his real intent. But the predatory CEO is certainly not the only villain in this story. According to *Bedrag*'s creator, the show deals with "the way financial structures link us all", and how they have "almost become like a force of nature, but a man-made one" (Cranswick). In other words, the series articulates "an underlining fear that the power of the state has been subsumed by global capitalism", while suggesting that "society *does* exist and is made up of millions of individuals that are all intimately connected, a world where a single action can cause a never-ending series of ripples across a whole community" (Creeber 24). This idea clearly emerges in the title sequence, built on aquatic images and the double meaning of "leaking". The first shots are set inside Energreen's headquarters and centre on a glass half full of water that slowly starts to ripple and leak. When a drop hits Claudia's face, viewers understand that the whole building is being flooded, while the people inside carry on with their tasks. This "upward movement creates a 'sinking' effect . . . suggesting a threatening or a dangerous mood", which is not limited to the capitalist elite at Energreen (Bruhn 70). In the subsequent shots, more water keeps leaking, coming out of a car that Nicky is fixing in his garage and flooding Mads' and Claudia's apartments. The water is unstoppable and cannot drain away, submerging everything and everyone, drowning again what little has managed to emerge. The last scene focuses on the main detective, now completely underwater, and his vain attempt to swim against the current and catch the investigation files that keep slipping through his fingers.

The limited role assigned to the detective, which mirrors his limited agency over such vast and polymorphous crime, allows the series to spend more time with the criminals, exposing the patterns and consequences of their actions, and the degree of personal responsibility. In the opening sequence, scenes of a TV interview with Energreen's CEO bragging from London about the high standards of his company are intertwined with shots from a crime scene in Copenhagen, where the police find the corpse of a Ukrainian worker at the company's wind park. Mads documents the poor living conditions and the brutality of the contracts of these foreign labourers, but the only result of his investigation is their sudden dismissal and the transfer of responsibility from Energreen to its contractors. In episode four, Claudia is sent to one of Energreen's partners, a solar panel manufacturer in the town of Fredericia. The show spends a lot of time with its workers, who are mostly Danish citizens who have devoted their entire lives to the factory's success and technological renewal. Even if the business is in good shape, its survival does not match Energreen's financial needs, and Alexander announces that the company must be closed and all workers dismissed.

Although most of the episodes are set in Denmark, Energreen is clearly presented as part of a larger, supranational system. Like *Karppi*, Nordic noirs are generally built around a defined location that specific iconic buildings make easily recognisable (Hansen and Waade 63). Viewers of *Bedrag* know that Copenhagen is the main setting, and they may have identified its offshore wind farm in the first episode. The protagonists' apartments feature key elements of Scandinavian design, like the well-known Danish chairs, that contributes to the social scrutiny of the series. But the main public locations where the story takes place, starting with Energreen's headquarters that also feature an always-visible world map, could be found in any major economic and financial city. Towards the end of the series, when Alexander, Claudia, and other colleagues travel from one European capital to the other to look for new investors, they move between visually identical spaces. If *Bedrag*'s Copenhagen does not have many distinctive elements, it is because it must appear as part of a system that goes far beyond the nation.

Energreen's involvement in the international energy business makes this systemic aspect even clearer. In episode three, Alexander takes part in the "2016 World Future Energy Summit" in Abu Dhabi and visits Masdar City, a complex built right outside the capital and completely sustained by renewable energy. While walking through the largest solar plant in the region – a controversial example of energy-efficient construction – the CEO's self-interested commitment to the renewable energy cause is exposed once again. As soon as he understands that Middle Eastern investors would not be attracted by Energreen's work on wind power, he organises the presentation of a superconductor material that aims at transporting energy without losses. When his chief scientist explains that the project is not yet ready and refuses to falsify the results of her research, Alexander criticises her lack of faith and replaces her. Finding a cheap and reliable way to store power is one of the actual challenges of energy research, but the company only sees it as a way to obtain foreign investments that would help expand its business and carry on its fraudulent operations. By questioning the way in which corporate power manages the energy system and deals with its technological and social challenges, *Bedrag* provides an effective critique of capitalist dogmas. Like *Karppi*, it highlights the pervasiveness of a "techno-utopian" public discourse and the risks of a merely technological perspective on energy futures, which is promoted by Alexander as well as the leader of the Emirates energy business.

During the presentation of Energreen's fallacious material, the narrative emphasises the almost magical character of this event. The technology behind this project is presented as so sophisticated as to be incomprehensible, while amazing the potential investors and making them overlook the unreliability of the product on display. In *Bedrag*, this unconditional faith in technological advancement is discussed as an example of capitalism's similarly unconditional faith in infinite progress and growth. Confronted with Claudia's doubts about Energreen's operations, Alexander always responds with the need for blind trust in the company's success, in their possibility of endless development. This is what he repeats until the very last sequence of the series, when he is in an unknown location outside Denmark after the company's fraud has been made partially public. While Alexander keeps fantasising about the economic possibilities of wind power, he is brutally killed by a gunshot that hits his glasses, symbolically destroying his world and future vision. *Bedrag*'s critique of a capitalist approach to energy rests on the belief that

> placing one's faith in environmental change in a market system built around growth and profit, endless expansion, and the bottom line, and one, furthermore, premised in a fundamental way on disavowing or negating the value of natural systems, is questionable.
>
> *(Szeman and Boyer 7)*

As a consequence, the series seems to prove the incompatibility between renewable energy and capitalist thinking; in Szeman's and Boyer's words, "there may have been coal capitalism and oil capitalism; there *cannot* be solar or wind capitalism" (7).

Bedrag's ending further reinforces this critique, since it provides no real sense of closure or return to order. Despite highlighting the impact and scale of the crime from the very beginning, Mads and his team must surrender to the impossibility of bringing all the criminals to justice. In the end, Alexander manages to escape, Claudia autonomously decides to turn herself in, and the other more powerful members of the company's board are left untouched. Moreover, Alexander's sudden death reveals a criminal international network fully active and completely unseen by the detectives. The failure of the police exposes the failure of Denmark's national system, which is represented as fragile and unprepared, led by corrupted public figures and untouchable capitalist businessmen that work together to produce social inequality. At the same time, the series makes clear that the flows in the system are also ascribable to everyday people. This approach makes the identification of individual culprits as well as their punishment much more complicated. Instead of leading to a reassuring restoration of the social order, the ending produces even more chaos and tends toward extensive, collective guilt. In the world of *Bedrag*, criminals are not exceptions but examples that point to the real culprit – capitalism and its systemic production of social and environmental injustice. Although the series is more concerned with financial than environmental crimes, its structure is an excellent example of how to remodel crime fiction to accommodate more collective forms of crime and guilt. The feeling of innocence granted by the traditional return to the status quo is replaced by a widespread sense of complicity and responsibility, which implicates the characters as well as the viewers.

The Question of Individual and Collective Responsibility

According to Imre Szeman and Dominic Boyer, narratives that confront energy issues should try to "first, grasp the full intricacies of our imbrication with energy systems . . . and second, map out other ways of being, behaving, and belonging in relation to both old and new forms of energy" (3). *Bedrag* and *Karppi* are certainly less ambitious in their discussion of viable alternatives to the culture behind the present energy system since the environmental issues are ancillary elements to broader narrative concerns. In *Karppi*, the ecocrimes work primarily as a red herring that postpones the solution of a more traditional murder, a violence that has been hidden in a small Finnish village for years. In *Bedrag*, the focus on wind power business is a way to expose the world of multinational capitalism and its predatory agenda, especially in response to the 2008 financial and bank crisis.

Despite their limitations, the two shows depict some of the fundamental characteristics of the international energy system and the transition to renewable energy. In doing so, their narratives make energy infrastructures visible. Energy production and distribution sites have been hidden from the public gaze in Western countries for a long time, and the same tendency seems reaffirmed with renewables. The offshoring of wind energy, for instance, is often seen as a way to "avoid public resistance by placing windfarms in locations that are far away from human habitation and therefore less visible" (Van der Horst 139). *Karppi* and *Bedrag* use wind turbine parks as locations for key plot twists and treat them as aesthetic elements of the landscape. In *Bedrag*'s opening sequence, the wind turbines are framed by a rosy sky at dawn and a shiny sea on the horizon. When *Karppi*'s two detectives travel to Germany, giant wind turbines tower above the main characters, like intermediaries between sky and earth, while their mobile shadows complement the visual value of the scene.

By bringing out the aesthetic qualities of technological infrastructures, fictional narratives can transform them into significant cultural objects that stimulate emotional responses and push viewers to question their position within the energy system and the processes that sustain it (Jørgensen and Jørgensen 8). According to Axel Goodbody and Bradon Smith, this is a fundamental step to expose "hidden mechanisms of power and social hierarchy" and reveal "the inappropriateness of the ways in which we tend to think about energy generation and consumption and about our relationship with the material world more generally" (14). In both TV series, the exposure and scrutiny of energy infrastructures become an opportunity to develop a more complex understanding of contemporary energy culture, its bond to capitalist practices, and their social and environmental consequences. While it is true that "the expansion of capital could not have occurred in the absence of oil as a hegemonic form of energy" (Szeman, "How to Know About Oil" 193), these narratives seem to be alerting us to the perils of using renewable energy to perpetuate the same unsustainable political, economic, and cultural systems of yesterday and today. Thanks to their depiction of ecocrimes and social inequality, *Bedrag* and *Karppi* make abundantly clear that new forms of energy can only produce old forms of injustice and crime when they are exploited to power the global capitalist system. Without really speaking to the future or imagining an alternative society, both series represent the immobility of the present and its resistance to change. However, the two shows take opposite paths in their response to this awareness and the distribution of individual and collective responsibility.

Karppi's more traditional narrative structure, built on the predominant role of the detective and her ability to identify the criminal and bring him to justice, demonstrates that the main crime is an exception in a well-functioning social system. As in classic detective fiction, the anomaly is successfully corrected, the social order restored, and viewers are reassured of their individual innocence (Moretti 138). To achieve this goal, the narrative shifts the focus away from ecocrimes and separates them from the actual investigation. Despite witnessing the illegal storage of hazardous waste, *Karppi* prioritises a murder that is much more circumscribed and therefore easier to solve. By assigning a private motive to the main crime and by relegating environmental crimes to a foreign country, the series reaffirms an anthropocentric perspective and avoids any form of shared guilt. Finland's political, economic, and social systems are shown as themselves victims of external predatory interests, ensuring the nation's collective innocence. Following Borges's famous definition of detective narratives, *Karppi* seems to be "safeguarding order in an era of disorder" (499). But in doing so, it misrepresents the global and temporal scale of crimes against the environment – what Rob Nixon influentially called slow violence – and represses environmental guilt (Rugg 597).

Like *Karppi*, many other examples of detective fiction struggle to properly depict the environmental crisis and related cultural issues. Crimes that transcend the individual and extend across space and time do not match some of the formal and thematic characteristics of more traditional detective fiction (Hollister 1012): the identification of crime with murder and the subsequent spotlight on human victims; the often private, traumatic context behind the criminal's behaviour; the privileged position granted to the detectives and their limited perspective; the need for a quick and single solution that marks the triumph of justice and the restoration of the status quo; and a concept of justice that disregards the more-than-human world. According to Marta Puxan-Oliva, the most successful examples of crime fiction that deal with environmental concerns are those that push the limits of their own genre conventions (369). Even without a real focus on ecocrimes, *Bedrag*'s thematic and formal choices build a narrative structure that can accommodate the complexity of the environmental crisis and question the historical and cultural practices behind it. The financial fraud at the centre of the plot makes the series embrace a broader concept of crime that translates into a web of different criminal actions with extensive repercussions. *Bedrag* succeeds in depicting the various facets of this complex crime by diminishing the role of the detective and adding three more protagonists. The time the narrative spends with the criminals allows for a deep exploration of their various degrees of responsibility, the impact their actions have on the entire society, and their link to global capital.

As Andrew Pepper writes about crime fiction that aims to address multinational capitalism, criminality in *Bedrag* can clearly be seen "not merely as individualised but more importantly as systemic" (358). The national system, which ensures the restoration of the social order in *Karppi*, is here as much ineffective as impotent. *Bedrag*'s Denmark fails not only because it is unable to defend its people from violence and injustice, but also because it finds itself enmeshed in the systemic problems it is supposed to solve. National borders are no guarantee of protection and innocence: the crimes of global capital manage to cross them and connect seemingly distant spaces. The failure of the police investigation is key in this process because it points to more, unseen culprits, and to a form of collective guilt. Since many of the criminals are presented as such from the beginning, the real mystery viewers are asked to unravel in *Bedrag* is their own role in capitalist violence and the resulting social and environmental forms of injustice. In other words, the series attempts to represent the parts we play in the Anthropocene, or better, in the Capitalocene. The focus on the international energy business and its dependency on global capital offers a particularly privileged perspective on these entangled histories, given that "the geological agency of humankind is directly linked to its ability to mobilize energy" (Bergthaller 128). A return to order is impossible in *Bedrag* because everyone participates in systemic problems no individual can solve alone, not even the most gifted of detectives. Viewers, as a result, are not reassured but pushed to perceive themselves as part of this system and to question their own behaviour and degree of responsibility. "Everyone is guilty and shared guilt is a common bond" (215), Deborah Bird Rose writes in her article "Anthropocene Noir". And it is precisely this awareness that "remove[s] us from that singular position of spectator" and reminds us that "in the midst of all we cannot choose, we do still make choices" (219).

Note

1. "Nordic Energy Research", www.nordicenergy.org/, accessed 20 June 2021.

Bibliography

Bergthaller, Hannes. "Energy." *The Anthropocene. Key Issues for the Humanities*, edited by Eva Horn and Hannes Bergthaller, Routledge, 2020, pp. 128–140.

Bird Rose, Deborah. "Anthropocene Noir." *Arena Journal*, vol. 41, no. 42, 2013, pp. 206–219.

Borges, Jorge L. "The Detective Story." *Selected Non-Fictions*, edited by Eliot Weinberger, Viking, 1999, pp. 491–499.

Bruhn, Jørgen. "Ecology as Pre-Text? the Paradoxical Presence of Ecological Thematics in Contemporary Scandinavian Quality TV." *Journal of Aesthetics & Culture*, vol. 10, no. 2, 2018, pp. 66–73, doi:10.1080/20004214.2018.1438729.

Cranswick, Amie. "Exclusive Interview with Follow the Money Writer Jeppe Gjervig Gram." *Flickering Myth*, 15 Apr. 2018, www.flickeringmyth.com/2017/04/exclusive-interview-with-follow-the-money-writer-jeppe-gjervig-gram/.

Creeber, Glen. "Killing us Softly: Investigating the Aesthetics, Philosophy and Influence of Nordic Noir Television." *The Journal of Popular Television*, vol. 3, no. 1, 2015, pp. 21–35, doi:10.1386/jptv.3.1.21_1.

Engelstad, Audun. "Framing Nordic Noir. From Film Noir to High-End Television Drama." *European Television Crime Drama and Beyond*, edited by K. Toft Hansen et al., Palgrave MacMillan, 2018, pp. 23–39.

Goodbody, Alex, and Bradon Smith. "Stories of Energy: Narrative in the Energy Humanities." *Resilience: A Journal of the Environmental Humanities*, vol. 6, no. 2–3, 2019, pp. 1–25, doi:10.5250/resilience.6.2-3.0001.

Hansen, Kim T., and Anne Marit Waade. *Locating Nordic Noir. From Beck to the Bridge*. Palgrave MacMillan, 2017.

Hollister, Lucas. "The Green and the Black: Ecological Awareness and the Darkness of Noir." *PMLA/ Publications of the Modern Language Association of America*, vol. 134, no. 5, 2019, pp. 1012–1027, doi:10.1632/pmla.2019.134.5.1012.

Jensen, Lars, and Kristín Loftsdóttir. *Crisis in the Nordic Nations and Beyond. At the Intersection of Environment, Finance and Multiculturalism*. Routledge, 2014.

Jørgensen, Dolly, and Finn Arne Jørgensen. "Aesthetics of Energy Landscapes." *Environment, Space, Place*, vol. 10, no. 1, 2018, pp. 1–14, doi:10.5749/envispacplac.10.1.0001.

Klitgaard Povlsen, Karen. "Gender and Geography in Contemporary Scandinavian Television Crime Fiction." *Scandinavian Crime Fiction*, edited by A. Nestingen and P. Arvas, Wales UP, 2011, pp. 74–81.

Leyda, Julia. "Petropolitics, Cli-Fi and *Occupied*." *Journal of Scandinavian Cinema*, vol. 8, no. 2, 2018, pp. 83–101, doi:10.1386/jsca.8.2.83_1.

Mittell, Jason. *Complex TV. The Poetics of Contemporary Television Storytelling*. New York UP, 2015.

Moretti, Franco. "Clues." *Signs Taken for Wonders. Essays in the Sociology of Literary Forms*, translated by Susan Fischer, David Forgacs, and David Miller, Verso, 1983, pp. 130–156.

Mrozewicz, Anna Estera. "The Landscapes of Eco-Noir: Reimagining Norwegian Eco-Exceptionalism in *Occupied*." *Nordicom Review*, vol. 41, no. s1, 2020, pp. 85–105, doi:10.2478/nor-2020-0018.

Nestingen, Andrew. *Crime and Fantasy in Scandinavia: Fiction, Film and Social Change*. Washington UP, 2008.

Oxfeldt, Elisabeth. "Framing Scandinavian Guilt." *Journal of Aesthetics & Culture*, vol. 10, no. 2, 2018, p. 1438725, doi:10.1080/20004214.2018.1438725.

Pepper, Andrew. "Crime Fiction and Global Capital." *The Routledge Companion to Crime Fiction*, edited by Janice Allan et al., Routledge, 2020, pp. 353–361.

Puxan-Oliva, Marta. "Crime Fiction and the Environment." *The Routledge Companion to Crime Fiction*, edited by Janice Allan et al., Routledge, 2020, pp. 362–370.

Redvall, Eva N. *Writing and Producing Television Drama in Denmark: From The Kingdom to The Killing*. Palgrave Macmillan, 2013.

———. "The Concept of 'Double Storytelling' in Danish Public Service TV Drama Production." *Ethics in Screenwriting. New Perspectives*, edited by Steven Maras, Palgrave MacMillan, 2016, pp. 33–55.

Rugg, Linda Haverty. "Displacing Crimes against Nature: Scandinavian Ecocrime Fiction." *Scandinavian Studies*, vol. 89, no. 4, 2017, pp. 597–615, doi:10.5406/scanstud.89.4.0597.

Sovacool, Benjamin K. "Contestation, Contingency, and Justice in the Nordic Low-Carbon Energy Transition." *Energy Policy*, vol. 102, 2017, pp. 569–582, doi:10.1016/j.enpol.2016.12.045.

Szeman, Imre. "How to Know about Oil: Energy Epistemologies and Political Futures." *On Petrocultures. Globalization, Culture, and Energy*. West Virginia UP, 2019, pp. 174–199.

———. "System Failure: Oil, Futurity, and the Anticipation of Disaster." *On Petrocultures. Globalization, Culture, and Energy*. West Virginia UP, 2019, pp. 93–112.

Szeman, Imre, and Dominic Boyer, editors. *Energy Humanities*. John Hopkins UP, 2017.

Van der Horst, Dan. "Justice in the Eye of the Beholder? 'Looking' beyond the Visual Aesthetics of Wind Machines in a Post-Productivist Landscape." *Environment, Space, Place*, vol. 10, no. 1, 2018, pp. 134–153, doi:10.5749/envispacplac.10.1.0134.

PART V

Energy, Globality and Circulation

28

"IT TASTED LIKE GASOLINE"

The American Roman Noir and the Oil Encounter in Elliott Chaze's *Black Wings Has My Angel* (1953)

Nathan Ashman

> When I was a young sprout, I had an ambition to become a cowboy and roam the range, shooting rustlers and rescuing good looking women. But I'm older and wiser now and I've learned there are more good looking numbers to be had riding around in a flash car with a pocketful of chips.
>
> *Martin M. Goldsmith, Detour, 1939**

The opening pages of Elliott Chaze's little-known roman noir *Black Wings Has My Angel* (1953) find protagonist and escaped convict Tim Sunblade coming to the end of a sixteen-week stint "roughnecking" on an oil drilling rig in the Atchafalaya River (7). The scene is ostensibly significant for what immediately follows, where Sunblade will use his newly earned capital to check into a hotel in Krotz Springs. Here he encounters the text's other central protagonist, an alluring yet mysterious sex worker called Virginia. The narrative then treads the path of a Bonnie and Clyde road novel, culminating in an audacious armoured vehicle heist that will set both characters on a fatalistic path towards oblivion. Yet it is both before and after the robbery, in the characters' various movements around a newly developing post-war American landscape of highways, gas stops and suburban subdivisions, that the text's opening passage takes on greater critical import, as Chaze presents us with a series of "aesthetic images" that conflate automotive movement with freedom, sovereignty and pleasure. Or, as Sunblade himself puts it, with "that very good feeling of going somewhere" (LeMenager 69; Chaze 13). This opening representation of oil extraction is therefore the first of many associated "industrial era" activities, such as driving, that are revealed within the pages of the text, all of which are seemingly "valorised" as means of "facilitating the body's self-extension toward other life" (LeMenager 69).

Yet these utopian "structures of feeling" are simultaneously counterbalanced by the otherwise tragic trajectory of the novel, as the characters' propulsive consumptive desires lead them not towards happiness and self-fulfilment, but disillusionment and death (Watson 90). Building on the work of Matthew Pangborn, what this chapter seeks to explore is how Chaze's novel, and the American roman noir more broadly, both ambivalently strive to give shape to a period of vast social, technological and topographical change, mapping, however obliquely, the very "transformation of twentieth century America by the power of oil" (Pangborn 781). We see this not only in the material geographies that the text reveals – a particular form of what Matthew Huber calls "auto-centric" space – but in its depiction of a commodified and alienated post-war subject trying, and ultimately failing, to navigate this new and bewildering landscape of desire (x).

DOI: 10.4324/9781003091912-34

Nathan Ashman

Where Is the Oil in Noir? The American Roman Noir and the Oil Encounter

In his essay "Where is the Oil in Modernism", Joshua Schuster points to a conspicuous disparity between the omnipresence of petroleum in the modernist era and its comparative absence in the canon of modernist writing. Despite being "everywhere" in the fabric of the surrounding culture – transforming the very "shape of the landscape with cars, roads, airplanes, military equipment, spawning suburbs, intensifying land speculation [and] producing new chemicals and plastics" – oil in its commodity form is, for Schuster, largely and curiously peripheral in modernist art (197). This recognition of oil as a kind of structuring absence can be equally and usefully extended to the field of noir fiction, particularly given the specific literary-historical and thematic symmetries between modernist writing and literary noir.[1] Lee Horsley is one of many critics who has positioned noir as a form of popular or – as Andrew Pepper puts it – "pulp modernism", noting a shared preoccupation with pessimism and existentialism, and also highlighting the ways that modernist themes and techniques helped shape noir formally, encouraging "the use of irony, non-linear plots, subjective narration and multiple viewpoints" (Pepper 60; Horsley 3). More significant is the fact that both noir and modernism narrate what Stephanie LeMenager has categorised as the "high point of petromodernity", a period where the concept of modern life "based in the cheap energy systems made possible by oil" became firmly entrenched (68). In the context of the United States, Matthew Huber terms this then emergent national imaginary "the American way of life", a specific social vision rooted in the precepts of "privatised social production, single-family housing and automobility" (xiv). Noir's "settings and atmosphere" are intimately connected to these infrastructures of petro-modernity, as indicated by its particular focus on a disaggregated world of "hotels, highways [and] flickering streetlights" (Pitt Scott).

Whilst it might therefore be claimed that there is a similarly surprising absence of oil in the American roman noir (as in modernism), we must also recognise that the 'problem of oil' is as much interpretive as it is representational. To paraphrase Joshua Schuster: if noir writers seemingly did not care much about the role of oil, neither have noir critics. To date, very little academic attention has been paid to the relationship between noir and petroleum culture, representing a conspicuous gap within the fields of crime fiction studies and the environmental humanities. Noir itself is a highly contested term, so it is perhaps unsurprising that critics who have tended to theorise the mode as an articulation of certain psychoanalytic structures (Joan Copjec, Slavoj Žižek, Mary Anne Doane), or as a set of mutating ahistorical 'affects' (Christopher Breu, Lee Horsley, Woody Haut), have overlooked the contextual implications of oil. It is with the more "historically minded" school of critics, those who are inclined to conceptualise noir fiction as either an "expression of post-World War II malaise or as a coded form of Depression-era social protest", that such an omission becomes a more substantial point of scrutiny (Breu 24).

In his book *America is Elsewhere: The Noir Tradition in the Age of Consumerism*, Erik Dussere locates noir fiction as a negative response to American consumerism, arguing that these works express a "newly urgent desire to rediscover a version of the nation imagined as authentic" (3). Dussere's discussion of what he calls "authenticity effects" in noir and hard-boiled fiction – representational literary strategies aimed at preserving American masculinity and America at large – offers a perceptive insight into post-war national identity in the context of mass consumption but fails to consider the ways that petroleum energy and petroleum products provided the material basis for this "geography of consumption" and for the very subjectivities produced therein (Dussere 4; Huber 64). A similar criticism can be extended to Sean McCann's *Gumshoe America*, which explores how mid-twentieth-century American crime writers attempted to grapple with the newly emergent paradoxes of New Deal liberalism yet overlooks the extent to which New Deal policies "set the stage" for a new form of "post-war American capitalism", one rooted in "mass petroleum production for mass petroleum consumption" (Huber 59). Other notable studies that have considered noir and pulp fiction either in the context of depression era politics (William Marling), or in relation to the rise of mass-market paperback publishing (Paula Rabinowitz), have likewise discounted how the material, commercial and imaginative landscapes of mid-twentieth-century American crime fiction were perceptibly transformed by the forces of oil capital.

The seeming dearth of criticism exploring the relationship between noir and early petroleum culture is, to an extent, expressive of a broader questioning of "how to know about oil" that has preoccupied environmental humanities scholars in recent years (Szeman 146). Indeed, one of the distinguishing features of "oil culture" is the way that it manages to "obscure the very source of energy which makes it possible in the first place" (Bergthaller, Grewe-Volpp & and Mayer 126). Whilst being imbricated at every level of society, culture and industry, oil is still mostly absent from the workings of everyday life, particularly in its material form. This goes some way towards explaining the relative scarcity of literary texts that have consciously taken oil as their primary subject of inquiry. In his now influential 1992 review of Abdelrahman Munif's oil quintet *Cities of Salt*, Amitav Ghosh was the first to observe that the "oil encounter" had up until then "produced scarcely a single work of note" (29). Ghosh points to Munif's text – an example of what he terms 'petrofiction' – as a rare outlier, one that simultaneously reveals how the broader, global history of oil is still "a matter of embarrassment verging on the unspeakable, the pornographic" (29).

Many scholars have subsequently interpreted the term 'petrofiction' as a distinct genre of 'oil conscious' novels that explicitly engage with the "physical stuff" of oil in their contents (Duckert 216). By working to elucidate a canon of texts that reflect deliberately on the oil encounter, the aim of such critics has been to highlight "the distortions and disjunctions, the forms of ignorance and bad faith which oil has engendered", whilst also to demystify the complex global and historical entanglements that have allowed, and continue to allow, for oil's predominance (Bergthaller, Grewe-Volpp & Mayer 126). Yet such an approach to 'reading for oil' also risks compressing the field of study into a limited corpus of 'actively' engaged texts and tends to overlook the more abstracted or undetermined cultural effects of petro-modernity. After all, Stephanie LeMenager suggests that "loving oil" – a phrase she uses to refer to our deeply embedded cultural attachment to the possibilities that oil grants us – has "a great deal to do with loving media dependent on fossil fuels" (66). Part of what makes oil such a destructive and all-encompassing organisational force is the way it has been supported by "overlapping media environments to which there is no apparent 'outside' that might be materialised through imagination and affect as palpable hope" (LeMenager 70). As LeMenager puts it: "To step outside of petromodernity would require a step outside of media, including the contemporary printed book" (64).

In response, critics such as Patricia Yaeger have sought to reorientate the work of the petrocritic, arguing that their job is not merely to constitute a new body of explicitly oil-conscious texts, but to "rechart literary periods and make energy sources a matter of urgency to literary criticism" (306). For Yaeger, the term 'petrofiction' should not be interpreted as a specific mode or genre, but rather as a "grand new periodising gesture", one that rethinks the twentieth and twenty-first centuries as a broader age of petro-modernity (Rubenstein 49). For the critic considering the relationship between energy, history and culture, this means uncovering the ways in which energy "invisibilities may constitute different kinds of erasures" (Yaeger 309). Pointing to Fredric Jameson's *The Political Unconscious: Narrative as a Socially Symbolic Act*, Yaeger outlines the possibility of reading for an "energy unconscious". She writes: "since fuel sources hover in the background of texts, if they speak at all, to pursue an energy unconscious means a commitment to the repressed, the *non-dit*, and to the text as a tissue of contradictions" (310 fn1). Whilst acknowledging that such a methodology risks reverting to "crude materialism", Yaeger maintains that by considering the ways in which energy sources "enter texts as fields of force that have casualties outside (or in addition to) class conflict and commodity wars", it might be possible to instantiate a "changing phenomenology that could re-create our ideas about the literary text's relation to its originating modes of production as quasi-objects" (309).

It is this kind of 'unconscious' signification that Matthew Pangborn seeks to uncover in his essay "Lessons in 'Bad Love': Film Noir and the Rise of the American Oil Regime in Edgar G. Ulmer's *Detour*", which represents the only sustained critical inquiry into the relationship between noir and petroleum culture to date. Drawing on the work of energy humanities scholars, the essay reads Ulmer's film, and the broader genre of film noir, as providing "a much-needed ground-level perspective on the efforts of industry and government to stimulate oil consumption by creating desires in a public struggling with the inherent paradoxes of new technologies" (780). Focusing specifically on Ulmer's "recognizably representative strategies of lighting and narration that identify its spectators with its mobile and criminal main character", Pangborn argues that the film takes its spectators "on a journey of self-exploration and even self-condemnation", all the while tracing the variegated consequences of the "new American oil regime" through the viewer's "entanglements in oil's disruptive information and media systems" (786). Pangborn also interrogates the film's "own shaping by oil's social, financial, and environmental networks", uncovering a "complex web of technologies, beliefs, and behaviours" that, from the film's very first frames, implicate its spectators in a "dirty secret the film otherwise struggles to bring to light" (786). The "shadowy and fated network" to which film noir alludes is not, in Pangborn's reading, some form of "abstract metaphysical contemplation" or generalised malaise catalysed by the contexts of economic depression and war, but what he describes as a "felt recognition" of the ways in which a "rapidly expanding network of extraction, distribution, and consumption was compelling Americans to remake their lives in dramatic ways that felt beyond their control" (780).

It is the 'felt recognition' of the oil encounter that I wish to investigate further in my reading of both *Black Wings Has My Angel* and the American roman noir. Firstly, I will suggest that the American roman noir reveals, in real time, the economic and material transformation of the American landscape as a direct consequence of New Deal policies to revive capitalism in the midst of the Great Depression. The New Deal ushered in a dramatic change of "everyday life", proposing "new imaginaries of living" that comprised "a whole set of practices and geographies that locked in ways of consuming energy" (Huber 41). This became expressed through a particular set of spatial arrangements: a "geography of mass consumption" (Huber 41). As will be shown in my reading of Chaze's novel, central to navigating this new and perplexing landscape of commodity desire was the automobile, which in the noir tradition, becomes inextricably bound up with fantasies of freedom, desire and autonomy. Yet these fantasies are also explicitly gendered, as indicated by the mode's propensity to depict women and cars as synonymous objects "constituted from, sustained by, and dependent on oil industries" (Devereux 163). Thus, the desire/danger dialectic that defines the figure of the femme fatale – represented in *Black Wings Has My Angel* by Sunblade's accomplice Virginia – becomes bound up not only with what Matthew Pangborn terms the "sexualisation" of the automobile, but also with a greater anxiety and terror regarding the possibility of a free and sovereign feminine subject (789).

Secondly, I will argue that it is noir's deeply embedded fantasy structure, typically characterised by the subject's "negative relationship to the law", that offers a particularly useful way of articulating oil's 'absent presence' in twentieth-century cultural production (Breu 25). Indeed, Sheena Wilson, Adam Carlson and Imre Szeman argue that "oil's constitutive relationship with modernity" can be rendered "culturally intelligible by way of the tragic narrative of Oedipal alienation, which governs both the education of masculine Petro-subjects and the objectification of women's bodies" (135). Here they point to oil capital as a kind of symbolic structure akin to the Law of the Father, one that is external to the subject but that determines both the shape of his/her identity and the very nature of his/her desires. Thus, it becomes possible to read the novels that comprise the tradition of the American roman noir as what Glenn Willmott terms "generic tragedies of Oedipal oil" because "they represent, in diverse cases, the destructive yet enlightening trajectory of human beings alienated both from each other and from their natural world, as these are commodified as resources to fuel the production of an imagined personal sovereignty" (193). Noir pessimism, in this sense, is not solely an expression of deep-seated and "transhistorical" psychoanalytic structures, but what Christopher Breu describes as a "contextually specific" response to the shifting social, cultural and financial climate of the United States following the desolation of the Great Depression (25). Yet whereas Breu positions noir's 'negativity' as being motivated by the advent of "corporate capitalism", I will suggest that it is responding to a more influential and impenetrable shift, one that literally *energised* this new era of corporate capitalism: the rapid rise of petro-modernity (25).

"Dames're Funny Machines": *Black Wings Has My Angel*, the Femme Fatale and the Erotics of Automobility

In the period immediately following the Second World War, American oil consumption – and the American economy at large – experienced an unprecedented boom. Between 1945 and 1950, the number of cars on the road doubled from 20 to 40 million, as an increasing number of families moved from densely populated urban centres to newly developed suburban subdivisions. This had a material impact on the growth of several interconnected and dependent commercial industries, with the total number of shopping centres, supermarkets, fast food restaurants and motor hotels rising rapidly and concomitantly with America's increasing thirst for oil. By the end of the decade, oil consumption had risen to 5.8 million barrels per day, a figure that would triple in the twenty-five years thereafter (McQuaid 54). Whilst cheap and abundant petroleum energy – combined with innovations in transport and technology – was the obvious factor behind this meteoric transformation of American life, the enmeshment of oil consumption in the advancement of a wider "cultural politics" had begun almost a decade earlier and formed part of the New Deal's attempts to reenergise capitalism during a period of economic depression (Huber 28). As Matthew Huber suggests, although the idea of "the American way" was already part of the cultural fabric prior to Roosevelt's election, the 1930s saw the "first widespread use of the postfix 'way of life'" in the evocation of a "particular set of lived practices and conceptions of freedom" (33). This 'new way of life' was based around "high wages, home ownership" and a set of newly developing "auto-centric suburban geographies predicated on the provision of cheap and abundant oil" (Huber 58).

The 1950s subsequently saw a "multiplicity of fractionated petroleum products" enter the marketplace, products that became inextricably intertwined with a set of consumptive habits rooted in "particular visions of freedom, domesticity and health" (Huber 72). Nothing gave shape to these ideologies more so than the automobile itself, which not only transformed the material geographies of everyday life, but reconfigured the basic assumptions and values that came to define the American twentieth century. These ostensibly utopian ideals of "unbridled mobility" and "personal expression" form one facet of what David Blanke describes as America's complex "automotive love affair" and appear prominently in the aesthetic and narratorial traditions of noir fiction and film. Mark Osteen is one of many critics to have noted a preoccupation with cars across the "noir cycle", pointing to a variety of films in which cars emerge as "overdetermined symbols" of characters' desires, as well as "signs of restored consumer power and renewed possibility – of a refurbished American dream" (184). With its doomed lovers, Edward Anderson's depression era noir novel, *Thieves Like Us* (1934), is pointed to as a prototypical example of this convergence between desire and unrestrained mobility, but one could easily extend this to a host of roman noirs that either feature automobility as a central or instigating plot point – *The Postman Always Rings Twice* (1934), *Detour* (1939), *Dark Passage* (1946), *Criss-Cross* (1949) – or where the very subject of mobility, and its limits, is placed into question – *They Shoot Horses Don't They* (1935), *In a Lonely Place* (1947). For Osteen, "noir's cars" come to represent "the propulsive aspirations of disenfranchised people who turn to crime", embodying the "possibility of social mobility through auto-mobility" in the post-war era (184). Yet, as most of these novels ultimately reveal, such fantasies are marked by a despotic underside, with the automobile transmuting from a symbol of liberation and self-fulfilment into something irrepressible and inescapable. It is through this typical narrative arc that the roman noir ambiently attempts to chart the tacit yet pervasive anxieties of a nation grappling with the transformative effects of a rapidly emerging petroleum culture, revealing that mobility (both automotive and economic) is rarely "evenly distributed", particularly along the axes of class, gender and race (Poll 192).

This trajectory is typified by Elliott Chaze's *Black Wings Has My Angel*, which centres on a characteristically fated relationship between escaped former convict Tim Sunblade and mob moll turned sex worker, Virginia. They first meet in the opening pages of the text, when Sunblade hires Virginia to mark (and celebrate) the end of an extended stint working on an oil rig in rural Louisiana. Initially disappointed by Virginia's appearance, preferring to have been fixed with "a big commercial blob of a woman" rather than the "slender poised thing with skin the colour of pearls melted in honey" that arrives at his door, Sunblade becomes quickly infatuated (11). The couple spend several days in the hotel room together, after which Sunblade decides to take Virginia with him on a road trip out West. In these early sections of the text, we begin to see the extent to which, in the noir tradition, automotive movement becomes heavily endowed with a range of attendant (and idealistic) values and beliefs, including self-determination, freedom and possibility. This is typified in the moment when Sunblade and Virginia first set out on their journey, crossing the "Red River Bridge" to the world beyond:

> I sailed my Mississippi tags over the iron railing and saw them hit the water with a splash, forty feet below. She watched me, leaning back in her leather-padded corner, smoking quietly. Nothing seemed to surprise her: the car, the tags, the business of taking an uncharted trip with an unknown man. The wind whipped her bright hair the way it does in the soft-drink advertisements, cooperatively, beautifully. The cross-stripes of tar on the white highway thumped faster and faster beneath the wheels until the thumping became a buzzing. The air was soft, yet not dead. And over all of it lay the very good feeling of going somewhere.
>
> *(13)*

Through the language of discovery – "uncharted", "unknown" – Chaze captures the promise, pleasure and adventure of the American highway, merging the traditional "frontier spirit" with the "worship of the machine as complex icon" (Priemeau 6). Tied into this is a palpable sense of renewal and self-determination, with Sunblade's dramatic discarding of his prison tags indicating a negation or erasure of both past and self. There is a phantasmagoric lexicon underpinning the images that Chaze presents to us, underscored by the description of Virginia's hair whipping in the wind like in "the soft drinks advertisements, cooperatively, beautifully". At once a picturesque or cinematic distillation of a moment of empowered automotive movement, what is also obliquely prefigured here is the text's broader commodification of femininity and desire in the pages to come, both of which become repeatedly and inextricably connected to objects of consumption.

In this instance, we are presented with the first of several associations, both metonymic and explicit, between the automobile and the female body. Indeed, submerged within these images of freedom and sovereignty is a tangible undercurrent of danger, speed, sex and annihilation, as typified by the description of the accelerating Packard thumping "faster and faster" across the tarmac, the air around it "soft, yet not dead" (12). It is a propulsive image, albeit one that subtly points to a steady loosening of Sunblade's control. What we can begin to trace here is the way in which "noir's obsession with dichotomous women characters" symbolically reflects the "standing of the car" in post-war American commodity culture (Pangborn 789). Whilst, on the one hand, driving encouraged a form of masculinist self-empowerment connected with a "somatic, even sexual exhilaration", on the other, it evoked concomitant ideas of danger and death, the fear of pushing one's desires to potentially fatal limits (Pangborn 789). As Paul Mason Fotsch aptly puts it: "[In noir], the automobile embodies the thrill and danger of [the] inevitable drive forward; it offers the possibility of greater power through mobility, but it also threatens to overpower the driver" (106).

These dialectical drives associated with automotive mobility permeate the early sections of *Black Wings Has My Angel*. Through its portrayal of a developing circuitry of roads and gas stations, as well as a vast and interconnected service economy of truck stops, motels, diners and bars, for large parts the novel can be seen to record a "variety of sensory and emotional values" associated with an emergent American oil culture (LeMenager 68). In one section, Sunblade and Virginia stop at a roadside barbecue stand where some "kind of engine turned the beef ribs over and over, like a bloody Ferris wheel, over the charcoal fire" (17). Sunblade describes the sensation of "slowly" eating the food, washing down "the greasy roasted meat with stingingly cold beer" and a cigarette. Virginia then leans over and kisses him for "a long time", her "lips cold and fresh and soft" (17). Roadsides are a recurrent feature of Chaze's text and, despite their inherent commerciality, are frequently naturalised and romanticised through images of somatic and carnal pleasure, emphasised here through the obvious conflation of eating and kissing. The stillness of the scene, as indicated by a concentration on the "slowness" of movement and action, also serves to consolidate the reader's perception on its affective and aesthetic dimensions. However, hinted at within the description of the "bloody Ferris wheel" of meat is the true function of the roadside as "combined theatre and amusement park", a site where a particular set of values are staged and reproduced (Belasco 105). In the same way, Chaze's novel itself vigorously "transforms the roadside into a vivid imaginary place", one capable of "generating affective investment" in a wider web of "petroleum infrastructures" (LeMenager 70–71).

Sustained within this imaginary of the roadside is the myth of the pastoral, the nostalgic fantasy of a life "separated from the distractions and corruption" of a mechanised and urbanised world (Hecht 12). We see this in another passage from the early sections of the text, when following a violent and erotic altercation that recalls scenes from Cain's *The Postman Always Rings Twice*, Sunblade and Virginia retreat into the mountains around Cripple Creek, Colorado, where they wait for their bruises and cuts to heal. Cooking off of hot coals, bathing in rock pools and freely navigating the "walloping stretch of space" that surrounds him, Sunblade valorises the "special feeling of freedom" he gets "up there on the rocks" (52). Comparing the independence and euphoria of standing "up in the sky and clouds" to the equally liberating sensation of "being able to move around", Sunblade claims to feel the same way about freedom as "an evangelist feels about religion" (31). Yet he also acknowledges that freedom is nothing without "the money to enjoy it", and that this, in itself, amounts to another "kind of religion, a very exclusive kind" (31). What we see here is the extent to which Sunblade's own desires and pastoral fantasies have already been commodified, and that the freedom he speaks of actually suggests a certain "post war ideology", one rooted in a form of "privatized, consumption driven social mobility" (Dussere 55). The surrounding and formerly idealised landscape of Cripple Creek is soon shown to be similarly tainted by these imbricated structures of capitalist accumulation, when Sunblade stumbles across a nearby and disused mineshaft called Katie Lewellyn. The mineshaft embodies the long prehistory of resource extraction on which modern America was built and thus further challenges Sunblade's pastoral fantasy. The memory of the shaft's "yawning mouth" – as well as the "insane forgotten blackness" that he glimpses within – will haunt Sunblade for the remainder of the text, and later play a significant role in his and Virginia's dramatic demise at the denouement (42,51).

Typically of noir, Sunblade's dreams of freedom and mobility (both spatial and socio-economic) are simultaneously burdened by an inescapable sense of anxiety, one that manifests most materially in his uneasy relationship with Virginia. Despite deciding to bring Virginia with him on his trip to Denver, Sunblade spends much of the early sections of the novel planning to ditch "her in the ladies john of some filling station" (17). He remains suspicious that, despite their intimacies, their relationship is still fundamentally transactional, an apprehension that his further compounded by Virginia's continued reference to a refrain from a country music song by Lefty Frizzell: "If You've Got the Money, Honey, I've Got The Time" (12). When Sunblade subsequently tries to leave her at a gas stop, he realises that she has pilfered the money he saved during his stint on the oil rig and returns to confront her. Following another explosive altercation, the two subsequently fall into an uneasy truce, with Sunblade deciding to bring her in on his plans to heist an armoured truck.

What is most striking about Virginia's positioning as a commodity item is the way that descriptions of her body and behaviours become intimately bound up with the aforementioned desire/danger dialectic of the automobile. Indeed, we can see the sexual thrill invested in the possibilities of automotive movement the first time that Sunblade "[gives Virginia] the wheel" following their stop at the roadside barbecue stand, where he observes her "hand[ling] the hydramatic shift without self-consciousness . . . [feeding] the heavy car the gas in a nice soft gush" (18). Chaze's description is permeated with phallic symbolism and the language of sexual climax, as both car and woman are conflated as objects of, and routes towards, desire. For Cecily Devereux, this is demonstrative of an ingrained cultural impulse to "repeatedly and insistently" represent "both cars as women and women as cars and thus, equally and interchangeably, as commodities", not only in the sense of metonymically "associating women with cars" as has historically been seen in media advertising (as in Figure 28.1), but of literally "constructing them equally as products" (163). Devereux argues that this correlation is therefore not simply a matter of "metaphor" but is instead a "foundational symbolic interchangeability" rooted in a "cultural system that situates its meanings in the late nineteenth- to early twenty-first-century business of oil" (163). In other words, it is petroculture that drives "the representation and performance of femininity", interpellating women in "time and space" and maintaining "their auto-immobility across multiple registers" (165). Crucially for Devereux, what makes petroculture's "mobilization of the standard commodity figures of femininity so significant in the aligning of women and cars in media and popular representation" is its solidifying of a particular "symbolic economy", one that is contingent on women "not having the capacity to be self-propelling or self-determining: not to be driving but to be driven, not to be subject but object" (163).

These conservative gender anxieties become central to noir's expression of a particular masculine erotics of automobility, embodied most readily through the liberated yet lethal figure of the femme fatale. As Jonathan Munby suggests, the recasting of "female independence" in this form – as "home-breaking, avaricious, a sexual predator, and user of men" – is constitutive to the mode's ideological orientation and plays a substantial role in demonising "women's desire for autonomy" in the post-war period (193). Thus, as well as sexualising and commodifying Virginia via this symbolic interchange between the female body and the automobile, *Black Wings Has My Angel* also represents the attendant angst and "horror" that arises when "the 'normal' is subverted and the car/woman takes on subjectivity and drives herself" (Devereux 170). Whilst Sunblade continually valorises Virginia's skill behind the wheel, referring to her as the "peerless driver of Packards", the potential power she holds through such autonomy becomes an increasing source of agitation and uncertainty as the text progresses (120). This is again channelled into the language of the novel, which remains rooted in the same rhetorical interchangeability between the female body and the automobile. In one passage, Sunblade describes Virginia's lovemaking as "mechanically splendid, yet as though the performance was the result of some remote control and did not really involve her" (24). Ostensibly a compliment, latent within this description is the language of mechanisation, the sense of something cold, depersonalised and threatening, of something outside of Sunblade's control. This rhetoric surfaces again later in the text, when Sunblade and Virginia hitch a ride with a trucker. After Virginia praises the trucker's proficiency behind the wheel, he responds by discussing, in an unblinkingly associative fashion, his wife's "female trouble" (180). Bemoaning that he can't fix his wife in the same way he can his truck, where he would "just go in there and tear out the whole rear end and be done with it", he concludes that "dames're funny machines" (181).

Figure 28.1 'Chevrolet Advertisement'. From *Evening Star*, November 23, 1947, Page 6, Image 97.
Source: Public Domain. Courtesy of Library of Congress, Washington, DC.

Intrinsic to these descriptions of the female body (in conflation with the automobile) is the masculine desire to master, objectify and control, all stemming from the attendant fear of an economically autonomous and sexually liberated female subject. Thus, it is possible to position the gendered antagonisms that define the noir cycle – antagonisms that, Christopher Breu argues, reveal an ideological post-war project to reprivatise women's roles through the "aggressive reassertion" of male privilege in the public sphere ("Radical Noir" 201) – as a latent symptom of the broader production of masculine subjectivities, and the control and commodification of the female body, in petroculture. We can again see the significance of oil in the material and social production of post-war American life – and how this manifests in noir's anxious gender contentions – in the mid-section of the text, when Tim and Virginia begin to actively plan for the heist. To allow them time to prepare, and to make them appear as "respectable and accounted-for as anyone in the city of Denver", the couple temporarily settle in a quiet suburb, where Sunblade begins working as a power shear operator at a "big dusty plant" on the south side of the town (70). To maintain an image of propriety, Virginia reluctantly takes on the role of the "healthy and conscientious and insatiate . . . young bride", dutifully watering the front yard and "kissing [Sunblade] off at the door" every morning in "a flowered cotton dress" (70, 77). Before long, Sunblade is progressively interpellated into this fantasy of post-war suburban life, not only growing to appreciate the "honest labour" that his work demands, but also the ideals of family and private ownership that come along with it. In one revealing section, Sunblade sits in his car, wistfully "looking at the little brick bungalow" and thinking how "it was the only home of [his] own [he'd] ever had" (94, 11).

What is ambiently pointed to here is not only the extent to which petroleum energy fuelled the spectacular suburbanisation of America in the post-war period, but how suburbanisation itself became central to reasserting the "centrality of traditional gender roles" (Rosser 139). Reviewing a range of "petroleum advertisements" from the 1950s – advertisements that marketed the convenience of petroleum-based products to modern living – Matthew Huber argues that such commercials worked to promote "a very particular vision of home replete with imagery structured through uneven relations of race, gender and class". Or, to put it more bluntly: "Men work, women stay at home, and *everyone is white*" (85). It is this image of post-war conformity that Chaze presents us with in the middle section of the text, albeit one that is increasingly destabilised as the narrative progresses. Characteristic of the femme fatale as an embodiment of masculine anxieties regarding female empowerment, Virginia steadily rebels against these privatised gender constraints, telling Sunblade in a dissatisfied rage: "Damn you, Tim, I'm not taking much more of this. . . . Plan or no plan, I'm not going to squat here in this thus and such and scrape the mildew off me" (77).

Following an incident a few pages later, where an intoxicated Virginia soaks Sunblade with a watering hose in front of their onlooking neighbours, she will subsequently disappear, only to be later discovered by Sunblade boozing at a local bar in the company of "a fat Filipino" (97). Sunblade quickly moves to reassert his white male authority, first hitting the man and then violently jerking Virginia off her stool. Virginia's rejection of Sunblade's patriarchal suburban fantasy prefigures the patterns of alienation and disillusionment that will come to permeate the final sections of the text, as both characters struggle to comprehend the extent to which their own subjectivities and aspirations, as well as the natural world around them, have already been subsumed and commodified within a privatised, petrocapitalist organisation of desire. This pervasion of the text by the signs of petroleum culture is neatly crystallised in a section just before the heist, when Sunblade stops for a coke at a Tuscany bar but finds himself unable to enjoy it. As he tells the reader: "It tasted like gasoline" (116).

Sunblade's attempts to know and master Virginia – an Oedipal trajectory that is fundamental to noir's specific fantasy structure – will typically end in tragedy and disaffection, becoming a synecdoche for the similarly abstracted and estranging nature of oil as a social organising force (or narrative). In other words, it is the language of Oedipal alienation (so constitutive to the workings of noir) that offers a particularly apposite framework through which we can begin to articulate the noir subject's anxious relationship with industrial modernity.

"A Terrible Fascination": Noir and the Ontology of Oil

Following several months of planning, Sunblade and Virginia will eventually carry out the heist, successfully stealing the armoured truck and murdering a guard in the process. They later drive the vehicle (concealed within a customised trailer) to Cripple Creek, dumping it, along with the guard's body, into the bottomless depths of Katie Lewellyn: the same abandoned mineshaft Sunblade had stumbled across earlier in the text. Taking off with their loot – which amounts to around $88,000 – Sunblade and Virginia later land in New Orleans, where they ingratiate themselves with the city's social elite and embark on a hedonistic gorge of booze, food and sex. Yet, in time, Sunblade grows dissatisfied with this world of depthless consumption and abundant material goods, finding that he has "so much stuff" that he can't "concentrate on any of it long enough to enjoy it" (138). This is typified by the enormous "sunken tub" that sits in the bathroom of his lavish apartment, an item that had always "been part of the dream" (144). But, two months in, he realises the tub "mean[s] nothing to [him], no more than the Exakta camera and the rest" (144). Instead, he finds himself fantasising about the first night he met Virginia in Krotz Springs, where he took "a wonderful bath in an old-fashioned tub with the rim of rolled iron" (144). Sunblade speaks of his old clothing in a similarly nostalgic fashion, claiming that the new "bench made" shoes he buys at "sixty dollars a crack" never feel as good or "as easy as the cheaper ones" (136).

It is Sunblade's increasing inability to control Virginia that causes the greater anxiety, as she embarks on a string of sexual assignations with "wealthy men" and other prominent social climbers. Sunblade begins to resent the money, to detest the "petulant, craving business of sweating over wealth" (136). He also grows to despise Virginia, who he perceives to have transformed from an "elegant adventuress with plenty of guts and imagination into a candy-tonguing country club Cleopatra who nested in bed the whole day long and thought her feet too damned good to walk on" (147). It is through these images of unsatisfied consumer desire that we can begin to see the broader cultural effects of what Stephanie LeMenager refers to as "petrotopia" – the now "ordinary" American landscape of "highways, low density suburbs, strip malls, fast food and gasoline islands and shopping centres" that Chaze reveals to us elsewhere in the text (75). While petrotopia is ostensibly a utopian projection, representing "itself as an ideal end-state, the service economy made flesh", it is one driven more by "economic and political will" than any egalitarian or democratic intent (LeMenager 75; Scott 185). Thus, as Heidi Scott suggests,

> When everyone is a consumer within a petrotopia, new problems emerge and the salable utopian image becomes tread-marked with the realities of its false design: not to make people happy, but to make them continually less happy, that is, continually consuming.
>
> *(Scott 185)*

Unfulfilled and dejected, Sunblade pines for the days before he was rich, where it was "only a matter of hanging onto life, a good, rugged, animalistic, instinctive thing that kept [him] hard on [his] toes" (136). He quits New Orleans shortly after, bringing a reluctant Virginia on a pilgrimage to his hometown of Masonville, Mississippi (where we learn that Sunblade's real name is Kenneth McLure). It is whilst driving past the house of his childhood sweetheart, Nora Hickman, that Sunblade spots a police cruiser following close behind in his rear-view mirror. A frantic car chase ensues, culminating in a bloody shootout that leaves both officers dead. Abandoned by a panicked Virginia, Sunblade is forced to flee on foot, only to be picked up and arrested by another police cruiser shortly after. Now a cop killer, he is brutally beaten and tortured by investigating officers but refuses to give Virginia up. His loyalty is repaid pages later, when Virginia seduces and incapacitates one of the prison guards before springing Sunblade from his cell. Now on the run and hounded by the law, Sunblade begins to rediscover the same sense of freedom and possibility that he held prior to executing the armoured car heist. This is crystallised in a moment in the immediate aftermath of their escape, when standing on the side of the road, the "big diesel trucks whooshing by on the highway", Sunblade is struck again by that familiar feeling of "going somewhere, going somewhere new" (179).

Yet, the final section of the narrative is distinguished by the movement backwards rather than forwards, as Sunblade and Virginia return to the near deserted Cripple Creek, renting a small cottage on the outskirts of town. Despite their attempt to recapture the feelings of freedom and escape that they felt up in the mountains towards the beginning of the novel, Virginia soon grows depressed, talking "more and more frequently of dying" (188). Sunblade, meanwhile, finds his mind preoccupied with thoughts of the old Katie Lewellyn mineshaft, which he feels is pulling him "closer and closer to it" with each passing day (191). This soon develops into an obsession, as he becomes plagued by the "absolutely senseless feeling" that if he gazes down into the blackness of the shaft, he will "see something down there", something seemingly primordial, revelatory or devastating. As Sunblade puts it: "[M]aybe this something would ease my mind, or maybe it would drive me stark raving mad, but whatever, I wanted to look. Like a kid who in the middle of the night thinks there's something under his bed" (192).

Associated with anxiety and estrangement, the mine becomes "the sign of a pure otherness, of a human capacity for alienation (misrecognition of home and kin) that is its tragic, oedipal *hamartia*" (Willmott 193). Whilst, as Willmott suggests, this is ostensibly an "inward and imaginative capacity" – of the type and kind that typifies the tragic or fatal narrative trajectory of protagonists in noir – the mineshaft also holds "an outward and material significance", one which focalises the "unrepresentable, absent cause of this alienation in modern society itself" (193). The absent cause, in this case, is the abstracted cultural and economic network of an emergent petro-capitalism, whose disaggregated sites of extraction and consumption operate to distort straightforward causal relations. Thus, whilst Katie Lewellyn is an abandoned gold mine, rather than an oil well, it metonymically recalls the drilling rig from the opening scene of the text and thus points to the imbricated histories/industries of resource extraction that have shaped (and continue to shape) modern America. It is here, I argue, that *Black Wings Has My Angel* begins to reveal an element constitutive to the roman noir more broadly. That is, that the widescale and variegated effects of industrial modernity are central to the mode's tragic imagination, and that these effects, in turn, work to draw noir's recognisable patterns of primordial alienation to the surface. As Glenn Willmott puts it:

> Oil is a synecdoche for a larger modernity . . . one that points inwards, toward the inhumanity latent in human society, toward a dangerously alienated self and the exercise of a ruthlessly indifferent sovereignty that is marked as timeless but awoken by a dystopian petroculture.
>
> *(191)*

In other words, noir alienation can be interpreted as a contextual response to a period of rapid development, one that indicates a tacit scepticism towards the freedoms and desires that petro-modernity purports to grant us.

The significance of the mineshaft as a symbolic representation of internal estrangement – as well as an external, material gesturing towards petroculture as "something larger and harder to define than a single object or event" (Pangborn 783) – is evidenced further at the climax of the text, when

Sunblade and Virginia finally surrender to its "hideous" magnetic force (Chaze 197). As they reach the summit, Sunblade describes how

> A slow sickening whirl seemed to move round and past us, as if the air itself had become so thick with evil as to be a tangible force. . . . [I felt] a blind disgust so intense, so very much of what it was, as to carry with it a terrible fascination. The ultimate horror is, for some unworldly reason, attractive. Hypnotic. For this reason you stare at the face of a leper . . . you are riveted by the scene of an automobile crash.
>
> *(197)*

As indicated by Sunblade's description, the mineshaft inhabits a space between epistemic polarities, inducing oscillating feelings of fascination and abjection, revelation and obscuration, desire and death. The result is a resistance to understanding or clear meaning, with Sunblade unable to give definitive shape to the object to which he refers. It is in this moment of unintelligibility that Chaze brings us closer to what Jennifer Wenzel describes as the "ontology of oil", which, she argues, can be characterised by an inherent resistance to the causative operations of narrative (212). There is something akin to the Freudian return of the repressed in the "deep subject" of the mine, a "hidden primal inhumanity" that gestures towards, and is brought to the surface by, the alienating mechanisms of petroculture (Chaze 200; Willmott 190). The shaft is anti-plot, a "hyperobject" vastly distributed in time and space that can only be glimpsed, rather than fully comprehended (Morton 1). Descriptions of it become permeated with the language of disease and pollution, with Virginia describing her own fixation to that of "a woman who's known for months she had cancer, and the doctor finally tells her it's there and tells her where to look to see it. And she must look at it but she can't" (198). Sunblade himself recognises this feeling of somatic encroachment, telling Virginia: "It's eating me too, baby" (199).

Eventually facing down into the ugliness, both characters are momentarily struck by a sense of release, as if by forcing themselves to look, they had somehow "rid [themselves] of some loathsome disease" (201). This elation is short lived, however, as Virginia stumbles backwards, her back "arch[ing] prettily" as she falls over the edge into the hole (203). A panicked Sunblade tries desperately to retrieve her from the jagged bow of rock on which she lands but is unable to find a rope or instrument long enough to reach down. After hours of failed rescue attempts, he awakens by the edge of the hole the following morning, unable to face up to the reality of Virginia's death. In the midst of fantasising that Virginia is elsewhere, perhaps "getting a massage at Mamie's", Sunblade spots four or five men "strutting" towards him over the ridge, among them Clell Dooley, the very same FBI agent who had "packed him off to Parchman in the dim days of poverty" (209, 135). In the closing lines, Sunblade is severely beaten by the men, all the while trying to ask them "if they had seen Virginia" (209).

Noir Ecologies: Oil, Noir and Narrating the Now

In characteristically noir fashion, *Black Wings Has My Angel* concludes with Sunblade's inevitable confrontation with the Law, as he becomes what Scott Loren describes as the dysfunctional "outlaw of the oedipal drama" (71). Epitomising the typical noir protagonist, "who finds himself trapped in a universe defined only by the impossibility of narrativization", what we see in Sunblade is an instituting image of the perennially alienated and estranged petro-subject, one who is incapable of articulating the elusiveness of oil as an object of narrative contemplation (Lehmann 109). The imaginative and tragic framework of the American roman noir not only taps into certain primordial or mythic desires and failures, but also strives, however unconsciously, to articulate a broader feeling of disorder and disaffection catalysed by a period of immense technological and social transformation. In this way, I argue that the roman noir can be seen to express something fundamental regarding the experience of the modern subject. After all, as Mathew Pangborn puts it: "noir's repetitive presentation of a story in which an actor becomes trapped by desire within a vast and mysterious system [is] perhaps the archetypal plot of petromodernity" (802).

In this way, the noir sensibility – defined by guilt, disorder and discontinuity – also speaks powerfully to our experience of 'the now'. In recent times, several scholars have drawn a persuasive connection between the anxious existentialism of noir fiction (and film) and the profound uncertainty of our ecological moment. As Lucas Hollister suggests, many of the questions that we ask when confronted with the Anthropocene's "destabilising frames of reference, its hyperobjects", are the very same questions prompted by the noir mode: "What is violence? How do we conceptualize networks and causalities of violence? How do we understand the distribution of victimization and guilt? What conclusions are we to draw from the violence that marks our political, economic, and social conditions?" (1012). Thus, whilst reconsidering the roman noir via its imbrication in the early U.S. oil regime offers, in the first instance, a valuable way of thinking about the "aesthetic logic of oil" in twentieth-century cultural production, it also substantiates noir's continuing relevance as a means of narrating that which defies narration: be it the opacity of global energy systems, the bewildering forces of capital or the density of the climate crisis (Perks 254).[2] This does not require a necessarily pessimistic or wholly doomed worldview either. Isak Holm, for instance, uses the term "noir optimism" to refer to a particular way of rethinking the "scales of our stories and their causal logics", one that could enable "new ways of attaching to the present" (24). In other words, the roman noir can not only help us articulate where we have been – knowing how oil is imagined in fiction is, after all, fundamental to an understanding of "what and who we might become without it" (Szeman 163) – but its narratorial frameworks can also provide a structure through which to shape our understanding of the now, and perhaps, even, to imagine different kinds of futures.

Notes

1. The terms 'literary noir' and 'noir fiction' are used synonymously here (and elsewhere) with what Andrew Pepper has labelled the 'American Roman noir'. Pepper's term refers to a specific cycle of novels published in the United States from the 1930s to the 1950s that share similar thematic and political concerns, notably relating to, as Pepper outlines, "the corrosive effects of money, the meaningless and absurdity of existence [and] anxieties about masculinity and the bureaucratisation of contemporary life" (60).
2. More certainly be done in examining the relationship between petroleum culture and the roman noir, particularly as regards to the material production and distribution of pulp fiction via mass market publishing.

Bibliography

Belasco, Warren. "Commercialized Nostalgia: The Origins of the Roadside Strip." *The Automobile and American Culture*, edited by David Lanier Lewis et al., U Michigan P, 1983, pp. 105–122.

Bergthaller, Hannes, et al. "Introduction to Petrofictions." *Green Letters*, vol. 23, no. 2, 2019, pp. 125–129, doi: 10.1080/14688417.2019.1654243.

Blanke, David. *Hell on Wheels: The Promise and Peril of America's Car Culture, 1900–1940*. UP of Kansas, 2007.

Breu, Christopher. *Hard-Boiled Masculinities*. U Minnesota P, 2005.

———. "Radical Noir: Negativity, Misogyny, and the Critique of Privatization in Dorothy Hughes's *in a Lonely Place*." *MFS Modern Fiction Studies*, vol. 55, no. 2, 2009, pp. 199–215, doi:10.1353/mfs.0.1607.

Chaze, Elliott. *Black Wings Has My Angel*. New York Review Books, 2016.

Devereux, Cecily. "'Made for Mankind': Cars, Cosmetics, and the Petroculture Feminine." *Petrocultures: Oil, Politics, Culture*, edited by Sheena Wilson et al., McGill-Queen's UP, 2017, pp. 162–186.

Duckert, Lowell. "Coal/Oil." *The Cambridge Companion to Environmental Humanities*, edited by Jeffrey Jerome Cohen and Stephanie Foote, Cambridge UP, 2021, pp. 214–228.

Dussere, Erik. *America is Elsewhere: The Noir Tradition in the Age of Consumer Culture*. Oxford UP, 2014.

Fotsch, Paul Mason. *Watching the Traffic Go By: Transportation and Isolation in Urban America*. U Texas P, 2009.

Ghosh, Amitav. "Petrofiction: The Oil Encounter and the Novel." *New Republic*, vol. 206, no. 9, 1992, pp. 29–34.

Goldsmith, Martin M. *Detour*. Open Water Publishing, 2014.

Hecht, Roger. "American Pastoral." *Encyclopaedia of the Environment in American Literature*, edited by Brian Jones and Geoff Hamilton, McFarland, 2014, pp. 12–14.

Hollister, Lucas. "The Green and the Black: Ecological Awareness and the Darkness of Noir." *PMLA/ Publications of the Modern Language Association of America*, vol. 134, no. 5, 2019, pp. 1012–1027, doi:10.1632/pmla.2019.134.5.1012.

Horsley, Lee. *The Noir Thriller*. Palgrave Macmillan, 2009.

Huber, Matthew T. *Lifeblood: Oil, Freedom, and the Forces of Capital*. U Minnesota P, 2013.

LeMenager, Stephanie. *Living Oil: Petroleum Culture in the American Century*. Cambridge UP, 2014.

McCann, Sean. *Gumshoe America: Hardboiled Crime Fiction and the Rise and Fall of New Deal Liberalism*. Duke UP, 2000.

McQuaid, Kim. *Uneasy Partners: Big Business in American Politics, 1945–1990*. John Hopkins UP, 1994.

Morton, Timothy. *Hyperobjects: Philosophy and Ecology After the End of the World*. U Minnesota P, 2013.

Munby, Jonathan. *Public Enemies, Public Heroes: Screening the Gangster from Little Caesar to Touch of Evil*. U Chicago P, 1999.

Osteen, Mark. "*Noir's Cars*: Automobility and Amoral Space in American Film Noir." *Journal of Popular Film and Television*, vol. 35, no. 4, 2008, pp. 183–192, doi:10.3200/jpft.35.4.183-192.

Pangborn, Matthew. "Lessons in 'Bad Love': Film Noir and the Rise of the American Oil Regime in Edgar G. Ulmer's *Detour* (1945)." *Journal of American Studies*, vol. 55, no. 4, 2020, pp. 780–814, doi:10.1017/ s0021875820001085.

Pepper, Andrew. "The American Roman Noir." *The Cambridge Companion to American Crime Fiction*, edited by Catherine Ross Nickerson, Cambridge UP, 2010, pp. 58–71.

Perks, Samuel. "'He Can't Throw Any of His Coal-Dust in My Eyes': Adventurers and Entrepreneurs in *Victory's* Coal Empire." *Conrad and Nature*, edited by Jeffrey Mathes McCarthy et al., Routledge, 2018, pp. 252–268.

Pitt Scott, Harry. "Offshore Mysteries, Narrative Infrastructure: Oil, Noir, and the World-Ocean." *Humanities*, vol. 9, no. 3, 2020, doi:10.3390/h9030071.

Poll, Ryan. "The Rising Tide of Neoliberalism: Attica Locke's Black Water Rising and the Segregated Geographies of Globalization." *Class and Culture in Crime Fiction: Essays on Works in English Since the 1970*, edited by Julie H. Kim, McFarland, 2014, pp. 175–200.

Primeau, Ronald. *Romance of the Road: The Literature of the American Highway*. Bowling Green State UP, 1996.

Rosser, Sue Vilhauer. *Women, Science and Myth: Gender Beliefs from Antiquity to Present*. ABC-CLIO, 2008.

Rubenstein, Michael. "Petrocriticism." *Futures of Comparative Literature: ACLA State of the Discipline Report*, edited by Ursula K. Heise, Taylor & Francis, 2017, pp. 49–50.

Schuster, Joshua. "Where is the Oil in Modernism?" *Petrocultures: Oil, Politics, Culture*, edited by Sheena Wilson et al., McGill-Queen's UP, 2017, pp. 197–213.

Scott, Heidi C. M. *Fuel: An Ecocritical History*. Bloomsbury, 2018.

Szeman, Imre. "How to Know about Oil: Energy Epistemologies and Political Futures." *Journal of Canadian Studies*, vol. 47, no. 3, 2013, pp. 145–168.

Watson, Jay. *Fossil Fuel Faulkner: Energy, Modernity, and the US South*. Oxford UP, 2023.

Wenzel, Jennifer. "Petro-Magic-Realism Revisited: Unimagining and Reimaging the Niger Delta." *Oil Culture*, edited by Ross Barrett and Daniel Worden, U Minnesota P, 2014, pp. 211–225.

Willmott, Glenn. "Oil Tragedy as Modern Genre." *Petrocultures: Oil, Politics, Culture*, edited by Sheena Wilson et al., McGill-Queen's UP, 2017, pp. 187–198.

Wilson, Sheena, et al. "On Petrocultures: Or, Why We Need to Understand Oil to Understand Everything." *Petrocultures: Oil, Politics, Culture*, edited by Sheena Wilson et al., McGill-Queen's UP, 2017, pp. 3–20.

Yaeger, Patricia. "Editor's Column: Literature in the Ages of Wood, Tallow, Coal, Whale Oil, Gasoline, Atomic Power, and Other Energy Sources." *PMLA/Publications of the Modern Language Association of America*, vol. 126, no. 2, 2011, pp. 305–326, doi:10.1632/pmla.2011.126.2.305.

29

OIL AND THE HARD-BOILED

Petromobility, Settler Colonialism and the Legacy of the American Century in Thomas King's *Cold Skies*

Alec Follett

In his award-winning historical work, *The Inconvenient Indian* (2012), Cherokee author Thomas King writes passionately about the bituminous sands of Northern Canada:

> The Alberta Tar Sands is an excellent example of a non-Native understanding of land. It is, without question, the dirtiest, most environmentally insane energy-extraction project in North America, probably in the world, but the companies that are destroying landscapes and watersheds in Alberta continue merrily along, tearing up the earth because there are billions to be made out of such corporate devastation. The public has been noticeably quiet about the matter, and neither the politicians in Alberta nor the folks in Ottawa have been willing to step in and say, "Enough," because, in North American society, when it comes to money, there is no such thing as enough.
>
> *(219)*

Here, King draws attention to the environmental and social impact of increasingly inaccessible and destructive oil deposits, as well as Canadians' ongoing attachment to the energy source. This concern with energy extraction and Indigenous peoples is a topic King has revisited throughout his career, especially in his celebrated literary novels. *Green Grass, Running Water* (1993) centres on a dam that threatens to disrupt a river and the Blackfoot community's Sun Dance that is intertwined with the riparian ecosystem, while *The Back of the Turtle* (2014) focuses on pipeline and oil sands projects that devastate the environment and an Indigenous community.

However, in this chapter I want to turn to King's lesser-known mystery novel, *Cold Skies* (2018), for the way it engages with North America's oil entanglements, both at the level of plot at the level of genre. *Cold Skies* is the third in a series of six novels that follow Cherokee detective Thumps DreadfulWater. Thumps, a former California cop, moves to Chinook Montana to escape his traumatic past and develop his career as a landscape photographer, but is inevitably dragged back into solving crimes. In *Cold Skies*, King's reluctant hard-boiled detective is asked to investigate two suspicious deaths. He soon learns that the deceased are co-owners of a company that has just developed an innovative oil deposit and aquifer mapping technology that was undergoing testing on the nearby Blackfeet reservation. While this new oil technology is at the centre of the investigation, oil permeates the novel from the gas-powered vehicles that variously excite and antagonise characters to the motels, airports, and other infrastructure that make petromobility possible.

DOI: 10.4324/9781003091912-35

In this chapter, I trace how the contemporary attachment to oil in the novel is perpetuated by a desire to extend twentieth-century American culture, or what Stephanie LeMenager calls the "American century", and how this desire is firmly rooted in the logic of settler colonialism (4). To tangle matters further, I will suggest that the hard-boiled mode is firmly entrenched in the American century and is therefore implicated in perpetuating a certain aesthetic investment in oil. And yet, through playful adaptation of the genre's conventions, such as the mobile detective, the red herring, and the restorative function of the investigator, King writes hard-boiled fiction that simultaneously exposes our entanglements with oil and ultimately exceeds the oil-bound limitations of the mode. Consequently, the novel suggests that, to comprehend our current relationship with oil, we must understand and ultimately rewrite our relationship with the American century and with genre.

Hard-Boiled Fiction, Oil, and the American Century

Despite *Cold Skies*' pointed social commentary, not to mention its expertly crafted dialogue, the novel has received very scant critical attention to date, largely owing, I would suggest, to an entrenched association between crime fiction and lowbrow culture. Critic Richard Bradford explains that, notwithstanding the clear aesthetic and social value of the crime novel, the genre "as a whole (is) still treated as a sub-species of mainstream literature" (115). Such highbrow-lowbrow distinctions between literary fiction and crime fiction have no doubt impacted the reception of King's DreadfulWater mysteries. As Roshaya Rodness puts it: "Although one of the staples of his artistic career is arguably his experimentation with genre, King's literary fiction and historical prose are often foregrounded while his genre fiction is placed in the background or displaced (62)". Rodness argues that King's mysteries are downplayed to preserve "his image as a literary author" and to help maintain "a certain prestige of iconic Canadian literary output" (63). Yet neglecting King's mysteries is especially limiting for critical conversations about his long-standing interest in energy extraction and settler colonialism because, as I argue in this chapter, the conventions of the hard-boiled mode allow King to generate his most compelling written exploration of oil dependency and colonial violence.

Although attending to oil in literary works is urgent and illuminating, critics within the energy humanities have underscored two major challenges related to reading for and writing about it: 1) oil is hard to make visible; 2) oil is hard to know. Indeed, Graeme Macdonald explains that, while there is increasing academic attention being directed towards oil's "violent visibility", there is a simultaneous acknowledgement that oil remains largely imperceptible in daily life and in cultural texts (293). Oil's lack of visibility, due in part to the fact that it is regularly consumed far from extraction zones, is one of many reasons why "there is very little 'petrofiction' that explicitly addresses" its complex histories and ongoing effects (Bergthaller, Grewe-Volpp & Mayer 125–26). Consequently, critics are attempting to make oil more visible by identifying, cataloguing and analysing works of "petrofiction", a term coined by Amitav Ghosh in the early 1990s and now commonly used to refer to a field of texts "that straightforwardly engage with the physical stuff of oil in their contents" (Duckert 216). However, as Lowell Duckert explains, "the idiom can also explore those imaginations sprung more obliquely from the oil-obsessed and oil-powered age known as Petromodernity: the gasoline, plastics, and the 6,000 other petroleum-based products consumed on a daily basis" (Duckert 216). This has prompted critics such as Imre Szeman to argue that commonplace understandings of energy sources, including oil, are still reductive and inadequate. Rather than simply knowing oil as an energy source that fits within existing social structures, he suggests that oil should be understood as an organising force that dictates society materially, politically and culturally:

> [Oil] shapes . . . [societies] in every possible way and at every possible level, from the scale of our populations to the nature of our built infrastructure, from the objects we have ready to

> hand to our agricultural and food systems, and from the possibility of movement and travel to *expectations* of the capacity to move and interact, not to mention the plastics used to encase our smartphones and other high-tech devices.
>
> *("How to Know" 147)*

Aligned with Szeman's argument, one that suggests a more encompassing understanding of the ways oil shapes society, Patricia Yaeger highlights the need to "rechart literary periods and make energy sources a matter of urgency to literary criticism" (306). For Yaeger, the term 'petrofiction' is not a genre or set of stylistic features, but a form of periodisation that forces us to rechart the twentieth and twenty-first centuries as the age of petro-modernity. Until we come to an understanding of the ways oil has shaped, and continues to shape, politics and culture (both locally and globally), it will not be possible to imagine different kinds of futures free from such dependencies.

In this chapter, I am interested in how the hard-boiled mode might be useful to our understanding of oil and its cultural effects. A more robust understanding of oil can be gained by engaging with genres, such as the hard-boiled, that emerged during what Stephanie LeMenager has categorised as the "high point of petromodernity", and that retain popularity in the contemporary moment (67). Through plot and form, the twentieth-century hard-boiled grapples with the pleasures and violences of the American century, all of which are shaped by oil, and these concerns are the legacy that contemporary manifestations of the mode must grapple with. By focusing on the hard-boiled, I build on Jennifer Wenzel's interrogation of the relationship between genre and oil, particularly her questioning of the way that *"different kinds of texts . . . work against or contribute to oil's invisibility"* and knowability ("How to" 158). This chapter also takes part in emerging critical interest in the relationship between crime fiction and oil, which has developed alongside a more general interest in crime fiction and ecology prompted by Jo Lindsay Walton's and Samantha Walton's influential special issue in the ecocritical journal *Green Letters*.

While critics of crime fiction and oil, including Sharae Deckard and Stephanie LeMenager, undertake compelling comparative approaches, juxtaposing texts from different periods and or locations, in this chapter I work alongside critics who attend to the conventions of the hard-boiled and to American contexts, because a more robust understanding of the American century, and its media, is also essential to understanding the contemporary reliance on oil. For example, Glenn Willmott reads early hard-boiled author Raymond Chandler's *The Big Sleep* as a kind of "oil tragedy", where oil becomes the

> synecdoche for a larger modernity . . . one that points inwards, toward the inhumanity latent in human society, toward a dangerously alienated self and the exercise of a ruthlessly indifferent sovereignty that is marked as timeless, but awoken by a dystopian petroculture.
>
> *(191)*

Nathan Ashman reads Ross Macdonald's 1973 novel, *Sleeping Beauty*, in a similar fashion, examining the ways in which oil dependency "shapes a larger framework of buried lies, concealed passions and terrible crimes" within the text (51). Mary Stoecklein considers the relationship between oil and American settler colonialism in Tom Holm's *The Osage Rose* and Linda Hogan's *Mean Spirt*; and Harry Pitt Scott considers the conventions of crime fiction and petromobility, especially the detective's hypermobility and regular movements across oil's infrastructure (4). Together, these critics are turning to crime fiction as a productive site through which to "name and frame energy", to borrow Imre Szeman's phrase ("Afterword" 390). By examining the connections between crime fiction and oil, they are signalling to scholars in the energy humanities that detective fiction is vital for what it reveals about the cultures that produce it. And yet, there is still much work to be done in thinking through the relationship between the hard-boiled mode and oil, particularly the extent to which the hard-boiled is intertwined with the oil-driven pleasures and injustices of the twentieth century and how the legacy of the American century contributes to contemporary oil dependency.

Whilst the power of oil ostensibly derives from the wide range of industrial and technological innovations that it made possible – from cars to plastics to a multitude of other oil derived products – Matthew T. Huber argues that at the core of oil's transformative effects was its empowerment of a certain "way of life" that emerged alongside the automobile in the early decades of the twentieth century, and that resulted in immense changes to the country's infrastructures and settlement patterns (58–59). Describing America's ongoing imbrication in ideas of freedom, movement and autonomy, he writes:

> Oil is primarily about powering a certain kind of mobility characterised by an individuated command over space. . . . Oil is a powerful force not only because of the material geographies of mobility it makes possible but also because its combustion often accompanies deeply felt visions of freedom and individualism.
>
> (*x-xi*)

Oil's ability to move Americans, both physically and emotionally, alongside the many other pleasures and material benefits that emerged and developed through the twentieth century, has resulted in a certain affective investment in the possibilities that oil allows, or what LeMenager terms a "love" of oil (18). However, oil's benefits were not shared equally among all Americans and often worked to exacerbate certain structural inequalities. As Ryan Poll writes of the twentieth-century United States: "Oil enables, prioritizes, and celebrates mobility (coded as freedom), but this mobility is not evenly distributed; rather the roads are still organized by the colour line" (192). In other words, oil directly harmed many minority communities. Indigenous peoples, in particular, were often adversely impacted by oil through extraction projects that resulted in the dispossession and degradation of land (Gilio-Whitaker 55, and Clarsen and Veracini 893). Petromobility also reinforced settler colonialism on an ideological level. Georgine Clarsen and Lorenzo Veracini argue that early automobility was used to "*disavow* actual indigenous presences" through drivers who imagined their car travel as "re-enactments of earlier colonial explorations" across unpeopled spaces (894).

The violence of oil continues for Indigenous communities in the twenty-first century as they face the uneven impacts of oil-fuelled climate change and the disproportionate negative effects of resource extraction projects. Oil's contemporary violence is exemplified in the extraction of the bituminous sands of Northern Alberta and its transnational transportation via pipelines. This oil project results in multiple harms from the contamination of local waters that threatens Indigenous health and traditional practices to the incursion of extraction and transportation pipelines that undermine Indigenous self-determination in their territory (Laboucan-Massimo 81–2). The "economic, social, and biological cost" of oil is increasingly understood by Americans, and yet the country's love affair with oil flows on, largely because it is unextractable from a certain 'way of life' that has come to define the American twentieth century (LeMenager 15). Explaining the desire to bring twentieth-century American "ideological, stylistic, military, and economic expressions of modernity" into the future, LeMenager argues that "the continuation of this American century is what's at stake in the 'race for what's left'" (4). Despite the many violences caused by oil, the expressions of modernity forged in the twentieth century – especially individualised mobility – are so compelling that they drive contemporary commitments to oil.

The hard-boiled mode of American crime fiction is, I argue, particularly entangled with American oil culture, emerging, as it did, from the oil-shaped pleasures and violences that defined the inter and post-war period. First realised in the United States of the 1930s through the adaptation of the "frontier hero" of an earlier, colonial era to the new, modern America (Scaggs 64), the hard-boiled emerged out of what LeMenager terms "the high point of petromodernity" (68). The relationship between oil and the early hard-boiled novel is most clear through the changing relationship that detectives and criminals forge with newly emergent transportation infrastructures. Jason Vredenburg highlights how early hard-boiled novels departed from previous crime fiction, which was often set in and around trains and the rail system. He argues that early hard-boiled novels worked through developing anxieties associated with automobility, which gave criminals the freedom of unconstrained mobility, especially before police surveillance technology developed ways to reveal criminals' movements (74). Yet these anxieties also become wrapped up with certain desires and pleasures, as indicated by the unrestrained movement allowed to the newly mobile private investigator.

Drawing on Raymond Chandler's *The Big Sleep*, Vredenburg argues that the detective "Marlowe gets around town, mainly by automobile, and his mastery of that terrain is essential to his success" (114). Vredenburg continues to explain that the detective's hypermobility, which emerged in the early days of the genre, became "typical" of the mode (114). Indeed, Sean McCann argues that by "navigating the furthest corners of the metropolis", the hard-boiled detective "ranges across its social and geographical terrain, tying the disparate features of the urban landscape into a legible map" (47). Moreover, in addition to the detective's and criminal's ability to navigate space through automobility, Harry Pitt Scott writes that at the core of the hard-boiled-adjacent genre of the noir is "petromodernity's infrastructure: hotels, highways, flickering streetlights and eerie hinterlands, ports and warehouses, and so on" (4). Consequently, we can begin to see how the hard-boiled, from its early to contemporary manifestations, is inextricably intertwined with twentieth-century petroculture. Although beholden to the twentieth century (and to oil) through the familiar tropes of individualised hypermobility, embedded urban infrastructures, and the detective's mastery and control over space, the mode is still particularly capable of reimagining itself. As Scaggs states: "the general tendency of hard-boiled fiction to replicate, explore, and even interrogate its own conventions allows the entire sub-genre to be appropriated for a variety of ideological, formal, and generic purposes" (78), including feminist revisionism of the 1980s by authors such as Sue Grafton and Sara Paretsky and more recent subversions of the genre's white gaze by authors including Walter Mosley and Steph Cha. In doing so, writers of the hard-boiled simultaneously draw on and subvert early tropes and conventions so their works speak to questions of justice that are meaningful to their present moment. Through the mode's ongoing investment in, and subversion of, twentieth-century American petro-modernity, the contemporary hard-boiled becomes an important site through which to understand the legacy of the American century and to investigate our current and deeply imbricated relationship with oil.

Investigating Petromobility and Resource Extraction in *Cold Skies*

If Beth Mooney, the coroner of King's fictional town Chinook, could conduct an autopsy on the American century, she would find a body bloated with oil. Oil in the gut drank down like Sheriff Duke Hockney's "soft tar" coffee (417) – oil that energised and empowered, but also destroyed. *Cold Skies* writes an obituary for the American century, examining the joys, the sorrows, and the structures of violence that enabled it. As a work of petrofiction, it juxtaposes oil as an abstracted commodity – one that is invisibly entrenched in American culture and media and intimately bound up with attendant ideas of freedom, pleasure, and power – with oil as a material substance that is inseparable from a settler-colonial extractivist ethos and the accompanying dispossession of Indigenous land. By underscoring the disproportionately felt pleasures and violences of oil, the novel makes legible contemporary oil dependency as a nearly insurmountable obsession that grew out of twentieth-century American culture, media, and economics. King first addresses the legacy of twentieth-century petro-modernity in the text's opening chapter, where he not only introduces readers to the crime scene that sets Thumps, and the plot in motion, but also to the complicated relationship between oil and pleasure.

Through the character Bob Tatum, who is travelling by airplane and automobile for work, the opening chapter considers the challenges of contemporary petromobility in contrast to the perceived joys of car culture. Tatum's commute involves two uncomfortable flights and a frustrating interaction with the car rental agency. His plane from Great Falls to Chinook suffers "a mechanical problem and had sat on the tarmac for almost five hours" (1). The in-flight food is unappealing, and his seatmate is awful. For Tatum, the damaged airplane becomes an unreliable, slow-moving, and confining mode of transportation. Tatum is not alone in these sentiments, as is suggested by the third person omniscient narrator who asserts that a delayed plane is "always a crowd-pleaser" (1). Tatum's commonplace business travels suggest that while petromobility is incredible for the ways it can move people across vast distances at great speeds, it is also disappointing for the ways it often moves passengers at a much slower and much more uncomfortable pace than advertised.

Tatum's experience at the rental car agency is also not as advertised. Tatum reserved a compact car but was "upgraded" to an SUV. Reflecting on this unexpected switch, Tatum thinks: "'upgraded,' . . . was just the expression car rental companies used to make you feel as though you had won something, as though getting a larger, clumsier gas guzzler was first prize in a lottery" (2). Tatum's concern with driving a large "gas guzzler" gestures to a frustration that his rental vehicle will cost more to fuel and will be more cumbersome to drive. His disappointment with the rental and with his flight are framed as individual inconveniences, however, and can thus be understood through Matthew T. Huber's notion that petro-modernity has generated an expectation that the individual can move freely and easily (xi). Despite facing these commonplace frustrations, there is little to suggest that Tatum's individualised inconvenience may develop into a deeper understanding of what has been called the "fundamental antagonisms" of automobility, antagonisms tied to concerns about oil's finitude, environmental damage and the logistical challenges of creating infrastructure capable of moving so many people (Böhm, Jones, Land, and Paterson 9–10). Instead, Tatum bears the inconveniences as if they are the exception to the norm, however unlikely this may be.

Despite Tatum's frustration, he is quick to reassociate petromobility with freedom and excitement. Shortly after reflecting on his flight and rental car, he muses,

> If he had his way, he'd drive a sports car, something agile and quick. . . . Maybe when he retired and the kids left home, he'd get a used Jaguar or better yet a Corvette, an older model with red and white tuck-and-roll upholstery.
>
> *(3–4)*

Cars may be essential to business trips but, as critic Simon Orpana writes, "they are also vehicles of fantasy, desire, and enjoyment" (80). By shifting from Tatum's frustration with petromobility to his enjoyment of classic gas-powered cars, King strikes at one of the core challenges of energy transition: oil generates a certain, highly mobile, and often pleasurable way of life that is deeply engrained in the American consciousness. Attending to the pleasures made possible by oil is important, argues Warren Cariou, for what it reveals about our attachment to oil:

> Most artists, activists and scholars who engage with the politics of petroleum are focused on its negative consequences – and for very good reasons – but sometimes they tend to forget the role of pleasure in all this. Aesthetic pleasure and other more elemental kinds. The pleasure of playing in the mud, for example, or the pleasure of stepping on the accelerator and feeling a surge of acceleration. If we didn't derive multiplicitous forms of pleasure from our relationship to petroleum, we would find it much easier to disengage from it now that we have a better understanding of how dangerous it is, not only for individuals working in close proximity to it but also for much of the life on this planet.

For Tatum, oil is pleasurable because of its connection to mobility. Through his emphasis on "an older model" Corvette, Tatum firmly associates the joy of automobility in the past, specifically in the mid-twentieth century when Corvettes were most popular. And yet, rather than leaving the pleasures of oil in the past, he desires to extend these indulgences into the future through his intention to buy a Corvette when he retires. Through a complicated logic that places the pleasures of oil in both the past and in the future, but not in the present, Tatum sidesteps the contemporary antagonisms of petromobility and reaffirms his commitments to oil as a locus of pleasure. For Tatum's reasoning to work he must rely heavily on a belief in American modernity of the twentieth century, and he must turn to that belief to bolster his ongoing commitment to oil. Here King lays important groundwork for questions that are taken up and turned over as the novel progresses, particularly regarding how oil produces pleasures and at whose expense.

Like Tatum, Thumps has a complicated relationship with petromobility; however, Thumps's must be read in light of the history of the character of the hard-boiled detective who typically relies on the automobile to master geographical space and crime, thereby validating this new auto-centric landscape. Thumps is certainly reliant on his Volvo to follow leads and finds detective work without his car to be uncomfortable. Following local Blackfeet friends who lead him to a new crime scene, Thumps walks along a deer trail and "scramble(s) to keep up" (84), and after complaining about the long walk he is informed that there is an "easier way" to their location as "the highway is just over there" (86). Although his friend explains that there "is nothing like a good walk" (86), for Thumps, the detective, bipedal mobility is not as convenient as automobility, which he typically relies on to solve crimes. However, Thumps is not entirely like the conventional detective who uses the automobile for fast and convenient movement. Instead, Thumps is reluctantly mobile and unable to effectively use the car to master space. His decades-old Volvo is unreliable, but he is uninterested in buying a newer model. The car fails to turn over in cold weather, and other mechanical problems impact Thumps's plans. Reflecting on driving the broken car to escape his past in California, the narrator states:

> Thumps . . . headed north and east for no better reason than that's the way the roads ran. Just outside Chinook, the Volvo had thrown a fuel pump, and Thumps wound up stuck in the town. Frost was already in the air and, as he waited for a replacement to be shipped from Salt Lake, an early blizzard had come down from Canada. The roads were closed for the better part of a week, and by the time they had been cleared, he realized that he had no place to go. So he stayed. Everybody died somewhere, and, for Thumps Chinook had looked to be as good a place as any. *(71)*

From the north and east direction of the highway to the broken fuel pump, road closures, and slow-paced shipment of replacement parts, King draws attention to the variegated aesthetics of petroculture. However, by emphasising Thump's struggle with cars and roadways, King creates a character who departs from the conventional hard-boiled detective, one who ordinarily traverses spaces by car with unrealistic ease and success. Thumps is depicted as a passive driver who accepts that cars break down, that bad weather can close roads, and that shipping goods takes time. Here, King's oil aesthetics thus present a counter image to the purported pleasures of petromobility and to the national mythology of highways as an infrastructure of convenience and adventure.

Moreover, King underscores how settler-colonial infrastructure and mobility has historically impacted Indigenous peoples, as he has Thumps recount the history of the removal of over "thirteen thousand" Cherokee in 1838, all of whom were placed in "concentration camps" then forced to walk west on a network of routes or trails: "More than four thousand Cherokee died on the two-thousand-mile trek. Two deaths for every mile" (205). As in the nineteenth century, the twentieth century and contemporary infrastructures of petromobility make possible settler-colonial culture and economy, or as Ojibwe scholar Winona LaDuke and geographer Deborah Cowen writes: "infrastructure is the *how* of settler colonialism" (245). With this colonial history in mind, King contrasts Thumps' passive mobility with that of the traditionally hypermobile hard-boiled detective figure who masters crime and geography, as well as with other characters in the novel, like Bob Tatum, who similarly champion the aesthetic pleasures of automobility.

In an explicit gesture to the difference between Thumps and the hypermobile detective of the genre, Thumps considers whether or not he should investigate the motel room that he is watching. The narrator states:

> As he sat in his car, he tried to remember the various techniques for breaking into a room. If the lock was as old as the motel, he might be able to get in with a credit card, though that always worked better in the movies than it did in real life. There was a trick with a car jack where you spread the frame until the door swung open. He could give that a try. If he had a jack. But he didn't . . . . And what would he do once he got inside? Tear up the carpet? Pull down the ceiling tiles? . . . God. No. Best to forget all this cloak and dagger stuff, drive home, and put the groceries away. He could make a pot of coffee, eat toast with a little jam, sit on the sofa, and read a book. Have a nap.
>
> *(244–45)*

Here, King juxtaposes the reluctant Thumps with the active and mobile fictional detectives who, in their desire to solve the case, turn to "car jacks" and other everyday objects to break doors. Thumps, who uses his car as a place to sit and think, and who desires to use his vehicle to shuttle him away from the crime scene, inverts the actions of the active detective who typically relies on the automobile to move toward, rather than away from, their sites of investigation. It is telling that Thumps would rather "read a book" than be an active and mobile detective. King could have written Thumps as a character with a newer, more powerful car that he uses to move swiftly and easily between spaces. However, by doing the opposite, King continues to adapt the detective archetype through what Julia Breitbach describes as a self-conscious and "humorous rewriting of the hard-boiled mode" (95). Humorous though it may be, King's attempt to subtly rewrite these traditions and tropes is significant to his broader questioning of the mode's historical entanglement with oil. If, as LeMenager writes, the twentieth-century love affair with oil "has a great deal to do with loving media dependent on fossil fuels", then King's critical, contemporary engagement with the tropes of mobility – that first emerged in the 1930s and developed through the century – works toward breaking up this affair with oil (66).

Stephanie LeMenager suggests that 'falling out of love' with oil is challenging precisely because it permeates all forms of cultural production. As she puts it: "[Oil] has supported overlapping media environments to which there is no apparent 'outside' that might be materialised through imagination and affect as palpable hope" (70). There is certainly 'no outside' of oil (or its infrastructures) for Thumps, who travels by car to a motel to investigate a murder that seems connected to the creation of a new oil mapping technology. There is also 'no outside' of oil for a genre like the hard-boiled, which, as discussed, is inextricably connected to the developments of petroculture. However, through a strategy of confronting oil at the level of plot and genre, King writes a work of hard-boiled petrofiction that subtly interrogates our relationship with the substance, questioning whether contemporary readers still require or desire a detective story with a hypermobile detective whose success and ethos is tied to the freedoms and pleasures of petro-modernity. If stories allow certain ethics "to exist and flourish" (King 164), then what energy ethics might emerge from a story like *Cold Skies*, one that reflects on the cultural and literary legacy of that oil-soaked American century? For King, to rethink our relationship with oil requires rethinking our relationship with the American century and with the forms of storytelling that it enabled. As King writes in *The Truth About Stories*: "want a different ethic? Tell a different story" (164).

King establishes that the freedoms and pleasures of the automobile and airplane are engrained in American culture – and media – and thus contribute to contemporary oil dependency; however, he also makes clear that this ubiquitous petromobility often occurs at the expense of Indigenous communities. King subtly links earlier forms of settler-colonial mobility to petromobility by way of Tatum who thinks "the rental cars were lined up like horses at a hitching post", and through the name of The Wagon Wheel motel, where one of the crimes occurred (3). However, while animal-powered modes of transpiration were essential to nineteenth-century settler mastery of the West, the emergence of oil-powered transportation, and its infrastructures, accelerated settler-colonial mobility and ushered in new reasons for corporations and the state to dispossess Indigenous peoples from their lands. In *Cold Skies*, King is aware of the long history of settler colonialism but is acutely interested in how oil enabled a new era in settler-colonial mobility and control over Indigenous communities and their lands. He explores these unevenly felt violences that are necessitated by petromobility through the novel's primary crime, which centres on an oil mapping technology.

The mystery revolves around the death of James Lester and Margo Knight. Lester and Knight are co-owners of Orion Technologies, which has developed a "revolutionary" oil deposit and aquifer measuring technology that will "change the energy resource industry" (124–27). The development of this new technology is made possible by ongoing colonial control over Indigenous land. Orion Technologies had been testing their resource analysis mapping technology on "tribal land" called Bear Hump, leased from the Blackfeet whose reservation is adjacent to Chinook (85). However, the company's use of tribal land is complicated, since the agreement was made with the state rather than the Blackfeet who are in the legal process of regaining the land, which had been wrongfully taken when "the terms [of the original treaty] were unilaterally changed by the U.S. Senate" (204). Through the settler state's legal mechanisms, Indigenous land is used, and often damaged, by the resource extraction industry that fuels the American way of life. And yet, as Chief Claire Merchant's battle for the title to the land suggests, questioning the legality of these projects may offer some restitution for Indigenous communities and may even disrupt resource extraction. Consequently, the attention to crime that is central to the hard-boiled mode becomes a tool through which King interrogates the intertwined processes of settler colonialism and resource extraction.

King considers the legality of resource extraction on Indigenous territory through the character Randall Boomper Austin, who is the primary suspect in the murder case. Austin, owner of Austin Resource Capital, is presumed to be in town because of his interest in Orion's resource analysis mapping technology. Austin is certainly interested in the technology, which he eventually buys from Orion Technologies. However, he is more interested in the leased land that comes with the purchase of Orion's technology. Late in the novel, it is revealed that Austin has discovered a deposit of the prized gemstone red beryl and is in Chinook to supervise a discrete mining operation. Austin's beryl mine exists in a legal grey zone. He can mine the beryl because he owns the lease to the land, despite the fact that it is soon to be legally returned to the Blackfeet Merchant. The ethics of his mining operation are also dubious, since he is excavating beryl at the same time he proposes to sign an agreement stating he would not draw water from the aquifer for bottling. In addition, he tells Merchant and Thumps he is

> even going one step further. . . . I'll keep the monitoring stations in play for now so I can recheck Orion's findings, but I'm going to begin to take them out, one by one, and return the land to its natural state.
>
> *(329)*

Austin answers the question of whether his actions are wrong when he states "guilty" after Thumps confronts him about the mine (448). What is more, Austin's confession and word choice shifts his actions from the realm of ethics to criminality, even if the mine is not a 'crime scene' by typical legal standards.

The crime of mining Bear Hump is further emphasised when its ecological impact is identified. On their way to investigating the mine, Blackfeet characters Moses and Cooley lament the diminishing tall native grasses that have been outgrown by invasive short grasses. At the mine, which Austin has hidden with sod, they see "six perfectly straight mounded swatches of new-growth short grass, thick lines drawn on the prairies" (291), which contribute to the destruction of this sensitive grassland ecology. By settler-colonial standards, where the damaging mines are associated with grand projects like open-pit oil sands mines or the mountaintop coal mines, the damage caused by Austin's mine is minor. However, for the Blackfeet, the disturbance to the grassland is both ecologically and culturally damaging because the Blackfeet have a traditional relationship with the plants and grassland ecosystem that Piikani scholar Betty Bastien describes as an "alliance" (82). For Jennifer Wenzel "the urgent task" of the environmental critic is "to understand its [literature's] role in calculating what counts as 'nature', 'environment', 'crisis', or even 'human'" (*The Disposition* 17). Because of the close yet often complicated relationship between ecological harm and settler laws and ethics, I would add to Wenzel's list, who or what counts as a victim, what counts as a crime, and to who? Following Wenzel's mode of inquiry, King's use of the hard-boiled mode, with its focus on "the failures of the legal and social order to protect decent people from elite predation and criminal abuse" (McCann 56), becomes a valuable way to articulate, from an Indigenous standpoint, the criminality of settler resource extraction projects.

If King uses the hard-boiled mode to raise questions about criminality, he must do so through an obligation to, and a subtle disruption of, the mode's conventions. As expected, King creates a crime for his detective to investigate with a cast of suspects and plot twists. In this conventional murder mystery, Austin is the primary suspect until a plot twist near the end of the novel reveals the murderer as Oliver Parrish: a serial killer and colleague of Lester's and Knight's who, through a series of slip ups, is at threat of being discovered by his business associates (424–5). Austin's role as the suspect is emphasised through red herrings spread throughout the novel. Thumps' friend, Archie Kousoulas, exclaims Austin "murdered two people!" (126). He continues, "Austin wants the technology. Lester and Knight won't sell it. Bam! He has them killed" (129). Then, on learning from Merchant that Austin may be interested in buying Orion's technology for the lease so he can bottle water from the aquifer, Thumps thinks: "There it was. A piece of the puzzle that hadn't been available until now. 'Boomper Austin'" (206). And finally, Thumps sums up the clues later in the novel: "Austin and/or Cruz [Austin's bodyguard] were the most likely suspects. Austin had a strong interest in Orion's technology" (306). These red herrings draw Thumps, and the reader, away from Parrish. The red herring, alongside the list of suspects and unexpected twists and turns results in a conventional murder mystery that makes the novel familiar to readers of genre fiction. However, King's formulaic use of conventions, specifically the red herring, also becomes a tool that shines attention on the crime of resource extraction. King subtly inverts the red herring, with the 'false lead' actually setting Thumps on the right path, albeit for a different crime. In doing so, King strikes a delicate balance between writing within the hard-boiled mode and giving weight to the subject of resource extraction on Indigenous land.

As a detective, Thumps plays a unique role in addressing the semi-legal, though highly unethical, crime of mining red beryl on reservation land. Through his careful attention to Austin's actions and to the sod-covered Bear Hump, Thumps builds a case and receives an admission of guilt from Austin in front of Merchant. Catching Austin in such a public manner ensures that he can no longer mine the stone undetected, but this case also gestures to the possibility of other ethically questionable dealings that Austin undertakes in his global oil and resource extraction businesses, dealings that are not investigated at this time and thus do not result in an admission of guilt. However, in this instance at least, Austin must work with Merchant to create an extraction "agreement that is mutually beneficial"

(448). King does not provide readers with details of the agreement, but the narrator's emphasis on Merchant's good mood due to "a spirited discussion with Boomper Austin about the price of red beryl and the cost of deceit" implies that the agreement is lucrative for the Blackfeet (452). Here, Thumps follows in the footsteps of the hard-boiled detectives who, Scaggs writes, "aim to restore and maintain the social order" (61). Thumps solves the crime, restores order for the Blackfeet, and wins over Merchant's attention. King writes a hopeful ending, with Merchant's successful business deal standing in contrast to the many times Indigenous communities have experienced extraction projects on their land without their consent or without receiving proper compensation.

Nevertheless, King's ending is not entirely positive when read in light of Walton's and Walton's suggestion that "Now, in an era of widescale environmental crisis, the detective's reassuring and restorative functions must, once again, be reconsidered" (2). Thumps is only able to help restore order within a settler-colonial framework, where Merchant's options are limited to Austin extracting the beryl for free or for a fee. In their discussions after Austin is exposed, Merchant does not indicate that she might oppose the mine. Rather, she only suggests that generating a "mutually beneficial" deal is going to require "more coffee" (448–49). Merchant is in an uncomfortable position that is familiar to many Indigenous leaders who are confronted with resource extraction projects. As Dina Gilio-Whitaker (Colville Confederated Tribes) states:

> Things get ethically complicated when Native nation governments willingly choose to engage in resource extraction – especially fossil fuels – given the environmental harm they cause, both in the extracting and in the production of climate-changing greenhouse gases. The need to escape poverty and assert sovereignty, weighed against cultural obligations to protect land, forces tribes into what can seem like an impossible double bind. It is a realm of difficult choices that exist beyond the binaries of black and white and right and wrong, necessitated by the unforgiving and unrelenting demands of capitalism.
>
> *(69)*

Given this harsh reality, Thumps' restoration of order is both comforting and troubling. It is comforting because it stops the crime against the Blackfeet and allows Merchant to strike a deal with Austin, but it is troubling because it does not stop the crime against land, plants, and culture. It also gestures to similar outcomes that occur in negotiations about oil extraction projects on Indigenous land, which allow petro-economies to continue. However, through this uncomfortable conclusion, more important questions are raised about resource extraction and criminality, particularly in relation to ideas such as justice and victimhood.

King's Blackfeet characters are stuck in a settler-colonial framework, in which extraction projects will continue to fuel the expressions of modernity, like cars and highways, that emerged in the twentieth century. In many ways, King's characters, to borrow LeMenager's phrase, "loathe to disentangle" themselves from oil, yet they also loathe to be entangled in it (7). They are haunted by the legacy of the oil-soaked American century, which simultaneously delivers the pleasures and antagonisms of petromobility and the financial reward and unevenly distributed environmental damage of resource extraction. It is through this rich treatment of the American century and his playful adaptation of the hard-boiled mode that King expose oil's continued influence in the early twenty-first century.

Despite oil's pervasiveness, however, King writes moments that resist and critique it at the levels of both theme and plot. There is the sense, within the novel's narrative and conventions, that stuck though we are, we need not be stuck in the future. Indeed, as Szeman writes, "looming in the background of any discussion of oil" – including *Cold Skies* – "are *finitude* and *the future*" ("Afterword" 392). Oil is both finite and unsustainable. Something will come next. For King, however, what comes next does not reside solely in the imaginative work of *Cold Skies*; it rests, instead, in the minds of readers and writers. In *The Truth About Stories*, King emphasises the reader's role in the storytelling process. In a refrain that ends each chapter's story, King states

> Take [the story]. It's yours. Do with it what you will. . . . But don't say in the years to come that you would have lived your life differently if only you had heard this story. You've heard it now. (29)

King's exhortation proposes that readers have the agency to take whatever it is they have learned or felt, if anything, from his stories and live accordingly. For King, then, the creative work of life is to grow and change with the help of good stories. It is urgent and empowering work, especially in this era of needed energy transition. Of course, the storyteller has the responsibility to tell what Cherokee writer and scholar Daniel Heath Justice calls "stories that heal" rather than "stories that wound" (1). It is a responsibility that King upholds through his decision to write *Cold Skies*, rather than to give up on genre fiction and return to his critically acclaimed literary and non-fiction writing after the self-proclaimed "bust" of the first DreadfulWater novel ("Another Interview"). In *Cold Skies*, King writes within and against the hard-boiled mode, exposing how the hard-boiled is invested in oil while attempting to rework it for political ends. In doing so, King demonstrates how the hard-boiled mode might reinvent itself in the early twenty-first century as a mode that heals rather than wounds. Fortunately, the hard-boiled, as *Cold Skies* reaffirms, is a malleable mode capable of doing the challenging and often uncomfortable work of confronting our complicated relationship with oil.

Bibliography

Andrews, Jennifer, and Priscilla L. Walton. "Revisioning the Dick: Reading Thomas King's Thumps Dreadful Water Mysteries." *Detecting Canada: Essays on Canadian Crime Fiction, Television, and Film*, edited by Jeannette Sloniowski and Marilyn Rose, Wilfrid Laurier UP, 2014, pp. 101–122.

Ashman, Nathan. "Hard-Boiled Ecologies: Ross Macdonald's Environmental Crime Fiction." *Green Letters*, vol. 22, no. 1, 2018, pp. 43–54, doi:10.1080/14688417.2018.1431139.

Bastien, Betty. *Blackfoot Ways of Knowing: The Worldview of the Siksikaitsitapi*. U Calgary P, 2004.

Bergthaller, Hannes, Christa Grewe-Volpp, and Sylvia Mayer. "Introduction." *Green Letters*, vol. 23, no. 2, 2019, pp. 125–129, doi:10.1080/14688417.2019.1654243.

Böhm, Steffen, Campbell Jones, Chris Land, and Mat Paterson. "Introduction: Impossibilities of Automobility." *Against Automobility*, edited by Steffen Böhm, et al., Blackwell Publishing, 2006, pp. 3–17.

Bradford, Richard. *Crime Fiction: A Very Short Introduction*. Oxford UP, 2015.

Breitbach, Julia. "Rewriting Genre Fiction: The Dreadful Water Mysteries." *Thomas King: Works and Impact*, edited by Eva Gruber, Camden House, 2012, pp. 84–97.

Cariou, Warren. "Petrography." *Warrencariou.com*, www.warrencariou.com/petrography.

Clarsen, Georgine, and Lorenzo Veracini. "Settler Colonial Automobilities: A Distinct Constellation of Automobile Cultures?" *History Compass*, vol. 10, no. 12, 2012, pp. 889–900, doi:10.1111/hic3.12015.

Daxell, Joanna. "The Native Detective à la King." *Beyond Comparison: Actes du 5ième colloque en littérature canadienne comparée, Université de Sherbrooke*, edited by Roxanne Rimstead et al., Topeda Hill, 2005, pp. 51–59.

Duckert, Lowell. "Coal/Oil." *The Cambridge Companion to Environmental Humanities*, edited by Jeffrey Cohen and Stephanie Foote, Cambridge UP, 2021, pp. 214–228.

Ghosh, Amitav. "Petrofiction: The Oil Encounter and the Novel." *New Republic*, vol. 206, no. 9, 1992, pp. 29–34.

Gillio-Whitaker, Dina. *As Long as the Grass Grows: The Indigenous Fight for Environmental Justice, from Colonization to Standing Rock*. Beacon Press, 2019.

Huber, Matthew T. *Lifeblood: Oil, Freedom, and the Forces of Capital*. U Minnesota P, 2013.

Justice, Daniel Heath. *Why Indigenous Literatures Matter*. Wilfrid Laurier UP, 2018.

King, Thomas. *The Truth About Stories: A Native Narrative*. Anansi, 2003.

———. "Another Interview with Thomas King." Interview by Jordan Wilson. *Canadian Literature*, 8 Oct. 2009, https://canlit.ca/another-interview-with-thomas-king-october-2009/.

———. *The Inconvenient Indian: A Curious Account of Native People in North America*. Anchor Canada, 2013.

———. *Cold Skies*. HarperCollins, 2018.

Laboucan-Massimo, Melina. "From Our Homelands to the Tar Sands." *Downstream: Reimagining Water*, edited by Dorothy Christian and Rita Wong, Wilfred Laurier UP, 2017, pp. 81–88.

LaDuke, Winona, and Deborah Cowen. "Beyond Wiindigo Infrastructure." *South Atlantic Quarterly*, vol. 119, no. 2, 2020, pp. 243–268, doi:10.1215/00382876-8177747.

LeMenager, Stephanie. *Living Oil: Petroleum Culture in the American Century*. Oxford UP, 2014.

Macdonald, Graeme. "'Monstrous Transformer': Petrofiction and World Literature." *Journal of Postcolonial Writing*, vol. 53, no. 3, 2017, pp. 289–302, doi:10.1080/17449855.2017.1337680.

McCann, Sean. "The Hard-boiled Novel." *The Cambridge Companion to American Crime Fiction*, edited by Catherine Ross Nickerson, Cambridge UP, 2010, pp. 42–57.

Orpana, Simon. *Gasoline Dreams: Waking up from Petroculture*. Fordham UP, 2021.

Pitt Scott, Harry. "Offshore Mysteries, Narrative Infrastructure: Oil, Noir, and the World-Ocean." *Humanities*, vol. 9, no. 3, 2020, doi:10.3390/h9030071.

Poll, Ryan. "The Rising Tide of Neoliberalism: Attica Locke's *Black Water Rising* and 'The new Jim Crow'." *Class and Culture in Crime Fiction: Essays on Works in English Since the 1970s*, edited by Julie H. Kim, McFarland, 2014, pp. 175–200.

Rodness, Roshaya. "Thomas King's National Literary Celebrity and the Cultural Ambassadorship of a Native Canadian Writer." *Canadian Literature*, no. 220, 2014, pp. 55–72, doi:10.14288/cl.v0i220.192604.

Scaggs, John. *Crime Fiction*. Routledge, 2005.

Stoecklein, Mary. *Native American Mystery Writing: Indigenous Investigations*. Lexington Books, 2019.

Szeman, Imre. "How to Know about Oil: Energy Epistemologies and Political Futures." *Journal of Canadian Studies/Revue d'études canadiennes*, vol. 47, no. 3, 2013, pp. 145–168.

———. "Afterword." *Fueling Culture: 101 Words for Energy and Environment*, edited by Imre Szeman et al., Fordham UP, 2017, pp. 389–394.

Vredenburg, Jason. *Motorcars and Magic Highways: The Automobile and Communication in Twentieth-Century American Literature and Film*. University of Illinois at Urbana-Champaign, PhD Dissertation, 2013.

Walton, Jo Lindsay, and Samantha Walton. "Introduction to Green Letters: Crime Fiction and Ecology." *Green Letters*, vol. 22, no. 1, 2018, pp. 2–6, doi:10.1080/14688417.2018.1484628.

Wenzel, Jennifer. "How to Read for Oil." *Resilience: A Journal of the Environmental Humanities*, vol. 1, no. 3, 2014, pp. 156–161, doi:10.5250/resilience.1.3.014.

———. *The Disposition of Nature: Environmental Crisis and World Literature*. Fordham UP, 2020.

Willmott, Glenn. "Oil Tragedy as Modern Genre." *Petrocultures: Oils, Politics, Culture*, edited by Sheena Wilson et al., McGill-Queen's UP, 2017, pp. 187–196.

Wilson, Sheena, Imre Szeman, and Adam Carlson. "On Petrocultures: Or, Why We Need to Understand Oil to Understand Everything Else." *Petrocultures: Oils, Politics, Culture*, edited by Sheena Wilson et al., McGill-Queen's UP, 2017, pp. 3–20.

Yaeger, Patricia. "Editor's Column: Literature in the Ages of Wood, Tallow, Coal, Whale Oil, Gasoline, Atomic Power, and Other Energy Sources." *PMLA/Publications of the Modern Language Association of America*, vol. 126, no. 2, 2011, pp. 305–326, doi:10.1632/pmla.2011.126.2.305.

30

"THE WHOLE WORLD . . . WAS A GIGANTIC PRISON"

Climate Crisis and Carceral Capitalism in Rachel Kushner's *The Mars Room*

Megan Cole Lyle

A few years after the twenty-something protagonist of Rachel Kushner's 2018 novel, *The Mars Room*, has been incarcerated for committing a violent crime, she begins to plot an elaborate escape into the California countryside. Inspired by her prison's exuberant GED tutor – a Henry David Thoreau acolyte who frequently espouses the healing potentialities of nature – protagonist Romy Hall grows consumed by dreams of breaking out of prison and making a new life among the trees. However, throughout the narrative, Romy realises that the salvific countryside of her fantasies is conspicuously and rapidly vanishing. Beyond the walls of the semi-rural prison Romy inhabits, there is nothing left resembling a pastoral paradise – instead, there is only the bleak reality of the Central Californian landscape in the early aughts. Where once lay lush expanses of nature immortalised by the likes of John Muir, there now remains only a "no-man's land" marred by thick smog, industrial agriculture, and oil derricks as far as the eye can see (Kushner 20).

As the novel progresses, the entanglements between carceral capitalism and environmental declension become increasingly explicit. In chapters interwoven with the primary storyline, Kushner incorporates direct excerpts from two idiosyncratic intertexts: the environmentally focused portions of "Unabomber" Ted Kaczynski's manifesto and an array of strikingly parallel sentiments from Thoreau's *Walden*. Both texts, Kushner suggests, famously characterise nature as a site of retreat, regeneration, and refuge from industrial modernity and its concomitant social ills. However, by juxtaposing these nature-reverent passages with stark descriptions of rural America as it stands in the early twenty-first century – drought-stricken, highway-riddled, oil-suffused, and studded with prisons – the novel highlights the hollowness of Thoreau's and Kaczynski's idealisations of the natural world. The petroleum-soaked reality of the Golden State's contemporary landscape haunts these canonical articulations of nature-as-refuge. As a result, throughout *The Mars Room*, Kushner suggests that the omnipresence of carceral petro-capitalism has eliminated nature as a marginal space of possibility altogether, foreclosing certain forms of environmentalist and abolitionist agency and leaving Romy and her milieu mired in cruel optimism. In her zealous conviction that the natural world might liberate her from incarceration, Romy sublimates the realisation that the natural environment itself has become a casualty of carceral petro-capitalism and is thus highly unlikely to save her the way it "saved" Thoreau and Kaczynski.

Through *The Mars Room*'s extended engagement with avatars of the primarily white and male-dominated environmental literary tradition in the United States, Kushner suggests that the popular

DOI: 10.4324/9781003091912-36

notion of nature as a liberation – as the "free" alternative to the carceral state, industrialism, and myriad other social ills – demands radical revision in the age of petro-capitalism. I contend that, throughout the novel, Kushner complicates the enduring masculinist American fantasy of nature-as-freedom by arguing that anthropogenic environmental destruction and the carceral state are both constitutive parts of the same capitalist matrix and to escape from one arena of domination is merely to enter another. Drawing upon Ruth Wilson Gilmore's and Nicole Fleetwood's meditations on the geographical and ecological dimensions of the prison-industrial complex, I argue that contemporary "crime petro-fictions", like *The Mars Room*, demonstrate the ideological limitations of the American environmental literary tradition. Such novels illuminate the striking convergences between carceral and ecological disaster, interrogating the efficacy of agency and escape in a world circumscribed by carceral petro-capitalism. Finally, I suggest that "crime petro-fictions" provide a unique lens through which to examine the twinned "slow violence" of carceral capitalism and of the climate crisis. Given that American prisons are increasingly rural, incarcerated people are uniquely positioned to witness the deterioration of "wilderness" spaces and to notice that sites of "pristine refuge" are now often sites of incarceration, industrialisation, and ecological degradation. Although such sites of socio-ecological violence are often shunted out of public view, Kushner demonstrates that they remain perfectly legible to the incarcerated people who are forced to inhabit them.

"Ecophilia" and the Politics of Pastoral Escape

As the field of ecocriticism has broadly demonstrated, ascribing a stable set of ideological meanings to the American natural landscape is an inherently impossible task. For ecocritic Lawrence Buell, the American idea of nature has become a "tangled ideological palimpsest" over the course of several centuries, an accruement of affective association that are multifaceted and even contradictory (5). Nature is a slippery referent that has alternately signified freedom, conquest, horror, riches, beauty, hope, hatred, and all permutations thereof. As Raymond Williams famously put it, "the idea of nature contains, though often unnoticed, an extraordinary amount of human history" (67). That is, nature signifies precisely what rhetors wish that it might signify, such that "the idea of nature is the idea of man" (Williams 71).

Despite its multiple valences, nature-as-freedom has long been among the most enduring – if most complicated – tropes across the American environmental literary tradition. Matthew A. Taylor has named this relatively romantic approach to the natural world "ecophilia" and has called it the dominant environmental affect of the post-industrial age. For Taylor, an "ecophilic" orientation is characterised by the idealistic reverence of nature as a sublime source of self-realisation. Nature, for "ecophilic" writers, exists as an unblemished space that is both extricable from societal issues and somehow capable of solving them.

In excerpts starkly juxtaposed against the otherwise carceral backdrop of Kushner's *The Mars Room*, ecophilia broadly characterises the affective orientations of both Thoreau and Kaczynski. In the mid-nineteenth century, following the publication of "Civil Disobedience" and *Walden*, Thoreau first began to weave together the strange American constellation of incarceration, isolation, individualism, and the natural environment – a constellation that inaugurated "the notion of androcentric pastoral escape" as "the great tradition within American literary naturism", per Buell (25). Thoreau, who went to prison for refusing to pay taxes to a federal government supporting slavery, extolled the virtues of living a relatively isolated life in nature, where one might be fundamentally insulated from the ills of society. Thoreau famously declared that nature constituted a realm apart from slavery and social inequality, avowing "that Nature has been partner to no Missouri Compromise" (*Slavery* 108). Taking flowers as metonyms for nature writ large, Thoreau once wrote:

> I scent no compromise in the fragrance of the water-lily . . . In it, the sweet, and pure, and innocent are wholly sundered from the obscene and baleful . . . The foul slime stands for the sloth

> and vice of man, the decay of humanity; the fragrant flower that springs from it, for the purity and courage which are immortal.
>
> *(108–109)*

For Thoreau, nature represented the last bastion of refuge – and autonomy – in a rapidly industrialising, unjust world. The natural landscape, he insisted throughout his life, was innately separable from "the decay of humanity" in its myriad manifestations.

More than a century later, between 1978 and 1995, wilderness recluse Ted Kaczynski continued this Thoreauvian dedication to nature in his own warped fashion, bombing 26 people whom he believed to be complicit in the industrial desecration of the environment. However, as Kaczynski notes in his diary, belying his superficial concern for environmental preservation was the desire to exact "personal revenge on those who deprive or threaten to deprive [his] own autonomy" (Kushner 275). The precise motives of Thoreau and Kaczynski diverge significantly, but Kushner implies throughout *The Mars Room* that – in the simplest terms – both men went to prison for their dedication to the natural environment, their commitment to self-reliance, and their deep resentment towards an industrialising social order with which they vehemently disagreed. To varying extents, their "ecophilia" led them both to abandon society for the solace offered by nature.

If *The Mars Room* seems an odd locus for intertexts by the Unabomber and the ur-Transcendentalist, it is because critical reception of the novel has focused primarily on its carceral politics, glossing entirely over its ecological undercurrents. If they have addressed *The Mars Room*'s environmental entanglements at all, most critics have framed Kushner's invocation of Thoreau's and Kaczynski's environmentalism as a persistent reminder that nature represents a world beyond carceral capitalism, and that, therefore, there is hope for those trapped within its brutal matrix. One critical interpretation by Amanda Holmes Duffy insists that "the strip club is not the real world, San Francisco is not the real world, and the prison system is not the real world" – but nature *is*, in fact, the real world. Another critic, Jo Livingstone, argues that through Romy's brief interactions with the natural environment, she is able to "sense the absolute difference between being locked up and being free", and therefore to maintain some level of optimism. However, I argue that *The Mars Room*'s extended meditation on Thoreau and Kaczynski suggests just the opposite: that the notion of nature as reprieve – as the "free", "real world" foil to the carceral state, industrialism, and infinite other social ills – demands radical revision. I suggest that critics of *The Mars Room* have fundamentally overlooked the novel's representation of carceral logic as spatially inescapable. Carceral logics cannot be circumvented by escaping the bounds of surveillance through the natural world, Kushner demonstrates, since the transcendence of social ills through nature has always been a mere illusion, foreclosed to all but the most privileged.

Carceral Omnipresence and the Limits of Nature-as-Salvation

As *The Mars Room* begins, Kushner's cast of women incarcerated in California's Central Valley – either a lush agricultural paradise or a stark industrial landscape, depending on perspective – briefly toy with "ecophilia", dreaming half-heartedly of a semi-pastoral escape into the vast almond orchards lining the prison yard, a feat in which Romy eventually (if only temporarily) succeeds. Shackled into a bus on the way to prison, Romy muses:

> In our closeness to the scumbly land beyond the meshed window, I longed for reality to twist itself like a bag and tear a hole from the twisting, rupture the bag and let me out, release me into that no-man's land.
>
> *(19–20)*

Her dreams provide a brief respite from the bleak dreariness of the bus ride and from her dire circumstances in general – Romy, a young mother and former sex worker, is serving more than two consecutive life sentences for murdering her stalker, and thus craves all the positive distractions she can get. However, she quickly brings herself back down to Earth: "I looked out the window and saw nothing but nature's carpet of rocks and shrubs darting past in an endless bumpy scroll", she laments (20). Indeed, outside the window, "there wasn't much to look at" beyond the omnipresent evidence of a desolate, car-centred landscape – "underpasses and on-ramps, dark, deserted boulevards", an empty world wherein "no one was on the street" (3). From the land in which she fleetingly seeks salvation from prison's realities – "for reality to twist itself like a bag and tear a hole from the twisting" – she finds only monotonous, "endless" desolation. As the bus makes its way to the Stanville Women's Correctional Facility, landscapes of hope begin to falter, becoming landscapes of petroleum-soaked desolation.

If the road to prison is bleak, the prison itself is a nightmare. Stanville, the fictional Central California city that houses Romy's correctional facility, lies at the nexus of compounded social and environmental crises. The city, "synonymous with its prison", not only boasts the "highest percentage of minimum-wage workers in the state", but myriad ecological ills: the "water is poisoned", the "air there is bad", and temperatures frequently reach the triple digits Fahrenheit (91). The "burnt smell of synthetic fertilizer" permeates the air, as the prison lies in the middle of "big agriculture" country, with almond groves as far as the eye can see (92). Fossil-fuelled, water-intensive industrial agriculture – and its incompatibility with human life – is a constant motif. As the bus draws closer to the prison, Romy muses that "we were surrounded by agriculture. I saw no human beings working in the fields. The fields were abandoned to machines" (29). The fog that constitutes the atmosphere "wasn't even real fog. It was haze from crop dusters that sprayed the almond fields surrounding the prison" (221). As fervently as Romy and her fellow incarcerated women attempt to romanticise the landscape – to find some pastoral paradise into which they might mentally retreat – it is impossible to ignore the omnipresent signifiers of petro-capitalism and its dire ecological impacts, both in the environment beyond the prison and finally within the prison itself.

After their demoralising bus trip to the correctional facility, the women are reintroduced to ecophilia when they meet the prison's GED tutor, Gordon Hauser. A recent hire at Stanville Women's Correctional Facility, Gordon abandoned his graduate studies in literature just before completing his dissertation on "Thoreau's image of a spiritual molting season, of a new man, the fateful concept of an American Adam" (52–53). In a mock-Thoreauvian gesture, Hauser lives alone in a cabin on the rural outskirts of the prison town, committing himself to personal philosophies of "reverence of nature" and "self-reliance" (92). He is a naïve, starry-eyed optimist who often attempts to proselytise the healing power of literature and nature to his incarcerated students, including Romy, to whom he becomes close over the course of the novel. To himself and to his students, Gordon often lauds the hyper-autonomous Thoreau and philosophises about his muddled ideological throughlines with Kaczynski. He compares the former's resentment of "the mass of men [who] lead lives of quiet desperation" to the latter's desire, "ideally, to break down all society into very small, completely autonomous units" through "the destruction of the worldwide industrial system" ("Excerpts" 16). Although Gordon condemns Kaczynski's violent tactics and strives to separate his radical ideological positions from his those espoused by Thoreau, his Transcendentalist hero, Gordon does occasionally sympathise with Kaczynski's stated desire to escape from societal malaise by immersing himself in nature. Gordon implicitly encourages his incarcerated students to do the same in a misguided attempt to relieve them of the agonies and indignities of life in prison.

However, as a free white man, Gordon struggles to comprehend that these aspects of a deeply American ethos – the volatile and oft-contradictory blend of radical autonomy, nature-worship, and anti-industrialism – are inaccessible to most of his students. The Stanville Women's Correctional Facility is composed mainly of incarcerated women and trans men, many of whom are mothers, sex workers, young, pregnant, Black, Brown, Indigenous, or some intersection thereof. Despite its heavy presence throughout the novel, Gordon's Thoreauvian fantasy of being "reborn unfettered and sinless", with nature and individualism as midwives, has always been denied to them (Kushner 53). Though Romy and her milieu attempt to liberate themselves through ecophilia, as the narrative progresses, they settle into the realisation that ecophilia was never designed for them.

Despite the increasingly apparent futility of providing "ecophilic" hope to Stanville's incarcerated women, Gordon encourages his imprisoned literature pupils to seek solace and self-discovery in the natural world, an orientation that his own literary education has instilled in him. If his students could only "learn to think well, to enjoy reading books" about nature and liberation, then "some part of them would be uncaged", Gordon insists to himself (49). While teaching John Steinbeck's *The Red Pony* to his GED class, Hauser notices the women comparing

> the mountains in the book and the ones they could see from main yard. They seemed afraid of the mountains, which surprised Gordon. He figured they'd regard the mountains as freedom, the one thing they could glimpse of the natural world.
>
> *(187)*

Gordon is confused by their indifference to the idea of nature as representative of "freedom" beyond the prison walls, ignoring the fact that for them, the distinction between "free" and "unfree", or "within" and "beyond", is tenuous at best. The women's deprivileged positionalities allow them to see what Gordon cannot – that the outside world is merely an extension of their carceral surroundings and that nothing is untouched by the carceral logics that have entrapped them. As Nicole Fleetwood explains, the work of incarcerated artists, writers, and thinkers has long "challenge[d] the 'inside' and 'outside' logic of carcerality" and illuminated the carceral logics suffusing institutions and spaces beyond the prison proper (6). Further, as Dylan Rodríguez puts it, "the emergence of imprisonment as a central 'constitutive logic' of the American social/racial formation" means that the binarism of "inside" and "outside", as well as "free" and "unfree", has become much less tenable for racialised and gendered subjects in the twenty-first century than it has ever been for the likes of Gordon or his white, male literary idols (39).

Just as carceral studies scholars have long maintained that the delineation between "inside" prison and "outside" prison is rapidly vanishing, ecocritics have suggested that the distinction between nature and non-nature is disappearing at a similarly alarming rate – if it is not already gone. By the turn of the twenty-first century, when the events of *The Mars Room* take place, any semblance of "pure, untouched" American wilderness – already a specious concept deeply constituted by colonial logics[1] – had been compromised by rampant ecological exploitation. As Bill McKibben famously put it in his 1989 book, *The End of Nature*, in the post-industrial era, there remains no real distinction between "pure" nature and nature transformed by human intervention. Since the advent of fossil fuel combustion, "we have changed the atmosphere, and thus we are changing the weather" (McKibben 58). Further, McKibben argues that "by changing the weather, we make every spot on earth man-made and artificial. We have deprived nature of its independence, and that is fatal to its meaning. Nature's independence is its meaning; without it there is nothing but us" (58). He laments, "we have ended the thing that has, at least in modern times, defined nature for us – its separation from human society" (64). As there is no "inside" and "outside" with relation to carceral capitalism, there is no "inside" and "outside" with relation to anthropogenic environmental crisis. Just as the whole world has been touched by carcerality, the whole world has been touched by "petro-modernity" and its concomitant ecological effects. Thus, there is no unsullied place toward which to escape from either carceral capitalism or ecological collapse – which are themselves increasingly intertwined, as the environmental neglect and devastation surrounding Stanville suggests within the novel.

The only incarcerated woman in *The Mars Room* who earnestly believes in the separability of carceral logic from the natural world is the widely reviled white supremacist inmate, Laura Lipp. To her cellmates' chagrin, she often reminisces aloud about her pastoral hometown, Apple Valley, California – a "wonderful place" where "you can practically smell the apple blossoms and hear the honeybees" (6). Lipp, recounting her origins to her unenthused cellmates, explains,

> "The town's got history though," she said, carefully moving away, out of fist-swinging range, just in case. "It was a fine place and it's gone downhill. We used to be cowboy country. All the country and western people came because of Roy Rogers. He had a museum with a big display of all his fishing lures. He owned the Apple Valley Inn. My father took us there for Sunday dinners. It was a carefree time. There weren't problems like now".
>
> *(159)*

Of course, such an idyllic, problem-free, "cowboy country" could exist only in the mind of someone like Lipp, for whom it is possible to believe in the existence (if only nostalgic or theoretical) of a "fine place" imbued with natural splendours beyond the purview of state surveillance, poverty, incarceration, and exploitation. For Lipp's cellmates, there remains no hope of an untainted, natural "outside" that is not intricately intertwined with carceral logics and the extractive, capitalist imperatives that underpin it.

As McKibben explains, even if such "natural purity" had ever been a viable concept, it has certainly been compromised by the age of scarcely checked oil spills, unprecedented levels of greenhouse gas emissions, and the myriad natural disasters they catalysed globally. Lipp and Gordon may be unable to recognise this, but the majority of the inmates in Stanville Women's Correctional Facility cannot help but notice it everywhere they look. While en route to the prison, for instance, Romy registers the sheer scale of the Capitalocene[2] and is struck by the indistinguishability between natural beauty and industrial disaster when she notices an alarming dark mass on the horizon:

> Somewhere deep in the Central Valley, the sky still dark, I looked out the window and saw two massive black shadows looming up ahead. They looked like dark oily geysers fluming upward on the side of the highway. What terrible thing was spewing into the sky like that, filling it with soot? They were huge black clouds of smoke or poison.
>
> I had read about a gas leak, about pounds of pollution issuing into the sky in Fresno or someplace. When gaseous quantities are measured in pounds you know there's trouble.
>
> Maybe this was some kind of environmental disaster, crude oil that had burst its underground pipe, or something too sinister for explanations, a fire burning black instead of orange.
>
> As our sheriff's department bus approached the giant black geysers, I got a close-up glimpse.
>
> They were the silhouettes of eucalyptus trees in the dark.
>
> Not an emergency. Not the apocalypse. Just trees.
>
> *(28–29)*

For Romy, instead of representing a pastoral escape, the eucalyptus trees beyond the prison bus's windows resemble a "terrible. . . . Disaster", "too sinister for explanations", and remain firmly ingrained within the same matrix of capitalist brutality that has engendered the structural conditions for mass incarceration. Further, these "black shadows of the apocalypse" seemed "dusty and sad . . . collect[ing] dust year after year, load[ing] up with dirt and car exhaust" (30). Capital's brutal regimentation and subsumption of nature appears, dressed in carceral rhetoric, everywhere Romy turns. It appears when Romy passes "a truck that held many rows of stacked metal cages" containing "turkeys, crammed so that they had to bow their long necks", the wind "pulling out their features, which flossed the highway in white flecks" (31). It appears once more when Romy remembers an anecdote about how, after early farmers accumulated too many sheep to keep track of them by name, they invented the concept of counting, turning their flocks from pastoral set pieces to calculable agricultural commodities. She ties the idea explicitly to her experience as a faceless and fungible inmate: the experience, for the sheep, was "like prison, from a name to a number", Romy muses (113). Again and again, nature is appropriated in the name of carceral capitalism, which contributes directly to environmental declension.

Even Gordon's Thoreauvian reveries are eventually pierced by the realities of industrial agriculture and its eerily carceral aesthetics. From his cabin in the woods, Gordon waxes poetic about "his solitary life, his acquired taste for valley beauty", but admits that his cherished vistas are, "to the untrained eye, not beautiful" (217). He deals with a water supply constantly poisoned by uranium and with foxes and other animals who veer uncomfortably close to his property as a result of being forced out of their natural habitats by industrial agriculture and highway construction. The oak trees surrounding his cabin are "drought-desiccated . . . their jagged little leaves coated in dust" (183). In the distance, the landscape is marred by "oil pipeline and derricks", the air is sweltering and "heavy with the smell of fertilizer", and the evidence of big agriculture is everywhere, "just power lines and almond groves in huge geometric parcels all the way to the prison" (184). In a moment of clarity, he explains that the valley beyond his door

> was a brutal, flat, machined landscape, with a strange lemonade light, thick with drifting topsoil and other pollutants from farm equipment and oil refineries. It was a man-made hell on earth but then again a real valley, with mountain ranges on either side. It was the size of industrial agriculture, scaled for that. It was difficult to imagine what it had looked like before it was farmed. It was hard even to imagine what it had looked like farmed in the old-fashioned way, by people. Machines shook the almond trees in synchronous violence.
>
> *(217)*

Not unlike the prison nestled within it, the farmland is bureaucratised, polluted, regimented, and fraught with "violence" – a "man-made hell on earth" inextricably caught in the gears of industrial production, Gordon admits, before attempting to resurrect the pastoral fantasies he has internalised through years of literary study ("but then again a real valley, with mountain ranges on either side", he demurs).

Romy later echoes similarly bleak sentiments, but jettisons any of Gordon's tendencies to romanticise. After finally – and, it turns out, very briefly – succeeding in her plan to escape from the prison into the almond orchards surrounding it, Romy observes,

> Everything was straight here. I was inside a giant grid; empty of people, but made by people. The whole world, at least this one, the Central Valley, from the mountains to the western horizon, was a gigantic prison. Orchards and power lines instead of razor wire and gun towers. Unmanned, and man-made.
>
> *(316)*

Romy articulates what Gordon can only tentatively intimate: that the ecophilic fantasy of nature-as-salvation, of some beautiful and ideologically vacant nature-beyond-culture, beyond-prison, has been subsumed by carceral logics and is susceptible to the same bureaucratic desolation as the prison itself. There is no clean "inside" and "outside" delineation between nature and prison, or between the "real world" and the carceral world. They are all intertwined in the same ideological matrices reflected in the writings of both Thoreau and Kaczynski – which, despite their apparent dissimilarities, rest on the same libertarian foundation of radical individualism, autonomy, and anti-sociality.

Critics of *The Mars Room* have assumed, with Gordon, that Thoreau and Kaczynski represent opposite ideological poles or at least that Kaczynski's resemblance to Thoreau is merely pale, perverse, and extreme. In her review of the novel, Abigail Deutsch interrogates the line between the figures and asserts that their juxtaposition points toward "how ill-equipped we are to determine who is breaking rules – to say nothing of what, precisely, the rules should be". Gordon is particularly "ill-equipped", in this sense, to the point of refusing to acknowledge any similarities between the figures: "True both lived in one room huts", Gordon writes at one point to an old friend from graduate school, "but I don't see much connection between them" (92). Gordon wonders again, later, "Why was Thoreau Thoreau, while Ted Kaczynski was Ted?" (272). He refuses to consider that their ideological similarities are multitudinous, their dispositions separated not by kind but by degree. Kushner herself acknowledges that the art of James Benning – whose works explicitly compare the two figures, "making discernable a multitude of contacts between Thoreau and Kaczynski's beliefs, political viewpoints, and experiences of seclusion" (Ault 1) – "directly inspired [her] thinking on Henry David Thoreau and Ted Kaczynski" in the novel (Kushner 338). Thoreau's and Kaczynski's ideologies assume the same gendered and racialised assumptions of nature as a site for individualist development and self-possession and build upon a set of American convictions intrinsically tied to extraction, autonomy, and – ironically, for Kaczynski – the very same ideological bedrock that promotes the industrial exploitation he claimed to despise. Thoreau famously devoted himself specifically to "deliberate" living, often focusing his energies inward rather than into the social sphere. In "Civil Disobedience", for instance, Thoreau objects primarily not to the social evil of slavery itself, but to the injunction to pay taxes toward institutions that he does not support. "It is not a man's duty, as a matter of course, to devote himself to the eradication of any, even the most enormous wrong", Thoreau insists, "but it is his duty, at least to wash his hands of it, and, if he gives it no thought longer, not to give it practically his support" (393). As Michael Meyer pithily put it in an introduction to Thoreau's canonical works, Thoreau "pledged allegiance not to the Republic but to the individualism for which he stood" (7). As Benning – and Kushner – illustrate, the leap from Thoreau's "humble pursuits of . . . autonomy in nature and for freedom from institutionalised power, two tenets that resonate through the core of American individualism", to Kaczynski's "horrific insurgence" based upon the same principles, is not outrageously far (Benning). As Kushner highlights in an excerpt from Kaczynski's diaries, his mass violence in the name of anti-industrialist, natural preservation was, in his own words, motivated solely by "personal revenge on those who deprive or threaten to deprive [his] own autonomy", explicitly tying natural preservation to radical individualism (275).

Gordon, the novel's mouthpiece for ecophilia in the Thoreauvian and Kaczynskian vein, illustrates how intimately their brand of naturism is intertwined with the sort of individualism and anti-sociality that cannot benefit his incarcerated pupils. While stargazing outside of his cabin one night, he notes that

> the way the stars flickered unevenly, from bright to dimmer to brighter, reminded him of car headlights. A car at night, moving along a tree-lined road, lights shining intermittently. But stars were wondrous, and headlights could be sinister. Stars were nature. Cars were unknown human intent.
>
> *(95)*

Gordon positions "human intent" as the explicitly toxic element corrupting an otherwise "wondrous" tableau, rejecting the social sphere in favour of total self-possession. People are the enemy, he maintains. Nature is their innocent victim.

However, for Romy and the women incarcerated with her, it is impossible to see nature as an extra-social space, evinced in part by Romy's failure to successfully escape into pastoral solitude at the novel's climactic conclusion. Using a prison-yard brawl as cover, Romy sneaks through an opening in the fence and tears into the almond orchards alone, into a natural space "where people are gone" and "the night falls upward, black and unmanned" (334). However, she is almost immediately apprehended and reminded of the impossibility of "escaping" the social sphere and capitalist, carceral logic into the "safe place" represented by trees and solitude (333). For the chronically surveilled, policed, and incarcerated – an increasingly large segment of the population in the age of late capitalism – Kushner suggests that there remains no viable hope of escaping into some secluded spot in the woods. As William Cronon puts it, the Romantic idea of wilderness as harbouring a "flight from history. . . . an escape from responsibility, the illusion that we can somehow wipe clean the slate of our past and return to the tabula rasa that supposedly existed before we began to leave our marks on the world" is, at best, a false hope (80). Promoting such hope serves only to reproduce cultural forms that structurally reinforce mass incarceration, by advocating for individualisation and escape rather than communal revolution.

Echoing this resistance to "escape" narratives, Ruth Wilson Gilmore argues that any effective resistance to mass incarceration must trouble the apparent borders of "inside" and "outside", urban and rural, and institutional and beyond. She explains that the "breadth" of carceral logic "belies the common view that prisons sit on the edge – at the margins of social spaces, economic regions, political territories, and fights for rights" (11). Further, "the government-organised and -funded dispersal of marginalised people from urban to rural locations suggests both that problems stretch across space in a connected way and that arenas for activism are less segregated than they seem" (ibid.). Rural landscapes are not only included within the American carceral matrix, Gilmore insists; they are the very loci of this matrix. There are no geographical or symbolic exemptions to capitalism or to the carceral regime, as Gordon's preoccupation with Thoreau suggests; rather, any remaining "natural" spaces have been co-opted, ensuring that, as Romy puts it, "the whole world . . . [is] a gigantic prison" (316).

"Slow Violence" and Revolutionary Futures

To suggest that the nature-as-escape paradigm demands revision, or that carcerality is spatially inescapable, is not to suggest that dismantling the prison-industrial complex is an impossible endeavour or that environmental affects have no part to play in liberation. Rather, as Nicole Fleetwood details through her examination of the art of incarcerated artist Todd (Hyung-Rae) Tarselli, nature can sustain hope through the invocation of alternate spatiotemporalities but cannot by itself dissolve carceral logic. Tarselli, who has spent years in solitary confinement, collects fallen maple leaves within the prison yard and paints photorealistic animals on them: "[scenes] from nature painted on a piece of nature created in a prison cell" (7). Fleetwood explains that "the life of the leaf as organic matter is part of the art, with holes on the surface from decomposing" (ibid.). Tarselli does not envision nature as true salvation or escape but uses its aesthetics as a means to "materialize the conditions out of which prison art emerges: penal space, penal matter, and penal time" (ibid.). The dissolution of the leaves on which he paints reminds viewers of the oppressive temporality of the carceral system, underscoring the fact that "natural" freedom remains foreclosed to incarcerated people. Still, connection to the natural world remains a revolutionary act, by virtue of being forbidden. Tarselli's nature-based artistic practice parallels a scene in *The Mars Room*, in which Gordon smuggles California poppy seeds into the prison for a student who gardens to pass the time. When the student first receives the seeds, she begins to cry, thanking Gordon profusely for the "God shot" (190). Gordon explains, "She'd replanted it, watered it. She told him she watched the sky and waited for birds to excrete seeds, and germinated them secretly in wet paper towels. Rules were such that no plants were supposed to grow", but her rebellious gardening practice sustains her (190). Environmentalism and nature-based art will not liberate her, of course – gardening in the prison is still gardening *in the prison*. However, developing close connections with nature – not in the idealised, Thoreauvian vein but in its blunt reality – might help her cultivate a revolutionary consciousness, like Tarselli's, that she may not have accessed without it.

Indeed, by the end of the novel, Romy and her fellow incarcerated women have begun to comprehend the massive scope of carceral cruelty and industrial environmental destruction – a violence "too sinister for explanations" – both of which eclipse the kind of interpersonal criminal violence they have committed themselves, often in self-defence or to secure basic needs. In her cell, Romy meditates on the difference between the kind of violence that landed her in prison and the kind of social and environmental violence ravaging the world beyond the correctional facility's walls:

> The word violence was depleted and generic from overuse and yet it still had power, still meant something, but multiple things. There were stark acts of it: beating a person to death. And there were more abstract forms, depriving people of jobs, safe housing, adequate schools. There were large-scale acts of it, the deaths of tens of thousands of Iraqi civilians in a single year, for a specious war of lies and bungling, a war that might have to end.
>
> *(265)*

Still, "according to prosecutors, the real monsters" were inmates like Romy and her cellmates. Romy muses, "In the primitive part of the mind, violence was body-to-body, punching and clubbing and cutting. Those people went to prison. Were not offered any kind of mercy" (265). However, the multinational corporations poisoning the air and water in Stanville and beyond, funding interminable wars, and manufacturing climate crises are pardoned – almost without exception – by virtue of their wealth and anonymity. Only after witnessing the scope and severity of environmental and carceral violence in Stanville does Romy understand why her individual act of "violence" was criminalised with two life sentences even though it pales in comparison to the structural violence being committed all around her.

In Rob Nixon's parlance, Romy is articulating the particular cruelty of "slow violence", a spatiotemporally distributed violence that does much more damage than Romy or her fellow incarcerated women have ever done. Slow violence, for Nixon, is "a violence that occurs gradually and out of sight, a violence of delayed destruction that is dispersed across time and space, an attritional violence that is typically not viewed as violence at all" (2). As opposed to the "spectacular", "instantaneous" violence for which people are imprisoned, slow violence is "incremental and accretive, its calamitous repercussions playing out across a range of temporal scales" (Nixon 2). Though Nixon applies the concept of slow violence primarily to climate change and environmental degradation, it is equally applicable to the carceral system, which, as Angela Davis has written, is similarly invisible: "The prison is one of the most important features of our image environment", she explains, and "this has caused us to take the existence of prisons for granted" (18). Whether in relation to carceral capitalism or the climate crisis, slow violence is notoriously difficult to represent and therefore to address. Nixon asks,

> How can we convert into image and narrative the disasters that are slow moving and long in the making, disasters that are anonymous and that star nobody, disasters that are attritional and of indifferent interest to the sensation-driven technologies of our image world?
>
> *(3)*

In answer, Kushner – and fellow authors of "crime petro-fictions" – suggest that incarcerated people might possess a uniquely sharp ability to apprehend and represent both environmental and carceral slow violence.

Slow violence is perpetuated because it often occurs out of sight, at a distance. Indeed, Nixon blames "capitalism's innate tendency to abstract in order to extract, intensifying the distancing mechanisms that make the sources of environmental violence harder to track" (41). However, incarcerated people are first-hand witnesses not only to "abstracted" carceral violence, but to "abstracted" environmental violence, given that – according to the U.S. Census Bureau – "a disproportionate share of inmates and prisons are located in rural areas" across the country (Porter et al 1). These rural areas, like Kushner's fictionalised Stanville, exist within the country's depopulated margins. These marginal spaces house not only unsightly prisons, but landfills, sites of industrial pollution and toxic waste, big agriculture, and the myriad other unpleasant by-products of "petro-modernity". America's "untouched" Thoreauvian landscapes – previously assumed to be sites of "pure nature" – are, in the twenty-first century, home to prisons and pollution. Thus, novels set in prisons – themselves often located in geographies most heavily ravaged by the effects of climate change, at the rural margins of U.S. society – and narrated by incarcerated people are uniquely positioned to testify that the "rural margins" are no longer some natural paradises into which anyone might escape – instead, they are sites of environmental and carceral "slow violence" that remain invisible to Americans who can afford to live elsewhere.

As fervently as Stanville's incarcerated women try to adopt ecophilia and hope for some natural paradise into which to escape, Romy and her milieu ultimately realise that the structures of carceral petro-capitalism are so extensive that environmental disaster *and* prison are everywhere. Individual escape á la Thoreau and Kaczynski has always been foreclosed to them. This realisation – and the solidarity and revolutionary fervour it inspires – may be the only thing to save them in the end. Alone, Romy fails to escape from the prison – but in community, the novel offers glimmers of insurrectionary optimism.

If there is indeed any hope to be found in the natural world, Kushner suggests, through her frequent allusions to Thoreau and Kaczynski, it cannot be autonomous, hyper-individualist, and antisocial, but must be rooted firmly in the social landscape. Within and beyond the prison context, isolation is a tool of oppression and domination. Kushner underscores this, illustrating that social ties – to her fellow inmates and to her young son, Jackson – are the only ties that truly sustain Romy throughout the novel. In one of the novel's most optimistic scenes, Romy recounts a memory from when Jackson was 5 years old, at the park with herself and her mother, watching autumn leaves fall off the trees. Romy and her mother express their melancholy at the changing seasons, but Jackson is "enchanted" at the convergence of nature, sociality, and cyclical futurity represented by the leaves:

> "All that beauty and for nothing," my mother said. "They'll fall off tomorrow."
>
> "But after they fall off," Jackson said, "the tree will grow new leaves, Grandma, and then those will turn colors, like these ones." It would keep happening over and over, Jackson said, through all the years. The leaves falling off meant new ones were coming. My mother looked at him like she wondered what planet he was from.
>
> *(175)*

Romy explains that the foundation of Jackson's optimism is that, in his mind, "there were always people. People taking care of other people" (176). Within the climate change-ravaged, carceral landscape of the twenty-first century, this social hope – and not the Thoreauvian or Kaczynskian fantasy of some rebellious escape into nature and solitude – may constitute all the hope in the world.

Notes

1. See Dina Gilio-Whitaker's "The Problem With Wilderness" in *As Long as Grass Grows: The Indigenous Fight for Environmental Justice, from Colonization to Standing Rock,* Beacon Press, 2020.
2. On the distinction between Eugene Stoermer's and Paul Crutzen's term, "Anthropocene", and its later iteration, "Capitalocene", popularised by Andreas Malm and Jason W. Moore among others, see *Anthropocene or Capitalocene?: Nature, History and the Crisis of Capitalism,* edited by Jason W. Moore. Moore argues that the idea of the Anthropocene focuses on consequences rather than causes of environmental declension, and thereby ignores the "naturalized inequalities, alienation, and violence inscribed in modernity's strategic relations of power and production" (4–5).

Bibliography

Ault, Julie, and James Benning. *(FC) Two Cabins by JB*. A.R.T. Press, 2011.

Benning, James. "Stemple Pass by James Benning." *Pleasure Dome*, 7 Mar. 2014.

Buell, Lawrence. *The Environmental Imagination: Thoreau, Nature Writing, and the Formation of American Culture*. Harvard UP, 1995.

Cronon, William. "The Trouble with Wilderness; or, Getting Back to the Wrong Nature." *Uncommon Ground: Rethinking the Human Place in Nature*, edited by William Cronon. W.W. Norton, 1995, pp. 69–90.

Davis, Angela Y. *Are Prisons Obsolete?* Seven Stories Press, 2003.

Deutsch, Abigail. "On 'The Mars Room' Fiction in Review." *The Yale Review*, 1 July 2018, yalereview.org/article/fiction-review-mars-room.

Duffy, Amanda Holmes. "Book Review: 'The Mars Room: A Novel' by Rachel Kushner." *Washington Independent Review of Books*, 3 May 2018, www.washingtonindependentreviewofbooks.com/index.php/bookreview/the-mars-room-a-novel.

"Excerpts from Letter by 'Terrorist Group,' FC, Which Says it Sent Bombs." *The New York Times*, 26 Apr. 1995, p. 16.

Fleetwood, Nicole. *Marking Time: Art in the Age of Mass Incarceration*. Harvard UP, 2020.

Gilio-Whitaker, Dina. "The Problem with Wilderness." *As Long as Grass Grows: The Indigenous Fight for Environmental Justice, from Colonization to Standing Rock*. Beacon Press, 2020.

Gilmore, Ruth Wilson. *Golden Gulag: Prisons, Surplus, Crisis, and Opposition in Globalizing California*. U California P, 2007.

Kaczynski, Ted. "Bombing in Sacramento: The Letter." *The New York Times*, 26 Apr. 1995.

Kushner, Rachel. *The Mars Room*. Scribner, 2018.

Livingstone, Jo. "Here and Now: The Bold, Vivid Worlds of Rachel Kushner's Novels." *The New Republic*, 7 June 2018.

Meyer, Michael. "Introduction." *Walden and Civil Disobedience*, edited by Henry David Thoreau, 1854. Penguin Classics, 1983.

Moore, Jason W., editor. *Anthropocene or Capitalocene?: Nature, History and the Crisis of Capitalism*. PM Press, 2016.

Nixon, Rob. *Slow Violence and the Environmentalism of the Poor*. Harvard UP, 2015.

Porter, Sonya R., John L. Voorheis and William Sabol. "Correctional Facility and Inmate Locations: Urban and Rural Status Patterns." *CARRA Working Paper Series, U.S. Census Bureau Center for Administrative Records Research and Applications*, July 2017, www.census.gov/content/dam/Census/library/working-papers/2017/adrm/carra-wp-2017-08.pdf.

Rodríguez, Dylan. *Forced Passages*. U Minnesota P, 2006.

Taylor, Matthew A. "The Nature of Fear: Edgar Allan Poe and Posthuman Ecology." *American Literature*, vol. 84, no. 2, 2012, pp. 353–379, doi:10.1215/00029831-1587377.

Thoreau, Henry David. *Walden and Civil Disobedience*. 1854. Penguin Classics, 1983.

———. *Slavery in Massachusetts*. 1854. Cambridge UP, 2012.

Williams, Raymond. "Ideas of Nature." *Problems in Materialism and Culture*. Verso, 1980.

31

READING DONNA LEON AS MEDITERRANEAN ECO-NOIR

Valerie McGuire

Critics have long understood detective fiction as emerging alongside the rise of the nation-state. The classic theorisation locates the genre's origins with the development of industrial urbanism, "and with the obliteration of the individual's traces in the big-city crowd" (Benjamin 43). The thick descriptions of cities that feature in the earliest examples of the genre "promised to overcome the unreadability of modern urban centres" (King 1238). A fixation on the urban produced a convention about famous serial detectives as embodying the cities within which they operate, e.g., Sherlock Holmes as synonymous with Victorian London, Philip Marlowe with hard-boiled Los Angeles, and so on. But what happens when localised sleuths start to confront a globalising world? According to *New Yorker* book critic Clive James, detective fiction's generic use of setting lingered on as a postmodern relic of its former ability to decode modernity. James laments that contemporary crime writers tend to "shift the reader's involvement from the centre to the periphery: to the location. In most of the crime novels coming out now, it's a matter not of what happens but of where".

For James, the fiction of Donna Leon is a fine emblem of international detective fiction's evaporation into "an unusually potent combination of atmosphere and event". It exemplifies how today's detective novels "are essentially guidebooks" (James). In its three decades of circulation, Donna Leon's bestselling Commissario Brunetti series has produced no less than thirty titles, a long-running German television adaptation, an associated cookbook, and lastly, an actual guidebook. It is easy to see why some might disparage Leon's fiction as mere armchair travel.

Yet, on the publication of the twenty-fifth novel of the series, the *Guardian* newspaper christened Leon an "Eco-Detective" writer (Rustin). Despite the apparent fairy tale quality that has tended to be a more well-known hallmark of the series, the theme of ecological corruption has also played a prominent part of its imaginative world. As Leon has stated elsewhere: "it might be better to ask me how I control [the theme of environmental destruction] not taking over the novels entirely" (Leon, 2020). The embattled M.O.S.E. project to install a series of mobile gates at the edge of Venice to regulate the inflow of water from the Adriatic Sea regularly features in the series. Drafted into law in 1987 but not initiated until 2003, the seven-billion-euro project has morphed into a signifier of the Italian state's propensity for clientelism, poor decision-making, government mismanagement (as of this writing in 2022, the project remains unfinished), and with construction of the gates having already damaged the local ecosystem, a symbol of the vulnerability of the lagoon's unique environment (Leon "On Italian Culture"). In Venice, world ecological crises overlap with local politics. The city has been the stage for a dramatic battle between political elites who promote global tourism

 DOI: 10.4324/9781003091912-37

and local democratic activists who protest against the environmental havoc wreaked by cruise ships (Guaraldo). What better location than Venice to rehabilitate detective fiction's traditional investment in the relationship between the individual and the state while incorporating questions of ecology and sustainability.

This chapter argues that the intensely local nature of Leon's Commissario Brunetti is but a red herring for the global scales of ecological (in)justice that it underlines. As the historian Piero Bevilacqua has suggested, Venice can be taken as a "planetary metaphor" for the uncomfortable cohabitation between nature and non-nature. Brunetti appears intensely local and oriented toward the Venetian setting, and unlike his counterpart Aurelio Zen (by the British-American author Michael Dibdin), the detective rarely leaves the immediate environs of the lagoon (Farina). But the commercial success of the series crosses way more than the Alps. Translated into more than thirty-five languages and with a large international and multilingual following, Leon seems to make an appeal to the possibilities of obtaining a "glocal" readership by converting an iconic genre of the nation-state, the detective novel, into vehicle for reaching a larger, global audience (Segnini). If the reader of the Brunetti series delights in finding series' next dead body floating in the Grand Canal, this is partly because they experience Venice in the series as a world heritage city that is itself under perilous threat. American characters regularly inhabit the series and some of them, like the opera diva Flavia Petrelli, appear as serial characters alongside Brunetti and his family, pointing to the inherently transnational matrix of the genre (Allan, Gulddal, King, and Pepper 2020). The success of Leon's 'worldly' detective novels pivots on a strong political messaging about ecology and the environment. Brunetti's Venice is not the city of canals, fog, and moonlight – as it has been in the Western literary imagination for hundreds of years. Leon's Venice is an act of hybris, a mixture of land and water. It is an amphibious city of a hundred little islands suspended in a lagoon that is highly exposed to not only the threat of climate change but to decades of polluting practices and the residues of industrial petro-modernity (Iovino)

Leon draws our attention to "obscured histories of a landscape" that are "often dismissed as backdrops to human activity" (Walton 115). The Porto Marghera and its controversy recurs in several of her novels. Created during the fascist dictatorship as part of an effort to modernise Venice and revive its status as a commercial port in the Mediterranean, the Porto Marghera became the site of a violent experiment in industrial modernity. In the 1930s, the industrial park produced aluminium alloys, iron, and steel products and chemical agents, including the sulphur mustards used in Italian colonial wars of aggression against Libya and Ethiopia in the 1930s (Iovino). When Italy entered the plastic era in the 1950s, the industrial plant became the home of production of petrochemicals, chlorine, PVCs, and some of the most dangerous and carcinogenic chemicals known to humankind. It was not until the 1980s when a case was brought to the Italian state that the full extent of the Porto Marghera as a landfill for toxic waste was revealed and only after decades of corporate efforts to withhold epidemiological research conducted by Italian and international researchers from the public (Casson). In *Through a Glass Darkly*, Leon's requiem to the dying artisanal glass industry, Brunetti's investigations lead him back to the fundamental criminality of the Porto Marghera, and further, the ways in which industrial crimes evades forms of justice normally meted out by the state. The crime of dumping of residuals from glass production pales in comparison with the much bigger crime of the illegal and continual dumping of industrial waste from the Porto Marghera that happens in the plain view.

In other novels, nature and its forces emerge as characters more powerful than their human counterparts – such as in *Acqua Alta*, in which the titular high tides flood a Renaissance palazzo and isolate the novel's villain, effectively submerging the evil that has disrupted the moral order. Critics that see Leon's fiction as an exercise in cliché overlook that the politics of location can tell histories of repression and domination and suggest discursive patterns of hierarchy and subjugation (Said).

Brunetti is a self-confessed "Eurosceptic", and his investigations tend to underscore the byzantine and ineffectual nature of institutions of European governance. Other characters embody the historic strength of environmental movements in Italy. Almost all of Brunetti's immediate family and many of his colleagues in the police turn to vegetarianism at some point in the series. The novels thematise how the economic demands of growth capitalism and the ongoing promotion of tourism to Venice conflict with local movements to protect its ecology and natural environment. Leon is deeply attuned to the Italian context where pride over the environment and the landscape, the *bel paese*, has produced a strong tendency among Italians toward ecological activism (Cesaretti 126). In fact, the series might well be housed under the label 'Mediterranean noir' and seen as part of a growing corpus of international detective fiction that "highlights unity over divisions, violence and discrimination" and seeks to counter the rhetoric of Fortress Europe with a sense of the "uprooted geographies" of the Mediterranean (Pezzotti 1) Immigrants sometimes figure into the sub-plots as wrongful suspects, mirroring the way that Italian culture and media has often linked immigrants with criminality (McGuire). Like other authors of Mediterranean noir, Leon participates in a critique of "exclusionary" European policies and is part of "a subgenre that showcases criminal organisations and beautiful landscapes" (Pezzotti 1). Leon's ecological 'work' interrogates European futures, all the while expanding the circle of inclusion to non-human matters too: animals, waters, trace elements, the air we breathe, and the soils we traverse.

However, Leon has experienced fierce criticism for vetoing an Italian translation of the series, a choice motivated by a desire to protect her anonymity (although she has since moved to Switzerland she travels regularly to Venice). An Italian critic, Ranieri Polesi, famously accused Leon of exploiting stereotypes – of rendering Italians as "not having changed since the era of Machiavelli"[1] – starting an Internet buzz about whether the real reason for refusing an Italian translation had to do with the tenor of her writing. The desire to keep her distance from the public is not unlike the well-known case of Elena Ferrante, whose identity has escaped all journalistic inquiry contributing to what is known in the United States as 'Ferrante fever'. Yet if Leon's success has been married to a debate about the presence of stereotypes, it is worth considering the exact nature of Leon's engagements with the generic conventions of crime fiction. As Andrew Pepper has argued, noir fiction developed as a critique of the totalising imperatives of the detective novel and to illustrate a "set of contradictions that ensue when the state takes control of the justice system" (*Unwilling Executioner* 1). Pepper is heavily indebted in his own theorisation of the crime novel to a set of Italian theorists – Antonio Gramsci, Toni Negri, and Giorgio Agamben – all of whom come out of an Italian tradition of critiquing the state as highly problematic thanks to "the interpenetration of state and civil society" (*Unwilling Executioner* 3). Seen from this perspective, Leon's Brunetti might not so much be a stereotype as a quintessential representation of the "unwilling" or "ambivalent" executioner of noir fiction. He is a (prototypically Italian) detective that may solve the crime but resist the disciplining requirements of the state. Brunetti recoils at the thought that he must inspect the pockets of victims: "he hated the first invasion of the privacy of the dead, this first awful imposition of the power of the State on the peace of the departed" (Leon, *Strange Country* 11). Brunetti "had, years ago, ceased caring about the perpetual fraud committed against the State" (*Strange Country* 22).

What's more, Leon's use of stereotypes might be meaningfully understood within the tradition of the golden age of Anglo-American crime fiction that often took on exotic settings and locations. Leon's Brunetti series clearly owes much to the Agatha Christie cosy-style murder mystery, with its defined set of suspects and the ways in which detective uses rationality to solve the crime, but also in its transnational orientation of the setting. Like Christie, Leon's novels present American and other European characters and "articulate a geopolitical concern with American power and influence" (King, 2018: 9). As Stewart King has recently argued, in Christie, "national attributes or stereotypes distract readers from solving the mystery" (9). Accompanying the Anglo-Euro-American presence in Italy is an 'outside' financial and economic hegemony that has challenged the territorial boundedness of the Italian state and its justice system – one of the salient concerns of crime fiction in the neoliberal era.

A stereotype about Italy as a certain kind of corrupt state may also serve as an analogy for the corruption of the state globally when it comes to ecology. As Pepper further observes, the contradictions that noir traditionally investigates have been further complicated in the contemporary era and by the expanding power of global capital, or the "deterritorializing impulses" of the modern state (*Unwilling Executioner* 208). Here again, Italy proves again a useful point of reference, for the country has historically been a country where Euro-American political and economic influence interfere with national politics. During the 1950s, it was closely surveilled as a proxy for the Cold War, with the United States fearing it was an incubator for Eurocommunism. More recently, Italy's handling of issues such as migration and health contagion (during the coronavirus outbreak) have set the stage for wider international policies. As Giorgio Agamben remarks: "Italy is a sort of political laboratory where new technologies of governance are tested" (41). Italy as "political laboratory" is an idea that might well apply to the place of Italian organised crime – also known as the ecomafia – in cheaply disposing of Europe's toxic waste.

Toxic Italy: *Death in a Strange Country* (1993)

In the second novel of the series, *Death in a Strange Country*, Leon pits her everyman detective-hero, the congenial Commissario Brunetti, against such an ecomafia. The body of an American public health inspector is found floating in a Venetian canal "strangely shrivelled and white, a sure sign he had been in the water a long time" (*Strange Country* 8). The death precipitates a mystery and investigation centred around the secrecy of toxic waste while revising the spatial and temporal containment of the classic locked-room murder mystery – to which in other ways the novel playfully nods. It resituates Italy as not so much a stereotypically dysfunctional and contained landscape, but one in which foreign influences, special interests, and local corruption experience deterritorialisation. Not only is the state itself unmoored from its territorial expression, but power is not embodied in the conventional archetype of the villain.

The word "strange" in the title refers to the fact that the American victim has died in a foreign country, but it also points to the foreign influences unsettling both the Venetian setting and the police procedural. As first revealed through some associations with straight teeth and physical fitness, readers learn that the victim is a soldier from the American military base in Vicenza, which curtails Brunetti's jurisdiction, forcing the hero to pursue an investigation that is not pro forma: for example, the detective removes drugs from the victim's apartment to evidence that there has been a cover-up. A second death, that of the American soldier's ex-girlfriend – also made to look drug-related – shows the lengths to which invisible circles of power will go to conceal what the two victims had discovered. An illegal hazardous-waste dumping site near the base is the cause of a rare and life-threatening disease afflicting small children. A third death, which ensues when Brunetti continues to pursue an investigation outside of his command, underscores the vastness of the network of corruption behind the crime. Faced with a chain of collusion so deep it circles back to his own immediate family, Brunetti is unable to pursue a legal punishment. He is silenced about the real crime of the novel – the Italian state's involvement with disposing of illegal toxic waste – and faces what Massimo Carlotto calls "strategic alliances between entrepreneurs, financial policing bodies, politics and organized crime" (Carlotto 12).

To no small extent, then, the novel resists the traditional denouement of the police procedural in which Brunetti's public revelation of the truth will lead to a judicial process and the restitution of order. The Italian government's lack of oversight and its capitulation to monied interest for personal gain suggest a spirit of free-market capitalism gone haywire. Leon offers us her own take on what Andrew Pepper refers to as "capitalist noir" (228). While there is resolution to the micro-conflicts in the novel, the larger issue of the way a state's concern for its citizens' safety and security is subordinated to the demands of growth capitalism produces a sense of "ongoing-ness" in which the narrative offers no real resolution to the crime it presents (Pepper, "Complex Crime Fiction" 130). In the end, Leon has it both ways, and leaning on stereotypes of *omertà*, the Sicilian mother of the novel's final victim extracts her own justice by killing the villain with her own hands. But this justice is importantly outside the state underlining further the state's collusion with environmental crimes.

By the same token, Leon overturns conventional myths of Italy as the *bel paese* and notions of Italian cultural autonomy free from 'foreign' influence (Nerenberg). Indeed, if Leon were solely relying on Italian stereotypes, she might well have set the novel – or at least the origins of the novel's villains – in Southern Italy, long held as the Italian repository of lawlessness and criminality. The Venetian setting in *Death in a Strange Country* serves instead as a setting for the revealing of the toxic shadow of industrial modernity in Europe. While Brunetti searches for culprits to a set of murders, he discovers a multifaceted history of the production of PCB chemicals (polychlorinated biphenyls). Commonly used as coolants within refrigerator and air-conditioning systems, but also present within plastics, hydraulic fluids, and lubricants, PCBs ceased to be produced in 1977 because of health effects associated with exposure. With a half-life of up to fifteen years, PCBs are difficult and costly to dispose of. Although intent on finding murderers, Brunetti maps a shadow economy that preys on the need for cheap and illegal disposal in bulk, with the German and American governments colluding to dump PCBs – in this case leftover waste from a Cold War military industrial complex. The density of the cover-up utterly frustrates the traditional modes of police inquiry, and Brunetti faces a constellation of dots that do not add up to culpability: "he tried to think of a way all of this could be made to connect, hoping that it would lead him to whoever it was had pushed the needle into the doctor's arm" (*Strange Country* 185). Brunetti detects in its place a kind of violence that is much more insidious and far more difficult to trace than murder. The crime and grief wreaked on victims of PCB chemicals is akin to the slow violence described by Rob Nixon: it "occurs gradually and out of sight, a violence of delayed destruction . . . incremental and accretive, its calamitous repercussions playing out across a range of temporal scales" (2).

The Italian setting still acts as a character in the novel. But it highlights not the *bel paese*, or Italy as a tourist destination, but a particular historical moment which revealed Italy to be a middle passage for the 'poison ships' (*navi di veleni*) of European toxic waste (see also Past, 2016). In 1988, authorities found over two thousand drums of toxic waste in the Niger Delta. The barrels contained PCBs as well as dimethyl formaldehyde and asbestos fibres; they leaked and contaminated the soil and groundwater leading to an outbreak of disease in a small fishing village of Nigeria. The Koko incident sparked an international outcry and led to the first international regulations on hazardous waste as well as a pan-African debate about "toxic colonialism" (De Majo). Investigative journalists further discovered that the barrels originated with Italian oil companies. As Leon describes in *Death in a Strange Country*, the incident proved the way in which environmental crime often crosses all sovereign borders and reinforces a set of uneven power relations with more powerful states exploiting weaker ones:

> For decades, the Third World had been the rubbish dump of the industrialized nations, taking in shiploads of toxic substances, which were scattered around their pampas, savannahs, and placed there in exchange for current wealth, no thought given to the future price that would come due and be paid by future generations. And now, with some of the countries of the Third World refusing any longer to serve as the dumping ground of the First, the industrial countries were

> constrained to devise systems of disposal, many of them ruinously expensive. As a result, fleets of phantom trucks with false manifests travelled up and down the Italian peninsula, seeking and finding place to unload their lethal cargoes.
>
> *(188)*

The issue of toxic waste disposal in Italy has been featured in the international media for two decades now and has engendered a set of cultural representations in both fiction and non-fiction – the best known of them being Roberto Saviano's 2008 expose, *Gommorah*, revealing the Neapolitan mafia's globalisation and involvement in a variety of industries including waste management.[2] Within the Italian context, it is widely agreed that Southern Italy is often the dumping ground of illicit industrial waste from the North in what is a new incarnation of an age-old disparity between Northern and Southern Italy – what Antonio Gramsci described as Italy's internal colonialism (Gramsci). Leon points to Italy's role within a larger European context, often doing the dirty work of northern, more industrial countries, like Germany, where Green party politics have had some victories (in their rejection of nuclear power, for example), disguising the fact that corporations still succeed in intimidating governments, and that black economies still prey on the demand for cheap ways of disposal. The real culprit of *Death in a Strange Country*, then, is the very society that often treats environmental crime as a *fait accompli* – a dirty fact of industrial modernity that needs to be buried out of sight. Brunetti winces when he sees the title of a book by Italy's Green Party, *Global Suicide*, thinking it to be hyperbole. But the messaging of the novel suggests just such a noir position of resignation – of global suicide – when faced with the magnitude of eco-corruption. Notably, the novel ends with Brunetti in a cascade of tears as he discovers that his father-in-law has business ties with the powerful figures behind the crime. Brunetti realises he is complicit in the systems which will damn us all – as in the classic formulation of the resigned or noir position, we are, like Philip Marlowe, "part of the nastiness" (Chandler 250).

Beyond the Tourist Maze: *Through a Glass Darkly* (2006)

The majority of Italy's 46.4 million tourists each year make a stop in Venice (Hom 27). Tourism has so overcome contemporary Venice that it has been described as "the first postmodern city, selling no product other than itself and its multiple images" (Davis and Marvin 3). Leon's refusal to translate the series into Italian has encouraged the view that her books are above all for a foreign market that consumes the multiple – and meaningless – images of Italy that circulate globally. It could be tempting to label Leon a hypocrite who decries ecological catastrophe but capitalises on Venice's evergreen popularity as a tourist destination. Yet it's worth considering that what is being pedalled is not destination Venice, but a meditation on the "politics of location" created by a bio-genetic age known as the 'Anthropocene' (Braidotti 51) If Venice is an amphibious city, then it is also a place where tourists may take stock of their current condition by witnessing the conflict between nature and culture as it once was before the 'Anthropocene'.

In her 2006 novel, *Through a Glass Darkly*, a crime that takes place at a glass factory on the island of Murano sets the stage to bring together these various threads, i.e., tourism, globalisation, and the environment. As Leon describes in detail over the course of the novel, glass making in Murano has long stood for the great beauty of artisanal production in Venice. As one character pointedly remarks, "What's more Venetian than Murano glass?" (*Glass Darkly* 160). But in the ever-complicated post-industrial present, Murano's glass industry fights a multi-front battle against extinction: Venetian glassmakers are at risk of eventually being made obsolete thanks to competition with cheap glass knockoffs imported from China, that's if the Chinese do not succeed first in purchasing Italian factories outright. The costs of gas required to fire the glass at extreme temperatures have risen so high as to make profits anaemic, and finally, drawing the ire of the glass-factory owners, are the high fees demanded by special firms to safely dispose of the residual acids generated in glass production (*Glass Darkly* 226–27). In short artisanal industries are being whittled away by the combined forces of EU bureaucracy and global neoliberalism.

On the one hand, Leon critiques the fantasy that Italy "remains a country of *piccole industrie* (small industries) and family-based capitalism in which the social relations between people trump the economies of scale" (Hom 24). She shows, indeed, the very postmodern condition of these industries which face extinction while still producing remarkable value for the circulation of Venice's image as a memento of an economy of a different era. On the other hand, she introduces the idea that concern for the environment trumps even the desire to sustain culture in our perilous posthuman futures. But the conflict between these two ideas proves to be the novel's great red herring. It is not the struggling owner of one of the glass factories that is the culprit but his prosperous neighbour – an aspiring politician who has built his reputation as a vocal opponent against the ongoing abuse of environmental law by the petrochemical plant in nearby Porto Marghera. Leon complicates the narrative by revealing that the murder is committed to protect the image of a politician as an environmentalist. How can the Laguna be saved if its politicians are so easily corruptible?

Through a Glass Darkly begins with the image of "No Global" protests that have turned rough and aligns with the political groundswell in Italy at the time of its publication. In Italy, a strange alliance has occurred between populism and environmentalist movements – as best articulated in the M5S or Five Star Movement whose platform includes public water, sustainable transport, sustainable development, technology, and environmentalism alongside anti-immigration. The entire cast of characters familiar to the series is swept up in a "contagion" of environmentalism:

> After Signorina Elettra left, Brunetti asked himself, as would someone from the Disease Control Centre, in which direction the arc of ecological infection was now likely to be passing: whether from her to Vianello or from the Inspector to her. His imagination was seized for a moment by this image, and he found himself wondering what risk of contagion he experienced by working in such proximity to them and when he might begin to feel the first symptoms.
>
> (Glass Darkly *163)*

Brunetti must resist the contagion of environmentalism, of course, because to embrace it would mean to recognise the inadequacy of the state and to live with cognitive dissonance as a member of its forces. As he observes about his friends, their ardent environmental positions do not align with their careers as bureaucrats:

> Given the sincerity of their beliefs, why then did Vianello and Signorina Elettra work for the police force, when they could be working for some sort of environmental protection office? For that matter, why did any of them continue to work for the police?
>
> *(163)*

Leon seems to ask questions about the viability of political solutions to environmental crimes. The novel's noir position of resignation is also one in which the sense of hypocrisy it engenders in all subjects is unavoidable.

The police inspector's crisis of identity mirrors the larger story of the novel about the victim who has repeatedly sent information to the Italian police asking it to investigate the poisonous toxins at the glass factory, believing that these are responsible for the birth defects of his children. Not only do his calls go unheeded, but he is thought to be a madman. In yet another off-books case, Brunetti finds his way back to the origin of the crime through a series of books and papers left by the victim: Dante's *Inferno*, several tomes on environmental law and the rules for dumping in the lagoon, and then a book about toxicity, *Industrial Illness*. Brunetti deduces the coordinates for the location in Venice where the glass-factory waste was illegally being dumped and where Brunetti discovers all manner of contaminants. The victim was not a madman but someone with an acute sense of perception. At the heart of *Through a Glass Darkly*, then, is a discussion of how having an ecocritical awareness means recognising the impossibility of reversing our impact on the environment. Like the Pauline epistle from which the novel draws its title, "For now we see through a glass, darkly, but then face to face. Now I know in part; but then shall I know, even as also I am known" (1 Cor. 13:12), the novel implies that if we are nearer to an end-of-times, we also may see our corruption and our failures more clearly.

Conclusion: A Corruptible Lagoon

In his seminal essay coining the term "Petrofiction", Amitav Ghosh paralleled the twentieth-century affair with Petro-modernity to the colonial encounter of the spice trade of the nineteenth century: "oil is clearly the only commodity that can serve as an analogy for pepper" (Ghosh 29). If Ghosh first described a "barren" literary homage to oil and attributed "muteness" on the subject to a variety of factors (corporate secrecy, that oil was discovered in regions that did not have a great literary tradition, and the fact that the territory of oil, with its plethora of global actors, is multilingual), a growing corpus of ecocritical studies suggests a much longer list of works of fiction that have dealt with the topic – if peripherally – since the global oil shocks of the 1970s (MacDonald, 2013). Moreover, while oil has powered a post-industrial revolution of sophisticated technologies such as automation and precipitated the disappearance of embodied labour, it has also produced a complex and amorphous residue in our urban landscapes and natural environments. As Elena Past describes, the Mediterranean offers a lens through which to address the question of waste and the remains that lurk "at home" of the oil revolution elsewhere. The poison ships that troll the Mediterranean disposing of radioactive waste, plastics, and chemical arms underline the way that out "anthropogenic litter" is not a national, regional, or exclusively human problem but one that, like the sea itself, courses through history and "geological layers, through political formations; carrying tourists, migrants, drones, military convoys, and documentarians" (Past 381).

If we respond to the ecocritical call to consider culture as not inseparable but as mutually permeable with nature, we look naturally toward Venice, a city so thoroughly penetrated by the sea, one that reminds us of the ways in which humans are not only in relationships with one another but also with the natural environment. When Napoleon invaded in 1797 and the city was partitioned between the French and Austrians, the Venetian empire, which had lasted almost eleven hundred years, disappeared almost without remark. From that point forward, the city became an emblem of its constitutive hybridity, both its corruption by nature and by the Ottoman East. Precisely this corruption made Venice, in 1910, the ideal place for Filippo Marinetti, founder of the Futurist movement, to stage his coup of Parisian modernism. Marinetti saw in the city the possibilities of palingenesis, a rebirth and a reawakening of the old imperial and maritime Venice of the past ("that can dominate the Adriatic Sea, that great Italian lake") and a way forward out of the crisis of modernity that was gripping not just Italy but all of Europe. As the symbol par excellence of decadence ("a bed whose bottom has been staved in by caravans of lovers"), Venice, the city of "reeking canals" and "leprous and crumbling palaces", could startle the world from its moral turpitude and slumber (Marinetti). For Marinetti, the sense of the city's immanent corruption made it the ideal setting to marshal a crisis of modernity for a new radical, avant-garde project of art as action (Scappettone, 2014).

Perhaps a similar sense of corruption makes Venice an ideal site for the moral crises of eco-noir, as it is explored by Donna Leon. As crime writer and critic Megan Abbott has argued, noir emerges "when certain structures of authority don't make sense any longer" – like, for example, the failure of states to take prompt action against pollution, toxicity, and the impact of carbon emissions on our ecosystems. Yet if Abbott focuses on the historical or temporal frames, we might think instead about noir from an ecocritical perspective and how certain environments might likewise be generative. There are many Anglo-American authors writing detective fiction set in Italy – so many, in fact, that "Italian Expat Crime Fiction" constitutes its own subgenre, with Leon arguably its most successful representative.[3] While canonised as very much a writer of commercial or "pulp" crime fiction and drawing scant critical attention, Leon seems nevertheless to have tapped into an age-old vein of Venice as both a symbol of corruption and decay.[4] Moreover, if Leon's target readership is largely Anglo-American, then the Venetian setting of her novels operates not just as a sentimental destination, but one that calls on travellers to have a modernist experience of the city. Her novels offer up, in a familiar cartography, the city as almost a series of Proustian madeleines, moments of sudden intimacy with a remote past. It can be no accident that Henry James is the favourite author of Brunetti's clever wife Paola. James visited and briefly resided in Venice nine different times between 1869 and 1907 and wrote extensively about his experience. The city appears as a setting in many of his novels about Americans in Italy and his love affair with the Laguna resulted in one work of illustrated travel literature, *Italian Hours* (1909), itself a literary Baedeker of sorts. Like Edith Wharton, who also set several of her novels in Venice, James mines the singularly hybrid backdrop – its Renaissance architecture, waterways, and interplay of light, fog, and shadow – to meditate on the relationship between our moral interiors and our exterior environments and raise questions about the ways in which "material and moral decadence accompany each other" (Hallamore 30). Taking a cue from James, the Venetian setting in Leon's fiction is an opportunity to uncover the relationship between environmental degradation and political decay and decline.

Interestingly, Leon's novels never outright reference the potential threat of climate change to Venice nor hint at a catastrophe scenario in which rising seas submerge the city, drowning two thousand years of world culture and history. Instead, her novels raise subtler issues of how environmental (in)justice has changed our minds and our spirits, how the corruption of our exterior environments has led to an inevitable cognitive dissonance within our interior lives. While science fiction has long been held as the genre naturally equipped to narrate dystopian futures in which our cities and planet are uninhabitable, commercial or "pulp" crime fiction is increasingly viewed as adapting perfectly to debates surrounding the climate crisis (Walton, 2018; King, 2021). By leaning on some of the generic conventions of noir fiction, such as the probing of the problem of moral responsibility and the reversal of the detective novel's mission to assign culpability and mete out justice, Leon suggests that corruption is now endemic to all societies. And yet, it is hard to imagine Leon achieving such results without drawing on a rich symbolic repertoire of Venice as a *locus amoenus* for nature's corruption by culture. Leon helps to articulate a related set of ecocritical questions about what happens to one of the central elements of the detective fiction genre, that is, a strong sense of setting, when it focuses on environmental imaginaries. Her subtle reversal of setting as readable and comprehensible in the classic detective novel to one that is mired in secrecy and illegibility may define the contemporary Mediterranean eco-noir.

Notes

1. In the original, *"Donna Leon parla degli italiani come se ancora fossimo ai tempi di Machiavelli."* Ranieri Polesi, "Passaggio in Italia, frontiera del Male." *Corriere della Sera*. July 26, 2009.

2. While Saviano's *Gommorah* is doubtless the best-known discussion of the corruption of waste management by the Italian mafia, it is by no means the only cultural example of it. The documentary film, "Wasting Naples" (2009) or the popular Imma Tatarani detective series by Mariolina Venezia set in Matera, Italy.
3. See http://italian-mysteries.com/.
4. A few exceptions have considered her inter-textual use of opera (Rye, 2014) or the ways that she foregrounds cuisine within the series (Anderson, Miranda, and Pezzotti, 2018).

Bibliography

Abbott, Megan. "A Conversation with Megan Abbott." *The Sewanee Review*, 2018, Summer, thesewaneereview.com/articles/a-conversation-with-megan-abbott.

Agamben, Giorgio. *Where Are We Now? The Epidemic as Politics*. Rowman and Littlefield, 2021.

Allan, Janice M., et al., editors. *The Routledge Companion to Crime Fiction*. Routledge, 2021.

Anderson, Jean, et al., editors. *Blood on the Table: Essays on Food in International Crime Fiction*. McFarland Et Company, Inc., 2018.

Benjamin, Walter. *Charles Baudelaire: A Lyric Poet in the Era of High Capitalism*. Translated by Harry Zohn, Verso, 1997.

Bevilacqua, Pero. *Venice and the Waters: A Planetary Metaphor*. Donzelli, 1995.

Braidotti, Rosi. *The Posthuman*. Polity, 2013.

Carlotto, Massimo, and Jean-Claude Izzo. "Eulogy for Jean-Claude Izzo." *Total Chaos*, translated by Howard Curtis, Europa Editions, 2013.

Casson, Felice. *La fabbrica dei veleni: Storie e segreti di Porto Marghera*. Sperling & Kupfer, 2007.

Cesaretti, Enrico. *Elemental Narratives: Reading Environmental Entanglements in Modern Italy*. Pennsylvania State UP, 2020.

Chandler, Raymond. *The Big Sleep*. Penguin, 2005.

Davis, Robert, and Garry Marvin. *Venice, the Tourist Maze: A Cultural Critique of the World's Most Touristed City*. U California P, 2004.

De Majo, Claudio. "Italy's Poison Ships: How an International Trade of Hazardous Waste Sparked a Grassroots Struggle for Environmental Justice." *Environment & Society Portal, Arcadia*, no. 43, 2020.

Farina, William. *Italian Crime Fiction in the Era of the Anti-Mafia Movement*. McFarland & Company, 2020.

Ghosh, Amitav, "Petrofiction: The Oil Encounter and the Novel." *New Republic*, vol. 206, no. 9, 1992, 29–34.

Guaraldo, Emiliano. "Resisting the Tourist Gaze. Art Activism against Cruise Ship Extractivism in the Venice Lagoon." *Lagoonscapes: The Venice Journal of Environmental Humanities*, vol. 1, no. 1, 2021, pp. 101–124, doi:10.30687/lgsp//2021/01/008.

Hallamore, Ann Caesar. "Americans Writing Venice: Edith Wharton and Henry James." *A Venetian Miscellany*. U Warwick P, 2016, pp. 27–32.

Hom, Stephanie Malia. *The Beautiful Country: Tourism and the Impossible State of Destination Italy*. U Toronto P, 2013.

Iovino, Serenella. "From Thomas Mann to Porto Marghera: Material Ecocriticism, Literary Interpretation, and Death in Venice." *Handbook of Ecocriticism and Cultural Ecology*, edited by Hubert Zapf, De Gruyter, 2016, pp. 349–367.

James, Clive. "Blood on the Borders." *The New Yorker*, 2 Apr. 2007, www.newyorker.com/magazine/2007/04/09/blood-on-the-borders.

James, Henry. *Italian Hours*. Houghton Mifflin, 1909.

King, Stewart. "E Pluribus Unum: A Transnational Reading of Murder on the Orient Express." *Journal of Detection*, vol. 36, no. 1, 2018, pp. 9–19.

———. "*Crimate* Fiction and the Environmental Imagination of Place." *The Journal of Popular Culture*, vol. 54, no. 6, 2021, pp. 1235–1253, doi:10.1111/jpcu.13083.

Leon, Donna. *Death in a Strange Country*. Grove, 1993.

———. *Acqua Alta*. Grove, 1996.

Leon, Donna, and CrimeReads. *Through a Glass Darkly*. Grove, 2006.

———. "Donna Leon on Italian Culture, Environmentalism, and Her Long-Running Series." *CrimeReads*, 10 Apr. 2020, crimereads.com/donna-leon-on-italian-culture-environmentalism-and-her-long-running-series/.

MacDonald, Graeme. "The Resources of Fiction: Energy, Oil, Literature." *Reviews in Cultural Theory*, vol. 4, no. 2, 2013, pp. 1–25.

Marinetti, Filippo. "Against Passéist Venice." *Futurism: An Anthology*, edited Lawrence Rainey et al., Yale UP, 2009, pp. 67–70.

Mattaruco, Giada. "Gialli Veneziani Recenti." *Quaderni veneti*, vol. 5, no. 2, 2016, pp. 5–26.
McGuire, Valerie. "Crimes of Diction: Language and National Belonging in the Fiction of Amara Lakhous." *Journal of Romance Studies*, vol. 15, no. 2, 2015, pp. 1–21, doi:10.3167/jrs.2015.150201.
Nerenberg, Ellen. *Murder Made in Italy: Homicide, Media, and Contemporary Italian Culture*. Indiana UP, 2012.
Nixon, Rob. *Slow Violence and the Environmentalism of the Poor*. Harvard UP, 2011.
Past, Elena. "Mediterranean Ecocriticism: The Sea in the Middle." *Handbook of Ecocriticism and Cultural Ecology*, edited by Hubert Zapf, De Gruyter, 2016, pp. 368–384.
Pepper, Andrew. *Unwilling Executioner: Crime Fiction and the State*. Oxford UP, 2016.
———. "'Complex' Crime Fiction and the Politics of Ongoing-Ness: Don Winslow's War against Endings." *Crime Fiction Studies*, vol. 1, no. 1, 2020, pp. 130–145, doi:10.3366/cfs.2020.0011.
Pezzotti, Barbara. *Politics and Society in Italian Crime Fiction: An Historical Overview*. McFarland & Company, 2013.
———. "Towards a Definition of Mediterranean Noir or Crime in the Mediterranean: Mediterranean Noir or Mediterranean Crime Fiction?" *Belphégor: Littérature populaire et culture médiatique*, vol. 20, no. 1, 2022.
Polesi, Ranieri. "Passaggio in Italia, frontiera del Male." *Corriere della Sera*. 26 July 2009.
Rustin, Susanna. "Donna Leon: Why I Became an Eco-Detective Writer." *The Guardian*, Guardian News and Media, 15 Apr. 2017, www.theguardian.com/books/2017/apr/15/donna-leon-interview-commissario-brunetti-earthly-remains#:~:text=Cultured%2C%20shrewd%2C%20honest%20and,by%20faking%20a%20heart%20attack.
Rye, Marilyn. "Mozart on Her Mind, Venice in Her Heart, Mysteries in Her Pen: Donna Leon's Commissario Guido Brunetti Series." *Bloody Italy: Essays on Crime Writing in Italian Settings*, edited by Patricia Prandini-Buckler, McFarland, 2014, pp. 133–164.
Said, Edward. *Orientalism*. Vintage, 1977.
Scappettone, Jennifer. *Killing the Moonlight: Modernism in Venice*. Columbia UP, 2014.
Segnini, Elisa. "Andrea Camilleri's Montalbano and Elena Ferrante's *L'amica geniale*: The Afterlife of Two 'Glocal' Series." *The Translator*, vol. 27, no. 3, 2021, pp. 254–270.
Walton, Samantha. "Studies in Green: Teaching Ecological Crime Fiction." *Teaching Crime Fiction*, edited by Charlotte Beyer, Palgrave MacMillan, 2018, pp. 115–130.

32

THE CIRCULATION OF GLOBAL ENVIRONMENTAL CONCERNS

Local and International Perspectives in the Verdenero Collection and Donna Leon's Crime Fiction

Aina Vidal-Pérez

Introduction

Prior to 2006, the Milan-based publisher Edizioni Ambiente used to produce annual reports by Italy's top environmental organisation, Legambiente, with little audience success. That year, in the hope of raising awareness among Italians regarding social and environmental degradation – and having noticed crime fiction's popularity in Italy – the publishing house contacted Italian crime novelists and brought them on board for a project in which both an environmental association and an environmental publishing house would suggest subject matters for crime novels. The alliance appropriately called this collection Verdenero and, up until 2012, managed to publish almost twenty titles portraying a miasmic Italy: a territory that, contrary to the international imaginary, is home to putrid waters, sewage, ashes, and waste. In 2011, the group published Francesco Aloe's *Il vento porta farfalle o neve* ('The wind brings butterflies or snow'), a novel dealing with the humanitarian and environmental catastrophe ensuing from the collision between a ferry and an oil tanker anchored three miles off the port of Livorno. The novel also contains a criminal plot to cover up the burgeoning trafficking of weapons and toxic waste at said port. Six years later, in an entirely different venture, the Swiss publisher Diogenes Verlag published *Earthly Remains* by best-selling author Donna Leon. In this novel, the twenty-sixth title in her crime fiction series featuring the famous Commissario Guido Brunetti, the police officer faces a large toxic waste management corporation endangering the Venetian ecosystem, the Mediterranean Sea, and certain areas abroad. Despite dealing with similar global concerns (water pollution, toxic waste trafficking, collusion between local administrations and international corporations, and North-to-South exploitation), Aloe's and Leon's novels have followed diverging paths: while *Il vento porta farfalle o neve* was not translated and only circulated within Italian borders, Leon's *Earthly Remains* was almost simultaneously translated to German, Catalan, Spanish, Dutch, French, Polish, and Portuguese, but, strikingly, not to Italian.

In my exploration of the relations between crime fiction and ecology, I suggest taking on a comparative analysis of the small, independent publishing house Edizioni Ambiente, with Aloe's *Il vento porta farfalle o neve* (2011) as a case study, and the global publishing phenomenon of Donna Leon's crime fiction, with a focus on *Earthly Remains* (2017). Though translated into more than twenty languages, Leon has barred the translation of her novels into Italian.[1] Aloe's novel, on the other hand, specifically aims to reach the audience that Leon excludes: the Italian reader who remains unaware of

DOI: 10.4324/9781003091912-38

the corrupt structures orchestrating the country and the world at large. Leon is a mainstream, global author whose crime novels have gathered international attention, while Aloe's circulation remains limited within Italian borders and participates in a publishing project concerned with raising awareness among Italians. Comparing the two will allow for a new reflection on crime fiction's modes of representation and circulation when dealing with global environmental concerns located in specific places in the world, such as Livorno and Venice. Taken together, these novels enrich the critical dialogue between the local and the global: first, they reveal the need for local and regional support – in this case, the Italian Mediterranean Sea and its shores and ports – to represent global phenomena, like environmental crime; second, both novels intentionally use crime fiction to reach popular audiences and raise environmental awareness at different scales; finally, both texts demonstrate the disparate scopes of circulation that novels face even when dealing with the same global issues, which casts light on certain novels' and authors' unequal mechanisms of internationalisation.

Ecocrime Fiction: From the Local to the Global

In recent years, the contemporary crime fiction genre has garnered increasing academic attention from a global perspective, particularly in two enriching directions: its study within the framework of world literature (which considers parameters involving circulation and the world book market and thematic issues like global capitalism) and its study from the interdisciplinary and environmentalist perspective of ecocriticism (given the substantial production of ecologically-oriented crime fiction works). However, only very recently have these perspectives been used to approach the study of the crime novel's involvement in environmental issues in the global era, that is, the ways that crime fiction as a genre is engaging with planetary emergencies like the global ecological crisis.

In a recent article, Neus Rotger and Marta Puxan-Oliva[2] posit that current literary criticism has detected a large set of novels engaging with global forces such as migration, war, capital flows, hyper-connectivity, and more frequently, the climate crisis. Appropriately, they note that scholars are connecting the novel to previously unacknowledged fields, from cross-disciplinary perspectives.[3] I would propose foregrounding ecology and ecocriticism as a field and discipline that can clearly contribute to local-global discussions regarding the market and the novel in the global era. Since Ursula Heise ("Sense of place") posited Ecocriticism as a promising approach to reconsider environment and representation, multiple theorisations on the scalar entanglement between the environment, the planet, and the novel have emerged. Joseph Keith's discussion of the "planetary novel", for instance, considers how the novel measures up against the representational challenges posed by the planetary scale of environmental crisis. While Amitav Ghosh argues that the novel cannot imagine the inhuman scales of ecological phenomena ("Petrofiction", "The Great Derangement"), Shouhei Tanaka points toward the ideas of "planetary petrofiction" to suggest how "energy inflects the organisation and mechanism of narrative itself" (193). The novel, then, is being discussed in terms of how suitably it can address specific environmental problems in relation to global issues concerning the survival of ecosystems. What kinds of fictions, then, can emphasise real agency in the face of planetary climate emergency? What representational limits must the novel face when it comes to environmental crisis?

Adam Trexler uses the term "Anthropocene fictions" to reflect upon the novel as "a privileged form to explore what it means to live in the Anthropocene moment" (27). Considering the links between climate change and narrative representations – with a specific focus on climate fiction – Trexler calls for formal innovation and conducts an analysis that can be very profitable for the crime fiction genre. In the same way as with climatic fiction, perhaps because of traditional beliefs around what constitutes "serious" or "literary" fiction, crime fiction novels had received scant attention from global literary and ecocritical studies. As Trexler posits, "academic criticism of contemporary fiction has been overwhelmingly focused on determining a literary canon deserving of serious study", leaving aside so-called "popular" authors (11). Ecocriticism's critical potential only started encompassing crime fiction studies a few years back, demonstrating the degree to which environmental awareness has permeated "commercial fiction".[4]

This chapter aims to contribute to the study of the global novel, as well as to an understanding of how crime fiction, more specially, is tackling the challenges of narrating phenomena like the environmental crisis. To do so, it takes on a twofold discussion that considers both formal and sociological aspects. Firstly, crime fiction has traditionally been studied through a set of prescribed typologies and formulae. This has shifted in recent years, however, as the genre's mutability, adaptability, and transgressiveness (Allan et al.), as well as its ability to integrate new shifts in transnational and global studies, have become an increasingly central facet of contemporary crime fiction criticism (Pepper and Schmid). Crime fiction that focuses on environmental crimes takes advantage of the genre's global transmission, whist managing to subvert prescribed formal functions with new strategies that contribute to a broader planetary imagination of the environment. In this sense, while, on the one hand, the crime fiction genre – because of its generic features, such as its involvement in social issues, attention to spatial representation, and realism – allows for the addressing of socio-environmental issues, on the other, the climate emergency works to transform its generic conventions. New forms of global environmental risk, and the ways they are perceived, seem to require new forms of expression and analysis. As a result, I will study the continuities and discontinuities of crime fiction tropes, focusing especially on issues of scale: the transformation of the notion of criminal agency, now depersonalised and decentralised by globalising forces distributed among local and state administrations, criminal organisations, and large international corporations; the depiction of other-than-human victims such as whole ecosystems; and the construction of a plot that expands the human-centred understanding of felony.

Secondly, the study of the ecocrime novel's global reach and its literary uses demands a sociological approach, one that considers the material conditions of these fictions' production, circulation, and reception. While crime narratives have garnered enormous attention in recent years, the dominant approaches have focused on national or regional scales, in addition to the hegemonic Anglo-American canon. Recent contributions are nevertheless studying crime fiction from a broader perspective, beyond the national or regional (see Matzke and Mühleisen; Krajenbrink and Quinn; Anderson et al.; King; Pepper and Schmid, Nilsson et al.; or Stougaard-Nielsen). Drawing from these global approaches, this chapter aims to analyse, through the two aforementioned case studies, how crime fiction has been used to deal with environmental and humanitarian issues. Through a combined approach, I will analyse both the formal and narrative strategies that are used to frame tensions between the local and the global, on the one hand, and the genre's disparate circulation and reception in the (unequal) international book market, on the other.

Global Crimes in Global Waters

In its English version, *Earthly Remains* opens and closes with a map of the Venetian lagoon. Although Venice – and its surroundings – comprises the space where the Venetian-to-the-core Brunetti investigates criminal cases, the *commissario* decides to vacation at a country home in Sant'Erasmo, one of the *laguna*'s multiple islands. He hopes to enjoy solitude, bathe, read, sunbathe, and paddle through the Adriatic waters so he can keep his head away from the *questura*. There, he meets a former rower called Davide Casati, who is obsessed with unearthing the cause behind the mass death of his beloved bees, which live in hives all over the marsh. One day, Casati's corpse appears floating in the waters of the lagoon. This marks the beginning of the police investigation, the resolution of which suggests that Casati was murdered by a hired assassin from GCM Holdings, one of the large industrial corporations located in the Porto Marghera industrial area (in front of Venice) that specialises in the transport, trafficking, and dumping of highly toxic waste.

Significantly, the first corpse to be discovered in this novel is not human, but rather those of the bees that have been struck down by the lagoon's high levels of environmental pollution. Corruption in the urban environment is not what catalyses the police plot in Leon's *Earthly Remains*, but rather corruption of the natural environment. Leon manages to portray how the polluted waters of the lagoon filter in like tentacles from the small canals of Venice and Marghera towards the various islands of the lagoon, opening up to the cold waters of the Adriatic as they head to the Mediterranean and finally end up in African countries like Nigeria. From the initial cartographic representation, the reader can imagine the expansion and scope of environmental pollution stemming from large-scale industrial activity.

In *Il vento porta farfalle o neve* the criminal investigation is conducted not by a commissioner or a policeman, but by a criminal, who narrates in the first person. Fratello, the main character, is a *'Ndrangheta* killer whose boss has asked him to eliminate two former collaborators. Fratello carries out the mission with Nino, but they soon become enmeshed in an extraordinarily complex conspiracy that forces them to investigate an international criminal plot that dates back twenty years and involves toxic waste and arms trafficking, the ascendancy of US military bases, the embezzlement of humanitarian aid campaigns in African countries, and the generalised corruption of local, state, and international administrations. All these criminal activities oscillate around a central and real tragic event: the Moby Prince disaster.

This catastrophe of international resonance took place in the port of Livorno, Tuscany, on the shores of the Tyrrhenian Sea. On April 10, 1991, the Moby Prince ferry collided with the Agip Abruzzo oil tanker. Hundreds of tons of Iranian oil spilled into the sea and into Moby Prince itself, causing a huge fire and eventually killing one hundred and fourty passengers who were charred or killed from gas poisoning. That night, other ships operating in the port of Livorno ended up being wrapped up in investigations related to international arms and waste trafficking. The ships' activity, camouflaged as humanitarian, was entangled with the accident. Francesco Aloe reconstructs this episode from Italy's past through Fratello, who ends up discovering Italy's historical tradition of eliminating – or disappearing – hazardous substances by burying them in southern Italy, shipwrecking the materials in offshore waters, and expelling or trafficking said waste in territories such as Lebanon, Somalia, Nigeria, and the Sahara.

Despite the scarcity of critical studies on the matter, the Mediterranean Sea as a space that is vulnerable to territorial exploitation given the dynamics of global capitalism has also gained attention within the crime genre over the last three decades.[5] Indeed, we may find a previous, well-established critical tradition dedicated to the study of crime fiction in the Mediterranean: the so-called "Mediterranean noir". Studied – and criticised – by many crime fiction and Mediterranean literature scholars and editors (see Barba; Reynolds; Smyth; Sánchez Zapatero and Martín Escribà; Turnaturi; Kinoshita), this subgenre includes realistic and critical criminal narratives that offer a glimpse of society's covert workings in places typically characterised by the radiant sun, blue skies, and crystal-clear waters. Though Leon's and Aloe's novels might fall within the parameters of Mediterranean noir, I suggest amplifying this regional understanding of the Mediterranean to a global scale in order to shed light on the tensions between the various scales that organise the contemporary spatial experience: the local, the state, the regional, and the global. The Mediterranean noir framework is not broad enough to analyse a growing number of novels that describe a *mare* that is no longer *nostrum* in the regionalist sense of the word, but rather speak of a Mediterranean with fundamental role within the dynamics of global power. How do literary representations of the Mediterranean make visible these complex, overlapping tensions between the local and the global?

I propose an analysis of the Mediterranean environment that not only expands our traditional imaginary of the sea, but also of the crime genre itself. In this sense, I use the term "narrative environments" put forward by Marta Puxan-Oliva as a way of considering traditional narrative space in historical terms, "not only as part of a fictional world, but as informed by its context", while also examining "the multi-layered historical conceptions that form them as environments" ("Colonial Oceanic Environments" 3–4). This approach can prove very rewarding when studying the Mediterranean Sea as a narrative environment: a deep-time conception that, on the one hand, links the narrative dimension to the experienced historical one, and, on the other, foregrounds its planetary resonances. In a recent research project, Puxan-Oliva promotes the global approach in literary spaces by suggesting the notion of "global environments" as spaces narrated "in such a manner that can hardly be considered in terms other than global" ("Global Literary Environments"). Oceans and seas, but also airspace, the poles, deserts, swamps, and jungles – spaces that are vast and hard to control – house resources that are often exploited for global economic benefit and possess features that transform them into internationally contested, transition spaces. Both notions – narrative environments and global environments – are extremely helpful to analyse how contemporary literature can address the environment beyond administrative borders through a deep-time conception, and how fiction concerns itself with the complex, interdependent relationship between humans and nature. How, then, does the Mediterranean as a global narrative environment adjust to Timothy Morton's well-known notion of "hyperobjects", or "things that are massively distributed in time and space relative to humans" (2)? How are criminal imaginations of the sea expanding our ideas around the Anthropocene's environmental and human impact? The following analysis focuses on two novels that address the effects of neoliberal globalisation both in their localised manifestations and through a transnational notion of crime.

Donna Leon's *Earthly Remains*: Trust in a Global Genre

Donna Leon wrote her first Brunetti crime novel, *Death at La Fenice*, in the early 1990s. Now a collection of thirty titles with over two million copies sold, the series has been translated into thirty-five languages, inspiring a German television series and spin-offs including a cookbook and a historical guide to Brunetti's Venice. Leon has participated in significant literary arenas, such as the Frankfurt International Book Fair (September 2016), accruing a plethora of literary prizes and recognitions. In April 2017, *Earthly Remains* saw simultaneous publication in English (by Grove Atlantic in the United States), German (Diogenes Verlag), Catalan (Grup 62), and Spanish (Seix Barral). Just one year later, the novel was acquired for translation and distribution in Dutch (De Bezige Bij), French (Calmann-Lévy), Polish (Noir sur Blanc), Portuguese (Relógio d'Agua), and English (by William Heinemann in the United Kingdom). Vigorously publicised, the novel enjoyed instant commercial success.

Leon occupies a space of constant visibility in the written media. Leading newspapers reviewed *Earthly Remains* as "one of her best" (Marilyn Stasio for *The New York Times*), with mentions in the *New York Times* "Bestseller", *New York Times* "Top Ten Crime Novels of 2017", and *New York Times* "Book Review Editors' Choice". The most noteworthy journalistic praise Leon accrued was in an interview with Susanna Rustin for *The Guardian*, in April 2017, as the piece clearly outlined Leon's growing inclination towards environmental criticism. Leon emerges as a fervent defender of ecology, explicitly rejecting Donald Trump's environmental policies and expressing her agitation at "the passivity of people in the face of [global warming]". Significantly titled "Donna Leon: Why I became an eco-detective writer", the interview legitimised Leon's authority in narratively dealing with global environmental issues. Leon uses crime fiction as a genre to convey global environmental conflicts, but how does she appropriate and transform its conventions? A mainstream, global author with widespread popularity worldwide, how does she narratively deal with her international audience? To answer these questions, in what follows, I will study how *Earthly Remains* is internationally oriented through its poetics, narrative, and market strategies.

The truth is that Leon's narrative develops an accurate and attractive sense of place while addressing Venice and its surroundings for a global audience. Eva Erdmann, an expert in crime fiction's internationalisation, shows that the crime fiction book market seeks "a target readership interested in the cosmopolitan and focuses on the export of familiar *topoi* and on a desire to conquer unknown and exotic territories" ("Topographical Fictions" 275). Having lived in Venice for more than thirty years, Donna Leon seems bent on continually proving her privileged position when it comes to talking about the city. Descriptions of Venetian customs, places, food, transport, and idiosyncrasies abound. To reinforce her authority, Leon almost obsessively flaunts her knowledge of Italian, with constant untranslated allusions to objects, social codes, elements of nature, culinary terms, and hints of what she calls *Veneziano* dialect ingeniously finding their way into the narrative.

We might say that Leon constantly exercises ethnographic exhibitionism, producing an abundance of passages in which the descriptions of place, people, and customs approach touristic discourse.[6] As a matter of fact, the ethnographic exhibitionism in the novel stands out to the extent that it seems to constitute – following Erdmann – an investigation in itself, a parallel "cultural investigation", designed for the reader ("Nationality International" 25). Many descriptions involve navigation, as the plot develops in the waters of the lagoon:

> Brunetti looked into the water and saw floating a metre below them a *puparìn*. . . . Closest kin to the *gondola*, though a bit shorter, the *puparìn* was Brunetti's favourite rowing boat, responsive and light in the water.
>
> *(51)*

Since Brunetti's adventures mostly unfold in Venice, the fact that *Earthly Remains*' criminal plot is set in the waters of the *laguna* gives Leon the opportunity to deploy all her knowledge of this new place, with "[n]o safe pavements, streets, places with names: only the veins and arteries of the *laguna*, disappearing with the tide, returning when it withdrew", a dizzying scenario where the sun is Brunetti's "only sure way to tell direction" (64).

Stereotypes of Mediterranean scenarios and ways of life abound within crime fiction, as the "Mediterranean noir" label suggests. The strategy may also prove remarkably successful in attracting an international audience. In this regard, scholars such as Ghose, Erdmann ("Nationality International"), Rolls et al., and Maher have specifically argued that this exoticised imaginary of Venice's cultural context is intentionally market-driven and transforms these narratives into real cultural commodities. This is evident not only in the multitude of volumes that Leon has published on the adventures of her Venetian commissario but also in the marketing strategies for her publications. All the Brunetti series' covers, for instance, generally include three fundamental elements:[7] the author's name in large and visible typeface, sometimes bigger than the title itself; an image having to do with the plot (in *Earthly Remains*, the boatman on the *puparìn*'s silhouette over the waters of the lagoon), and, finally, a Venetian landscape or some other symbolic element of the city (in this novel, the skyline of St. Mark's Cathedral), blurred and mysterious. This front-cover structure is often replicated in the translations. Strikingly, the *Earthly Remains* covers do not reflect, neither in the original nor in the translations, what would seem to be the main motivation behind the criminal plot: concerns over an environment endangered by the trafficking and dumping of toxic materials.

Despite not being reflected in her book covers, Leon's narrative shows a strong will to lay out the planet's environmental risks and the political corruption that informs them. She achieves this through an expertly crafted narrative strategy that contrasts the beauty of an exoticised Venice with the endangered planet. This can be traced through three central characters. First, through Brunetti, whose idealised gaze over the lagoon gives way to his acknowledgement of its degradation, vulnerability, and corruption after discovering the traffic and dumping of toxic waste in the waters of Venice and abroad. Following the tenets of Mediterranean noir, Leon exploits the potential impact of the contrast between a paradisiac location and its veiled corruption. Brunetti's exoticising perspective turns to horror when he discovers that "[the corporation and its workers] dumped [the toxic waste] in the *laguna*" (268). From this moment on, stereotypical discourse coexists with the rhetoric of risk society. This is specifically obvious in a secondary character, Casati, who constantly alludes to human impact on nature:

> "Look at that . . . Everywhere, we've built and dug and torn up and done what we wanted with nature. And look at this," he'd said, turning to his right and waving out over the *laguna*, "we've poisoned this, too. . . . We've poisoned it all, killed it all".
>
> *(158)*

In her study of the cultural configurations surrounding our imaginaries of endangered species, Heise ("Imagining Extinction") points out how modern environmentalist thinkers have commonly sustained "the idea that modern society has degraded a natural world that used to be beautiful, harmonious, and self-sustaining and that might disappear completely if modern humans do not change their way of life" (7). While discussing this statement is beyond the scope of this article, Leon undoubtedly uses this fatalistic perspective regarding human impact on nature, exploiting a well-known, global Mediterranean imaginary and articulating it poetically through the powerful contrast between the tropes of *paradise lost* and *risk society*. She projects it through yet another character, Patrizia Minati, the soil scientist who analyses Casati's samples of bees and sludge in the laboratory, hence giving authority and scientific backing to the rower's suspicions. Minati, who used to work in Uzbekistan for FAO (Food and Agriculture Organisation of the United Nations), specifically in the desertified Aral Sea, analysed soil samples to try to find "a natural way to rebalance the soil" (179). She eventually discovered that "the cotton that was growing everywhere . . . had killed the sea" (182), but this discovery bothered a certain diplomat, leading to her dismissal from her job. Minati's experience in the Aral Sea serves as an omen for the Venetian lagoon, an expression of the-risk-to-come. Minati shares a risk discourse with Casati, announcing catastrophe and the end of nature as starting at the Venetian lagoon but moving outwards to planet Earth as a whole.

> "I'm seeing how peaceful and beautiful it is, how lovely the birds are, how perfectly it has evolved . . . And I'm watching it die.". . . She raised a hand and waved in the direction of the water. . . . "There are fewer birds – some species no longer come here to nest – there are fewer fish. I seldom see a crab in the water. The frogs are gone. The tides don't make any sense anymore. And . . ." she began . . . "the earth itself . . ." . . . "The earth itself is what, Signora?" the inspector asked. . . . "I thought, since you work with the soil, that you meant it literally, Signora, about the soil.". . . "No, I meant the Earth, the planet. I suppose I was going to say it's gone mad".
>
> *(185–186)*

This linguistic game between earth/Earth (soil/planet) takes the global urgency that Leon transmits through ecological disasters into account. With the transnational experience of the soil scientist in the Aral Sea providing evidence of a future at risk, Leon shares speculations on the ecological catastrophe that may occur in a possible, near future, not just in the Venetian lagoon but reaching other oceans through the polluted Mediterranean waters and via the freighters transporting waste containers to countries in the so-called Global South, such as Nigeria.

I suggest that Leon manages to translate the imminent threat of a catastrophic future in the Venetian lagoon and abroad into the present. She merges the traditional realism of crime fiction novels with the rhetoric of catastrophe, deploying a quasi-speculative narrative mode. In the 2020 essay "Catastrophic Form and Planetary Realism", Debjani Ganguly discusses how the distinction between speculative and realist fiction has to be reconsidered in the face of climate catastrophe. Catastrophic scenarios do not belong to the indefinite future, but to the real present time, hence demanding and stimulating narrative strategies capable of reflecting this tension. Ganguly urges literary theorists "to pay attention to mutations in realist fiction that encode not an imaginary future of humankind, but nonhuman planetary futures that are already being written into the earth's stratigraphy" (422). While rooted in reality and in the realm of evidence – a key feature of crime fiction novels – Leon enters the domain of the possible and teases the reader with the rhetoric of the risk-to-come, in a redefinition of the temporal trope of the "yet-to-come" that Jameson assigns to science fiction. This evidence of the possible is mainly embodied by the soil scientist, through whom Leon serves to legitimise her catastrophic gesture.

Earthly Remains' most remarkable aspect lies in precisely how Leon lays out the large-scale, low-intensity, deep-time environmental effects of climate catastrophe by locating them in a very specific ecosystem: the Venetian *laguna*. While the novel wields conventional crime fiction characters and largely narrates their stories on the human scale, they are simultaneously framed within global, deep-time scales. Leon speculatively uses crime fiction formulae to direct some of these conventions into an environmentalist discourse in which the victim is the whole Venetian ecosystem. The villain goes beyond the person to comprise a complex structure of Italian administrations, criminal organisations, international corporations, and multilateral corrupted businesses. The plot circles around two investigations. One of them, performed not by a commissioner or a detective but by a *Veneziano*, sustains a risk discourse through the imagination of an endangered Earth. Leon addresses global issues while constantly providing a vivid sense of place in a gesture that could come off as exoticising. It is precisely through these mechanisms of Mediterranean exoticising that Leon speaks to a (non-Italian) international audience, becoming a global writer who mediates between a Venetian reality and a pre-existent global readership eager for crime fiction. Her narrative successfully engages in a moment in which global thinking, discourses, and narratives are imbued with catastrophism more than ever before.

Verdenero Collection and *Il vento porta farfalle o neve*: From Report to Crime Fiction

The Milan-based publishing house Edizioni Ambiente is committed to the documentation of environmental and sustainable-development issues in partnership with Legambiente, Italy's main environmental organisation. This organisation dedicates itself to identifying land- and nature-related abuses and aggressions. The annual *Ecomafia report. Le storie e i numeri della criminalità ambientale* documents their findings and uses the term "ecomafia", coined by the journalist Enrico Fontana in 1994, to refer to environmental crimes that involve organised crime. However, the audience of these reports was once limited to previously committed sectors, such as local administrations, official agencies, and ministries. In 2006, when the editorial team reassessed its strategy to attract a larger audience and increase social impact around territorial and environmental problems, it came up with a solution: each report would inspire a crime fiction or detective novel. The new collection, appropriately called Verdenero, was thus born. As its director Marco Moro explains, the scarce success of the *Ecomafia* reports underscores a paradoxical deficiency "both in the communication and the perception of mafia phenomena and of their penetration and distribution throughout the Italian territory" (196). Transforming the reporting format to fiction emerged as a brilliant idea in a country where "noir is a genre that enjoys wide success and has a large audience" (196).

The editorial team thus decided to contact crime fiction authors and start publishing commissioned novels inspired by their environmental reports, with the author roster including well-known Italian crime fiction writers such as Sandrone Dazieri, Carlo Lucarelli, Giancarlo de Cataldo, and Massimo Carlotto, as well as collectives such as Wu Ming, Michael Gregorio, Kai Zen, and Francesca Vesco, along with less-consolidated authors such as Francesco Falconi, Patrick Fogli, Alice Audouin, Eraldo Baldini, and

Francesco Aloe. The more than twenty Verdenero publications depict an underground Italy made of concrete, trash, asbestos, ashes, and poison, sketching an image of a country that remains oblivious to what lies beneath the agricultural soil and the sand on beaches, below the sea, and in the rivers. In opposition to the *bel paese* rhetoric, Moro explains that these phenomena are perfectly suitable for crime fiction.

One of the authors in the collection, Francesco Aloe, is of particular interest.[8] *Il vento porta farfalle o neve* (2011) is his only collaboration with Edizioni Ambiente, which ranked fourth in La Feltrinelli classification in April 2011. Though the book secured numerous press reviews, Francesco Aloe was mainly interviewed in the local press, including *Corriere della Calabria*, *Gazzeta di Parma*, and *Corriere Fiorentino*, and in online environmental media outlets such as the magazine *GreenNews*. The author was also interviewed and reviewed by online media outlets dedicated to Italian crime fiction, such as *Noir Italiano* and *Thriller Magazine*.

Il vento porta farfalle o neve follows the tenets of the true crime genre: built upon documents and historical facts and contexts, the novel constantly points out correspondences between fiction and reality, inviting the reader to weigh evidence and testimonies, as well as to compare perspectives and interpret different pieces of information. In *Il vento porta farfalle o neve*, the historical events documented as having occurred stem from evidence compiled by the Associazione 10 Aprile, as well as throughout police investigations and judicial proceedings. These documents are woven into the novel in different ways, with paratexts bolstering the facts. The novel opens with a personal note from Aloe, who dedicates the novel "a 140 persone. Uccise" [to 140 murdered people], signalling to the audience that they will face a non-official recreation of the catastrophe. This note is followed by a preface signed by Luchino Chessa, son of the Moby Prince commander Ugo Chessa, who, together with his brothers, promoted the formation of the Associazione 10 Aprile Familiari Vittime Moby Prince in March of 1995.[9]

The narration is full of passages that interweave real reports with the fictional plot. In fact, the fight to acquire confidential information from official reports motivates the entire plot: Fratello's boss needs access to the data collected in official reports, recordings, testimonies, etc., over the years, but the data has been kept hidden in a microchip. An example of this interweaving of the reports in the narration lies in the moment when Fratello reviews his mafia boss's papers containing all the information his clan had managed to compile in relation to the events of 1991:

> A partire dal 1987 è attiva in Italia una lobby affaristico-criminale che gestisce il traffico di rifiuti tossici-nocivi e radioattivi, stupefacenti, armi, titoli di Stato falsificati e materiali strategici nucleari. Per quanto riguarda le scorie tossiche e radioattive, lo smaltimento può avvenire con tre distinte modalità: l'interramento in località del Sud Italia in vecchie cave o discariche, l'affondamento di navi normalmente in zone extraterritoriali o lo smaltimento presso paesi del Terzo mondo come il Libano, la Somalia fino al 1992, la Nigeria e il Sahara.
>
> *(272)*
>
> [Since 1987, a business-criminal lobby has been active in Italy that manages the trafficking of toxic-hazardous and radioactive waste, drugs, weapons, falsified government titles and strategic nuclear materials. As for toxic and radioactive waste, disposal can take place in three ways: burial in places in Southern Italy in old quarries or landfills, the sinking of ships normally in extra-territorial areas, or disposal in Third World countries such as Lebanon, Somalia until 1992, Nigeria, and the Sahara].[10]

Finally, the book closes with some annexes titled "Fonti" in which the author lists the sources that have him helped build the story: testimonies of the disaster, voice recordings, videos of the fire, legal medical reports, court sessions, and journalistic investigations in *Repubblica* and *Corriere de la Sera*, alongside other data, such as the chronology of events and the "Elenco delle vittime" (list of the victims). All these texts and paratexts intensify and bolster the novel's verisimilitude and sense of outcry, thus expanding on what I consider the dominant narrative strategy: insistence on fact.

In *Il vento porta farfalle o neve*, Fratello's first-person narrative voice accompanies the reader and guides the research process. Choosing the voice of a mafia killer considerably increases the story's credibility: Fratello, an experienced hitman, knows the dark businesses and shady procedures of the mafia and its networks like the back of his hand, but, at the same time, he remains sensitive and has an ever-so-slight sense of justice. The character's double condition strengthens the reader's trust and helps unfold the alarm discourse while emphasising the corrupt linkages between politics and economics that can be read on a scalar structure: from local politics to international deals. While *Earthly Remains* sustains what I termed a 'risk discourse' based on the imminent end of nature through an environment-centred approach, in *Il vento porta farfalle o neve*, the alarm discourse focuses on the revelation of present social corruption and citizens' lack of awareness. By casting light on the interests behind the world's main political and social agencies, and the international relations of power established by global capitalism, the novel stands as an eye-opener for Italian citizens. From time to time, the boss accuses Fratello of ignorance, contemptuously repeating: "Che minchia sai tu del tuo Paese? Niente" (What the fuck do you know about your country? Nothing), and "Tu non sai un bel niente del tuo Paese" (You don't know a thing about your country) (242–243).[11] Despite the deterritorialisation of its jurisdiction, Aloe insists on the nation-state as the leading organisational agent that can demand accountability for such crimes.

Il vento porta farfalle o neve takes on the arduous task of representing the Mediterranean Sea as a global space that is both a victim and facilitator of human and environmental crime. Aloe manages to trace an international network of influences, power, exploitation, fraud, and crime through an intricate plot that ultimately links – through the maritime circulation of toxic waste and weapons – international macro-operations to Italy, Spain, the United States, Somalia, the Sahara, Libya, Ireland, Iran, and Lebanon, together with military, judicial, political, and civil agencies from all over the world, from the American army in the port of Livorno to the Italian mafia that disposes of the charred remains of the Moby Prince. In addition, the author links various historical events, from the Gulf War to the identity struggle of the Tuaregs, to the 1992 Progetto Urano. All this is condensed in the port of Livorno, where these specific Mediterranean waters are depicted as a splendid global narrative environment, but not as Leon's exportable, deeply exoticised space.

Il vento porta farfalle o neve draws from facts to build an alarm discourse in its quest to raise readers' awareness of the systemic corruption of Italian authorities and complex global structures. Aloe's narrative is driven by facts, resulting in a hybridised form of novelistic crime and true crime that is enriched through paratextual devices. By combining fictional and non-fictional forms, Aloe succeeds in highlighting the extent to which certain criminal activities in contemporary Italy exceed the boundaries of the state. At the same time, and by structuring the plot around the evidence of real facts, Aloe centres the state as the main evader of responsibility. The novel uses the conventions of crime fiction to depict the complexities of globalisation and its criminal impacts on the shores of Italy and abroad through an anthropocentric alarm discourse focused on the human dimension of catastrophe in terms of justice and reparation. While producing a distinctive narrative form by complicating the linkages between the national and the international, Aloe's target audience is explicitly Italian, and Verdenero's main goal is not to compete with international publishing conglomerates, but to nurture a national narrative based on true facts in order to build awareness among the Italian citizenship around the mechanisms that underlie state activities.

Conclusions

Crime fiction has proven a suitable genre to deal with global environmental and humanitarian concerns and their implications on local, state, and regional scales. The analysis of *Earthly*

Remains and *Il vento porta farfalle o neve* shows that crime fiction's adaptability facilitates the representation of the complexities of climate catastrophe, especially with regard to issues of scale, as the local and the global find themselves in constant tension through the setting (needing specific places to contain the abstract notion of the global), the plot (with the apparently accidental deaths in specific contexts often revealing an intricate plot of international reach), the victim (human but also non-human, even including the ecosystem), and the crime (with human deaths often accompanied by environmental crimes).

The Mediterranean emerges as a global environment that is both victim and victimiser. As per the tensions outlined earlier, the crime is located in a global scenario where state responsibilities pretend to be offshored or exported. Leon's and Aloe's novels denounce the scalar perversion of a political, social, economic, and moral global system of waste transport, trafficking, and dumping. To do so, both novels point to specific seaports (Venice and Livorno) as local spaces in which massive international finance networks and global geopolitical operations, official and unofficial, are deployed. However, Leon emphasises the planetary dimensions of environmental catastrophe through a deep conception of time and space, sustained through her quasi-speculative narrative mode that connects with a global trend of leaning into catastrophist thinking. In this sense, Leon highlights the irrelevance of the state, eschewing its primary agency, and developing a deterritorialised perspective of felony. Meanwhile, Aloe exposes the massive scale of a humanitarian crime perpetrated in a single night, connecting it to previous felonies to ultimately sustain not an ecosystem-focused approach but an anthropocentric one. Through the hybridisation of fact and fiction, Aloe (together with Edizioni Ambiente's Verdenero collection) aims at bridging the gaps between state violence in Italy and the lack of awareness among the Italian citizenship, thus binding the novel within Italy itself.

Both Donna Leon and Francesco Aloe (as well as the Verdenero collection) use crime fiction not only because of its capacity to encompass global concerns, but because of its traditional audience success. Leon's success must be analysed in connection with the importance of literary agents, book fairs, and her successful adaptation of a globally celebrated genre with exotic Mediterranean *couleur locale*. Translated into more than thirty languages, Leon intentionally addresses her narratives to a solid, international crime fiction audience, using the genre's existing transnational scope to raise awareness among readers from all over the world, while setting the novel in a stereotyped Venice as a guarantor of circulation. In contrast, Aloe and the Verdenero collection target what Leon precisely excludes: the Italian audience. Their use of the crime mode to fictionalise real-world reports is based on an awareness of the genre's historical tradition and success among Italian readers. These two examples prove that crime fiction stands as a genre that, despite being tightly linked to literary market strategies, can call out environmental crime at different scales: Verdenero takes advantage of its local and state impact, while Leon maximises the genre's transnational exportability, paradoxically excluding impact on the local scale despite her novel's Italian setting.

Acknowledgements

I am indebted to Neus Rotger and Marta Puxan-Oliva for generously sharing ongoing findings and providing key recommendations.

Funding

This research was conducted with the support of the Secretariat of Universities and Research of the Generalitat de Catalunya and the European Social Fund, and the project "The Novel as Global Form. Poetic Challenges and Cross-border Literary Circulation" (Spanish Research Agency, PID2020–118610GA-I00/AEI/10.13039/501100011033).

Notes

1. Leon has repeatedly explained to the media that this bar is a way of remaining anonymous in Venice, where she lived for more than thirty years.
2. This excellent article provides the first in-depth state of the art on the global novel and points toward new lines of inquiry through two case studies. I highly recommend reading it to delve further into the emergence of the global novel as a genre, including theorisations and controversies.
3. As Rotger and Puxan-Oliva posit, novel-studies scholars are linking the genre from different disciplines such as ecocriticism, cosmopolitanism, and translation studies with previously unrecognised fields like human rights, migration, translation, and ecology.
4. In 2009, Murphy urged his ecocritical colleagues to go beyond "the canonically defined literary texts . . . to read more widely in the realms of popular genre fiction", as well as "to understand the degree to which environmental consciousness and nature awareness has permeated popular and commercial fiction" (119).
5. For instance, in most of their novels, acclaimed crime-fiction writers such as Manuel Vázquez Montalbán, Jean-Claude Izzo, and Petros Markaris deal with unequal urban development and the consequent gentrification of Mediterranean cities to satisfy international financial interests. Lesser-known writers on the international level, like Xavier Moret (*Tramuntana* 2011, *Formentera blues* 2019) and Eduard Palomares (*No cerramos en agosto* ['We do not close in August'] 2019) criticise the Mediterranean coast's speculative real-estate industry and mass international tourism. Massimo Carlotto and Marco Videtta (*Nordest* ['North East'] 2005) portray how mafia networks use Mediterranean waters to transfer and dump toxic waste, while Carlo Lucarelli (*Navi a perdere* ['Ships to be lost'] 2008) casts light on environmental pollution resulting from toxic material discharges and big-industry activity.
6. Brunetti has been considered a tour guide to Venice, of sorts, for the foreign reader (see Ghose, Maher).
7. For an in-depth study of the covers of criminal novels, see Louise Nilsson's chapter, "Covering Crime Fiction: Merging the Local into Cosmopolitan Mediascapes", in the volume she edited together with Theo D'haen and David Damrosch, *Crime Fiction as World Literature*.
8. A specialist in foreign languages and literature, he published in 2008 his first thriller, *Vertigine* (later translated into Spanish), with Lettere Animate, followed by *Il vento porta farfalle o neve*. He later published the climatic dystopia *L'ultima bambina d'Europa* (Alter Ego Edizioni) in 2017 and the surreal crime-fiction novel *Aspetta l'inverno* (Alberti Compagnia Editoriale) in 2019.
9. For further information, see the website www.mobyprince.it/lassociazione-10-aprile/.
10. My translation.
11. My translation.

Bibliography

Allan, Janice M., et al. "Introduction: New Directions in Crime Fiction Scholarship." *The Routledge Companion to Crime Fiction*, edited by Janice M. Allan et al., Routledge, 2020, pp. 1–9.

Aloe, Francesco. *Il vento porta farfalle o neve*. Edizioni Ambiente, 2011.

Anderson, Jean, et al., editors. *The Foreign in International Crime Fiction: Transcultural Representations*. Bloomsbury, 2014.

Barba, David. "La novela negra mediterránea." *Primer encuentro europeo de novela negra. Homenaje a Manuel Vázquez Montalbán*, edited by David Barba, Planeta, 2005, pp. 21–22.

Erdmann, Eva. "Nationality International Detective Fiction in the Late Twentieth Century." *Investigating Identities. Questions of Identity in Contemporary International Crime Fiction*, edited by Marieke Krajenbrink and Kate M. Quinn, Rodopi, 2009, pp. 11–26.

———. "Topographical Fiction: A World Map of International Crime Fiction." *The Cartographic Journal*, vol. 48, no. 4, 2011, pp. 274–284, doi:10.1179/1743277411y.0000000027.

Ganguly, Debjani. "Catastrophic Form and Planetary Realism." *New Literary History*, vol. 51, no. 2, 2020, pp. 419–453, doi:10.1353/nlh.2020.0025.

Ghose, Indira. "Venice Confidential." *Venetian Views, Venetian Blinds. English Fantasies of Venice*, edited by Manfred Pfister and Barbara Schaff, Rodopi, 1999, pp. 213–224.

Ghosh, Amitav. "Petrofiction: The Oil Encounter and the Novel." *Incendiary Circumstances: A Chronicle of the Turmoil of Our Times*. Houghton Mifflin, 2005, pp. 138–151.

———. *The Great Derangement: Climate Change and the Unthinkable*. U Chicago P, 2016.

Heise, Ursula K. *Sense of Place and Sense of Planet: The Environmental Imagination of the Global*. Oxford UP, 2008.

———. *Imagining Extinction: The Cultural Meanings of Endangered Species*. U Chicago P, 2016.
Keith, Joseph. "The Novel as Planetary Form." *The Cambridge Companion to the Novel*, edited by Eric Bulson, Cambridge UP, 2018, pp. 268–283.
King, Stewart. "Crime Fiction as World Literature." *Clues. A Journal of Detection*, vol. 32, no. 2, 2014, pp. 8–19.
Kinoshita, Sharon. "Mediterranean Literature." *A Companion to Mediterranean History*, edited by Sharon Kinoshita and Horden Purcell, John Wiley & Sons, 2014.
Krajenbrink, Marieke, and Kate M. Quinn, editors. *Investigating Identities: Questions of Identity in Contemporary International Crime Fiction*. Rodopi, 2009.
Leon, Donna. *Earthly Remains*. William Heinemann, 2017.
Maher, Brigid. "A Crook's Tour. Translation, Pseudotranslation, and Foreignness in Anglo-Italian Crime Fiction." *Perspectives on Literature and Translation*, edited by Brian Nelson and Brigid Maher, Routledge, 2013, pp. 145–158.
Matzke, Christine, and Susanne Mühleisen. *Postcolonial Perspectives: Crime Fiction from a Transcultural Perspective*. Rodopi, 2006.
Moro, Marco. "Room with a View (of a Landfill): The Making of Verdenero." *Italy and the Environmental Humanities. Landscapes, Natures, Ecologies*, edited by Serenella Iovino, Enrico Cesaretti, and Elena Past, U Virginia P, 2018, pp. 194–199.
Morton, Timothy. *Hyperobjects: Philosophy and Ecology after the End of the World*. U Minnesota P, 2013.
Murphy, Patrick D. *Ecocritical Explorations in Literary and Cultural Studies: Fences, Boundaries and Fields*. Lexington Books, 2009.
Nilsson, Louise, et al., editors. *Crime Fiction as World Literature*. Bloomsbury Academic, 2017.
Pepper, Andrew, and David Schmid. "Introduction." *Globalization and the State in Contemporary Crime Fiction*, edited by Andrew Pepper and David Schmid, Palgrave Macmillan, 2016, pp. 1–19.
Puxan-Oliva, Marta. "Colonial Oceanic Environments, Law and Narrative in Herman Melville's *Benito Cereno* and Juan Benet's *Sub Rosa*." *English Studies*, vol. 99, no. 4, 2018, pp. 426–441.
———. "Global Literary Environments, or the Global Discourse of Space in Contemporary Narratives." *Global Literary Studies. Key Concepts*, edited by Neus Rotger and Diana Roig-Sanz, De Gruyter, forthcoming 2022 [Accepted for publication. Manuscript courtesy of the author].
Reynolds, Michael. *Black and Blue: An Introduction to Mediterranean Noir*. Europe Editions, 2006.
Rolls, Alistair, Marie-Laure Vuaille-Barcan, and John West-Sooby. "Translating National Allegories: The Case of Crime Fiction." *The Translator*, vol. 22, no. 2, 2016, pp. 135–143.
Rotger, Neus, and Marta Puxan-Oliva. "The Novel after the Global Turn: Decentered Perspectives from the Spanish Literary Field." *Studies in the Novel*, vol. 53, no. 3, 2021, pp. 285–305, doi:10.1353/sdn.2021.0025.
Rustin, Susanna. "Why I Became an Eco-Detective Writer." *The Guardian*, 15 Apr. 2017, www.theguardian.com/books/2017/apr/15/donna-leon-interview-commissario-brunetti-earthly-remains. Accessed 29 May 2021.
Sánchez Zapatero, Javier, and Àlex Martín Escribà. "La novela negra mediterránea: crimen, placer, desencanto y memoria." *Pliegos de Yuste*, no. 13, 2011–2012, pp. 45–53.
Smyth, Edmund J. "Marseille *Noir*: Jean-Claude Izzo and the Mediterranean Detective." *Romance Studies*, vol. 25, no. 2, 2007, pp. 111–121, doi:10.1179/174581507x193009.
Stougaard-Nielsen, Jakob. "World Literature." *The Routledge Companion to Crime Fiction*, edited by Janice M. Allan et al., Routledge, 2020, pp. 76–84.
Tanaka, Shouhei. "The Great Arrangement: Planetary Petrofiction and Novel Futures." *MFS Modern Fiction Studies*, vol. 66, no. 1, 2020, pp. 190–215, doi:10.1353/mfs.2020.0008.
Trexler, Adam. *Anthropocene Fictions: The Novel in a Time of Climate Change*. U Virginia P, 2015.
Turnaturi, Gabriella. "The Invention of a Genre: the Mediterranean Noir." *New Perspectives in Italian Cultural Studies: The Arts and History*, edited by Graziella Parati, vol. 2, Rowman & Littlefield, 2012.

33

MAGIC SEEDS AND THE LIVING DEAD

Investigating Transnational Ecocrimes in Rajat Chaudhuri's *The Butterfly Effect*

Damini Ray

Functioning within the demands and constraints of popular fiction, crime fiction has been variously perceived as lowbrow, pulp, and formulaic writing. At the same time, it has enjoyed immense commercial success over the decades and remains one of the bestselling book genres across the world. As Raymond Chandler puts it in his essay, "The Simple Art of Murder" (1944): "The average detective story is probably no worse than the average novel, but you never see the average novel. It doesn't get published. The average – or only slightly above average – detective story does" (3). Popular genres such as crime fiction, science fiction, or fantasy have, for a long time, been considered unsuited to the processes of serious critical enquiry or social and political activism.[1] Commenting on this perceived demarcation between 'serious' fiction and genre fiction, Michael Chabon has stated: "For even the finest writer of horror or sf or detective fiction, the bookstore, to paraphrase the LA funk band War, is a ghetto" (21). Despite this ghettoisation of popular fiction, some of the major works of climate change fiction and eco-fiction in the last two decades have been written in one or a combination of these genres, for example, *The Windup Girl* (2009) by Paolo Bacigalupi; *Drive Your Plow Over the Bones of the Dead* (2009), an "eco-mystery" by Olga Tokarczuk; *Memory of Water* (2014) by Emmi Itäranta; or *American War* (2017) by Omar El Akkad. Critical scholarship has also refocused on examining how these genres, while being ostensibly market-driven, provide possibilities for disruptive thinking, critiquing accepted hegemonic structures, or playing with received notions of culture, class, gender, or the environment.

Many writers continue to choose popular modes to engage with ecological crisis, climate change, and human-induced environmental disasters in the Anthropocene, blending and bending genres as diverse as speculative fiction, fantasy, detective fiction, and historical fiction. In *The Great Derangement* (2016), Amitav Ghosh observes that climate change events are:

> peculiarly resistant to the customary frames that literature has applied to 'Nature': they are too powerful, too grotesque, too dangerous, and too accusatory to be written about in a lyrical, or elegiac, or romantic vein. Indeed, in that these events are not entirely of Nature (whatever that might be), they confound the very idea of 'Nature writing' or ecological writing: they are instances, rather, of the uncanny intimacy of our relationship with the non-human . . . we are confronted suddenly with a new task: that of finding other ways in which to imagine the unthinkable beings and events of this era.
>
> *(Ghosh 36)*

DOI: 10.4324/9781003091912-39

Environmental disasters in fiction run the risk of appearing as terrible but isolated incidents, or events that are too far removed from our common lives. This chapter explores how crime fiction offers certain unique investigative opportunities when looking into an ecological crisis, prompting the reader to interpret it in terms of real-time human actions and their consequences. Through a close ecocritical study of *The Butterfly Effect* (2018), written by Indian writer and environmental activist Rajat Chaudhuri, this chapter argues that the text locates climate change and environmental crisis within certain narrative and thematic frameworks of crime fiction and thereby poses critical questions about human culpability, justice, and ethics with regard to environmental transgressions that extend across nations. This also allows for a more immediate and urgent engagement with environmental change, which is inextricably linked with both personal and collective trauma in the novel.

While it has been variously categorised as 'biopunk' and 'climate fiction', *The Butterfly Effect* consistently resists strict generic labels and intricately interweaves timelines, places, and narrative styles. It presents an environmental thriller about the dystopian consequences brought about by genetic engineering, irresponsible competitive corporations, and autocratic governments that unfold over large parts of Asia: an underrepresented and oft-overlooked region when it comes to climate and ecofiction. In recent years, some climate fiction works have explored Asian settings, such as *The Last Children of Tokyo* (2014), written in Japanese by Yōko Tawada and translated into English by Margaret Mitsutani (published in 2018); *Latitudes of Longing* (2018) by Shubhangi Swarup; and *The Gun Island* (2019) by Amitav Ghosh. The continent has a major number of underdeveloped, developing, and poor nations with widespread problems of overpopulation, high-consumption lifestyles, and unsustainable development which make them especially susceptible to climate risk and environmental disasters. Yet it has not featured much in Western literary considerations of climate change. This chapter argues that the aspects of crime fiction that are strategically incorporated by Rajat Chaudhuri throughout the novel not only help probe deeper into the processes behind a massive environmental disaster but also allow him to explore its transnational nature. The detective, as this chapter proposes, has meaningful roles to fulfil in the face of ecological crisis and needs to be reimagined and re-evaluated within the context of ecofiction. The first section of this chapter aims to examine the character of Private Investigator Kar as an agent of enquiry into the complex web of interconnected events and factors behind the ecological calamity, as he travels through the different locales in the novel. The second section addresses some of the ecological perspectives presented in the text and evaluates Kar as an enduring witness to ecological crimes in the Darkland narrative of the future. The final section of the chapter looks into the ways in which environmental crime, violation, and transgression complicate, and intersect with, issues of social injustice, corporate greed, and state oppression in the text.

Ecofiction and climate change fiction are still emerging forms in Indian Anglophone writing, even though nature figures prominently in both the long oral and folk tradition of India and its written literature. Rajat Chaudhuri – who has been actively involved with the United Nations Commission on Sustainable Development (UNCSD) and has contributed to the United Nations World Human Development Report (UNDP, New York) – believes that stories have the potential to spark collective interest, debate, and concern over critical issues like climate change, environmental hazards, and sustainable consumption. He has also talked about the responsibility of artists and writers to engage with environmental concerns through collaborative and individual efforts:

> We need people in large numbers to believe and engage in efforts that address emissions. Such efforts cover protests, awareness, advocacy and other tools but for movements to gain traction we need a spark, we need momentum . . . a great cli-fi novel can supply the spark and also help sustain the momentum.
>
> *(Chaudhuri, "Q&A")*

Chaudhuri's preference for popular formats to amplify and advocate for climate concerns is evident in his other creative endeavours, such as his narrative contribution to the cli-fi videogame, *Survive the Century* (2021), where the player has to perform as the senior editor of the topmost news organisation in the world. In *The Butterfly Effect*, he boldly attempts to envision how the use of untested genetically modified paddy causes an unprecedented pandemic of rapid ageing that is accelerated by drastic climate change. These events soon lead to massive environmental displacement, an acute shortage of resources, and a dystopian restructuring of power in major parts of Asia. As argued by Stewart King, "*crimate* fictions frame the causes of human-induced climate change as a criminal act of which there is a victim and for which those responsible should be held accountable" (1237).

The book progresses through a fragmented but cyclical structure, with four major narrative strands set in different timelines that come together loosely at the end, without a firm closure. Captain Old, an off-duty policeman in the 'Darkland' of the near dystopian future, is tasked with locating and assassinating the leader of the dissenting Red Dawn organisation which has been trying to topple the dictatorship of the Supreme Guide; PI Kar takes on a case to track down a group of missing Indian tourists in South Korea; Tanmoy Sen, a talented geneticist tries to create a new kind of drought-resistant, nutrient-enriched paddy in GeneLab UK; the mysterious Miss Park is employed as a guide for the Indian tourists in South Korea. Chaudhuri tries to explore the myriad interconnections and far-reaching ramifications of ecocide in a capitalist, power-grabbing world by decentring his story and travelling to and fro in time, societies, cultures, and geographies.[2] As the title suggests, through a patchwork of individual narratives, the novel emphasises the ways in which apparently minor and isolated changes, incidents and actions can set off unprecedented repercussions in complex and chaotic environmental systems.

Investigating the Detective Figure

From the beginning, the novel consciously employs an investigative plot to draw connections between its disparate narrative fragments. Private Investigator Kar and Captain Old, who turn out to be the same character at the end of *The Butterfly Effect*, carry out a complex interrogative function through the present and future narratives. Kar is introduced to the reader as a self-occupied, cynical, and jaded detective in Calcutta, with a slightly delinquent assistant, Chaitanya. Chaudhuri's fascination with the character precedes this novel. The detective was first introduced in Chaudhuri's debut book *Hotel Calcutta* (2013). PI Kar's investigation into the mysterious disappearance of the group of Indian tourists sets off the truth-seeking mission and involves the reader in this shared quest to assemble and make sense of the series of disjointed events.

Initially, PI Kar performs some of the conventional storytelling functions of the detective figure by disclosing information or 'clues' and aspects of the mystery. Through his interrogation, it is revealed that Tara, the only missing tourist to have returned, has developed some sort of unfamiliar amnesia and lost major control of her cognitive functions. This is of course an early hint about the effects of the environment that the tourists have been exposed to and also a foreshadowing of the epidemic of accelerated ageing and amnesia that would spread worldwide in the future. In Captain Old's parallel narrative of the dystopian future, there are indications of this disease in the use of terms such as "memory pill", "youngblood", and "freshface", which are only elucidated at the end (*Butterfly Effect* 372, 373). However, instead of leading the reader toward the revelation of a criminal or the human agents behind the disappearance of the tourists, PI Kar's investigation increasingly moves away from the human drama and refocuses the narrative towards much graver crimes against the environment, of which this case of the missing tourists is only a tiny ripple effect.

The 'private investigator' is understood as a quintessentially solitary, individualistic, outlaw figure who is not a part of the socio-economic and familial order. John Scaggs further points out how his 'private' nature is "indicated in the first-person 'I' of the term 'PI'" (59). In the hard-boiled tradition, the PI lives and works for himself and by himself. Interestingly, as PI Kar moves from Calcutta to Seoul to the White Cloud Mountain in search of the tour group, he gets more and more entwined in the events and actions that not only imperil individual futures but global futures. In an important revision of the classic private investigator figure, PI Kar doesn't find the group of tourists by himself. Rather, he is sought out by Jiyoo Park and escorted to them so that he can take the quickly ageing tourists back to their homes. This deemphasises the focus on the individual prowess of the PI in finding answers, proposing the need for more collaborative and globally minded responses to environmental crisis. The journey to the site of the disappearance takes Kar from the highly modernised Seoul right into the heart of the dense forests of White Cloud Mountain. Marta Puxan-Oliva observes that the

> wilderness has long had an acknowledged place in crime fiction: By imagining isolated places such as these, where crimes remain hidden from human sight and outside social frameworks, narratives can induce fear, produce a sense of mystery and highlight the violation of legal frameworks, since those frameworks are mostly based in built environments.
>
> *(363)*

In crime fiction, the forest can serve as a setting where the victim is isolated from civilisation and easy to prey upon or as a hunting ground to stalk, overpower, or violate a weaker quarry. Olga Tokarczuk offers an interesting spin on this in *Drive Your Plow Over the Bones of the Dead* where animal hunters become the hunted and their dead bodies are found one after the other in the forest. The protagonist of the novel claims that these murders could have been perpetrated by animals. The wilderness is a primitive zone that lends itself well to imaginings of psychological or sexual deviance and their concealment. However, it can also offer refuge, sustenance, and opportunities for healing for those who are fleeing from aggressors, have not found legal and social justice, or who have been let down by civilisation.

Chaudhuri presents the forests of White Cloud Mountain as a powerful, "primeval", liminal space where the natural and the supernatural coexist "dense as the blackest dream" and alive with dangers (Chaudhuri, *Butterfly Effect* 305). Jiyoo Park finds solace, purpose, and a deep connection with this wilderness and keeps returning again and again to the forests and the community living in the hidden valley to escape the alienation and impersonality of the city. Her decision to bring the Indian tourists to visit this special place stems from her urge to share this communion with nature. The forests and the hidden valley of White Cloud Mountain do not serve as passive settings or backdrops for human crimes. These places rather become the sites of human transgression and environmental degeneration; they are the first victims of ecocide in the novel. It is there that PI Kar first finds evidence of the irreversible damage to the ecosystem caused by the genetically modified paddy that had been planted in the valley. He notes:

> The dull yellow weed seemed to have taken over the mountains. Wherever he looked the carpet, like dead skin, covered the floor of the forest . . . The pine and birch stood faithlessly amidst the sea of other worldly vegetation but something seemed to have drawn their sap. The trees looked weak, shriveled and febrile unlike the green canopied giants he had seen earlier.
>
> (*Butterfly Effect* 312–313)

It almost mirrors the degradation in the decrepit, maggot-ridden dead body of one of the tourists Kar comes across soon after. The genetically spliced rice affects most of the tourists – and the Korean settlers living there – with a kind of "runaway amnesia" and accelerated ageing that makes them too sick to return home by themselves (354). In an uncanny episode with Jiyoo Park and the ghost of her dead sister Jia, he is informed that these untested splices had been stolen from a British laboratory by North Korean agents and a sample of that paddy landed up with Jiyoo Park, a defector from North Korea. In her desire to give back to the settlement and her spiritual guide, Jiyoo had offered the 'magic seeds' in the belief that they would yield an improved, richer crop.

It is important to note the complex transnational context of the novel in terms of the shifting locations it reveals – from Darkland to Cleanland, the cities of Calcutta, Seoul, London, the British countryside, the dense forests of White Cloud Mountain or the prison camps of Pyongyang. In the face of media silencing, corporate and governmental pressure, and international cover-ups, PI Kar alone cannot expose or resolve the crimes that trigger the environmental crisis.[3] But he is also collectively implicated in them and not a neutral, outside observer. Kar's function as a detective in this regard does not extend to bringing about justice or averting the ecocide. Neither is possible. He can only attempt to make sense of the interrelations between human actions and events and the environmental disasters that follow:

> Unprecedented situation' and "All out efforts" were the two expressions that had become the stock-in-trade for politicians, but in the fog of the weather events and the cover-up efforts, the effects of the deadly Scourge went unnoticed or were misinterpreted till it was too late . . . Finally the shortage of food, the burden of disease, and the millions of migrants pouring across borders had precipitated the war.
>
> *(360)*

Chaudhuri makes it clear that no individual, not even the lone detective, can operate authoritatively by himself or remain unaffected amidst rapid and calamitous changes to one's environment. It is essential to review the 'private' investigator within this transcultural and transnational matrix of socio-political, economic, and ecological factors that act upon the collective because understanding "our ecological enmeshment shifts emphasis to the detective as a situated and imperfectly knowledgeable agent who must act nevertheless" (Walton 3).

Ecological Consciousness and Memory

The detective, through the course of his investigation, often has to interrogate and work with a vast array of people that cuts across class, wealth, professions, cultures, organisations, and sometimes even nations. In the process of extracting the truth (or fragments of the truth), he interacts with the police, specialised experts, the judicial system, criminals, informers, outcasts, and communities. This multiplicity of perspectives can offer valuable insights into how different people, or groups of people, conceive of ideas such as crime, justice, violence, and transgression. From an ecocritical point of view, this also provides the opportunity to examine different "cultural understandings of human-nature relations and environmental crisis" (Walton 2). In *The Butterfly Effect*, Kar acts as a contact point for a number of varying ecological perspectives that oppose, influence, and enhance his understanding of nature and humanity's relationship with the environment. In this sense, Kar's role as a detective functioning amidst environmental crisis is reimagined in a new capacity. In an initial conversation with the NIS officer Kim in Seoul, PI Kar is annoyed and surprised by one of Kim's theories that the tourists could have been lured into some unknown valley hidden deep within the mountains of the country. He is baffled by Kim's assertion that there exist "secret valleys in mountains around the world protected by the Masters . . . places where they say the righteous will be sheltered in times of great strife" (94). This idea of a mystical natural space contends with Kar's rational view of the world as ordered, mapped, accessible, and ultimately, anthropocentric.

PI Kar's investigative encounter with the 'Keun Sunim' at White Cloud Mountain in South Korea gives an insight into more nurturing environmental cultures and behaviours. He is a Buddhist monk whom Jiyoo Park describes as a master of the "Short Path" (Chaudhuri, *Butterfly Effect* 266). Through the Keun Sunim and the community he has fostered in the hidden valley, Chaudhuri underscores an environmental philosophy and an ecological consciousness that draws from some aspects of deep ecology. The Keun Sunim himself is deeply in harmony with the mountain and its mystical dense forests. PI Kar finds out how he helped build, sustain, and guide the small settlement in the secluded valley, initially with some of the wounded soldiers after the war. Over the years they were joined by weary defectors from North Korea who couldn't assimilate well and people who chanced upon the monk or the settlement and gained his trust. As their spiritual guide, the Keun Sunim fosters a shelter, a kind of commune that offers refuge, healing, and meaningful environmental and social connections to people who have suffered trauma, injustice, and alienation. The Keun Sunim's insistence on simple living and maintaining the traditional way of life in the valley, and his belief that the community can only sustain itself as long as it is private and small ("Too many people will spell the end of our experiment"), resonates with some of the basic principles of deep ecology (*Butterfly Effect* 353). Deep ecology argues that humans have interfered with, overused, and encroached upon nature and the non-human world for too long. Human societies will benefit from a decline in human population and a focus on life quality and low consumption, instead of an "increasingly higher standard of living" (Naess 29).[4]

In *The Butterfly Effect*, after the tourists arrive in the valley, they join and contribute to the community by participating in its culture, work, and responsibilities. The commune espouses a certain kind of environmental ethics; the settlers derive sustenance from the woods, grow most of their food, share resources sustainably, and cooperate and collaborate without aggressively modernising the settlement and disrupting the ecosystem. In this sense, it can be considered an 'eco-village': "an intentional, traditional or urban community that is consciously designed through locally owned, participatory processes in all four areas of regeneration (social, culture, ecology and economy) to regenerate their social and natural environments" (*What*). Significantly, the introduction of the untested, genetically modified paddy has a deeply destructive and irreversible impact on the carefully-preserved natural ecosystem of the valley and the people living in it. Environmental crimes are not just restricted to directly exploitative and damaging human activities like mining, deforestation, or pollution. They can also involve irresponsible experimentation with nature and food and the use of technologies that interfere with the natural processes of the environment. With regard to the novel, Rajat Chaudhuri has commented,

> climate threats have been growing in exponential proportions and dangerous technologies are being pushed into our lives and food plates without proper testing and without the use of the "precautionary principle," which is a foundational principle of the Biosafety Protocol.
>
> *(Chaudhuri, "Wild Authors")*

Kar plays an important role in both the present and future narratives as a witness to the ecocide that unfolds throughout the book and to the failure of global and national power systems in redressing it or restoring the environment. Though he cannot bring about justice as a detective, Kar's remembrance of past events, encounters, and connections between the ecological crisis and negligent human actions is particularly crucial in a world suffering from mass amnesia. In the dystopian future, he appears to be one of the lucky few in Darkland who are resistant to the disease of rapid ageing and amnesia. Kar recalls, "And still he had remained young while the others had withered away. He didn't even have to pay the gerontology cess. No doubt they loved to call him Captain Old" (361). The notion of the detective in the novel thus extends beyond his traditional epistemological function, becoming instead a custodian and keeper of individual and collective memories. While surrounded by mass apathy and forgetfulness, the task of investigating the truth and persevering it becomes an important ethical responsibility of the detective.

Amnesia is a common device in crime fiction and films and is often employed through unreliable witnesses or narrators, gaps in information, and missing identities. In Alfred Hitchcock's iconic *Spellbound* (1945), a psychoanalyst tries to help an amnesiac man who confesses that he might have committed murder. On the other hand, it is the secret agent himself who has psychogenic amnesia in *The Bourne Identity* (1980) by Robert Ludlum. Paula Hawkins explores how blackouts induced by binge drinking interfere with the protagonist's ability to discern the truth as she looks into a murder in *The Girl on the Train* (2015). On a deeper thematic and symbolic level, memory loss in crime fiction can be tied to struggles with physical as well as psychological trauma and abuse, the vulnerability of subjective realities, identity erasure, and the repression of systemic violence or persecution. The ecologically triggered disease of amnesia that plagues large parts of the world in the Darkland narrative facilitates the totalitarian order of the Supreme Guide. Official accounts of the War of the Great Basins and its aftermath have been easily distorted to suit the dictatorial propaganda; all media agencies are state-run, and public access to archives, libraries, and information centres is highly controlled and restricted by the state. It is through the retrospective account – the living memory of Captain Old – that most of the calamitous effects and events that followed the discovery of the tourists in the contaminated valley of White Cloud Mountain are presented. Adam Trexler notes that "climate dystopias attempt to shift the political and ethical calculus between present and future, indicating the necessity of immediate action . . . such dystopias trace the failure of politics to craft an effective response to climate change" (120). Most of the inhabitants of Darkland do not know or recollect anymore how the Scourge, the Superstorms, and the war came to pass. 'Memory pills' and 'solar amnesia boards' are used to mitigate regular memory loss. The act of remembering the past by the detective, therefore, becomes a kind of resistance to distorted official accounts and the repressive state ideology. Captain Old not only remembers crimes of the past but also bears witness to the terrible human degradation and victimisation that is mirrored by the environmental degradation in the capital city of Darkland. In this dystopian reality, he is constantly surrounded by

> the cacophonous orchestra of the diseased and the hungry left out on the streets to wither away. The bodies, with just a flicker of life, massed on pavements or abandoned in weed-infested parks . . . The mongrels of the city had grown red fangs and flesh eaters prowled the suburbs . . . The flame trees on both flanks that had been planted repeatedly for shade had withered but their carcasses remained erect.
>
> *(Butterfly Effect 48)*

Faced with an extreme scarcity of resources and lack of basic necessities like uncontaminated food or proper shelter, daily survival becomes the foremost priority for the 'living dead'. Kar works with the law-and-order system in Darkland, and he is asked to find and eliminate the leader of the rebel group trying to rouse the masses against the Supreme Guide's regime. In totalitarian states, policing, surveillance, and punitive action are not aimed at combating crime but are used to enforce state oppression and criminality.

The disappearance of collective memory can involve the erasure of micro-histories and shared histories, the loss of cultures and ways of life and major shifts in systems of knowledge. In Yōko Ogawa's Japanese science fiction novel, *The Memory Police* (1994), significant and trivial objects keep disappearing from an unnamed island, followed by collective loss of memory of the object in the inhabitants. The fascist Memory Police of the island rigorously wipe out all associations, information, and evidence of the existence of the object once it vanishes, thus ensuring its complete obliteration. The few people who retain their memories are hunted down by the Memory

Police. When birds disappear, for example, wider connotations with flight slowly disappear from the collective imagination and shared knowledge. *The Memory Police* explores the materiality of memory and the importance of collective memory in subjective meaning-making and in the preservation of knowledge.

In an ecologically ravaged dystopia, the scarf from South Korea gifted by the Keun Sunim, which Captain Old "had carried with him all these years" acts as a material signifier for the environmental consciousness and ecosophy of the monk and his community (*Butterfly Effect* 361). It is a reminder not only of a different time and place but of what has been lost and must be regained. This can be better understood if we look at "memory as multidirectional: as subject to ongoing negotiation, cross-referencing, and borrowing" (Rothberg 3). Astrid Erll has argued that memories 'travel' across "social, linguistic and political borders" by virtue of different processes (4).[5] They are capable of forming connections and of moving and mixing over time and space. The piecing together of the truth in crime fiction often depends a great deal on the memories of witnesses. Their testimonies also offer varying perspectives and understandings of the nature of the crime and its redressal. From an ecocritical point of view, preserving environmental consciousness is as important as remembering environmental injustice – both of which are carried out by the detective on behalf of the collective in this novel.

Captain Old's memory can allow him to mourn the loss of ecosystems and peoples, as he preserves the fragments of truth that he has painstakingly gathered journeying through nations, cultures, and communities. It also establishes a rational link among the confusing narrative strands and brings a sense of order to the array of seemingly disparate events and actions. To resist the erasure of the truth and the stripping away of selfhood and individual identity within a totalitarian regime like the Darkland dictatorship, it becomes imperative to retain personal and shared memories, especially for an environmentally displaced, amnesiac population.

Unmasking the Perpetrators of Borderless Ecocrimes

Though there are characters, such as the geneticist Tanmoy Sen or the North Korean agents, in *The Butterfly Effect* whose actions trigger and facilitate the devastating scourge, Chaudhuri does not pin the culpability for the ecological crisis on individual figures, a single evil corporation, or some easily identifiable antagonist alone. The environmental catastrophe brought about by genetic engineering and climate change is an indictment of human society's indifference towards climate concerns and aggressive consumption and its disregard for the environment. The lack of easy answers as to the identity of the 'real culprits' behind this catastrophe is intentional. The novel indicates the interrelations between individual negligence, self-serving corporate organisations, collective apathy, and autocratic power structures that cause the ecological disaster and its dystopian consequences. Amitav Ghosh points out that this collective accountability has mostly been eluded

> at exactly the time when it has become clear that global warming is in every sense a collective predicament, humanity finds itself in the thrall of a dominant culture in which the idea of the collective has been exiled from politics, economics, and literature alike.
>
> *(Great Derangement 80)*

Tanmoy Sen is not the hackneyed evil scientist creating weapons of mass destruction in Chaudhuri's novel. In fact, a major portion of the book is devoted to Sen's narrative which traces his attempts to develop a revolutionary kind of spliced paddy that can withstand droughts and carry essential nutrients and vitamins. Sen dreams of addressing food shortage and ending world hunger, but also experiences bouts of self-scrutiny and ethical doubt following his exchanges with the character

Henry David, who strongly advocates against aggressive scientific experimentation with natural processes. The novel is by no means a cautionary tale about the pitfalls of science. Rather, it highlights the risk of environmental catastrophe present in introducing potentially damaging technologies into the environment without rigorous testing, study, and observation.

Corporate greed, competition, and hopes of monopolising the market prevent Sante, the corporation that owns much of GeneLab UK and funds Tanmoy Sen's project, from involving the scientific community in observing the development of the splices and testing the outcomes. Profit-driven attitudes and irresponsible decision-making by industries, organisations, and big corporations are greatly responsible for environmental damage. Sen is asked by the Group CEO of Sante to create paddy seeds that can only be used for one season so that they can maintain high profits. The secrecy maintained about the project and Sen's exclusive control over the engineering of new splices spell doom when both the rice samples and Sen's computer containing all his research are stolen. As it turns out, the experimental paddy contaminates the land, causes disease, and cross-pollinates and spreads to other plant species and food sources. Crime fiction often depicts corporations brazenly covering up their crimes and mistakes and silencing media without having to face any punitive action. It reveals the extremely uneven distribution of power in neoliberal, capitalist structures. It is not surprising that Sante hushes up the circumstances surrounding Sen's suicide in the novel and excuses itself from the responsibility of tracking down the stolen seeds.

In *Sense of Place* (2008), Ursula Heise contends that the representation of climate change as a global phenomenon and threat "requires the articulation of connections between events at vastly different scales" (205). Since climate change unfolds over a long period and at different paces in different parts of the world, writers face the challenge of representing it in concrete terms and conveying the urgency of addressing it. Different writers have adopted different narrative strategies to convincingly situate climate change in fiction, such as by using multiple narrators spread out over time and space or by amplifying and focusing on the effects of extreme weather events in one particular local setting. Rajat Chaudhuri discusses how he chose to present climate change in *The Butterfly Effect*:

> While a single storm can have dramatic elements which can fit into a short story, and we have the example of movies like *Into the Storm*, communicating the widely distributed developments surrounding climate change as a phenomenon is far more difficult. To work around this problem, I built my story around something else, a fast progressing genetically modified food disaster, to guide the flow of the narrative, while disparate climate events unfolding in the background reinforced the disaster.
>
> *(Chaudhuri "Dystopia")*

Climate change affects regions unevenly, based on geographical location and socio-economic factors. Ecological crises can thus arise in areas or nations which are far away from the sites where environmental damage, high emissions, and ecocrimes have taken place. Therefore, it makes sense to look at environmental disasters across national borders and delve into the factors that render certain regions and nations particularly vulnerable to human-induced environmental disasters and climate-related risks. The narrative frameworks of crime fiction are especially suitable for this since they naturally allow the detective as well as the reader to follow the investigative trail and trace the motives, agents, and repercussions of ecocrimes over different areas and nations. By situating the narratives across different nation-states, especially within Asia, Rajat Chaudhuri has attempted to imaginatively assess how different regions in the continent may suffer and respond very differently when confronted with a far-reaching environmental crisis. Considering both the vast human population and the natural resources and habitats that Asia encompasses, the ramifications of climate change are alarming. The dystopian aftermath in the novel divides the majority of Asia and large parts of Europe into Cleanland

and Darkland. As is evident from the nomenclature, Cleanland has mostly been able to evade and mitigate the calamitous consequences due to "geographical and other reasons" (*Butterfly Effect* 371).

Countries with low economic development and poor infrastructure are much less equipped to deal with ecocide and do not have the means to recover from the tremendous damages caused by extreme climate events like floods, droughts, storms, or heat waves. Negligent governments that fail to address climate risks and prolonged social and political unrest also make human and natural ecosystems more susceptible. In *The Butterfly Effect*, environmental damage, disease, and the destruction of important ecosystems quickly lead to instability in these vulnerable regions, followed by conflict and violence. Food shortage and the rising scarcity of resources most intensely affect poor countries and communities whose livelihoods depend on nature or natural resources. The alarming rise in sea levels in countries like Japan, environmental damage, and the War of the Great Basins cause massive migration, especially in Asia, as many parts of the world become uninhabitable or terribly ravaged. The novel draws distinct correlations between societal crime and environmental conditions. Immigrants and refugees are often the most victimised and oppressed. Captain Old is approached by a pimp on the streets to try Japanese girls who are "fresh off the boat" and chides him, "What would these poor girls eat . . .?" (Chaudhuri, *Butterfly Effect* 29). From sex trade and child pornography to assault and theft, criminal activity of all kinds thrives in Darkland as violence, extreme deprivation, and dehumanisation are a daily reality.

The investigation plot in the novel is useful for looking into the intersections between human-induced environmental disasters and other forms of injustice and violence. In *The Butterfly Effect*, though the North Korean agents Jia and Eugene steal the genetically modified paddy from the lab and unknowingly initiate the food crisis and pandemic, they cannot be simply categorised as the antagonists – their narratives are tales of victimhood, not transgression. The twin sisters Jiyoo and Jia Park voice their long, traumatic history of state persecution, coercion, and brutality to PI Kar at White Cloud Mountain, as he tries to understand how the valley was contaminated. Their narratives elaborate on the repressive way of life in totalitarian North Korea, their attempts to flee the country, their family imprisonment, rigorous state interrogations, violence, terror, and the extremely dangerous predicament of defectors. The terrible hardships and torture that the family faces in the prison camps at Pyongyang and forced labour camps compel Jia to agree to work as a spy for the 'Fatherland'. The agents Eugene and Jia are summarily killed after bringing back the seeds. Though Jiyoo successfully escapes, the novel details the harrowing challenges and privation that she has to face as she tries to reach South Korea from China, where North Korean defectors are often forcefully repatriated. Totalitarian states are also less likely to be able to manage the effects of an environmental catastrophe. Representations of human rights violations are thus weaved into the representation of the steady violation of the environment in *The Butterfly Effect*, emphasising the need to probe their interrelationships.

Few literary genres have proliferated and evolved as rapidly and multifariously as crime fiction since its genesis in the mid-nineteenth century, as it has tried to adapt to the rapid change in the inner and outer lives and landscapes of the world. Rapid industrialisation and urbanisation, colonialism, the two major World Wars, and the digital boom have all manifested themselves in the myriad forms that crime fiction has branched into – from the Golden age locked-room mysteries and classic whodunnits to spy thrillers, police procedurals, legal dramas, and cyber-crime fiction. By closely inspecting certain narrative strategies, my reading of *The Butterfly Effect* indicates how a hybrid form like the environment thriller, which does not exclusively focus on an ecological crime narrative, can engage with and speculate on critical issues of climate change through the vehicle of crime and mystery. *The Butterfly Effect* raises important concerns about the uneven transnational impact of environmental disasters, the convergence of human and environmental injustice, and the significance of memory in ecological consciousness. An assessment of Kar in the novel shows that, in an ecologically threatened world, the investigator has an important function in examining, understanding, and remembering environmental crimes. He also reveals several perspectives and approaches to the environment, ranging from the exploitative and the unethical to the nurturing and the ecocentric. Rajat Chaudhuri ultimately incriminates both the individual and the collective in his novel and thus urges readers to re-evaluate their own efforts in mitigating or aggravating environmental degradation.

Notes

1. See Stowe for a fuller analysis of elitist critical approaches to popular fiction.
2. The Independent Expert Panel for the Legal Definition of Ecocide, June 2021, defined ecocide as "unlawful or wanton acts committed with knowledge that there is a substantial likelihood of severe and either widespread or long-term damage to the environment being caused by those acts". The panellists have petitioned for ecocide to be accepted as a crime by the International Crimes Court.
3. Rajat Chaudhuri has been awarded a number of travel grants and writing fellowships such as the Charles Wallace Creative Writing Fellowship, the Hawthornden Castle Trust Fellowship, and the Korean Art's Council-InKo Writers' Residency in South Korea to aid the writing of this novel.
4. See Naess for more insight on the eight principles of deep ecology proposed.
5. Erll has also discussed the concept of 'transcultural memory' and proposed it as a new methodological framework for memory studies.

Bibliography

Bacigalupi, Paolo. *The Windup Girl*. Hachette, 2012.

Beckbessinger, Sam, Nicholson, Simon, Trisos, Christopher. *Survive the Century* (Web Browser version), 2021, https://survivethecentury.net/.

Chabon, Michael. "Trickster in a Suit of Lights." *Maps and Legends: Reading and Writing Along the Borderlands*. McSweeney's, 2008, pp. 13–26.

Chandler, Raymond. "The Simple Art of Murder (An Essay)." *The Simple Art of Murder*, Vintage, 1988, pp. 2–9.

Chaudhuri, Rajat. *Hotel Calcutta*. Niyogi Books, 2013.

———. *The Butterfly Effect*. Niyogi Books, 2018.

———. "Dystopia or Utopia? Politics or Aesthetics? What Are the Challenges of Writing Climate Fiction?" *Scroll.in*, 13 Oct. 2019, scroll.in/article/940301/dystopia-or-utopia-politics-or-aesthetics-what-are-the-challenges-of-writing-climate-fiction.

———. "Wild Authors: Rajat Chaudhuri." *Artists & Climate Change, The Arctic Cycle*, 20 Jan. 2020, artistsandclimatechange.com/2020/01/20/wild-authors-rajat-chaudhuri/.

———. "Read: Q&A with Rajat Chaudhuri." *Writers Rebel*, 22 Oct. 2020, writersrebel.com/read-qa-with-rajat-chaudhuri/.

El Akkad, Omar. *American War*. Pan Macmillan, 2017.

Erll, Astrid. "Travelling Memory." *Parallax*, vol. 17, no. 4, 2011, pp. 4–18, doi:10.1080/13534645.2011.605570.

Ghosh, Amitav. *The Great Derangement: Climate Change and the Unthinkable*. U Chicago P, 2017.

———. *Gun Island: A Novel*. Farrar, Straus and Giroux, 2019.

Hawkins, Paula. *The Girl on the Train*. Penguin Books, 2021.

Heise, Ursula K. *Sense of Place and Sense of Planet: The Environmental Imagination of the Global*. Oxford UP, 2008.

Hitchcock, Alfred, director. *Spellbound*. Selznick International Pictures and Vanguard Films, Inc., 1945.

Independent Expert Panel for the Legal Definition of Ecocide. *Legal Definition of Ecocide Drafted by Independent Expert Panel*. Stop Ecocide International, www.stopecocide.earth/legal-definition.

Itäranta, Emmi. *Memory of Water*. Thorpe, 2015.

King, Stewart. "*Crimate* Fiction and the Environmental Imagination of Place." *The Journal of Popular Culture*, vol. 54, no. 6, 2021, pp. 1235–1253, doi:10.1111/jpcu.13083.

Ludlum, Robert. *Bourne Identity*. Orion Books, 2016.

Naess, Arne. *Ecology, Community and Lifestyle: Outline of an Ecosophy*. Cambridge UP, 1990.

Ogawa, Yoko. *The Memory Police*. Translated by Stephen Snyder, Pantheon, 2019.

Puxan-Oliva, Marta. "Crime Fiction and the Environment." *The Routledge Companion to Crime Fiction*, edited by Janice Allan et al., Routledge, 2020, pp. 362–370.

Rothberg, Michael. *Multidirectional Memory: Remembering the Holocaust in the Age of Decolonization*. Stanford UP, 2009.
Scaggs, John. *Crime Fiction*. Routledge, 2005.
Swarup, Shubhangi. *Latitudes of Longing: A Novel*. HarperCollins India, 2018.
Tawada, Yoko. *The Last Children of Tokyo*. Portobello Books, 2018.
Tokarczuk, Olga. *Drive Your Plow Over the Bones of the Dead: A Novel*. Penguin Random House, 2020.
Trexler, Adam. *Anthropocene Fictions: The Novel in a Time of Climate Change*. U Virginia P, 2015.
Walton, Jo Lindsay, and Samantha Walton. "Introduction to Green Letters: Crime Fiction and Ecology." *Green Letters*, vol. 22, no. 1, 2018, pp. 2–6, doi:10.1080/14688417.2018.1484628.
What is an Ecovillage? GEN Europe, gen-europe.org/ecovillages/what-is-an-ecovillage/.

INDEX

Note: Page numbers in *italic* indicate a figure on the corresponding page.